Benchmark Papers in Microbiology

Series Editor: Wayne W. Umbreit
Rutgers—The State University

PUBLISHED VOLUMES

MICROBIAL PERMEABILITY
John P. Reeves
CHEMICAL STERILIZATION
Paul M. Borick
MICROBIAL GENETICS
Morad Abou-Sabé
MICROBIAL PHOTOSYNTHESIS
June Lascelles
MICROBIAL METABOLISM
H. W. Doelle
ANIMAL CELL CULTURE AND VIROLOGY
Robert J. Kuchler
PHAGE
Sewell P. Champe
MICROBIAL GROWTH
P. S. S. Dawson
MICROBIAL INTERACTION WITH THE PHYSICAL ENVIRONMENT
D. W. Thayer
MOLECULAR BIOLOGY AND PROTEIN SYNTHESIS
Robert A. Niederman
MARINE MICROBIOLOGY
Carol D. Litchfield

Additional volumes in preparation

Benchmark Papers in Microbiology / 11

A BENCHMARK ® Books Series

MARINE MICROBIOLOGY

Edited by

CAROL D. LITCHFIELD

Rutgers—The State University

Dowden, Hutchinson & Ross, Inc.

STROUDSBURG, PENNSYLVANIA

Distributed by

HALSTED PRESS A Division of John Wiley & Sons, Inc.

To My Father

Benchmark Papers in Microbiology, Volume 11
Library of Congress Catalog Card Number: 76-15955
ISBN: 0-87933-076-7

78 77 76 1 2 3 4 5
Manufactured in the United States of America.

LIBRARY OF CONGRESS CATALOGING IN PUBLICATION DATA

Main entry under title:
Marine microbiology
(Benchmark papers in microbiology / 11)
Includes indexes.
1. Marine microbiology. I. Litchfield, Carol D.
QR106.M39 589.9'09'208 76-15955
ISBN 0-87933-076-7

Exclusive Distributor: **Halsted Press**
A Division of John Wiley & Sons, Inc.
ISBN: 0-470-98924-6

ACKNOWLEDGMENTS AND PERMISSIONS

ACKNOWLEDGMENTS

DANISH SCIENCE PRESS LTD.—*Galathea Report, Copenhagen*
Deep-Sea Bacteria

HARVARD UNIVERSITY—*Farlowia*
Marine Fungi: Their Taxonomy and Biology

THE UNIVERSITY OF CALIFORNIA PRESS—*Bulletin of the Scripps Institution of Oceanography*
A List of Marine Bacteria Including Descriptions of Sixty New Species

PERMISSIONS

The following papers have been reprinted with permission of the authors and copyright holders.

ACADEMIC PRESS, INC.—*Cryobiology*
The Effect of the Rehydration Temperature and Rehydration Medium on the Viability of Freeze-Dried *Spirillum atlanticum*

GEORGE ALLEN AND UNWIN LTD.—*Oceanography and Marine Biology Annual Review*
Effects of Hydrostatic Pressure on Marine Microorganisms

AMERICAN SOCIETY FOR MICROBIOLOGY
Applied Microbiology
Bacteriology of Manganese Nodules: II. Manganese Oxidation by Cell-Free Extract from a Manganese Nodule Bacterium
Bacteriological Reviews
The Question of the Existence of Specific Marine Bacteria
Journal of Bacteriology
Geographic Regularities in Microbe Population (Heterotroph) Distribution in the World Ocean
Indigenous Marine Bacteriophages
Nutrition and Metabolism of Marine Bacteria: II. Observations on the Relation of Sea Water to the Growth of Marine Bacteria
Separate Nitrite, Nitric Oxide, and Nitrous Oxide Reducing Fractions from *Pseudomonas perfectomarinus*
Studies on Bacterial Utilization of Uronic Acids: III. Induction of Oxidative Enzymes in a Marine Isolate
Taxonomic Relationships Among the Pseudomonads

AMERICAN SOCIETY OF LIMNOLOGY AND OCEANOGRAPHY—*Limnology and Oceanography*
Adenosine Triphosphate Content of Marine Bacteria
Bacterial Populations in Sea Water as Determined by Different Methods of Enumeration

Characteristics of a Marine Nitrifying Bacterium, *Nitrosocystis oceanus* sp.n.
Growth of Marine Bacteria at Limiting Concentrations of Organic Carbon in Seawater
High Nitrogen Fixation Rates in the Sargasso Sea and the Arabian Sea
Isolation of Yeasts from Biscayne Bay, Florida, and Adjacent Benthic Areas
The Production of Vitamin B_{12}-Active Substances by Marine Bacteria
Respiration Corrections for Bacterial Uptake of Dissolved Organic Compounds in Natural Waters
The Spread Plate as a Method for the Enumeration of Marine Bacteria

AMERICAN SOCIETY OF ZOOLOGISTS—*American Zoologist*
Isolation of Bacteria from the Corallum of *Porites lobata* (DANA) and Its Possible Significance

THE BIOLOGICAL BULLETIN—*The Biological Bulletin*
Marine Bacteria and Their Role in the Cycle of Life in the Sea

CAMBRIDGE UNIVERSITY PRESS
Journal of General Microbiology
Mechanism of the Initial Events in the Sorption of Marine Bacteria to Surfaces
Journal of the Marine Biological Association
Bacteria of the Clyde Sea Area: A Quantitative Investigation

ELSEVIER SCIENTIFIC PUBLISHING COMPANY—*Biochimica et Biophysica Acta*
Effects of Salt Concentration During Growth on Properties of the Cell Envelope of a Marine Pseudomonad

JOURNAL OF APPLIED BACTERIOLOGY—*The Journal of Applied Bacteriology*
A Determinative Scheme for the Identification of Certain Genera of Gram-Negative Bacteria, with Special Reference to the Pseudomonadaceae
The Preservation of Marine Bacteria

MACMILLAN JOURNALS LTD.—*Nature*
Actinomycetes in North Sea and Atlantic Ocean Sediments

MISAKI MARINE BIOLOGICAL INSTITUTE—*Bulletin of the Misaki Marine Biological Institute, Kyoto University*
The Basic Nature of Marine Psychrophilic Bacteria
Distribution, Taxonomy and Function of Heterotrophic Bacteria on the Sea Floor
Some Ecological Aspects of Marine Bacteria in the Kuroshio Current

PERGAMON PRESS LTD.—*Deep-Sea Research and Oceanographic Abstracts*
Distribution and Activity of Oceanic Bacteria
On the Production of Particulate Organic Carbon by Heterotrophic Processes in Sea Water
A Water Sampler for Microbiological Studies

THE SEARS FOUNDATION OF THE JOURNAL OF MARINE RESEARCH, YALE UNIVERSITY—*Journal of Marine Research*
Apparatus for Collecting Water Samples from Different Depths for Bacteriological Analysis
Studies on Marine Bacteria: I. The Cultural Requirements of Heterotrophic Aerobes

THE SOCIETY FOR EXPERIMENTAL BIOLOGY AND MEDICINE—*Proceedings of the Society for Experimental Biology and Medicine*
Factors Influencing the Plate Method for Determining Abundance of Bacteria in Sea Water

SERIES EDITOR'S FOREWORD

Although the sea constitutes an environment clearly greater in extent and in volume than that of land, certain factors tend to restrict the diversity of microbial forms found in it, compared to the enormous variety inhabiting a single pound of soil. Yet despite these almost homeostatic factors—temperature, salts, limiting nutrient—the sea contains a diverse and distinctive population whose nature and activity are of great interest and potential benefit to man.

Indeed, during the very early stages of microbiology, in the late 1880s, some very prominent microbiologists turned their attention to the sea. But the methods that had been developed for terrestrial environments applied only fitfully to a water-salt environment and new methods of study had to be worked out. Although new approaches are still being developed, the basic "classical" pattern was set in the 1950s. These methods have revealed a marine population distinct from land forms, related to each other as the fish to the mouse, but just as distinct as these two creatures. Adaptation to the sea, and to portions of its microenvironment, permitted changes in the structure of the microbial cell wall, in response to the requirements of pressure and temperature. But, although one of the main functions of microorganisms in the sea is the recycling of organic materials, the environment remains, to a large extent, singularly poor in nutrient, especially poor in available organic matter, especially restricted by temperature and salt content. Yet even this flora involves a wide variety of organisms—algae, of course, but also fungi, yeasts, bacteria, streptomyces, although it may be lacking in nitrogen-fixing forms. The pattern is at least similar to the patterns of activities found on land.

There is no question that marine microbiology is a science in its own right. The wealth of material available and the progress made in bringing it to a coherent science are aptly outlined in the selections made by Dr. Litchfield, and this service (collecting, sifting, and rediffusion of the literature) is of particular importance in the field of marine microbiology, where the literature sources are so scattered. The marine scientist is unlikely to have readily available the *Journal of General Microbiology, Applied Microbiology,* or even the *Journal of Bacteriology,* whereas the microbiologist is unlikely to have *Limnology and Oceanography, Deep-Sea Research,* or the *Journal of Marine Research.* It is this fusion of the special literature needed in this very special field which is a vital service to students of marine microbiology.

WAYNE W. UMBREIT

PREFACE

Although the *Challenger* expedition (1873–1876) put oceanography as a whole on a scientific basis, and the era 1855–1870 laid the foundations for scientific microbiology, the study of microorganisms indigenous to the sea was not actively pursued until 10 to 15 years later. This book is concerned with the meeting of these two sciences and the evolution of the resulting discipline, marine microbiology. The early pioneers, Certes, Regnard, Russell, Fischer, at first had to be concerned with simply demonstrating that bacteria, yeasts, and molds could be found in parts of the ocean not "contaminated" directly by land. Despite the prevailing opinions of their times regarding abiogenic oozes and the nonmicrobial nature of the seas, they persevered and laid the foundations for the investigations of today. Their papers still make exciting reading. In preparing our papers for publication, we should remember that Certes opened the entire field of marine microbiology with only a three-page paper in 1884!

It would be quite impossible and impractical to survey the entire field of marine microbiology in one volume. Because of the development of new and sophisticated techniques for sample handling, sample analysis, and the interpretation of data, an understanding of the role of microorganisms in oceans and sediments is now within reach. After many years of research effort, the answer to the question "What is distinctive about marine microbes?" also seems at hand. Therefore, I have chosen to emphasize these aspects of marine microbiology, to the exclusion of other equally worthwhile topics. The eucaryotic organisms are represented in this book only by the yeasts and fungi because of their appearance on "standard" bacteriological media and thus their heterotrophic metabolism, but protozoa and other equally interesting microorganisms have been omitted for reasons of space, not lack of interest.

After a short section on nineteenth-century studies in marine microbiology, a major emphasis has been placed in Part II on the solutions to the various problems in the sampling, cultivation, and identification of marine bacterial isolates. In Part III, we are concerned with investigations into the structural and physiological nature of marine bacteria.

Parts IV through VI comprise a selection of papers dealing with the area in which we can most expect major discoveries in the immediate future—the real or potential activities of microorganisms in the marine environment. Based upon the work described in some of these papers, we may soon expect to see a picture of microbial function based not on hypothesis, but on fact.

Part VII is concerned with a survey of the distributions of various types of microbes in the sea—fungi, yeasts, bacteria, and viruses. Moreover, certain papers have been included here to demonstrate the types of integrated studies that are the hallmark of marine microbiology in the 1970s.

On occasion, an author has so succinctly summarized a large body of literature that the student or future investigator can gain better insights into the field from this one paper than from the individual papers. When available, such an outstanding review paper has been included as a tribute to all those who provided the necessary information and to the individual who has the gift to clarify a complex field.

No author composes a book without help from many others. The early literature was made available through the combined efforts of many librarians and especially those in the reference division of the Rutgers University Library of Science and Medicine. For their persistence and friendly aid I am most grateful.

Additionally, many of the authors of included papers have kindly provided suggestions and helped me to place my ideas in a better historical perspective, especially Dr. C. E. ZoBell, to whom I am most grateful for insight and suggestions. Throughout the writing and final stages of manuscript preparation, Dr. W. J. Payne has remained a special source of information and encouragement. He has clarified some of my more obtuse statements and removed excess verbiage so beloved by authors. To Dr. Payne I extend my deepest personal appreciation.

Marine microbiologists have traditionally been a relatively small coterie of dedicated individuals willing to share their knowledge and experience and lend a helping hand to the newcomer. For marine microbiology, like the oceans we study, touches all continents and is truly an international endeavor. It is hoped that this book will further this spirit of cooperation and friendship as we continue our quest for an understanding of the microorganisms of the seas.

CAROL D. LITCHFIELD

CONTENTS

Acknowledgments and Permissions v
Series Editor's Foreword vii
Preface ix
Contents by Author xvii

PART I: THE FIRST MARINE MICROBIOLOGISTS

Editor's Comments on Papers 1 Through 3 2

1 **CERTES, A.:** On the Culture, Free from Known Sources of Contamination, from Waters and from Sediments Brought Back by the Expeditions of the *Travailleur* and the *Talisman*: 1882–1883 8
Translated from *C. R. Hebd. Séances Acad. Sci.*, **98**, 690–693 (1884)

2 **FRANKLAND, P., and G. P. FRANKLAND:** *Micro-organisms in Water* 11
Longmans, Green & Company Ltd., London, 1894, pp. 112–116

3 **BEIJERINCK, M. W.:** *Photobacterium luminosum*, a Luminous Bacterium in the North Sea 16
Translated from *Archives Neerl. Sci. Exactes Nat. Haarlem*, **23**, 401–415 (1889)

PART II: THE PROBLEMS OF THE SAMPLING, CULTIVATION, AND IDENTIFICATION OF MARINE BACTERIAL ISOLATES

Editor's Comments on Papers 4 Through 6 28

THE DEVELOPMENT OF SAMPLING DEVICES

4 **RUSSELL, H. L.:** Studies on Bacteria Living in the Gulf of Naples 34
Translated from *Z. Hyg. Infectionskr.*, **11**, 165–167, 167–175 (1892)

5 **ZOBELL, C. E.:** Apparatus for Collecting Water Samples from Different Depths for Bacteriological Analysis 44
J. Mar. Res., **4**(3), 173–188 (1941)

6 **NISKIN, S. J.:** A Water Sampler for Microbiological Studies 60
Deep-Sea Res. Oceanogr. Abst., **9**, 501–503 (1962)

Editor's Comments on Papers 7 Through 13 65

THE DEVELOPMENT OF CULTURAL AND ENUMERATION TECHNIQUES

7 **LLOYD, B.:** Bacteria of the Clyde Sea Area: A Quantitative Investigation 73
J. Mar. Biol. Assoc. U.K., **16**(3), 879, 881–900 (1929–1930)

8 **BUTKEVICH, N. V., and V. S. BUTKEVICH:** Multiplication of Sea Bacteria Depending on the Composition of the Medium and on Temperature *(abridged English summary and figures)* 93
Microbiol. USSR, **5**, 322–343 (1936)

9 **ZOBELL, C. E.:** Studies on Marine Bacteria: I. The Cultural Requirements of Heterotrophic Aerobes *(summary)* 97
J. Mar. Res., **4**, 69–70 (1941)

10 **CARLUCCI, A. F., and D. PRAMER:** Factors Influencing the Plate Method for Determining Abundance of Bacteria in Sea Water 99
Proc. Soc. Exp. Biol. Med., **96**, 392–394 (1957)

11 **JANNASCH, H. W., and G. E. JONES:** Bacterial Populations in Sea Water as Determined by Different Methods of Enumeration 102
Limnol. Oceanogr., **4**(2), 128–139 (1959)

12 **BUCK, J. D., and R. C. CLEVERDON:** The Spread Plate as a Method for the Enumeration of Marine Bacteria 114
Limnol. Oceanogr., **5**(1), 78–80 (1960)

13 **HAMILTON, R. D., and O. HOLM-HANSEN:** Adenosine Triphosphate Content of Marine Bacteria 117
Limnol. Oceanogr., **12**(2), 319–324 (1967)

Editor's Comments on Papers 14 Through 18 123

THE PRESERVATION AND IDENTIFICATION OF MARINE BACTERIA

14 **ZOBELL, C. E., and H. C. UPHAM:** A List of Marine Bacteria Including Descriptions of Sixty New Species 128
Bull. Scripps Inst. Oceanogr. Univ. Calif., **5**(2), 239–240, 246–247, 251–253, 280–281 (1944)

15 **SHEWAN, J. M., G. HOBBS, and W. HODGKISS:** A Determinative Scheme for the Identification of Certain Genera of Gram-Negative Bacteria, with Special Reference to the Pseudomonadaceae 137
J. Appl. Bacteriol., **23**(3), 379–390 (1960)

16 **COLWELL, R. R., and J. LISTON:** Taxonomic Relationships Among the Pseudomonads 150
J. Bacteriol., **82**, 1–14 (1961)

17 **FLOODGATE, G. D., and P. R. HAYES:** The Preservation of Marine Bacteria 164
J. Appl. Bacteriol., **24**(1), 87–93 (1961)

18 **CHOATE, R. V., and M. T. ALEXANDER:** The Effect of the Rehydration Temperature and Rehydration Medium on the Viability of Freeze-Dried *Spirillum atlanticum* 171
Cryobiology, **3**(5), 419–422 (1967)

PART III: THE STRUCTURAL AND PHYSIOLOGICAL RESPONSES OF MARINE BACTERIA TO THEIR SALINE ENVIRONMENT

Editor's Comments on Papers 19 Through 21 176

NUTRITIONAL NEEDS OF MARINE BACTERIA

19 **MacLEOD, R. A., and E. ONOFREY:** Nutrition and Metabolism of Marine Bacteria: II. Observations on the Relation of Sea Water to the Growth of Marine Bacteria 182
J. Bacteriol., **71**(6), 661–667 (1956)

20 **PAYNE, W. J.:** Studies on Bacterial Utilization of Uronic Acids: III. Induction of Oxidative Enzymes in a Marine Isolate 189
J. Bacteriol., **76**(3), 301–307 (1958)

21 **MacLEOD, R. A.:** The Question of the Existence of Specific Marine Bacteria 196
Bacteriol. Rev., **29**(1), 9–23 (1965)

Editor's Comments on Paper 22 211

IS THERE A STRUCTURAL RESPONSE BY BACTERIA TO THE MARINE ENVIRONMENT?

22 **BROWN, A. D.:** Effects of Salt Concentration During Growth or Properties of the Cell Envelope of a Marine Pseudomonad 217
Biochim. Biophys. Acta, **49**, 585–588 (1961)

Editor's Comments on Papers 23 Through 25 221

MARINE BACTERIAL RESPONSE TO IN SITU TEMPERATURES AND PRESSURES

23 **MORITA, R. Y.:** The Basic Nature of Marine Psychrophilic Bacteria 231
Bull. Misaki Mar. Biol. Inst. Kyoto Univ., **12**, 163–177 (Feb. 1968)

24 **ZOBELL, C. E., and R. Y. MORITA:** Deep-Sea Bacteria 246
Galathea Rept., Copenhagen, **1**, 139–154 (1959)

25 **MORITA, R. Y.:** Effects of Hydrostatic Pressure on Marine Micro-organisms 262
Oceanogr. Mar. Biol. Annu. Rev., **5**, 187–203 (1967)

PART IV: MICROBIAL CYCLING OF ORGANIC MATTER

Editor's Comments on Papers 26 Through 29 280

26 **PARSONS, T. R., and J. D. H. STRICKLAND:** On the Production of Particulate Organic Carbon by Heterotrophic Processes in Sea Water 288
Deep-Sea Res. Oceanogr. Abst., **8**, 211–222 (1962)

27 **HOBBIE, J. E., and C. C. CRAWFORD:** Respiration Corrections for Bacterial Uptake of Dissolved Organic Compounds in Natural Waters 300
Limnol. Oceanogr., **14**(4), 528–532 (1969)

28 **JANNASCH, H. W.:** Growth of Marine Bacteria at Limiting Concentrations of Organic Carbon in Seawater — **305**
Limnol. Oceanogr., **12**(2), 264–271 (1967)

29 **STARR, T. J., M. E. JONES, and D. MARTINEZ:** The Production of Vitamin B_{12}-Active Substances by Marine Bacteria — **313**
Limnol. Oceanogr., **2**(2), 114–119 (1957)

PART V: THE MICROBIAL ROLE IN THE NITROGEN CYCLE OF THE SEA

Editor's Comments on Papers 30 Through 33 — **320**

30 **DUGDALE, R. C., J. J. GOERING, and J. H. RYTHER:** High Nitrogen Fixation Rates in the Sargasso Sea and the Arabian Sea — **330**
Limnol. Oceanogr., **9**(4), 507–510 (1964)

31 **WAKSMAN, S. A., M. HOTCHKISS, and C. L. CAREY:** Marine Bacteria and Their Role in the Cycle of Life in the Sea — **334**
Biol. Bull., **65**, 138–146 (1933)

32 **WATSON, S. W.:** Characteristics of a Marine Nitrifying Bacterium, *Nitrosocystis oceanus* sp. n. — **344**
Limnol. Oceanogr., **10**(suppl.), R274–R289 (Nov. 1965)

33 **PAYNE, W. J., P. S. RILEY, and C. D. COX, Jr.:** Separate Nitrite, Nitric Oxide, and Nitrous Oxide Reducing Fractions from *Pseudomonas perfectomarinus* — **360**
J. Bacteriol., **106**(2), 356–361 (1971)

PART VI: SECONDARY ACTIVITIES OF MARINE BACTERIA

Editor's Comments on Papers 34 Through 37 — **368**

34 **DiSALVO, L. H.:** Isolation of Bacteria from the Corallum of *Porites lobata* (DANA) and Its Possible Significance — **377**
Am. Zool., **9**, 735–740 (1969)

35 **EHRLICH, H. L.:** Bacteriology of Manganese Nodules: II. Manganese Oxidation by Cell-Free Extract from a Manganese Nodule Bacterium — **383**
Appl. Microbiol., **16**(2), 197–202 (1968)

36 **MARSHALL, K. C., R. STOUT, and R. MITCHELL:** Mechanism of the Initial Events in the Sorption of Marine Bacteria to Surfaces — **389**
J. Gen. Microbiol., **68**, 337–348 (1971)

37 **SOROKIN, YU. I., T. S. PETIPA, and YE. V. PAVLOVA:** Quantitative Estimate of Marine Bacterioplankton as a Source of Food — **401**
Okeanologiya, **10**, 253–260 (1970)

PART VII: MICROBES IN THE SEA

Editor's Comments on Papers 38 Through 45 — **410**

38 **BARGHOORN, E. S., and D. H. LINDER:** Marine Fungi: Their Taxonomy and Biology — **431**
Farlowia, **1**(3), 395–401, 436–440, 452, 456, 459, 463–465 (1944)

39 **FELL, J. W., D. G. AHEARN, S. P. MEYERS, and F. J. ROTH, Jr.:** Isolation of Yeasts from Biscayne Bay, Florida, and Adjacent Benthic Areas **447**
Limnol. Oceanogr., **5**(4), 366–371 (1960)

40 **WEYLAND, H.:** Actinomycetes in North Sea and Atlantic Ocean Sediments **453**
Nature, **223**(5208), 858 (1969)

41 **SPENCER, R.:** Indigenous Marine Bacteriophages **455**
J. Bacteriol., **79**(4), 614 (1960)

42 **LISTON, J.:** Distribution, Taxonomy and Function of Heterotrophic Bacteria on the Sea Floor **456**
Bull. Misaki Mar. Biol. Inst., Kyoto Univ., **12**, 97–104 (Feb. 1968)

43 **KRISS, A. E., S. S. ABYZOV, M. N. LEBEDEVA, I. E. MISHUSTINA, and I. N. MITSKEVICH:** Geographic Regularities in Microbe Population (Heterotroph) Distribution in the World Ocean **464**
J. Bacteriol., **80**(6), 731–736 (1960)

44 **SIEBURTH, J. McN.:** Distribution and Activity of Oceanic Bacteria **470**
Deep-Sea Res. Oceanogr. Abst., **18**, 1111–1121 (1971)

45 **TAGA, N.:** Some Ecological Aspects of Marine Bacteria in the Kuroshio Current **481**
Bull. Misaki Mar. Biol. Inst., Kyoto Univ., **12**, 65, 67–68, 70–76 (Feb. 1968)

Author Citation Index **491**
Subject Index **505**
About the Editor **519**

CONTENTS BY AUTHOR

MARINE MICROBIOLOGY

Part I

THE FIRST MARINE MICROBIOLOGISTS

Editor's Comments on Papers 1 Through 3

1 **CERTES**
On the Culture, Free from Known Sources of Contamination, from Waters and from Sediments Brought Back by the Expeditions of the Travailleur *and the* Talisman*: 1882–1883*

2 **FRANKLAND and FRANKLAND**
Micro-organisms in Water

3 **BEIJERINCK**
Photobacterium luminosum*, a Luminous Bacterium in the North Sea*

The earliest interests in the microorganisms in the sea were centered around the question of the survival of the newly discovered waterborne pathogenic bacteria—the typhoid and cholera bacilli. Indeed, the first papers on bacteria in seawater by de Giaxa, and others, were published more or less as adjuncts to the "proper" study of pathogenic bacteriology (9). This is completely understandable when the reader recalls that the last half of the nineteenth century was an extremely active period during which the diseases that had caused so much grief historically were found to be due, not to a malevolent god, but to microorganisms. Hence, microbes provided an explanation and a direction that man could take to protect himself from plague, diphtheria, typhoid, and the like. Thus, it was natural for the first concerns with the microbial flora of water to be directed toward the hunt for these deadly pathogens.

Several major works appeared on aquatic microbiology in the late 1880s and 1890s; they described methods for the purification and analysis of water (15, 23, 37), the microbes that lived in fresh waters (12, 17, 34, 37), and the inability of *Bacillus* (or *Bacterium*) *typhosa*, especially, to survive for long in coastal waters (10). Those at the forefront of research were impatient with nonmedical microbiology, and many of the early workers felt that they had to "justify" their aquatic studies by relating them to the distribution of cholera vibrios or typhoid bacilli. We have still not escaped from this need to justify, for frequently we feel we must compare marine bacteria to *Escherichia coli* to obtain approval for our work.

The emphasis on aquatic pathogens was not totally limiting, though, for Certes had already published his articles on bacteria in the sea and ocean sediments and demonstrated their distinctive natures (7, 8). In 1884, he first reported that it was possible to isolate marine bacteria from the oceans, and his original paper is reprinted here in translation (Paper 1). Following collection, Certes, in fact, cultured aerobic heterotrophic bacteria on seawater media, and even used a high-pressure device that allowed him to cultivate the microorganisms at pressures up to 300 atm.

This work was followed by the work of de Giaxa (10) and Russell (35), which was summarized by Frankland and Frankland in 1894 (16). The relevant portion of their book is reprinted here as Paper 2. This couple spent their lives studying and writing about what today has come to be called environmental microbiology. They were certainly a husband and wife team that was 70 years ahead of its time!

Meanwhile, basic studies were also appearing in Monaco, where Prince Albert lent his royal presence to the study of marine bacteria (1). This undoubtedly encouraged Richard to consider marine bacteria and their contributions to luminescence and the recycling of organic matter in his extensive volume *L'Océanographie* (31). This is the first oceanographic monograph that recognizes bacteria, and the first to include an extensive discussion on the need for improved techniques for aseptic sampling.

The Plankton expedition from Germany in 1889 (14) provided a major stimulus to the fledgling studies on marine microorganisms and established Germany as the center for such studies until World War I. By 1912, Benecke (3) could include a major chapter of 37 pages on microorganisms in seawater, sediments, and plankton and discuss their probable roles in the sea. After the war, however, the center for marine microbiology shifted, and major investigations began to originate in Great Britain, where Drew (11) was investigating denitrifying bacteria and Lloyd produced his classic papers on the Clyde Sea area (see Paper 7) (24). At the Tortugas Laboratory, Drew caused a major storm with his contentions on the role of microorganisms in $CaCO_3$ precipitation in coral reefs. Part of his hypothesis was confirmed over 50 years later by DiSalvo (Paper 34).

Many early papers, especially those from Germany, were concerned with two aspects of marine microbiology: the nitrogen cycle (discussed in Part V) and luminescent bacteria. Studies of the latter subject, which began with Pfluger's extensive discussion of the biological nature of luminescence (30) and Fischer's isolation of luminescing bacteria (13), led Beijerinck to study their distributions, nutrition, and taxonomy. In Paper 3, Beijerinck was concerned with much more than the idea that bacteria could be luminescent and found on sandy beaches. In fact, I believe, he would have been surprised not to find them there! He wanted to understand their function, nutrition, and biochemistry. He believed that the light-emitting capacity must be related to the nutrition of the organisms as well as the oxygen requirements (2) and their seasonal distribution. Thus, he recognized the importance of temperature, light, seawater, and fish extracts in his pure culture studies as he attempted to duplicate *in situ* conditions. The proteolytic enzymes released by the organisms were also of particular interest. This eminent Dutch microbiologist recognized diversity within the species concept, and grouped his isolates and those he could obtain from other workers into defined but flexible clusters of species.

Studies on luminescent bacteria continued very actively until the 1930s, with publications on the physiology of bioluminescent bacteria coming predominately from the laboratories of Richter (32, 33), Molisch (27), Mudrak (28), and Fuhrmann (18) on the European continent. Across the Atlantic, Harvey and Johnson and co-workers were also investigating this group of microorganisms (5, 19, 21, 22). Indeed, from 1920 to 1940 there was such an active worldwide research effort on the mechanism of light production and the physiology and nutrition of the bacteria that until quite recently very few papers have been published since then. Newer studies are now beginning to unravel the mysteries of the light-producing mechanisms of the luminescent bacteria (4, 9, 26, 29, 36).

Although not as widely recognized, several aquatic research institutes in the Soviet Union were producing basic studies on the distribution and roles of microorganisms in the Arctic Ocean (20) and the Caspian and Azov seas (6). Then, in the 1930s, in the United States, similar studies were conducted by Waksman (38, 39) and ZoBell (40).

The end result of this early effort was a concentration of major research emphasis on the functional aspects of microbial life in the seas, and one ceased to be amazed that microorganisms could be found in saline environments. Today, in the 1970s, we are rediscovering many of the old truths and gaining new insights with more sophisticated and more sensitive analytical techniques, which provide the quantitative data that were frequently unavailable to Certes, Russell, and all the other early pioneers in marine microbiology.

REFERENCES

1. Albert, Prince of Monaco. 1905. Considérations sur la biologie marine. *Bull. Mus. Oceanogr. Monaco, No. 56*:1–13.
2. Beijerinck, M. W. 1889. Les bactéries lumineuses dans leurs rapports avec l'oxygène. *Arch. Neerl. Sci. Exactes Nat. Haarlem 23*:416–427.
3. Benecke, W. 1912. *Bau und Leben der Bakterien*. B. G. Teubner, Berlin, 650 pp.
4. Borkhsentus, S. N., Fomina, V. V., Vinokurdova, T. I., and Domanskil, M. N. 1969. Isolation and crystallization of enzymes concerned with the bioluminescence of marine bacteria. *Biochemistry USSR 34*:1216–1222.
5. Brown, D. E., Johnson, F. H., and Marsland, D. A. 1942. The pressure-temperature relations of bacterial luminescence. *J. Cell. Comp. Physiol. 20*:151–168.
6. Butkevich, V. S. 1938. On the bacterial population of the Caspian and Azov seas. *Mikrobiology USSR 7*:1005–1021.
7. Certes, A. 1884. Note relative à l'action des hautes pressions sur la vitalité des micro-organismes d'eaux et des sédiments du *Travailleur* et du *Talisman*. *C. R. Seances Soc. Biol. Fil. 36*:220–222.
8. Certes, A. 1884. De l'action des hautes pressions sur les phénomènes de la putréfaction et sur la vitalité des micro-organismes d'eau douce et d'eau de mer. *C. R. Hebd. Seances Acad. Sci. 99*:385–388.
9. Cline, T. W., and Hastings, J. W. 1974. Bacterial bioluminescence *in vivo*: control and synthesis of aldehyde factor in temperature-conditional luminescence mutants. *J. Bacteriol. 118*:1059–1066.
10. de Giaxa, Prof. 1889. Ueber das Verhalten einiger pathogener Mikroorganismen im Meerwasser. *Z. Hyg. Infectionskr. 6*:162–224.
11. Drew, G. H. 1911–1913. The action of some denitrifying bacteria in tropical and temperate seas, and the bacterial precipitation of calcium carbonate in the sea. *J. Mar. Biol. Assoc. U.K. 9*:142–155.
12. Fischer, A. 1897. *Vorlesungen über Bakterien*. Fischer, Jena, 186 pp.
13. Fischer, B. 1887. Bacteriologische Untersuchungen auf einer Reise nach Westindien. II. Ueber einen lichtentwickelnden, im Meerwassen gefundenen Spaltplez. *Z. Hyg. Infektionskr. 2*:54–95.
14. Fischer, B. 1894. Die Bakterien des Meeres nach den Untersuchungen der Planktonexpedition unter gleichzeitiger Berücksichtigung einiger älterer und neurer Untersuchungen. *Zentralbl. Bakteriol. 15*:657–666.
15. Fischer, F. 1891. *Das Wassen, seine Verwendung, Reinigung und Beurtheilung*. Springer, Berlin, 284 pp.
16. Frankland, P., and Frankland, G. C. 1894. *Micro-organisms in Water*. Longmans, Green, & Co. Ltd., London, 532 pp.

17. Frankland, P., and Ward, M. 1892. First report to the Water Research Committee of the Royal Society on the present state of our knowledge concerning the bacteriology of water with especial reference to the vitality of pathogenic Schizomycetes in water. *Proc. R. Soc. Lond. B Biol. Sci. 51*:183–279.
18. Fuhrmann, G. 1914. Uber Nahrungsstoffe der Leuchtbakterien. Ver. 85. *Vers. Ges. Dtsch. Nat. Ärzte Wein. 11*:638–639.
19. Harvey, E. N. 1915. The effect of certain organic and inorganic substances upon light production by luminous bacteria. *Biol. Bull. (Woods Hole) 29*:308–312.
20. Issatschenko, B. 1911. Erforschung des bakteriellen Leuchtens des Chironomus (diptera). *Biol. Lab. K. Bot. Gart. St. Petersburg*:31–43.
21. Johnson, F. H. Eyring, H., Steblay, R., Chaplin, H., Huber, C., and Gherardi, G. 1944–1945. The nature and control of reactions in bioluminescence. *J. Gen. Physiol. 28*:463–538.
22. Johnson, F. H., and Harvey, E. N. 1938. Bacterial luminescence, respiration and viability in relation to osmotic pressure and specific salts of sea water. *J. Cell. Comp. Physiol. 11*:213–232.
23. Konig. J. 1887. *Die Verunreinigung der Gewässer.* Springer, Berlin, 624 pp.
24. Lloyd, B. 1930–1931. Muds of the Clyde Sea area. II. Bacterial content. *J. Mar. Biol. Assoc. U.K. 17*:751–765.
25. Mace, E. 1891. *Traité pratique de bactériologie.* Libraire J.-B. Baillière et Fils, Paris, 740 pp.
26. Makemson, J. C. 1973. Control of *in vivo* luminescence in psychrophilic marine photobacterium. *Arch. Mikrobiol. 93*:347–358.
27. Molisch, H. 1904. Die Leuchtbakterien im Hafen von Triest. *Akad. Wissen Wein. Math.-Naturwiss. 113*:513–528.
28. Mudrak, A. 1933. Beitrage zur Physiologie der Leuchtbakterien. *Zentralbl. Bakteriol. 88*:353–366.
29. Nealson, K. H., Platt, T., and Hastings, J. W. 1970. Cellular control of the synthesis and activity of the bacterial luminescent system. *J. Bacteriol. 104*:313–322.
30. Pfluger, E. 1875. Über die Phosphorescence Nerwesender Organismen. *Arch. Gesamte Physiol. Mens. Tiere (Pfluegers) 11*:222–263.
31. Richard, J. 1907. *L'Océanographie.* Vuibert and Nony, Editeurs, Paris, 398 pp.
32. Richter, O. 1926. Bakterienleuchten "Ohne Sauerstoff." *Plant. Arch. Wiss. Bot. (Berlin) 4–5*:569–587.
33. Richter, O. 1928. Natrium ein notwendiges Nährelement für eine marine mikroärophile Leuchtbakterie. *Denkschr. Akad. Wiss. Wein. Math.-Naturwiss. Kl. 101*:261–294.
34. Roux, G. 1892. *Précis d'analyse microbiologique des eaux.* Libraire J.-B. Baillière et Fils, Paris, 404 pp.
35. Russell, H. L. 1893. The bacterial flora of the Atlantic Ocean in the vicinity of Woods Hole, Mass. *Bot. Gaz. 18*:383–395, 411–417, 439–447.
36. Srivastava, V. S., and MacLeod, R. A. 1971. Nutritional requirements of some marine luminous bacteria. *Can. J. Microbiol. 17*:703–711.
37. Tiemann, F., and Gartner, A. 1889. Die Chemische und Mikroskopisch-Bakteriologische. *Untersuchung des Wassers,* 3rd edition containing

Kubel-Tiemann's *Anleitung zur Untersuchung von Wasser.* Friedrick Viewieg und Sohn, Braunschweig, 715 pp.

38. Waksman, S. A. 1934. The role of bacteria in the cycle of life in the sea. *Sci. Mon. 38*:35–49.
39. Waksman, S. A., Carey, C. L., and Reuszer, H. W. 1933. Marine bacteria and their role in the cycle of life in the sea. I. Decomposition of marine plant and animal residues by bacteria. *Biol. Bull.* 65:57–79.
40. ZoBell, C. E. 1934. Microbiologicl activities at low temperatures with particular reference to marine bacteria. *Quart. Rev. Biol.* 9:460–466.

1

ON THE CULTURE, FREE FROM KNOWN SOURCES OF CONTAMINATION, FROM WATERS AND FROM SEDIMENTS BROUGHT BACK BY THE EXPEDITIONS OF THE TRAVAILLEUR AND THE TALISMAN: 1882–1883

A. Certes

This article was translated by Marcel A. Gradsten and Carol D. Litchfield, from "Sur la culture à l'abri des germes atmosphériques, des eaux et des sédiments rapportés par les expéditions du Travailleur *et du* Talisman*: 1882–1883,"* C. R. Hebd. Seances Acad. Sci., **98**, *690–693 (1884)*

According to all observers, dredges from great depth never yield either plants or animals in a state of decomposition. How can this fact be explained? Would there not be at the bottom of the sea microbes analogous to the ones which, under our eyes, are working daily on the transformation of organic matter into inorganic matter?

I hasten to say that the experiments which I have the honor to report to the Academy do not solve the problem. However, they reveal a certain number of facts to which I deem [it] worthwhile to draw the Academy's attention.

Starting about two years ago, these experiments consisted essentially of the culture, protected from contamination, of sediments brought back in 1882 by the *Travailleur,* and since October 1883 [of the culture] of water and sediments brought back by the *Talisman.* From more than a hundred flasks cultured with a drop of water or a particle of mud from great depths,[1] only four of the cultures in contact with air have not grown. The cultures *in vacuo,* on the other hand, without exception remain sterile up to now. Thus,, anaerobic microbes but not aerobic microbes would have to be absent at the bottom of the sea.

Seawater sterilized at 120 and 128°C is a component of most of these nutrient liquids. I use it with the addition sometimes of a large quantity, sometimes of only a few drops, of calf or chicken broth, hay infusion, milk, or albuminous broth. I also made use of the liquids

[1]Depth of soundings (m)

Travailleur	927	1015	1094	2660	3100	4557	5100
Talisman	500	1918	2638	2685	3175	3705	

of Raulin and of Cohn. Prior to inoculation the flasks were held for several days in an oven. In a word, as indicated by the title of this paper, none of the precautions recommended by Mr. Pasteur were neglected to avoid the introduction of atmospheric or other contaminating germs. This condition, a sine qua non for all experiments of this nature, is easy to realize in a laboratory, but [it is] much more difficult to achieve on board a ship [which is] in motion. Thus, guided by a sense of prudence that will certainly be approved of, I thought I ought not publish the results of the experiments carried out in 1882 with the sediments of the *Travailleur*. Thanks to Mr. Alph. Milne-Edwards, who was kind enough himself to supervise the details of these delicate operations, all possible sources of error on board the *Talisman* seemed to have been [overcome], even the most difficult one with respect to the water tubes. The tubes to be used had been flamed at 200°C beforehand. By an ingenuous device produced by Mr. Alph. Milne-Edwards they only opened under water, [and] at the desired depth and at the precise moment when by turning around the recording thermometer to which they were attached [they] broke their taper.

At the present state of the art, we know that it is, so to speak, impossible to identify clearly the species of microbes, be it from the standpoint of their physiological activities [or] above all from considering their shape. Therefore, I will limit myself to describing the media in which several organisms were obtained from the mud.

Growth is very slow in certain media; but, in general, [it is] faster in the incubator than at room temperature. For example, the hay infusions and Raulin's medium sometimes become turbid only after 9 to 10 days. The molds appear lastly and only in milk, very dilute broths, and in Raulin's medium, where they grow to the exclusion of all other organisms. In the milk they show up only several days after the bacilli and probably when this first culture has altered the composition.[2]

In the neutral media, only bacilli appear, mostly motile, rather long, very thick, with refractive bulky spores. Less frequently, one encounters a large type of vibrio in the shape of a club, these organisms giving the appearance of a single or double capsule, and finally some micrococci.

The cultures from water (500, 1918, 3975 m) present the particular feature that the microbes, always the same, are much smaller and more agile than those from the mud. They form a veil [pellicle] on the surface, which happens less frequently in the other cultures. These differences are noteworthy, but in view of the limited number of flasks inoculated with water, it would be premature to conclude from this that the organisms from water always differ from those from mud. I have not encountered ciliated or flagellated infusoria in any of these cultures. It was different with water from the Sargasso Sea, which was given to me by the Marquis de Folin and which I have cultured,

[2]These molds were sent to the exhibition of the *Talisman* at the museum.

protected from contamination, by the addition of a few drops of calf broth. In the flasks so prepared I found, besides the usual small rods and characteristic diatoms, numerous amoebae, remarkable for their extreme smallness and flagella, among which a small number [was] of a highly interesting and probably new kind.

By successive culturing, several pure cultures were obtained, that is, consisting of only one kind of organism. It was a big bacillus, quite abundant, and nearing sporulation. On my request, Professor Cornil was so kind [as] to inoculate guinea pigs. These inoculations have never affected the health of the animals, not even in massive doses; the slight inflammation that was produced at the site of the inoculation always promptly disappeared without a trace.

To sum up, as of now it is legitimate to concede that at great depths of the ocean, the water and the sediments contain germs, which, in spite of the enormous pressure that they undergo, do not lose the ability to multiply when they are placed in the right media at the right temperatures. Do these germs originate exclusively at the surface and are they slowly deposited at the bottom of the seas? Can they be regarded as physiologically distinct species from the ones we know already? At this time we are unable to give the answer, but we can at least try to resolve this difficult question by new experiments. Thanks to the kindness of Mr. Cailletet, who obligingly put at my disposal his ingenious devices, I am now undertaking a new series of experiments with cultures in which will be duplicated, as much as possible, the pressure and temperature conditions prevailing at great depth. These delicate experiments will require a certain amount of time, and this is one of the reasons that cause me to submit to the Academy the results of my first experiments.[3]

The experiments with cultures in air and *in vacuo* will also be continued with numerous sediments that I have not yet had time to utilize.[4]

[3]In an initial experiment we have recovered living, chlorophyll-containing flagellated infusoria held for several hours at a pressure of 100 atm and even in a few cases at 300 atm.

[4]These experiments have been carried out at the laboratory of Mr. Pasteur, to whom, as well as to his regular co-workers, Messrs. Chamberland, Roux, and Loir, I would like to express my thanks.

2

Reprinted from pp. 112–116 of P. Frankland and G. P. Frankland, *Micro-Organisms in Water*, Longmans, Green & Company, London, 1894, 532 pp.

MICRO-ORGANISMS IN WATER

P. Frankland and G. P. Frankland

[*Editor's Note:* In the original, material precedes this excerpt.]

Sea-water.—The bacterial contents of sea-water is a subject of obvious interest, but one on which but very little information was until recently available. Even now, practically, the whole extent of our knowledge is based upon the results obtained by two investigators, De Giaxa[1] and Russell,[2] both of whom conducted their experiments in the Gulf of Naples.

De Giaxa, whose investigations only incidentally included some determinations of the number of bacteria

[1] *Zeitschrift für Hygiene*, vol. vi. 1889, p. 186.

[2] *Ibid.* vol. xi. 1891, p. 177.

in sea-water, found as many as 298,000 in a c.c., but the sample was taken only 50 metres from the entry of the Chiatamone Canal (an excessively foul water containing the sewage of a large part of Naples); at a distance of 350 metres the number fell to 26,000, whilst at 3 kilom. only 10 were discoverable.

Russell,[1] whose researches in this particular direction were much more extended, examined samples taken from depths of 75 to 800 metres, at distances of 4 to 15 kilom. from the shore, and found from 6 to 78 microbes in 1 c.c. taken from the surface, and from 3 to 260 at various depths below.

The following table, whilst showing the particular distances from land as well as the various depths at which the samples were taken, brings out very clearly that, whilst the total number of bacteria in sea-water is comparatively small, there appears to be no marked decrease in the numbers corresponding to the greater distance from land of the water selected for examination, all the samples, be it observed, having been collected at a distance of upwards of 4 kilometres from the shore, and thus beyond the reach of any littoral influences.

The Distribution of Bacteria in Sea-water both Vertically and Horizontally (Russell)

Depth of the sea-bottom at places of collection	Distance from land	Number of bacteria in 1 c.c. of water taken from the surface	Number of bacteria found in 1 c.c. of water taken at various depths							
			75 m.	100 m.	150 m.	200 m.	250 m.	300 m.	500 m.	800 m.
m.	km.									
75	4	64	57	—	—	—	—	—	—	—
100	6	22	3	5	—	—	—	—	—	—
150	9	8	—	—	10	—	—	—	—	—
200	11	26	—	260	—	112	—	—	—	—
250	10	15	—	—	—	—	10	—	—	—
300	11	78	—	20	—	—	—	5	—	—
500	15	6	—	—	—	—	53	—	23	—
800	6	30	—	—	—	—	—	—	—	3

[1] *Zeitschrift für Hygiene*, vol. xi. 1891, p. 177.

Thus the nearest point to the land at which a sample was taken was 4 kilom., where the sea had a depth of 75 metres, and hence the chance of any disturbance in the normal bacterial contents of this sea-water from accidental contamination from the shore is very slight; and even if any such source of error should arise, it would at most be only likely to take place during or after violent storms blowing from the coast.

Examination of sea-mud revealed the presence of very large quantities of bacteria. It was found, however, that the numbers steadily diminished up to a certain point with the greater depth at which the samples were abstracted, this being especially noticeable in the immediate vicinity of the coast. After 250 m. and up to 1,100 m. no further important reduction was observed.

Number of Bacteria found at various depths in 1 *c.c. of sea-mud and* 1 *c.c. of sea-water respectively* (Russell)

Depth in metres	Water	Mud
50	121	245,000
85	57	285,000
100	10	200,000
140	10	70,000
200	59	70,800
250	31	27,000
300	5	24,000
400	30	22,000
500	22	12,500
825	31	20,000[1]
1,100	—	24,000[1]

[1] Russell states that these higher figures must not necessarily be regarded as indicating a larger number of bacteria as being present at the greater depth, but are to be attributed rather to slight *local* variations, as the samples were collected over as wide an area as possible.

As regards the varieties of organisms present, it was ascertained that more than half appeared only to belong to this mud, and were not discoverable in the sea-water itself; three individual microbes in particular were especially characteristic of this mud, as much as 35 per cent. of the total number of colonies found on gelatine-plates consisting of these three forms (see pp. 454–456).

Russell has more recently [1] extended his investigations to an examination of the sea-water and mud in the vicinity of Wood's Holl, Massachusetts. The number of microbes present in these more northern and cooler waters was markedly less at this point than in the Mediterranean. The slime from Buzzard's Bay yielded an average of 10,000 to 30,000 bacteria in 1 c.c., which Russell says represent but a small fraction of those present in the Mediterranean mud at equal depths. Here again two species were found to be specially prevalent in the water, together with two or three other forms occasionally met with. The mud also contained these two prevailing water forms, but another form, an indigenous slime bacillus, occurred in such large numbers as to make up from thirty to fifty per cent. of the whole quantity present.

Samples of mud were also obtained, about 100 miles from the shore at the depth of 100 fathoms, on the edge of the great continental platform skirted by the Gulf Stream. These samples are the farthest from land that have ever been bacteriologically examined, and bacteria were present in large numbers; moreover, the two prevailing species present were identical with those obtained near the shore at Wood's Holl.

Russell mentions that the *Cladothrix intricata* (see p. 517) was only rarely met with, whereas in the Mediterranean mud it was frequently found.

On comparing these results with those already referred to for fresh-water lakes, it will be seen that the distribution of bacteria in the two cases is substantially similar, both the lake and the ocean at a

[1] *Botanical Gazette*, vol. xvii. p. 312. 1892.

distance from land being characterised by poverty in bacterial life.

Having now obtained some idea of the bacterial contents of various waters, we must direct our attention to a consideration of some of the numerous methods which may be adopted for removing them.

3

PHOTOBACTERIUM LUMINOSUM, A LUMINOUS BACTERIUM IN THE NORTH SEA[1]

M. W. Beijerinck

This article was translated by Marcel A. Gradsten and Carol D. Litchfield, from "Le Photobacterium luminosum, *bactérie lumineuse de la Mer du Nord,"* Arch. Neerl. Sci. Exactes Nat. Haarlem, **23**, *401–405 (1889)*

To describe a luminescent bacterium not recorded to date, I deem it necessary, in view of the imperfect state of the literature on this group of organisms,[2] to give first what is known about the occurrence of the different types of luminous bacteria. Because it was impossible for me to set up distinctions sufficiently important to permit generic separation of the five (or six) species that I have studied, and, on the other hand, because the need is felt for possibly designating these very interesting organisms simply by one name, I venture to assign them all to the genus *Photobacterium*. The different species found up to now are the following:

1. *Photobacterium phosphorescens*, the ordinary luminescent bacterium, nonliquefying, from phosphorescent fish. There are some doubts about the legitimacy of this generally used name (cf. *Pflüger Archiv.*, vols. 10 and 11, 1874 and 1875; Lassar, *Pflüger's Archiv.*, vol. 21, 1880; Ludwig, *Zeitschrift für Mikroskopie*, vol. 1, 1884; Tilanus, *Tijdsch. v. Geneeskunde*, vol. 2169, 1887; Forster, *Bacteriol. Centralblatt*, vol. 1, p. 337, 1887).[3]
2. *Photobacterium indicum*, a luminescent bacterium from the Caribbean Sea, discovered and described by Fischer (*Zeitschr. f. Hygiene*, vol. 2, p. 54, 1887) under the name of *Bacillus phosphorescens*.
3. *Photobacterium fischeri*, a luminescent bacterium from the Baltic Sea, discovered and described by Fischer (*Bacteriol. Centralblatt*, vol. 3, p. 105, 1888).
4. *Photobacterium luminosum*, named here for the first time.

To these four species must be added two species from the Baltic Sea, not described yet and closely related to *P. fischeri*; one of them slightly dissolves gelatin, whereas the other one does not liquefy it.

[1]*Editor's Note:* See p. 22 for Drawings 1–4, referred to throughout the text of this article.
[2]Cf. Macé, *Traité Pratique de Bactériologie*, p. 585, Paris, 1889.
[3]From the standpoint of priority, the name *Micrococcus phosphoreus* Cohn should be used for the ordinary luminous bacterium (*Verzameling van stukken betreffende het Geneeskundig Staatstoezicht in Nederland*, 1878, p. 126), but usage has decided in favor of *P. phosphorescens*.

On closer examination they will perhaps be recognized as variants of *P. fischeri*.

The reasons which lead me to believe that all[4] the species mentioned above should be combined into one genus are as follows:

1. They all grow better or even exclusively when the food contains 3.5 percent sea salt or isotonic ratios of other mineral salts.
2. They lose their [light-emitting] power by the addition of 2 percent or more glucose to their food, [because] they form then an acid and take on very peculiar shapes.
3. Peptone is their main source for nitrogen intake; they receive their carbon from very dilute solutions of glucose, levulose, maltose, galactose, calcium lactate, and above all from glycerol, and this assimilation is accompanied by the production of light.
4. They develop in a neutral or slightly alkaline medium, and a trace of acid is sufficient to extinguish the light.
5. They never form spores; they can all be brought by culture to motile stages, which swim toward the sources of oxygen, under certain conditions taking on the shape of spirilla and of vibrios.
6. None of them secretes diastatic or inverting enzymes. Hence, soluble starch, cane sugar, and milk sugar can neither serve for food nor for the production of light, because these substances as such are not oxidized.
7. All give a continuous light spectrum between the D and G lines, hence, in the yellow, green, and blue region.

Besides these analogies, there exist, however, several important differences among *P. phosphorescens*, and the other species (I have here particularly in mind numbers 2 and 4, since I have not yet completely studied the others). These differences consist primarily in the unique ability of *P. phosphorescens*, among all these species, to ferment glucose, levulose, maltose, and galactose in the absence of air, and secondly, that *P. phosphorescens* does not secrete a proteolytic enzyme as readily as the others (except for one of the two forms not yet described). Later, differences in the shape of these species will be mentioned. After this brief introduction, I pass now to my actual subject, the description of *P. luminosum*.

At the end of the summer of 1888, I studied repeatedly with great care the phosphorescence of the sea between Katwijk and Scheveningen [the coast off South Holland]. I do not know whether the phenomenon is always exactly the same as during the period I am referring to. I assume that along the coast line in question, and perhaps also farther north up to Nieuwe-Diep, one could always observe [it] to about the same extent that I saw last year. The main points of what I saw are as follows.

[4]I have studied in detail and for a long time species 1, 2, and 3, but for a briefer time, species 4, for which Prof. Fischer provided me the material, and which I had myself once isolated from a sea fish. The two other forms, not described, are also in my possession, thanks to the kindness of Mr. Fischer.

The actual cause for the light emitted by the surf and the sand of the shore must be searched for in the microscopic animals and bacteria. In fact, some coelenterata (of rather considerable size) such as *Cydippe pileus* and *Phialidium variabile* are phosphorescent. Certain species of *Sertularia* and of *Obelaria* can also usually be found along the coast of the North Sea and the Phialidies, particularly during the hot summer evenings, and can be seen by the thousands on the beach. However, all these forms of varying size contribute only a very little to the light of the surf spread over large areas. The microscopic animals that I observed belonged mostly to some kinds of Crustacea and Dinoflagellates, which I was unable to define; there was also a small amount of *Noctiluca miliaris.* These are the animals to which must be attributed the characteristic scattering of sparks by the waves as they unroll over the sand, as well as the brilliant scintillation caused by an abrupt movement or by the falling of raindrops in the puddles of water on the beach. Also, the lighting itself due to these animal forms was plainly local; and the mat glow extending over the foam of the surf, even on cursory examination, must be related to a different cause.

As a cause I was able to identify, after some research, a form of bacterium new to science.[5] Before describing this species, which I named *Photobacterium luminosum,* I should mention that not only the seawater but also the sand on the shore, to the extent that it is subject to the action of the tide, is, so to say, completely penetrated by the culture, and that the peculiar, luminous aureole formed around each imprint of the foot, so well known to all those who take a walk on our beaches in the evening, is due to the presence of *P. luminosum.*[6] Besides, I once happened to isolate this bacterium in a plaice, bought in the market in Delft, which had become luminous in my laboratory. Hence, the possibility of encountering this bacterium deeper in the interior of the country cannot be entirely excluded.

The easiest way to obtain *P. luminosum* as pure culture is the following. In the darkness of the evening after a hot summer day, look for one of those water puddles which the retreating tide leaves on the beach and which has been heated by the rays of the sun. At the edge and at the bottom of such a puddle, the sand is extremely rich in luminescent bacteria and eminently suitable for inoculation. A mixture composed of an extract of fish in seawater with the addition

[5]From the descriptions of eyewitnesses I must infer that at a southern point in Holland, far from the coast, never close to the shore, the phosphorescence of *Noctiluca miliaris* can be observed in all its splendor; in Scheveningen, Katwijk, and Zandvoort this spectacle never presents itself, if I have understood correctly my sources of information. Is this perhaps related to the circumstance that at the above localities the water of the North Sea contains less salt?

[6]The speed with which our bacteria multiply in the sand, evidently at greater speed than in seawater, it perhaps explained by the increase of the proportion of salt in the sand, and inevitable result of desiccation at ebb tide.

of 0.5 percent peptone and coagulated with 7 percent gelatin is especially recommended as the nutrient medium. Luminous sand then is mixed with boiled seawater, and a needle dipped in this mixture is used to draw lines on the gelatin. [Alternatively] the seawater covering the sand may be poured onto the gelatin medium and subsequently the gelatin is drained until its surface is dry. After only 24 hours at summer temperatures, brightly luminescent colonies can be seen here and there within the lines or in the surface growth covering the gelatin. By making from them, in the usual manner, new cultures in the same medium, a multitude of isolated colonies could be secured with a second run.

Photobacterium luminosum is one of the bacteria that liquifies gelatin quite strongly (Drawing 4a). When this fusion takes place in the presence of much nitrogeneous material, for example, 0.5 percent of asparagine and 0.5 percent of peptone, no fetid products develop; when, on the other hand, the nitrogen supply is limited, a putrefaction process sets in, evidently at the expense of the gelatin. In the following manner, it can be easily demonstrated that the liquefaction is due to a particular substance. *Photobacterium luminosum* grows rapidly and glows weakly in a medium similar to the one described above when the gelatin is replaced by agar. If, in addition to this agar medium, a little coagulated egg white, finely powdered fibrin, or coagulated blood serum is added, and if on this cloudy plate a wide line of *P. luminosum* is traced, a large, cleared region, which is due to the dissolving of the albumin introduced into the agar, is seen within a few days, forming around and below the bacterial culture. On digging a small cavity into this cleared area and replacing the agar removed by gelatin, the liquefaction of the latter is immediately observed. Microscopic examination shows that at the spots where these events occur, not a trace of bacteria exists; hence, they must depend on the presence of an easily diffusible enzyme.

As with the other photobacteria, *P. luminosum* develops only in a neutral or slightly alkaline medium. A minute amount of acid suffices to prevent completely growth and the production of light. This fact is remarkable, because under unfavorable cultural conditions, for instance in the presence of glucose, our bacterium itself secretes small quantities of acid that cut short the luminous emission. But this property is also common to other luminescent bacteria that I have studied. On the gelatin [prepared] from the broth of peptonized meat and on blood serum, *P. luminosum* does not grow at all. However, if we add 3 or 3.5 percent sea salt, potassium chloride, or magnesium chloride, the above media become quite suitable for cultivation, and the light can attain the same intensity as on the fish extracts. Based upon experiments with *P. phosphorescens*, I am convinced that everything is dependent on the proper osmotic pressure. I actually found that this species can emit light and even maintain growth in very different inorganic salt solutions, as long as they are isoosmotic with a 3 percent sodium chloride solution. This is contrary to the quite different effect of organic substances, such as glycerol and sugar. For that reason, it must be concluded that the phosphorescence of meat, potatoes, and other organic materials, of which the literature provides numerous data, can only be attributed to the presence of *P. luminosum,* if these products contain a sufficient proportion of salt. This reasoning applies also to other known luminous bacteria (*P. phosphorescens, P. indicum,* and the bacteria from the Baltic Sea, *P. fischeri*).

The shape of *P. luminosum* varies greatly, depending upon the nature of its food (Drawing 4b and d). If it contains little nitrogen and carbohydrate, our bacterium is generally very small and its shape resembles cholera vibrios. Here and there spirilla of various length are seen, along with the small rods. When observed attentively during the course of their existence, we sometimes see them fractionate into short vibrios. The particular curvature of the spirilla is the reason that their fission products result in curved rods. This is what led Koch, in the analogous case of cholera spirilla, to name his vibrios "comma bacilli." Spirilla and vibrios move rapidly, looking for free oxygen at the edge of the preparations. Growth without free oxygen is impossible regardless of the conditions.

This is the place to describe briefly the differences in shape existing among the four species described up to now. In the presence of the extreme polymorphism to which they are all subjected, it is, however, impossible to cite constant morphological differences, recognizable under all circumstances. Under ordinary culture conditions, that is, under bright lighting and on not too rich a medium, the following is generally observed. The ordinary nonliquefying *P. phosphorescens* (Drawing 1a and b) forms rounded or somewhat irregular pouches, often containing a darker inclusion that multiplies by independent binary fission at the same time as the bacteria; in rarer instances, there are small rods of varying length, double rods or diplococci, some of which, at any rate, swim slowly in various directions. The thickness of the small rods is about 0.5 μm, and the length about 1 μm; the dimensions of the globular cells are very variable, fluctuating between 0.5 and 2 μm. In the presence of glucose, and particularly in the simultaneous presence of glucose and aspargine, the small rods and globules swell up considerably, and, due to the internal formation of an acid, completely lose their luminescent properties. Under these circumstances, some small rods appear to show branching and take then the "bacteroides" shape peculiar to *Bacillus radicicola* from the tubercles of Papilionaceae.

Photobacterium indicum, the strongly liquefying luminescent bacterium of the Caribbean Sea (Drawing 3a and b), closely resembles *P. lumi-*

[7]The excavation of the surface is probably linked to the removal of water from the colony by the gelatin that surrounds it.

nosum as to the exterior aspect of the colonies, but its luminescent intensity is much greater and the hue of the fluid of the colonies (the fluid has a slightly concave surface[7]) is more of an ashen gray. The shape of this fast-moving bacterium is that of a straight small rod with rounded extremities; only rarely are individuals seen that curve inward and approach the shape of spirilla.

Photobacterium indicum and *P. luminosum* behave in about the same manner toward oxygen; the motile stages search for the sources of this gas. Without oxygen there is no or almost no growth, nitrates are not reduced at all, and indigo blue is reduced only with difficulty. The optimal temperature for the growth and luminescent functioning of *P. luminosum* lies around 25 to 28°C, which is lower than for *P. indicum,* whose optimum lies around 30 to 32°C according to my experiments. Hence, the speed of development of *P. luminosum* is notably greater than that of *P. indicum*. However, it is impossible to state a general rule in this respect, because the (optimal) temperatures in question depend upon the nature of the food and upon the proportion of salt. I will return later to the changes in shape to which these two bacteria are subject under the influence of glucose mixed into their food.

For the moment, I wish to say a word about the shape of the bacillus of the Baltic Sea, *P. fischeri* (Drawing 2a and b). While sharing its proteolytic power with the two previous species, *P. fischeri* is easily distinguished from them by the small dimensions of the vibrios and the small rods which swim with agility. These small rods are only 0.1 to 0.3 μm thick and about 1 μm long, and I never saw among them longer filaments or spirilli, which are so characteristic for *P. indicum* and *P. luminosum,* at least in young cultures. The exterior aspect of the colonies, the luminous power of which equals that of *P. indicum,* surpasses that of *P. luminosum,* and yields only to that of *P. phosphorescens,* is very peculiar because of the profound excavation (Drawing 2a), which we have already mentioned. Besides, our bacterium is easily recognized by the fact that its cultures continue to produce light for a long time, without any renewal of nutrients. In this respect, it exceeds by four to six weeks the other two liquefying species, which in seven to twelve days are usually extinguished, and it thus equals *P. phosphorescens*. Its light does not hve the beautiful bluish-green hue so typical for the last-named three species. It draws more toward the orange, and its luster is less lively. In this respect, *P. fischeri* resembles very much the two species not yet described but mentioned above, species that also have a close affinity to it as regards the shape of the bacteria themselves.

Nonluminescent and Nonliquifying Cultures

Once in a while, although in general seldom, the inoculations carried out with aged cultures of *P. indicum, P. luminosum,* and *P. phos-*

Forms of Luminous Bacteria and of Their Colonies

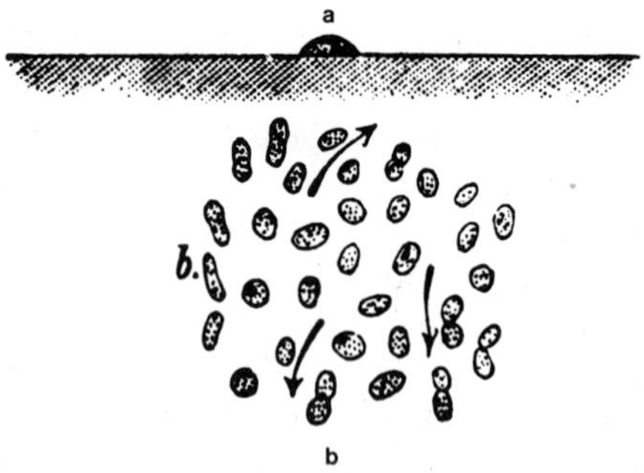

Drawing 1 *Photobacterium phosphorescens,* the ordinary luminous bacterium. (a) Ordinary colony on gelatin from a culture. (b) (800) Ordinary shape of the bacterium from colony *a,* seen in sea water; the arrows indicate the mobility.

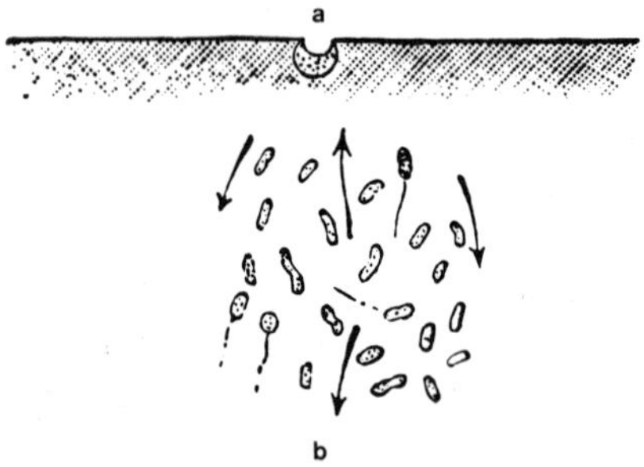

Drawing 2 *Photobacterium fischeri,* luminous bacterium from the Baltic Sea. (a) Colony causing gelatin flow with concave surface. (b) (800) Ordinary form in seawater.

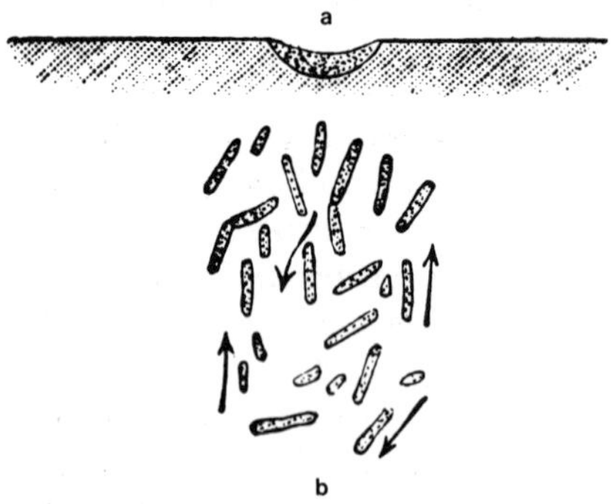

Drawing 3 *Photobacterium indicum,* luminous bacterium from the Caribbean Sea. (a) Colony. (b) (800) Form in seawater.

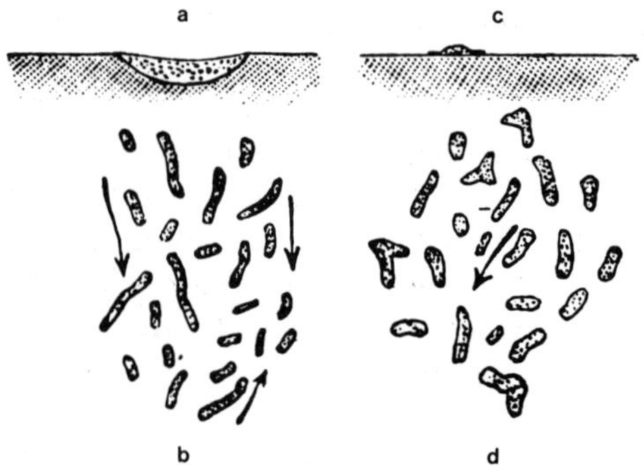

Drawing 4 *Photobacterium luminosum,* luminous bacterium from the North Sea. (a) Ordinary colony causing flow. (b) (800) Shape of the bacteria from colony *a* seen in seawater. (c) Colony not causing flow, under the influence of glucose. (d) (800) Shape of bacteria from colony *c* seen in seawater.

phorescens give rise to colonies with litle or no luminosity at all. Sometimes, the luminous power of the entire seeding is uniformly weakened; in other cases, only a few individuals have undergone this modification. On carrying out new inoculations with reduced or nonluminescent colonies, the newly acquired property proves hereditary; however, on further repetition of the operation, one frequently succeeds in restoring the original luminescent power. In some instances, notably with *P. phosphorescens,* I have seen successions of inoculations derived from nonluminescent colonies become spontaneously luminescent again after some time. The phenomenon of weakening, just discussed, is much more often seen with *P. indicum* and *P. luminosum* than with *P. phosphorescens;* when inoculating from tube to tube, one is thus constantly exposed to the decreasing luminescent power of the cultures. To keep it at the same level, a selection in the inoculation of the colonies must be effected from time to time. It is certain that the exhaustion of the culture medium or, perhaps more exactly, the extended action of the secretion products of the bacteria without sufficient input of nutrient material, determine in one way or another the change in question. But as to the why and how of the phenomenon, there is no hypothesis at the moment. Naturally, the case makes one think of the production of the two forms, well characterized by their colonies, in the putrefaction organism, *Vibrio proteus,*[8] or of the separation of *Micrococcus prodigiosus* into pigmented and nonpigmented individuals,[9] or perhaps of the loss of virulence in contaminators. Obviously, this matter is important, since it has bearing on the general problem of heredity, and luminous bacteria will perhaps be more useful in examining this question than other species.

The preceding lines had already been written when Prof. Fischer from Kiel, to whom I had sent some time ago my *P. luminosum,* announced that my cultures had produced two very different forms of luminescent bacteria. One of them was denoted by Mr. Fischer as "in the form of the hay bacillus," the other one resembling more "the bacillus of anthrax" as in the form of "a Medusa head"; by inoculation both forms were recognized as stable. Since I am quite sure that the bacteria from my shipment came from one colony only and the latter from a single individual, a process must have taken place analogous to that with *Vibrio proteus.* Very probably, the nature of the food must have played a role here, since at my place no change has occurred similar to that just discussed, and undoubtedly the nutritive mixtures employed are never exactly the same at the various bacteriological laboratories. Along with *P. luminosum,* I sent Mr. Fischer weakly luminous cultures of *P. indicum* and *P. phosphorescens;*

[8]Gruber, *Bacteriol. Centralblatt,* T. V, p. 345, 1889.
[9]Schottelius, Untersuchungen über den *Micrococcus prodigiosus, Festschrift für Kölliker,* Leipzig, 1887.

yet, with him, these cultures fully regained their light-emitting power, while he was not able to observe its weakening. Moreover, I ascertained great differences in stability among the poorly luminous forms; I actually own an *almost* completely dark state of *P. indicum*, and this state is very stable; when moved to different culture media, it remained dark; not a single colony showed any tendency to return to the former state.

Besides the loss of luminescent power due to hereditary influences, there exists still another effect of the same type, which is purely temporary and which can be caused at will by modifying the food in certain ways. This type of weakening in luminosity is always accompanied by a prompt arrest in growth, a more or less complete loss in liquefying power, and a most peculiar change in the shape of the small rods and of the vibrios. Its cause must certainly be sought for in the penetration into the interior of the organism of particular substances. As such, I identified certain sugars, primarily glucose, and then, to a lesser extent, levulose and maltose; aspargine too exercised an analogous action. The amounts of these substances needed to stop the emission of light or liquefaction are minimal; the addition of 1 percent of glucose or 1 percent of asparagine is ample. The light of liquefying luminescent bacteria by no means comes forth, as sometimes claimed, from a secreted substance; as in *Phosphorescens*, it is linked to the living matter itself, whereas the liquefying power depends on the secretion of an enzyme analogous to trypsin.[10] It is, therefore, evident that the penetration of glucose or asparagine into the substance of the bacterial body must cause a twofold disturbance. In fact, nothing favors the assumption that the enzyme, as long as it is enclosed in the body of the bacterium, acts as a combustible. In all probability, the light-emitting power is incidental to the respiration of oxygen; the energy developed during this process, which by ordinary organisms is given off as heat, is here partially transformed into light.

The hypothesis put forth by Mr. Radzzewsky,[11] to wit that the emission of light would depend on the preliminary formation of some special matter, and subsequently accepted by Mr. Bütschli,[12] does not seem to me very probable. It is certain, at the least, that the aldehydes which I have tested, such as lophine, amarine, hydrobenzamide,

[10]The proof that bacteria endowed with peptonizing action do not secrete, as is usually claimed, a pepsin, but a substance resembling the enzyme of albumin, secreted by the pancreas, results from the transformation occurring only in a neutral or slightly alkaline reaction. With pepsin, on the other hand, the presence of free acid is required. The proof is further substantiated by the nature of the products formed, among which, besides peptones, tyrosine and leucine are found, which are absent in true peptic processes.

[11]Ueber die Phosphorescenz der organischen und organisirten Körper, in *Liebigs ann.* 203, 1880, p. 305.

[12]*Protozoan*, p. 1004.

bitter almond oil, trimethylene oxide $(CH_2O)_3$, and acetaldehyde,[13] give rise, even in small amounts, to the extinction of the light. The conclusion that phosphorescence depends upon free and not combined oxygen derives from the way *P. phosphorescens* acts. In this species, growth and partition can actually take place by fermentation, even in a medium completely deprived of oxygen, where indigo white does not turn blue to any degree. But in that case, not the slightest trace of light is observed. The ability of hydrogen peroxide to support phosphorescence is explained by the violent decomposition of this compound under the influence of luminescent bacteria.

[13]All these substances were donated to me by Prof. van't Hoff.

Part II

THE PROBLEMS OF THE SAMPLING, CULTIVATION, AND IDENTIFICATION OF MARINE BACTERIAL ISOLATES

Editor's Comments on Papers 4 Through 6

4 **RUSSELL**
Studies on Bacteria Living in the Gulf of Naples

5 **ZOBELL**
Apparatus for Collecting Water Samples from Different Depths for Bacteriological Analysis

6 **NISKIN**
A Water Sampler for Microbiological Studies

THE DEVELOPMENT OF SAMPLING DEVICES

When standing on the deck of a ship and looking down into the depths of the sea, one knows that microscopic life is there, waiting to be studied. But how may one recover samples of water or sediment from the desired location and know that the samples are uncontaminated by the overlying water? How does one culture these mysterious creatures; what are their requirements for growth; will they even grow in the laboratory at all? What kinds of microorganisms are there? Do they fit into the classical taxonomic schemes? These were among the many questions that were asked and had to be answered before any progress in our understanding of the role of microorganisms in the sea could be achieved. Intuitively, one says that the microbes must be both the same and different. After all, they are microorganisms and hence possess certain basic properties; but because of where they live, they must have different requirements for growth, perhaps even specialized physiological mechanisms to account for the effects of temperature, pressure, and salinity. Before seeking answers to these questions, techniques had to be developed for the aseptic recovery of marine samples.

Although disputed by some even today (9), others (6, 25) have demonstrated the importance of aseptic sampling devices to protect from both bacterial and fungal contamination. Each early investigator was on his own in regards to sampling equipment and cultural techniques. Although Certes (5) maintained a sterile water sample, no detailed description of it has been found. Besides being among the first marine microbiologists, Russell, in Paper 4, recognized the need for the aseptic sampling of seawater; he therefore developed a simple large test tube with a rubber stopper closure. The rubber stopper contained a small sealed glass tube whose end was broken by a messenger. Russell also described a sediment sampler with a small cutting edge at one end and a screw-top that could be placed on the opposite end to ensure sterility once the sample was recovered. By the use of a sterile cork borer and knife, he recovered sediment from the center of the core; and since he found more bacteria here than in the immediately overlying water, he concluded that the sediment was not seriously contaminated by outside sources. Today, over 80 years later, we have not really improved on this technique for obtaining sediment for microbiological analysis. The only currently available sterile sediment sampler was designed for use in shallow water (60 ft) and results in a thoroughly mixed surface sediment–water interface(24). By this procedure it is not possible to obtain more than 4 to 6 cm of sediment.

Several workers are currently developing sediment samplers that would permit aseptic recovery of mud cores in which the core liner has not been contaminated by passage through the water. Some are even working on ways of maintaining the *in situ* pressure so that at last we can analyze the importance of pressure on the growth and reproduction of deep-sea marine microbes.

Unlike the comparative neglect in regards to the development of sediment samplers, the technological evolution of bacteriological water samplers has been continuous since Russell's time (19, 20). Until recently, two predominant types had evolved centered around (1) improvement in Russell's sterile bottle, and (2) the evacuated bulb technique. Among the changes in Russell's design (19) are the improvements in frame construction and capillary tubing advanced by Abbott in 1912 (1) and Wilson in 1920 (26) after experiencing difficulties with the wooden frame support that Russell employed. An early extension of the Russell design was introduced by Abbott, who eliminated the glass inlet tube and instead had the rubber stopper removable at the desired depth by means of a separate line. The sealed sterile bottle could be lowered to the desired depth and the cap removed by this second line; after an appropriate time was allowed for filling, the cap was reinserted, and the closed bottle returned to the surface. The obvious disadvantage, which made further refinements necessary, was the difficulty of adequately replacing the stopper, besides the continual problem of having a second line over the side of the ship, which could then become fouled.

A simplification of the closure system was proposed by Bertel in 1911 (3), who used a messenger-activated system. However, his bottle was unfortunately a nickel-plated cylinder, which proved to be toxic to the collected bacteria. Portier and Richard (15) proposed an all-glass ampule with a capillary tube along with a metal cover, which permitted sterile sampling and closure of the opening to the chamber as the unit was returned to the surface. The all-glass Wilson water sampler was only slightly modified and used by Lloyd.

At about this time, Matthews (10) modified the metal cylinder of Otto and Neumann (12) by devising a deep-sea glass sampler. The ends of the container were covered with washers; when submerged, the sampler, which was filled with 95 percent alcohol to kill any contaminating bacteria, was lowered to the desired depth, the ends were opened, and, as the alcohol rushed out, the container was filled by in-rushing seawater. The ends were then closed and the sampler returned to the ship. It seems likely, now, that the residual disinfectant probably resulted in lower counts of the bacteria in seawater, or that the disinfectant might have leaked out under pressure and thus negated its function. Also, the fragility of the apparatus undoubtedly discouraged its widespread acceptance. This device was modified by the insertion of a piston and the substitution of a metal chamber by Young et al. (27). In any event, the Matthews sampler was not widely accepted, and glass tube samplers were replaced by the evacuated tube device (20), which was cheaper and easier to prepare; it was used extensively by the early workers (8, 11, 13, 16).

Most early designs were really not suitable for deep-sea microbiological sampling because of fragility, difficulties in filling the containers, and small sample volume. Therefore, in 1941, ZoBell introduced a modification of both the glass bottle and the evacuated tube. In Paper 5 he describes the modification of the glass capillary attached to a rubber pressure tube, which in turn is attached to the rubber stopper inserted in a citrate of magnesium bottle. It was the positioning of the inlet tube and the bottle holder that made this a more acceptable design for deep-sea sampling. The entire apparatus could be sterilized, it was easy to construct, and, except for the inlet tube, it was reusable. Today, with only minor modifications, this type is commercially available as the J-Z sampler. Also in Paper 5, ZoBell describes the use of an evacuated bulb that could be attached to a sterile sealed tube and used for sampling at greater depths than was possible with the early designs. Slight changes, especially in the attachment of the inlet tube to the bulb, have since been made, but these are also a commercially available product today.

Perhaps, though, the most important aspect of Paper 5 was ZoBell's report on the toxicity to marine bacteria of the metal Nansen bottles, It has often been standard practice by oceanographers to obtain water samples for both microbiological and chemical studies using these open ended, nonsterile, metal bottles. Bedford (2) in 1931 pointed out the toxic effects of 60-min exposure to the "Prince Rupert" bottle described above. ZoBell extended the work of Raadsveld (17) on the oligodynamic effects of heavy metals on microbial growth and demonstrated that even a 5-min exposure to the metal bottle could be toxic to many marine species. Although disputed by some workers (9), others have confirmed both the toxicity and the contamination that can result from the use of Nansen bottles for the collection of bacteriological samples (18, 22, 23).

Despite the advancement in techniques for the aseptic recovery of water samples, one major drawback to the J-Z system has been the small volume of the water sampler. The ideal system would allow one to determine microbial populations or microbial activity as well as analyze for various nutrients, all on the same sample of seawater. This approach provides more meaningful comparative data, but it requires a larger volume of seawater than the previously described samplers permit. Consequently, the Niskin sampler was developed to obtain aseptically up to 2 liters of seawater at any depth, without leakage, in a sterile disposable bag (Paper 6). In the few years since the Niskin sterile water sampler has become commercially available, it has achieved widespread acceptance and is generally considered the sampler of choice for collecting large volumes.

Other highly specialized techniques have also been developed by various investigators. However, these are usually designed for a more limited specific purpose, such as the "microlayer" sampler of Sieburth (21) for surface layer samples. The bubble scavenging technique for concentration of cells when population densities are low (4), the exotic capillary tubes of Perfil'ev (14), and the syringe-dialysis bag for small-quantity anaerobic samples (7) are all examples of special adaptations of equipment designed for obtaining samples in which specific activities or properties of marine bacteria may be demonstrated. Several of these procedures are discussed more extensively in the following sections.

Undoubtedly, as more complex and penetrating questions are asked about microorganisms in the oceans, each investigator will have to modify or develop the sampling techniques necessary for his purpose. On a routine basis, though, the J-Z and Niskin samplers for seawater appear to fill most current needs for aseptic water samples. Future advances in sampling technique must be made in the methods for obtaining aseptic sediment samples and for the maintenance of both seawater and sediments under ambient temperatures and pressures to permit the study of more realistic microbial interactions and functions.

REFERENCES

1. Abbott, A. C. 1915. *Principles of Bacteriology*. Philadelphia, 617 pp.
2. Bedford, R. H. 1931. The bactericidal effect of the "Prince Rupert" sea water sampling bottle. *Contrib. Can. Biol. Fish. 6*:423–426.
3. Bertel, R. 1911. Ein einfacher Apparat zur Wasserentnahme aus beliebigen Meerestiefen für bakteriologische Untersuchungen. *Biol. Zentralbl. 31*:58–61.
4. Carlucci, A. F., and Williams, P. M. 1965. Concentration of bacteria from seawater by bubble scavenging. *J. Cons. Perm. Int. Explor. Mer. 30*:28–33.
5. Certes, A. 1884. Sur la culture, ã l'abri des germes atmosphériques, des eaux et des sédiments rapportés par les expéditions du *Travailleur* et du *Talisman*: 1882–1883. *C. R. Hebd. Seances Acad. Sci. 98*:690–693.
6. Holmes, R. W. 1958. *Scripps Cooperative Oceanic Productivity Expedition, 1956*. U.S. Dept. Inter. Fish Wildl. Serv. Spec. Sci. Rept. Fish. 279.
7. Jannasch, H. W., and Maddux, W. S. 1967. A note on bacteriological sampling in seawater. *J. Mar. Res. 25*:185–189.
8. Johnston, W. 1892. On the collection of samples of water for bacteriological analysis. *Can. Res. Sci. 5*:19–28.
9. Kriss, A. E., Mishustina, I. E., Mitskevich, I. N., and Zemtsova, E. V. 1967. *Microbial Population of Oceans and Seas*, K. Seyers (trans.) and G. E. Fogg (ed.). St. Martin's Press, New York, 287 pp.
10. Matthews, D. J. 1913. Deep sea bacteriological water bottle. *J. Mar. Biol. Assoc.* 9:525–530.
11. Miquel, P., and Cambier, R. 1902. *Traité de bactériologie pure et appliquée*. Paris, 372 pp.
12. Otto, M., and Neumann, R. O. 1904. Ueber einige bakteriologische Wasseruntersuchungen im Atlantischen Ozean. *Zentralbl. Bakteriol. 13*:481–489.
13. Parsons, P. B. 1911. Apparat zur Entnahme von Wasser aus grösser Liefe. *Zentralbl. Bakteriol., Abt. II, 32*:197–207.
14. Perfil'ev, B. V., and Gabe, D. R. 1969. *Capillary Methods of Investigation of Microorganisms*. University of Toronto Press, Toronto.
15. Portier, P., and Richard, J. 1907. Sur une méthode de prélèvement de l'eau de mer destinée aux études bactériologiques. *Bull. Inst. Oceanogr. No. 97*:2–4.

16. Praum, L. 1901. Einfacher Apparat zur Entnahme von Wasserproben aus grösseren Tiefen. *Zentralbl. Bakteriol., Abt. I, 29*:994–996.
17. Raadsveld, C. W. 1934. The oligodynamic effect of metals and metal salts. *Chem. Weekbl. 31*:497–504.
18. Rodina, A. G. 1972. *Methods in Aquatic Microbiology,* R. R. Colwell and M. S. Zambruski (trans.). University Park Press, Baltimore, Md., 461 pp.
19. Russell, H. L. 1892. Untersuchungen über im Golf von Neapel lebende Bakterien. *Z. Hyg. Infectionskr. 11*:165–206.
20. Russell, H. L. 1892. Bacterial investigation of the sea and its floor. *Bot. Gaz. 17*:312–321.
21. Sieburth, J. McN. 1965. Bacteriological samplers for air-water and water-sediment interfaces. *Ocean. Sci. Ocean Eng.,* pp. 1064–1068.
22. Sieburth, J. McN. 1971. Distribution and activity of oceanic bacteria. *Deep-Sea Res. Oceanogr. Abst. 18*:1111–1121.
23. Sorokin, Yu. I. 1964. On the problem of the method of microbiological sampling in the sea in the light of recent developments in marine microbiology. *Okeanologiya 4*:349–353.
24. Van Donsel, D. J., and Geldreich, E. E. 1974. Bacterial bottom sampler for freshwater sediments. Preprint. Environmental Protection Agency, Washington, D.C.
25. Willingham, C. A., and Buck, J. D. 1965. A preliminary comparative study of fungal contamination in non-sterile water samplers. *Deep-Sea Res. Oceanog. Abst. 12*:693–695.
26. Wilson, F. C. 1920. Description of an apparatus for obtaining samples of water at different depths for bacteriological analysis. *J. Bacteriol. 5*:103–108.
27. Young, O. C., Finn, D. B., and Bedford, R. H. 1931. A deep sea bacteriological water bottle. *Contrib. Can. Biol. Fish. 6*:417–422.

4

STUDIES ON BACTERIA LIVING IN THE GULF OF NAPLES

H. L. Russell

This article was translated by Marcel A. Gradsten and Carol D. Litchfield, from pp. 165–167, 167–175 of "Untersuchungen über im Golf von Neapel lebende Bacterien," Z. Hyg. Infections., **11**, *165–206 (1892)*

INTRODUCTION

While substantial progress was made during the last decade in the study of freshwater bacteria, investigators paid little attention to seawater species. The reason for this is probably the lack of practical significance of the latter for hygiene and for the fast developing pathological bacteriology. Nevertheless, seawater species seem to offer a productive area for investigation. The whole field of bacteriology is a new science and every investigation that promises to throw more light on the morphology and biology of these organisms deserves to be carried out with care. This study is not only of systematic interest but leads also to problems of a different type. It is possible that a thorough investigation of all species to be considered, and particularly those from the depths of the sea, may throw some light on the hitherto unresolved problem of the decay of organic matter in deep sea. The peculiar conditions under which the species of the deep sea exist also present problems that are of interest from a systematic as well as from a physiological standpoint. The investigation presented here was carried out in the spring and during the start of the summer of this year[1] and should not be considered completed; however, the results obtained seem to me of sufficient importance to justify the publication of the pages to follow.

My studies should yield a quantitative estimate and qualitative description of the bacteria living in the deeper water layers as well as of those from the mud from various depths. Although, an institute as well equipped as the Zoological Station offers all the facilities to carry out a long series of investigations, the shortness of time at my disposal may explain why my studies are not more exhaustive. The geographical location of the Zoological Station is also particularly favorable for the exploration of species living at greater depths. While the Gulf of Naples has expanses at considerable depths, the sudden and sharp dip of the bottom of the sea near the island of Capri particularly favors

[1]At the bacteriological laboratory of the Zoological Station at Naples.

the study of the deeper layers of the sea. At this point the bottom of the sea sinks within a few kilometers from land to a depth of 1500 m. Most of the observations were carried out at lesser depths, since it had to be regarded as best first to get acquainted with the easily attainable species that were living there before turning to the investigation of the deeper water layers. Research at greater depths was engaged in from time to time, whenever an occasion presented itself, in order to determine the generality of the distribution of bacteria.

I feel indebted to privy councillor Prof. Dr. Dohrn, and wish to express my sincere gratitude for the invariable kindness with which he put everything at my disposal that could advance my endeavor; I would also like to thank Dr. Kruse, the head of the bacteriology department, for his advice and interest in my work.

As far as I could gather, there are no observations yet on the life of bacteria at greater depths of the sea; research on the surface layers of the seawater seemed to prove that bacteria play only a very minor role among the multitude of organisms living in the ocean. The prevailing opinion seems to be that, due to its salt content, seawater acts more or less as a disinfectant, being not only free of indigenous species of bacteria, but also capable of destroying fission fungi or at least of inhibiting their growth. This idea, which arises from the ability of salt water to preserve organic tissue, does not seem to have an adequate scientific foundation. Forster's[2] investigations on the effect of sea salt, even in concentrated solutions, on the viability of pathogenic germs show that the latter are able to live for a rather long time in salt solutions. De Giaxa[3] studied, here in Naples, the effect of sterilized as well as unsterilized seawater on the growth of certain pathogenic microorganisms. He found that they not only live but also grow in seawater free of pathogenic species; however, in seawater with the normal content of bacteria (he took his samples near the coast), after various periods of time, they succumb in the battle for existence to the saprophytes, which are better suited to the life in the sea.

The usual fission fungi present in freshwater and in the soil are destroyed by the action of seawater and its microorganisms. Sanfelice[4] found that the bacterial content of seawater rapidly drops with distance from shore; although he did not determine by qualitative analysis whether this decrease in bacteria occurred at the expense of bacteria brought in by waste products, he considered it a distinct possibility. The investigation that I carried out with surface water distant from the coast showed that it did not contain terrestrial bacteria.

[2]Forster, *Muenchener medicinische Wochenschrift*. 1889. p. 497.
[3]de Giaxa, *Muenchener medicinische Wochenschrift*. 1889. Vol. VI. p. 162.
[4]Sanfelice, *Estr. dal. Boll. della Soc. d. nat. in Napoli*. 1889.

The following paper primarily deals with bacteria from the deeper seawater layers and the mud beneath them. Analyses of the surface water were only made to check the content of the layers below it.

[*Editor's Note:* Material has been omitted at this point.]

APPARATUS FOR THE EXTRACTION OF MUD SAMPLES

Another apparatus was put together for obtaining mud from the bottom of the sea. The relative solidity of mud permits the use of simple instruments based on the principle of the borer; however, it is difficult to open and close an apparatus on the bottom of the sea. The apparatus constructed for this purpose probably satisfies all requirements, although it cannot be totally sterilized. At the least, I believe to have obtained by it mud samples free of foreign matter, containing nothing but what actually can be found in mud beds. The appended sketch may serve to facilitate understanding [Figure 2]. A

Figure 2

simple iron tube of small diameter (approximately 10 mm) picks up the mud. The lower end is trimmed so that it can more easily penetrate into the bottom, but [is] not so sharp that it breaks or bends when it comes upon a stony expanse. The upper end, equipped with a cap with [a] female thread, can be opened and closed at will. In its turn, this cap carries on top a valve closure with a rubber gasket whose construction is shown in the sketch. The valve closure is so arranged that it opens when the tube is lowered, which allows the water to flow through the tube; when pulling up the tube filled with mud, the valve closes as a result of water pressure, thus preventing further entrance of water and, on the other hand, escape of mud. The tightness of the valve closure can be increased by fitting an annular depression of the rubber bearing to an annular elevation of the cap. This precaution is, however, hardly necessary because the smooth valve plate already closes tightly enough to prevent [the] entrance of water when pulling up the apparatus.

The tube intended for taking up the mud is fastened to an iron rod heavy enough to hold the apparatus constantly in a vertical position and capable of pushing the apparatus deeply into the mud immediately after hitting the bottom. If everything is correct, the tube should fill almost completely or completely with mud of regular consistency by a single push at the bottom. Securing a mud sample by means of a tube of not more than 10-mm diameter is only troublesome when pure gravel sediment is present. The bottom of the sea is usually covered with an extraordinarily fine, somewhat sticky mud, and in this case, an even strong admixture of gravel is not obstructive for obtaining a bottom sampling. [However], from a depth of a few meters where the bottom consists of pure sand without mud, it is difficult to obtain material.

The material secured is now treated in the following manner. As soon as the apparatus comes out of the water, the cap is screwed off, and the mud is pushed so far outward, by means of a piston fitting the tube, that it protrudes sausage-shaped 2 to 3 cm over the lower end of the tube. Then this end of the small mud cylinder, which was in contact with the water when the apparatus was pulled up, is cut off with a sterilized knife, and a sterilized small brass tube of 2- to 3-mm diameter is bored into the center of the mud still remaining in the metal tube. In this manner a small amount from the center of the bottom sample in question is obtained aseptically and quickly. The mud always forms quite a solid, small column, in which, obviously, the bacteria have much less motility than in a liquid. Since no water can flow through the filled tube, the bacteria have no way of spreading except by diffusion, which in at most 5 minutes could have happened to only a very small degree.

METHOD OF INVESTIGATION

The usual methods for the investigation of water were employed. After thorough shaking for best distribution of the microorganisms, 0.5 ml was taken from the tubes that contained the seawater by means of a sterilized pipette, mixed with seawater gelatin, and the plates prepared in the usual manner. The plates remain for 48 hours in the humidity chambers in order to avoid as much as possible any impurities from the air. While counting the colonies, the plates were kept under sterilized covers, and the glass window of the counting apparatus was sterilized with mercuric chloride. By applying these precautionary measures, the number of contaminations was reduced to a minimum, and plates free of bacteria could be obtained for 10 days and longer without any contaminants. On soil investigations, which are also applicable for mud examinations, opinions differ as to which method yields the best results. Fränkel[5] claims that the best results are obtained by mixing the soil sample directly with the liquefied gelatin and employing the roller method of Esmarch. Beumer[6] and others say that the soil must first be shaken up in sterilized water to effectuate an even distribution of the microorganisms in the gelatin. It is my belief that not one and the same method should be always applied, but that the quality of the soil and other conditions should be considered before one or the other method has been selected. As Reimers[7] has already pointed out, a sandy soil, which disintegrates easily, mixes very well with gelatin, whereas a muddy, clayey soil first needs shaking in a thinner liquid before it can be adequately distributed in the gelatin.

In my sea-bottom investigations it was impossible, even with the help of a sterilized glass rod, to mix the more or less sticky, clayey mud with gelatin and to distribute it evenly in the latter. If the gelatin is shaken vigorously it traps many air bubbles, which makes a subsequent, accurate count impossible. For this reason I used the dilution method, which gave me favorable results, although this method requires more manipulations and consequently introduces new sources of error into the investigation. The plate method was used for water as well as for mud because of the uniformity of the investigation. In addition, the preparation of Esmarch roll cultures was made impossible by the presence of a species common to mud that liquefies gelatin very easily. Sometimes I had to add a drop of a disinfectant to delay as much as possible the development of this species. As a unit of measure, I used a certain volume because it was easier and in general as reliable to work with as with weights. By means of this unit of volume I could also compare the results of the mud investigations with the ones of water.

The unit of measure was prepared in the following manner. As mentioned before, the mud was removed from the iron pipe by means of a small brass tube (2 to 3 mm in diameter). The smallest unit from a cork borer set is particularly suitable for this purpose, because its sharp, knife-like edge enables it to penetrate the small mud column without pressing it together; this latter condition is necessary if the volume is to be used as the unit of measure. This instrument can also be quickly cleaned and sterilized in an open flame if repeated use is required. The content of the cork borer can be readily determined,

[5]Fränkel, this journal, Vol. I. p. 527.
[6]Beumer, *Deutsche medicinische Wochenschrift.* 1886. nr. 27.
[7]Reimers, this journal, Vol. VII. p. 307.

and a given part of its length can serve as the unit. The mud is removed out of the borer with the help of the stamper coming with the borer for its ordinary use. I employed as the unit a certain length of the small mud column, that is, 0.5 cm, which corresponded to a volume of 0.05 ml. By varying the length of the segment of the little mud column, any multiple of this unit can be readily obtained. The portioned off amount of mud was transferred to 100-ml Erlenmeyer flasks, each of which contained 25 ml of sterilized, distilled water, and vigorously shaken for several minutes. The liquid became usually more or less turbid, and when, during clearing, small lumps formed at the bottom, the fineness of the precipitate was increased by crushing and stirring with a sterilized glass rod. When the dispersion of the mud had sufficiently progressed, 0.5 ml of this suspension was withdrawn from the flask by means of a sterilized pipette and, in the same way as the water samples, used for the preparation of plates. In this manner, 0.001 ml became the unit in the mud investigation.

When looking for a unit, be it a volume or a weight, it is first necessary to determine approximately the number of organisms in a given quantity. Obviously, the unit chosen is much too large if the plates prove to contain thousands of colonies. In my opinion, the most satisfactory and precise results are obtained when the number of colonies on one plate amounts at most to a few hundreds. If the colonies are too crowded on the plate, it becomes very difficult to count even a small number of squares on the glass window, and estimates made on this basis always contain a substantial source of error. It is improbable from the outset that the various species contained in the material for investigation grow at the same speed, or that even the ability to partition is the same for single individuals of the same species. On a totally overgrown plate that had not been observed for 2 to 3 days, the slowly growing species were not noticed at all. It may be objected that heavy addition of water increases the chance for committing mistakes, but I believe that this objection can be entirely refuted if, on different plates of the same mud sample, about the same number of colonies is found, because this would indicate that the bacteria are quite evenly distributed throughout the whole diluting liquid.

Table 1, which gives mean values for the thinly poured plates, shows the increase in number of colonies on consecutive days. It is evident that high dilution is also an advantage in the determination of the species. In the beginning the growth of many forms, and particularly of the saprophytes, is not so characteristic as to make it possible to differentiate among the various species before the colonies have aged for several days. The customary method of water investigation yields, at the best, approximate values, as the true number of microorganisms present will be always greater than the numbers of the grown colonies. However, if one proceeds always in the same manner, the source of error remains constant, and the information so obtained will become more accurate.

Table 1

Date of preparation of plates	Material of samples	Depth (m)	Number of colonies on consecutive days			
			23/V.	24/V.	25/V.	26/V.
21/V.	Mud	15	140	232	275	280
			108	212	249	265
21/V.	Mud	85	102	241	290	293
			96	215	278	278
21/V.	Mud	100	58	162	193	212
			65	155	168	168
21/V.	Water	100/100[a]	4	8	11	15
			5	5	5	11
21/V.	Water	50/100	5	10	20	37
			1	6	17	20
21/V.	Water	85/85	19	29	32	37
			8	24	25	25
21/V.	Water	0/100	22	30	31	42
			21	32	41	42
21/V.	Water	0/85	51	66	70	98
			35	49	64	90

[a]In fractional numbers, the denominator means the depth of the sea at the place in question and the numerator the depth from which the water was taken.

The conditions at which the bacteria at the bottom of the sea and from the deep sea exist are so different from those offered to them by the ordinary culture methods, which are geared to freshwater species, that with this manner of investigation their percentage loss must be greater than for freshwater species.

In regular cultures one obtains, of course, the prevailing aerobic species. In our case, however, anaerobic or at least facultative anaerobic bacteria must be expected because of the great distance from the atmosphere. To ascertain the anerobic organisms on hand, cultures were prepared by three different methods: (1) in small Erlenmeyer flasks in which the air was replaced by water, (2) in test tubes with pyrogallol solution, as specified by Buchner, and (3) on glass plates covered by another sterilized glass plate, as recommended by Sanfelice.[8] The latter method of investigation was used most often

[8]Sanfelice, *Atti della R. Acc. Med. di Roma*. Anno. XVI. 1890. p. 378.

because of its simplicity, and to prevent the entrance of air, sterilized gelatin was poured around the culture. As a nutrient medium, gelatin with the addition of 1 percent of glucose or sodium formate was used. It is not acceptable to estimate the number of germs present per unit by adding up the colonies obtained from anaerobic and aerobic cultures because of the existence of bacteria that grow as well on the admission of as on the exclusion of air. These latter species can be observed not only in anaerobic and aerobic plates, but also in high stratum cultures, growing everywhere with the same abundance.

RESULTS OF THE QUANTITATIVE INVESTIGATION

In a quantitative analysis of the sea bacteria, those from the bottom of the sea as well as the ones from the surface layers have to be considered. A priori one might expect that gravity affects bacteria in the same way as any other matter, so that the number of bacteria increases in the deeper water layers and becomes highest at the bottom of the sea. A rather significant number of analyses carried out with mud from various depths and with the water located right over it always showed that the *bacterial content of the mud was much higher than that of the same volume of water above it.* Table 2 gives the results of these analyses.

Table 2

Depth (m)	*Number of microorganisms found in 1 ml*			
	Water	*As an average from samples*	*Mud*	*As an average from samples*
50	121	6	245,000	6
85	57	4	285,000	4
100	10	3	200,000	4
140	10	5	70,000	3
200	59	6	70,800	10
250	31	4	27,000	2
300	5	2	24,000	2
400	30	2	22,000	4
500	22	2	12,500	4
825	31	2	20,000	5
1100	—	—	24,000	5

On the basis of the results, one must ask for an explanation for this excess of mud bacteria. The question has various answers; the number of bacteria at the bottom increases by a gradual depositing of the bacteria from above, in a vegetative stage or in the form of spores, or by the growth and propagation of the bacteria indigenous to the mud, or finally by a combination of both factors. To check the correctness of the first assumption, it was necessary to get a good idea about the distribution of the bacteria in the total quantity of water.

It follows from the many experiments carried out with freshwater that local variations in the bacteria content are always so large that only an extensive number of analyses will permit one to draw a valid conclusion on its size. In any case, I have not carried out enough determinations to offer a final opinion; nevertheless, my results should suffice to indicate what we may expect, since I analyzed for their bacteria content water samples originating close under the surface, as well as next to the bottom and in the middle layers of the sea. The *Challenger* expedition proved that sea fauna divides into two relatively well separated zones, one living at the surface and the other one at the bottom strata of the sea. Accordingly, it would be interesting to find out to what extent this law applies to plant life. Obviously, the higher organized plants and those with chlorophyll cannot exist at a depth where light is missing. Therefore, the majority of the fungi are also unable to grow under such conditions. It is most probable that the larger part of the plants living at any depth of consequence belong to the schizomycetes and the other lower organized classes.

Table 3 shows not only the vertical but also the horizontal distribution of the bacteria. The figures represent the mean content of the microorganisms of all analyses made at the various depths. From these results, it is impossible to arrive at a uniform, vertical distribution of the bacteria. The number of fission fungi per milliliter is always so small, thus increasing the source of error to such an extent in the preparation of the plates, that a much more comprehensive series of investigations would be needed to be able to tell whether a regularity in the distribution exists. The table indicates that, contrary to the animal world, the bacteria do not decrease substantially at certain depths. With respect to the horizontal distribution in the surface strata, clear regularity ceases when exceeding the boundary posed by possible

Table 3 On the vertical and horizontal distribution of bacteria in the seawater investigated

Depth of sea (m)	*Distance from land (km)*	*Number of bacteria in 1 ml of surface water*	*Number of bacteria found at various depths (m)*							
			75	*100*	*150*	*200*	*250*	*300*	*500*	*800*
75	4	64	57	—	—	—	—	—	—	—
100	6	22	3	5	—	—	—	—	—	—
150	9	8	—	—	10	—	—	—	—	—
200	11	26	—	200	—	112	—	—	—	—
250	10	15	—	—	—	—	10	—	—	—
300	11	78	—	20	—	—	—	5	—	—
500	15	6	—	—	—	—	53	—	23	—
600	6	30	—	—	—	—	—	—	—	3

impurities from land bacteria. Sanfelice and de Giaxa both observed a rapid decrease of germs living at the surface depending upon the distance from land, but not beyond a distance of 3 km. Table 3, which covers analyses made up to 15 km from land, shows that the number of germs in 1 ml always remains relatively small, and that no significant decrease occurs with distance from land. As one can see, I made observations at a distance of at least 15 km from land only in water 75 m deep. Usually, impurities from land do not reach such distances, unless they are moved by heavy, sea-bound storms. Table 3 indicates furthermore that *the bacteria content of the bottom strata does not fluctuate according to a certain rule,* and that on the whole it is not particularly affected by change of depth. But we must also remember that the number of germs in 1 ml is too small to give accurate information about the influence of the conditions which change with the depth. As to the distribution of the bacteria of the sea, a great difference is noticeable.

5

Reprinted from the *J. Mar. Res.*, **4**(3), 173–188 (1941)

APPARATUS FOR COLLECTING WATER SAMPLES FROM DIFFERENT DEPTHS FOR BACTERIOLOGICAL ANALYSIS[1]

By

CLAUDE E. ZOBELL

Scripps Institution of Oceanography[2]
University of California
La Jolla, California

Prerequisite to quantitative studies on the occurrence and importance of bacteria in the sea or other natural waters is a satisfactory device for collecting water samples for bacteriological analysis from any desired depth. Such a device must be easily sterilized, susceptible to aseptic manipulation under the most rigorous field conditions, possible to operate at high hydrostatic pressures and it must be made of biologically inert materials. It is desirable that the bacteriological water sampler can be used on the standard hydrographic or sounding wire in tandem or multiple units to provide for the simultaneous collection of samples from different depths and in conjunction with Nansen, Ekman, or Allen bottles, reversing thermometers or other hydrographic apparatus on the same wire. Sturdiness of construction, convenience of operation and economy of manufacture are also important features.

A survey of the literature reveals that more than a hundred bacteriological water samplers have been described introducing new

[1] Assistance in the preparation of these materials was furnished by the personnel of Work Projects Administration Official Project No. 65-1-07-2317.

[2] Contribution new series No. 150.

principles or modifications, but they all have limitations which restrict their usefulness. Unfortunately the data obtained from the bacteriological analysis of water collected with many samplers are open to criticism or are of only historical significance because the containers could not be sterilized, they were subjected to possibilities of extraneous contamination or they were made of metals which exert a bactericidal effect. The majority of the samplers which have been described are suitable only for collecting surface samples or samples from shallow depths. Others are unreliable, excessively expensive, intricately complicated or otherwise unsatisfactory. The apparatus which is described below has given good results under various conditions, it is simple to construct and it is positive in operation at any depth.

HISTORICAL

Besides the numerous methods which have been described for collecting water samples from the surface, three different types of samplers have been used: (1) weighted glass bottle from which the stopper is removed by a string, spring or messenger at the desired depth, (2) cylinder to which water is admitted by opening valves or by withdrawing a piston and (3) partially evacuated glass receptacle to which water is admitted by breaking a capillary glass tube.

Johnston (1892), Heydenreich (1899), Abbott (1921), Whipple (1927), Zillig (1929), Eyre (1930) and others have described ingenious devices for removing the stopper from a sterile bottle after it has been lowered into the water. However, it is obvious that the operation of such samplers is restricted to relatively shallow depths because the hydrostatic pressure at greater depths prevents the removal of stoppers from empty bottles. (The hydrostatic pressure of water increases approximately one atmosphere, or 15 lbs. per square inch, for each ten meters of depth.) In fact, unless provisions are made to prevent their descent, rubber stoppers will be pushed into the bottle at 50 to 100 meters and eschewing this, the empty bottles will be crushed at depths of a few hundred meters.

Otto and Neumann (1904) collected samples in a metal cylinder the ends of which could be closed with rubber gaskets at the desired depth. A nickel-plated cylinder with cocks operated by a messenger was used by Bertel (1911). However, it is objectionable to send such a cylinder down open because it might become contaminated with extraneous material while descending through the water and it is virtually impossible to perfect leak-proof valves or cocks which operate with sufficient ease to permit a pressure-resistant cylinder to be sent down empty. Matthews (1913) sought to obviate this difficulty by filling a glass-lined cylinder with 95 per cent alcohol which provided for the sterility of the apparatus as well as for the equalization of pressure. One messenger was dropped to open the ends of the cylinder thus permitting the disinfectant to diffuse out, after which a second messenger closed the valves to entrap a water sample from the desired depth. Drew (1914) used this apparatus with expressed confidence to depths of 800 meters. Others have used it with phenol solution or other disinfectants. Besides the expense and inconvenience of operating this apparatus there is always a possibility of the disinfectant diffusing out prematurely through a faulty valve, or all of the disinfectant may not be washed out during the time the valves are opened and closed.

Young, Finn and Bedford (1931) fitted a phosphor-bronze cylinder with a brass piston which is withdrawn by a messenger-activated mechanism thereby aspirating a water sample. In order to eliminate the bactericidal effect of metals, Renn (unpublished) used a large ground-glass cylinder fitted with a solid ground-glass piston or plunger. It is doubtful, though, if such cylinder and piston arrangements can be made so that they are absolutely leak proof at great depths and it is especially difficult to exclude water from the orifice until the messenger is dropped. Unless it is closed right to the end, the orifice may admit a small amount of surface water which will be aspirated when the piston is withdrawn. This objection applies to Reyniers' (1932) glass cylinder which is ingeniously opened and closed by flexible rubber tubing.

Russell (1892) applied the unpublished evacuated glass-bulb capillary-tube idea of Massea by fitting a small flask with a glass tube the end of which was sealed hermetically. Provisions were made for breaking the end off the glass tube with a messenger thus permitting the ingress of water. Modifications of this capillary tube idea have been described, in most cases independently, by Praum (1901), Miquel and Cambier (1902), Portier and Richard (1906), Druse (1908), Parsons (1911) and Issatschenko (1914). In its most practical form Wilson (1920) used a large test tube fitted by means of a rubber stopper with a capillary glass tube the end of which is sheared off by a lever activated by a messenger. Gee (1932) attached it to an Ekman bottle, provided a long capillary tube to retard the sudden inrush of water when the tip was broken at great depths and used a thick-walled test tube with the neck constricted to prevent the descent of the rubber stopper into the test tube with increasing pressure. ZoBell and Feltham (1934) bent the glass inlet tube in such a way that it was broken at a point remote from possibilities of contamination by the messenger and simplified the breaking mechanism.

BACTERICIDAL EFFECT OF METALS

Alloys of copper, lead, nickel, silver, tin, zinc or other metals lend themselves readily to the manufacture of water sampling apparatus but it has been recognized for many years that all except the noblest metals have an oligodynamic effect (Raadsveld, 1934). After finding a reduction of 97 to 100 per cent in the bacterial population in 100 ml. quantities of sea water exposed to 2 square inches of bronze, nickel, brass and other alloys, Drew (1914) concluded that platinum is the only metal suitable for the interior of bacteriological water sampling apparatus. According to Bedford (1931) the bactericidal effect of the bronze and brass sampling bottle of Young et al. (1931) was inappreciable in sixty minutes but noticeable in ninety minutes.

Observations made at the S. I. O. during the last decade indicate that the less noble metals have a pronounced effect on the survival and activity of bacteria in samples of sea water. Under certain conditions most of the bacteria in sea water are killed within a few minutes when in contact with bright brass, bronze or other alloys containing copper, nickel, tin or zinc. Sea water itself stored in brass receptacles for a few hours becomes bacteriostatic for some bacteria although it enhances the growth of other bacteria. It is beyond the scope of this paper to consider the factors which influence the oligodynamic effect of heavy metals on the activity of bacteria in sea water but a few experiments will be summarized to show that metal containers for bacteriological water samples should be shunned.

Investigators on several oceanographic expeditions have analyzed bacteriologically water samples collected with Nansen, Ekman or other metal bottles. As a rule they observe that in spite of the fact that the bottles were not sterilized, few or no bacteria could be demonstrated in the water. Such observations have been interpreted as indicating that there are few or no bacteria in the sea. That such an interpretation is not always valid is proved by the data recorded in Table I which gives a protocol of a series of experiments in which sea water was stored in Nansen bottles.

Thoroughly cleaned Nansen bottles were filled with raw sea water. After different periods of storage at 22° C. samples were withdrawn for bacteriological analysis. Control water samples were stored in glass bottles for the same time. The bacterial population was determined by plating procedures using nutrient sea water agar (ZoBell, 1941). It will be observed from Table I that whereas the

TABLE I—Number of Bacteria per Cubic Centimeter of Sea Water After Different Periods of Storage at 22° C. in Glass Bottles and in Brass Nansen Bottles

	Experiment 1		*Experiment 2*		*Experiment 3*		*Experiment 4*	
Period of Storage	*Glass bottle*	*Nansen bottle*	*Glass bottle*	*Nansen bottle*	*Glass bottle*	*Nansen bottle*	*Glass bottle*	*Nansen bottle*
0	94	—	108	—	143	—	294	—
1 hour	86	30	92	25	137	80	253	86
2 hours	72	21	102	18	89	26	222	19
3 hours	90	15	110	10	118	25	266	4
5 hours	52	0	76	3	226	1	207	0
24 hours	446	6	539	0	590	1	890	2

number of bacteria in the control sea water in glass containers decreased slightly with storage and then started to increase, the bacteria rapidly disappeared from the sea water which was stored in Nansen bottles for a few hours.

Only when collected from appreciable depths is there a likelihood of the water being in the Nansen bottle for an hour or longer. However, further experiments summarized in Table II indicate that many

TABLE II—Number of Viable Bacteria in Sea Water After Different Periods of Storage in Different Receptacles at 22° C., the Number Being Expressed as the Per Cent of the Original Number Present

Time in minutes	*Nansen bottle No. 3*	*Nansen bottle No. 13*	*Nansen bottle No. 18*	*Nansen bottle No. 19*	*Glass bottle control*
0	100	100	100	100	100
5	89	94	85	69	103
10	77	82	74	62	95
30	71	78	56	40	96
60	42	70	51	36	92

bacteria are killed or rendered incapable of multiplication in Nansen bottles within a few minutes. The average results of from three to six different experiments with four different Nansen bottles are given. Some of the Nansen bottles had a greater bacteriostatic effect than others, probably due to differences in the metallic surfaces which were exposed to the water samples. It is generally recognized that brightly polished metallic surfaces are more bacteriostatic than those which are coated with oxides or other film-forming substances.

GLASS BOTTLES

After trying different sizes and kinds of flasks, test tubes and bottles it was found that citrate of magnesia (CM) bottles with a one-hole rubber stopper substituted for the patented stopper, or ordinary pop or beer bottles are the best for collecting water samples. They are economical, rugged in construction, of sufficient volume and they stand erect on their flat bottoms without the need of special racks. Moreover, unlike the test tubes used by Wilson (1920), Gee (1932) and ZoBell and Feltham (1934) there is less tendency for the stoppers to be pushed in by the hydrostatic pressure. This can be prevented completely by inserting a piece of 8 mm. glass tubing of such a length that its ends rest upon the bottom of the bottle and the lower end of the stopper.

The use of the specially bent capillary inlet tubes as devised by various workers are somewhat difficult to construct and they are easily broken during sterilization, transportation or handling. The use of a piece of pressure rubber tubing, as used by Schach (1938), provides for flexibility and greatly simplifies the construction of the piece of glass that is broken at the time a sample is collected. Combining the most desirable features of the samplers of Schach (1938) and ZoBell and Feltham (1934), we have a device which is simple to construct and positive in action.

By means of a No. 1 one-hole rubber stopper a CM bottle is fitted with a four inch piece of 5 mm. glass tubing bent at right angles one inch from the end. The longer piece of the resulting L goes through the stopper into the bottle and the shorter arm is fitted with a four inch piece of 1/8 inch heavy-wall rubber pressure tubing. This part of the apparatus can be used over and over again almost indefinitely. The piece of projecting rubber tubing is fitted with a four inch length of 4 mm. glass tubing hermetically sealed at the outer end in a flame. The assembled apparatus is sterilized in the autoclave with the stopper resting loosely on the neck of the bottle held in place by the extruding glass tube. As soon as the bottles can be removed from the autoclave, preferably while they are still full of steam, the stoppers are firmly seated exercising aseptic precautions. When cool the bottles will be from 50 to 90 per cent evacuated.

To collect a sample of water the bottle is clamped into the brass carrier illustrated in Figure 37. The rubber tubing is bent around at an angle of 180° and held in this position by the spring clamp T. C. which is operated with the thumb. When lowered into the water to the desired depth the messenger is dropped. The glass tubing

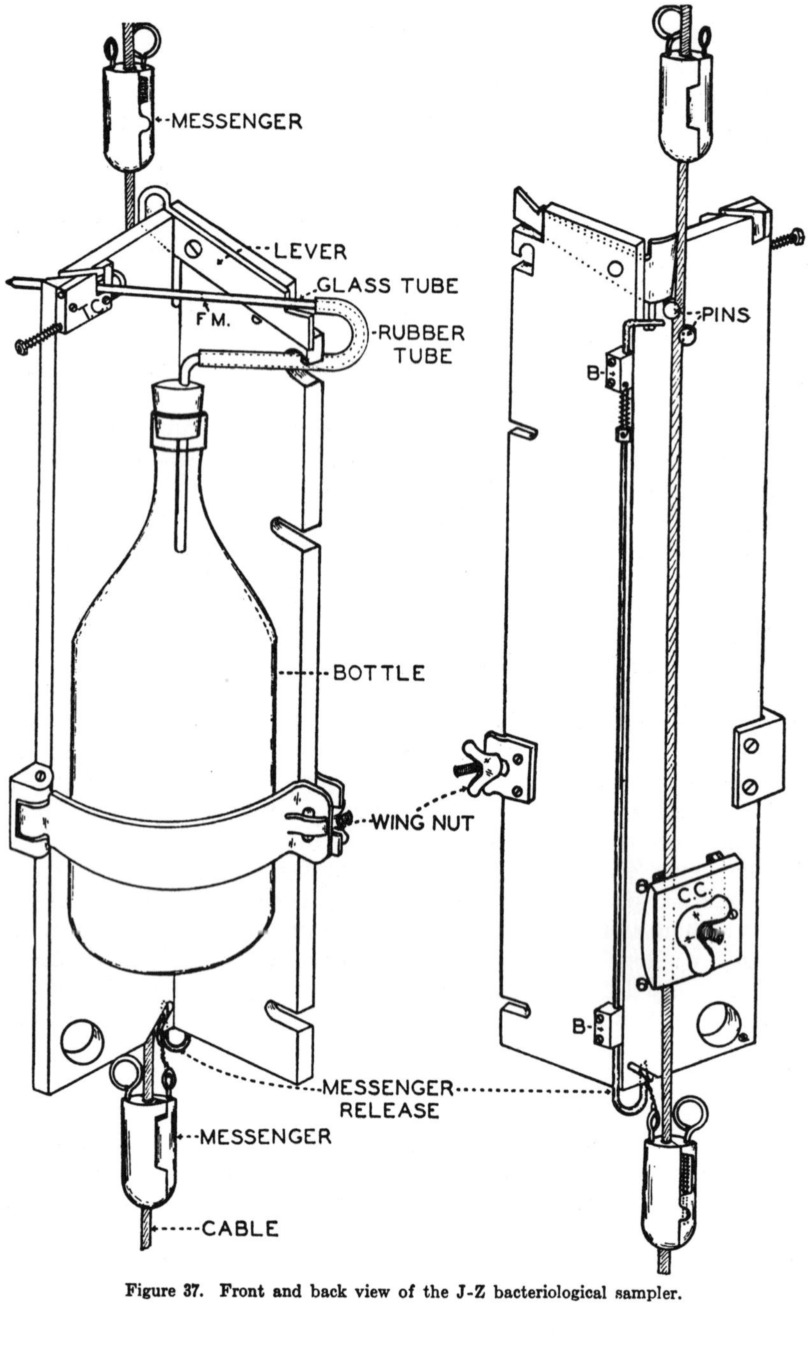

Figure 37. Front and back view of the J-Z bacteriological sampler.

is broken at a file mark midway between the clamp and the end of the lever when the messenger strikes the other end of the lever. Immediately the rubber tubing with the broken piece of the inlet tube flips out to assume a straight position again and a sample of water is aspirated three or four inches away from any part of the apparatus. Even at the instant it is broken the orifice is at least an inch from any part of the apparatus which might carry contaminating organisms, and there is no need for the operator to touch the inlet tube after it is sterilized.

The carrier is constructed of a piece of ¼ inch angle brass 2 × 2 inches wide and 11 inches long. The clamp for holding the bottle in place is made of a six inch length of ⅛ inch strap brass one inch wide bent in a semi-circle. One end is hinged to the carrier frame 4½ inches from the bottom and the other end is fitted with a bolt by which the brass strap can be tightened against bottles of different sizes with a wing-nut. A ⅛ inch notch ½ inch deep to accommodate the rubber tubing is cut over the top of the carrier, the bottom of the notch being 1½ inches from the top of the carrier frame. The notch is made somewhat larger at its inner end to help secure the rubber tube when pushed into the notch. The top corner is cut out as illustrated in Figure 38 to accommodate the inlet tube when the latter is bent into place. On the other side of the carrier a notch 3/16 inches wide and ¾ inches deep is cut at an angle of approximately 45° to accommodate the end of the glass inlet tube. The latter is held in place by the spring clamp made of a piece of brass rod bent U-shaped (T. C. in Figure 38).

The carrier is attached to the sounding wire, cable or rope by means of a wing-nut clamp on the back of the carrier. It is maintained in an upright position by placing the wire in the slot between the two pins P. P. When only one sample is to be collected there are simpler methods for attaching the carrier to the end of a rope, for example, but the sampler as illustrated is designed for rapid attachment to a hydrographic wire or cable.

The ⅛ inch brass rod shown in Figure 38 extending along the outer corner of the carrier is for releasing a messenger which is hooked on the bottom of the carrier. The upper end of the rod is bent over at an angle of 90° so it is engaged by the lever when the lever is hit by the first messenger. This forces the rod to slide down and in so doing the lower U-shaped end slides out of the 3/16 inch oblique notch cut in the bottom of the carrier. This releases the second messenger which then slides down the wire or cable to activate other apparatus attached below. The ⅛ inch brass rod is held in position by the bearings BB. These are constructed from pieces of brass ¼ inches thick and ⅝ inches square. After drilling a ⅛ inch hole along the edge of them they are fastened in place with two ⅛ inch machine screws although they can be soldered to the carrier frame. The brass rod is held in position by means of a spring.

This device with the attached glass bottle has proved to be satisfactory for collecting water samples when used singly or on a line in combination with other apparatus in lakes as well as in the ocean.

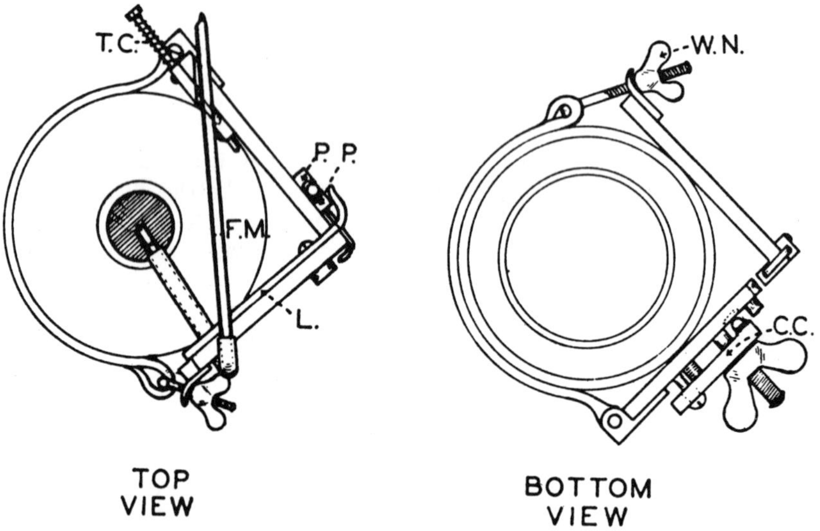

Figure 38. Top and bottom view of the J-Z bacteriological sampler with a citrate of magnesia bottle and its connections in place.

As many as eight of them fastened seriatum on the same line have made it possible to obtain samples from eight different depths in one cast. The repeated failure to recover test pigmented bacteria purposely applied to all parts of the carrier, the cable and the messenger proves that uncontaminated water samples can be collected under the most rigorous conditions. After a few words of instruction, uninitiated deck-hands and technicians alike can collect uncontaminated water samples with the apparatus suitable for bacteriological analysis.

In order to get a sample of water near the surface it is necessary to have the bottle partly evacuated. However, at depths exceeding 10 meters the bottle rapidly fills with water even if the bottle is not evacuated, the time required for the bottle to fill decreasing with depth due to the increasing hydrostatic pressure. Once equilibrium has been established, which requires from two to ten minutes after the inlet tube is broken, there is no possibility of water from higher levels entering the bottle as it is pulled toward the surface because the pressure within the bottle is greater than that in the water at higher levels. Actually there is a slight tendency for water or air to be forced out of the bottle as it is raised. Sometimes when bottles fitted with small capillary tubes are hauled rapidly to the surface they appear with a fine jet of water being ejected.

Although the hydrostatic pressure aids in the collection of water samples, it prevents the apparatus from being used successfully at all depths. Some of the CM bottles will be crushed by the hydrostatic pressure at depths between 200 and 300 meters. The trial immersion of over a hundred stoppered CM bottles tied in groups on the submerged sounding line showed that 7 per cent of them were broken at 200 meters, 29 per cent at 300 meters, 81 per cent at 400 meters and all of them were broken at 600 meters. Although most water samples for bacteriological analysis are collected from the principal biotic zone which rarely exceeds 200 meters, in exploring the abyssal depths of the sea, it is desirable to have a bacteriological sampler which will function at depths exceeding 10,000 meters. Bacteriologists have been seeking such an apparatus for several years.

A PRESSURE RESISTANT BOTTLE

None of the so-called "deep sea" bacteriological water sampling bottles has been designed to function at depths exceeding a hundred meters or so, with the possible exception of those filled with a disinfectant (Matthews, 1913, Drew, 1914).

As mentioned above, valves and pistons on metal bottles are not leak-proof as far as bacteria are concerned. Evacuated glass bulbs which are large enough to collect a satisfactory sample are broken by the high hydrostatic pressure encountered at greater depths. The best high pressure bottle we could find without going to the expense of having special ones fabricated was a 125 ml. thick-walled flask with a round bottom supplied by the Corning Glass Company. It was cracked at a depth of 1200 meters.

While unquestionably glass bottles holding 100 ml. or more could be fabricated to tolerate any pressure in the sea, the cost is excessive and, moreover, it is doubtful if evacuated bottles should be used at great depths for collecting water samples for bacteriological analysis. When the capillary inlet tube on such a bottle is broken at a depth of 1000 meters, for example, the bacteria aspirated with the water sample are subjected to an instantaneous pressure change of approximately 100 atmospheres, and correspondingly greater pressure changes at greater depths. The recovery of viable bacteria by Certes (1884), Carey and Waksman (1934) and others from depths near 5000 meters proves that neither the pressure encountered (approximately 500 atmospheres) nor its gradual release, effected by bringing the samples to the surface, killed all of the bacteria but it is indeterminate if some failed to survive this change. The work of Larsen et al. (1918) indicates that while certain bacteria tolerate pressures of 6000 atmospheres (about six times as high as any encountered in the sea), they are injured by the sudden release of pressures of 50 atmospheres or less depending upon the gas tension and other factors.

After experimenting with many types of devices to obviate the pressure factor for collecting samples at great depths, we have perfected the collapsible rubber bottle idea suggested to us by Dr. Austin Phelps of the Hopkins Marine Station. In its simplest form heavy rubber pear-shaped aspirator bulbs holding around 100 ml. are used (Figure 39). A 5 inch length of 1/8 inch heavy rubber pressure tubing is cemented in the 6 mm. opening of the rubber bulb with rubber cement. The rubber tubing is fitted with a 4 inch length of 4 mm. glass tubing with the outer end sealed in a flame, this part of the apparatus being the same as that used with the CM bottles described above.

The rubber bulbs can be sterilized in the autoclave after which the inlet tubes with the sealed ends are fitted aseptically with the bulb collapsed or depressed. At sea they can be sterilized by boiling, or in an emergency they can be sterilized by filling with 70 per cent alcohol or other disinfectant and then rinsed with sterile water after squeezing out the disinfectant, exercising aseptic technique. In practice we prepare and sterilize in convenient receptacles or paper-wrapped packets several hundred of the glass inlet tubes preparatory to embarking on a cruise. The rubber bulbs are sterilized by boiling or in a pressure cooker as needed, the bulbs being fitted with the glass inlet tubes while the former are still quite warm to lessen the likelihood of contamination and to get a better evacuation of the bulbs.

These rubber bulbs function in precisely the same way as the similarly fitted CM bottles. When the capillary inlet tube is broken by the lever activated by the messenger, the rubber bulb aspirates a water sample as it assumes its normal shape. It cannot be broken by the pressure regardless of whether it is collapsed or inflated with air because as it is compressed by the increasing hydrostatic pressure, the pressure will always be nearly the same inside and outside the bulb due to the flexibility of the rubber. Consequently there is very little pressure change as the water sample is aspirated.

The rubber bulbs are secured to the same carrier as is used for the CM bottles by means of an auxiliary collar made of a 1¼ inch length of brass pipe 1⅛ inches in inside diameter. A hole is drilled ⅜ of an inch from the edge of the collar and threaded to take a ¼ inch machine screw 1¼ inches long having a knurled or wing-nut head. Two nuts are placed on the screw as illustrated in Figure 39 to keep the collar and attached rubber bulb in the proper position in the carrier. The collar is slipped over the neck of the sterilized pear-shaped bulb which is ready for use. Then the bulb is placed in position on the carrier and fastened there by slipping the screw projecting from the brass collar into a ¼ inch slot which is cut ¾ inches deep 4 inches from the top of the carrier. It is secured by tightening the machine screw and in so doing the end of the screw presses into the neck of the rubber bulb for about ⅛ inch thereby preventing the bulb from dropping out under any conditions. The collars are made detachable to facilitate handling the water samples collected in the rubber bulbs as well as to facilitate the sterilization of the latter and their assembly.

The trial immersion of the assembled rubber bulbs to depths exceeding 4000 meters without breaking the inlet tube proved that they were leak-proof. When the inlet tube was broken by the messenger-activated mechanism, the bulbs came up filled with water. Although we have not had an opportunity to send them down to the greatest depth in the sea, there seems to be no reason why they will not be just as efficient at 10,000 meters as at 4000 meters.

With the brass collar removed the rubber bulbs stand in an upright position on their flat bottoms. The sample can be transferred to a sterile glass container by squeezing the bulb, or aliquot portions can be removed as needed. The water samples should be stored in the refrigerator until they can be analyzed, although to insure reliable results the water should be analyzed for its bacterial population as soon as possible to minimize the changes which accompany the storage of water samples (ZoBell and Feltham, 1934, ZoBell and Anderson, 1938).

Since the rubber is not entirely inert the samples should not be left in the rubber bulb any longer than is necessary. There are certain bacteria found in the sea which slowly attack rubber as well as the synthetic products, neoprene and isoprene, as manifested by

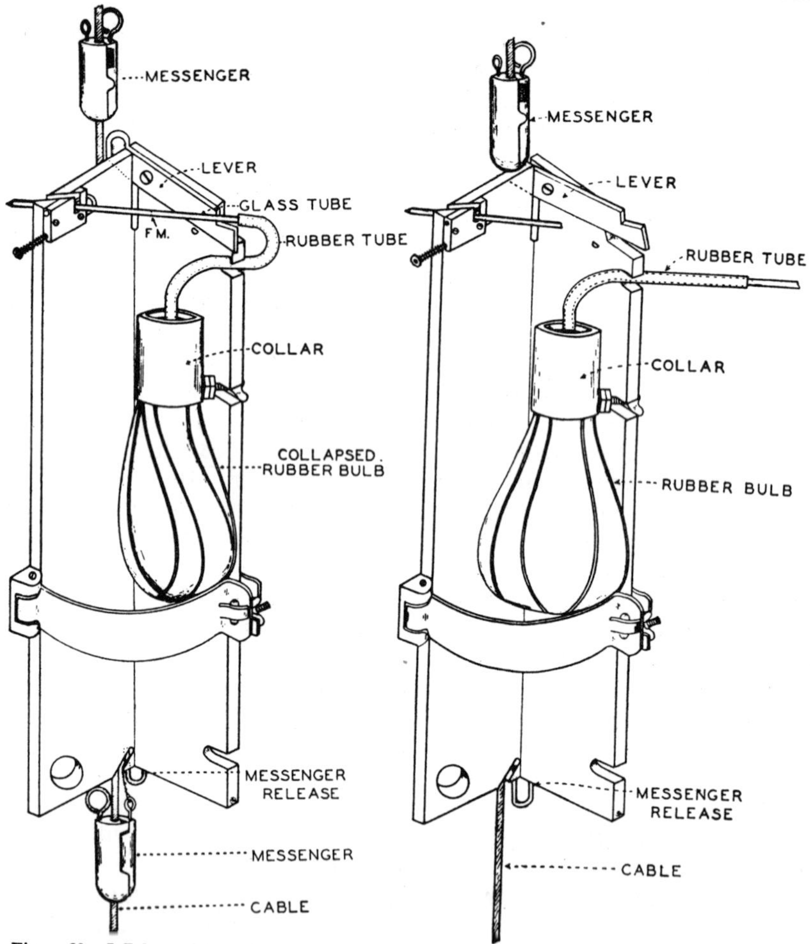

Figure 39. J-Z bacteriological sampler fitted with a pressure-resistant collapsible rubber bottle before and after collecting a sample of water. TC, thumb clamp, PP, pins for guiding cable; FM, file mark on glass inlet tube; L, lever which breaks tube; WN, wing nut on bottle clamp; CC, cable clamp for securing apparatus to cable.

oxygen consumption and the increasing bacterial population after several days. However, there is no evidence that the rubber is attacked for several days, so this possible source of error can be discounted provided the water samples are not left in the rubber bulbs for more than a few hours. The bacterial population and other properties of water samples of known bacterial content stored in rubber bulbs for 24 hours were not unlike those of water stored in glass receptacles under comparable conditions.

Several hundred water samples for bacteriological analysis have been collected from various depths in the sea proving the apparatus to be quite satisfactory. The sampler is known as the "J-Z" because of the activities of the S. I. O. mechanic, Carl I. Johnson, in perfecting the carrier.

Summary

None of the numerous samplers which have been described in the literature is entirely satisfactory for the collection of water samples for bacteriological analysis.

Containers made of copper, zinc, tin or nickel alloys are not suitable for the collection of samples of sea water for bacteriological analysis due to the inimical oligodynamic action of the metals. Many bacteria are killed and the sea water itself may be rendered bacteriostatic by exposure to the metals.

An apparatus is described which can be used on the standard hydrographic wire or cable for the collection of water samples aseptically from any desired depth in the sea. Multiple units provide for the simultaneous collection of samples from several different depths or the bacteriological sampler, known as the "J-Z", can be used in conjunction with other hydrographic sampling bottles and instruments. Glass bottles can be used to advantage for collecting samples at depths not exceeding 200 meters and collapsible rubber bottles are recommended for greater depths. High hydrostatic pressures do not interfere with the operation of the rubber bottles.

LITERATURE

Abbott, A. C.

1921. The Principles of Bacteriology, New York, Lea & Febiger, 10th Ed., pp. 650–652.

Bedford, R. H.

1931. The bactericidal effect of the "Prince Rupert" sea water sampling bottle. Contr. Can. Biol. and Fish., 6: 423–426.

Bertel, R.

1911. Ein einfacher Apparat zur Wasserentnahme aus beliebigen Meerestiefen für bakteriologische Untersuchungen. Biol. Centralbl., 31: 58–61.

Drew, G. H.

1914. On the precipitation of calcium carbonate in the sea by marine bacteria and on the action of denitrifying bacteria in tropical and temperate seas. Carnegie Inst. Wash., 5: 7–45.

Eyre, J. W. H.

1930. Bacteriological Technique. Wm. Wood & Co., New York, 3rd Ed., pp. 493–495.

GEE, H.
1932. Bacteriological water sampler. Bull. Scripps Inst. Oceanog., Tech. Ser., 3: 191–200.

HEYDENREICH, L.
1899. Einige Neuerungen in der bakteriologischen Technik. Zeitschr. f. wissensch. Mikroskopie, 16: 144–179.

ISSATCHENKO, B. L.
1914. Investigations on the bacteria of the glacial arctic Ocean. Monograph, Petrograd, 300 pp. (In Russian.)

JOHNSTON, W.
1892. On the collection of samples of water for bacteriological analysis. Canadian Rec. of Sci., 5: 19–28.

KRUSE, F.
1908. Beitrage zur Hygiene des Wassers. Zeitschr. f. Hyg., 59: 6–94.

LARSEN, W. P., HARTZELL, T. B. AND DIEHL, H. S.
1918. The effect of high pressures on bacteria. Jour. Infect. Dis., 22: 271–279.

MATTHEWS, D. J.
1913. A deep-sea bacteriological water-bottle. Jour. Marine Biol. Assn. U. K., 9: 525–29.

MIQUEL, P., AND CAMBIER, R.
1902. Traite de Bacteriologie pure et appliquée. Paris. 372 pp.

OTTO, M. AND NEWMANN, R. O.
1904. Ueber einige bacteriologische Wasseruntersuchungen im Atlantischen Ozean. Centralbl. f. Bakt., II Abt., 13: 481–489.

PARSONS, P. B.
1911. Apparat zur Entnahme von Wasser aus grösserer Tiefe. Centralbl. f. Bakt., Abt. II, 32: 197–207.

PORTIER, P. AND RICHARD, J.
1906. Sur une methode de prelevement de l'eau de mer destinee aux études bacteriologiques. Compt. rend. Acad. Sci., 142: 1109–1111.

PRAUM, L.
1901. Einfacher Apparat zur Entnahme von Wasserproben aus grösseren Tiefen. Centralbl. f. Bakt., I Abt., 29: 994–996.

RAADSVELD, C. W.
1934. The oligodynamic effect of metals and metal salts. Chemisch Weekblad, 31: 497–504.

REYNIERS, J. A.
1932. Apparatus for taking water samples from different levels. Science, 75: 83–84.

RUSSELL, H. L.
1892. Bacterial investigations of the sea and its floor. Botanical Gazette, 17: 312–321.

SCHACH, H.
1938. Ein Apparat zur Entnahme von Meerwasserproben aus der Tiefe für bakteriologische Untersuchungen. Jour. du Conseil, 13: 349–356.

Whipple, G. C.
1927. The Microscopy of Drinking Water. John Wiley & Sons, New York, Ed. 4, 585 pp.

Wilson, F. C.
1920. Description of an apparatus for obtaining samples of water at different depths for bacteriological analysis. Jour. Bact., 5: 103–108.

Young, O. C., Finn, D. B., Bedford, R. H.
1931. A deep sea bacteriological water bottle. Contrib. Canadian Biol. & Fish., 6: 417–422.

Zillig, A. M.
1929. Bacteriological studies in Lake Erie. Bull. Buffalo Soc. Nat. Sci., 14: 51–55.

ZoBell, C. E.
1941. Studies on marine bacteria. I. The cultural requirements of heterotrophic aerobes. Jour. Mar. Res., 4: 42–76.

ZoBell, C. E. and Anderson, D. Q.
1936. Observations on the multiplication of bacteria in different volumes of stored sea water and the influence of oxygen tension and solid surfaces. Biol. Bull., 71: 324–342.

ZoBell, C. E. and Feltham, C. B.
1934. Preliminary studies on the distribution and characteristics of marine bacteria. Bull. Scripps Inst. Oceanogr., Tech. Ser., 3: 279–296.

ACKNOWLEDGEMENT

Ideas from several of my associates at the Scripps Institution of Oceanography, the Limnological Laboratory of the University of Wisconsin, Hopkins Marine Station, and the Woods Hole Oceanographic Institution have contributed to the perfection of the bacteriological sampler. This help together with the material assistance rendered by Dr. Sydney C. Rittenberg and Catharine B. Feltham is gratefully acknowledged.

ERRATUM

Page 184, line 28: For more accurate information, see "Some Effects of High Hydrostatic Pressure on Apparatus Observed on the Danish GALATHEA Deep-Sea Expedition," C. E. ZoBell, *Deep-Sea Res. Oceanogr. Abst.* *2*(1):24–32 (1954).

6

Reprinted from *Deep-Sea Res. Oceanogr. Abst.*, **9**, 501–503 (1962)

A WATER SAMPLER FOR MICROBIOLOGICAL STUDIES

Shale J. Niskin

Institute of Marine Science, University of Miami

Abstract—A bellows type water sampler consisting of a spring-activated metal frame and a detachable two-litre, sterile, disposable container, is described. The device can operate at any oceanic depth.

INTRODUCTION

INVESTIGATIONS of the biology and distribution of marine yeasts and fungi have led to a need for an improved sampler for the aseptic collection of these organisms in the sea. Some of the problems with existing sterile samplers are : (1) depth restrictions (ZOBELL, 1941; SIROKIN, 1960; JENSEN and STEEMANN NIELSEN, 1953), (2) capacity limitations (ZOBELL, 1954; JENSEN and STEEMANN NIELSEN, 1953), (3) unreliable sampling at given depths (ZOBELL, 1954), and (4) shipboard storage, breakage, and resterilization of apparatus (SIROKIN, 1960). The described sampler was designed to overcome these difficulties.

GENERAL DESCRIPTION

The sampler operates like a spring-opened bellows. A hinged metal frame, opened by torsion springs, supports and expands a sterile, detachable container made of polyethlylene film. Water enters the latter via a rubber tube initially sealed with a glass tip. A mechanical system, mounted on the frame, is required (1) to break the glass tip seal, (2) to activate the sampler,(3) to release a messenger for tripping instruments below, and (4) to reseal the rubber tube after the container has filled. Details of the sampler are shown in FIGS. 1 and 2.

OPERATION

The sampler is assembled first by drawing the container between the latched, forked plates of the frame, the plate prongs entering pockets attached to the sides of the container. The rubber tube is then pulled through the open U-clamp and its glass tip is secured under the plunger rod cap (FIG.3). The sampler is then clamped to the hydrographic wire and a messenger attached. As many samplers as required can be placed at intervals on the wire.

Within five minutes after messenger impact, the container fills and automatically seals. To prevent air contamination after recovery, the rubber tube is sealed with a spring clip prior to its removal from the U-clamp. The container can then be pulled off the frame.

DISCUSSION

The sampler is designed to collect two litres of water in a sterile container at any oceanic depth. Larger volumes of water may be obtained by constructing a larger sampler.

The polyethylene containers can be gas sterilized (ethylene oxide) and compactly stored in quantity. They are inexpensive enough to be considered disposable, and their plastic construction obviates risk of metallic ion contamination.

*Contribution No. 428 from The Marine Laboratory, University of Miami.

A pronounced concavity of the unsupported sides of the container, observed upon recovery, provides a positive indication that the container has remained essentially undamaged and that the U-clamp has properly resealed the rubber tube.

The sampler is not adversely affected by high hydrostatic pressures. In order to prevent the rubber tube from being forced into its glass tip at great depth, a few cubic centimeters of sterile water are introduced into the glass tip before lowering (ZOBELL, 1954). Two types of glass tips, which obviate this precaution, have been tested recently at sea and are currently being evaluated.

The sampler has been extensively field-tested during deep-sea microbiological work, and has demonstrated its value in studies of this type. It is now being used routinely at sea.

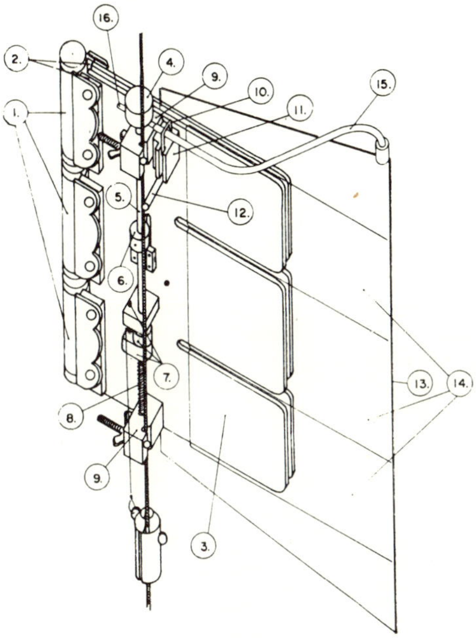

FIG. 1.

Assembled sampler attached to the hydrographic wire.

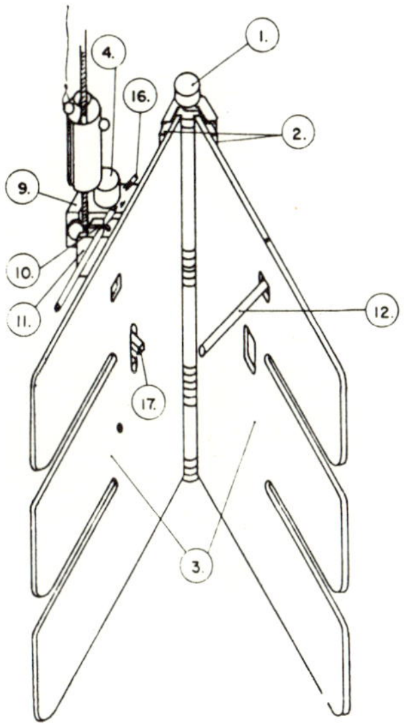

FIG. 2.

Open frame.

Legend to Figs. 1 and 2.

The frame, constructed of anodized aluminium and stainless steel, consists of two forked plates (3) riveted to shimmed (2) hinges (1) which enclose adjustable torsion springs. The mechanical system operates as follows : the plunger rod (5) is driven down by the impact of a messenger (released from shipboard or by a previously tripped sampler) striking the plunger rod cap (4) which breaks the glass tip seal (16) placed on the wire clamp mounting block (9). Simultaneously, the plate latch (17) is tripped by the latch disc (6) fastened to the plunger rod, and the messenger release pin is depressed against spring (8) compression, by a disc (7) also secured to the plunger rod, thereby releasing the attached messenger. As the torsion springs open the frame to an angle of 45°, the U-clamp releasing rod (12) is pulled from under the U-clamp (10) housed in the U-clamp mounting block (11). The released U-clamp reseals the rubber tube (15).

For economy and simplicity of construction, the container (13) is fabricated from 0·004 gauge stock tubular polyethylene film. When collapsed, it is trapezoidal in shape, which allows it to assume the form of a prism upon expansion of the frame.

The container is attached to the frame by means of plastic film pockets (14) (heat sealed to both of its sides) into which fit the prongs of the forked plates. In addition, this arrangement increases the efficiency of the bellows action by providing rigid support to both sides of the container.

The rubber tube (15) is inserted through a small hole cut in a corner of the container. A short length of elastic cord binds and seals the plastic film to the rubber tube.

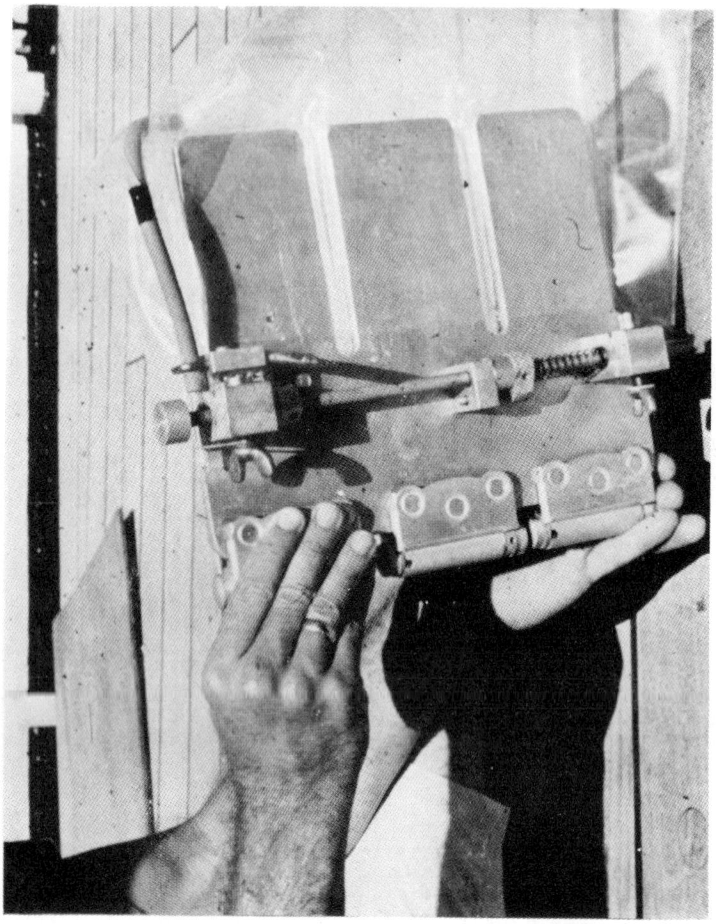

FIG. 3. Sampler ready for attachment to hydrographic wire.

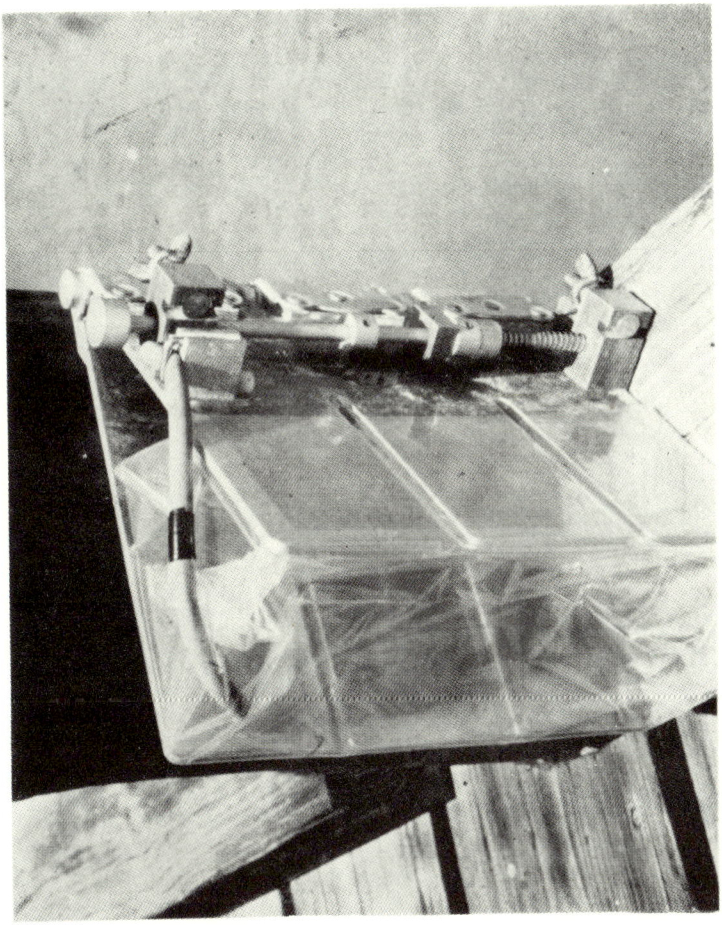

FIG. 4. Opened sampler. Note U-clamped fill tube.

Addendum

Since this article went to press, the following modifications, which obviates the rubber tube glass tip, has been effected.

The rubber tube is encased in a sealed plastic film sheath, the latter being formed as an integral part of the container. The sheathed rubber tube is pulled through the open U-clamp, and then doubled back so that the sheath, extending beyond the end of the rubber tube, can be stretched taut under the plunger rod cap to which is attached a cutting blade.

On messenger impact, the sheath is severed close to the end of the rubber tube, permitting water entry and freeing the rubber tube to spring away from the device. The possibility of the sampler itself providing a source of contamination is thus minimized.

Acknowledgements—The development of this device was supported in part by the Office of Naval Research, Contract Number 840 (01), under the supervision of Dr. F. F. KOCZY, Institute of Marine Science, University of Miami. The writer wishes to express his appreciation to Dr. SAMUEL P. MEYERS and his associates for their assistance throughout the course of the work. Further acknowledgement is made to the Velsicol Chemical Corporation for their aid in making the facilities of M/V VICCA available for deep-sea microbiological studies with this instrument.

REFERENCES

JENSEN, E. A. and STEEMANN NIELSEN, E. (1953) A water sampler for biological purposes. *J. de Conseil* **18**, 296–299.

SIROKIN, I. I. (1960) A sampler for the selection of samples of water for bacteriological analysis. *Bull. Inst. Reserv. Biol. Acad. Sci., U. S. S. R.* (6), 53–54.

ZOBELL, C. E. (1941) Apparatus for collecting water samples from different depths for bacteriological analysis. *J. Mar. Res.*, **4**, 173–188.

ZOBELL, C. E. (1954) Some effects of high hydrostatic pressure on apparatus observed on the Danish *Galathea* deep-sea expedition. *Deep-Sea Res.*, **2**, 24–32.

Editor's Comments on Papers 7 Through 13

THE DEVELOPMENT OF CULTURAL AND ENUMERATION TECHNIQUES

Once a marine sample is aseptically collected and brought to the surface, traditionally the questions to be resolved are what bacteria are in the sample and how many are there? The problem of estimating the total number of viable bacteria remains only partially resolved today. In considering this problem of isolation and enumeration, early investigators reasoned that the imitation of nature would probably yield the best results. Thus, Certes (10), Russell (34), Beijerinck (2), Fischer (13), Otto and Neumann (31), and Schmidt-Nielsen (35), without benefit of our extensive knowledge of the mineral requirements of these microbes, added seawater or 3.0 percent NaCl to their media. When cultivating bacteria from fish, they reasoned, a type of fish-protein extract would be the logical basis for the medium; and for the isolation of microbes from sediment samples, one should use mud extracts. Their difficulty in resolving the problem was primarily concerned with the level of organic enrichment, since they followed the lead, again, of the medical bacteriologists in using very high levels of organic enrichment. However, a pattern was established so that each succeeding investigator produced his own "ideal, complete" medium. This trend still continues, for despite our greater knowledge of marine bacterial metabolism, no one single medium will support the growth of all the types of microorganisms present or even of all the heterotrophs.

It is generally recognized, now, that seawater or a salts mixture is required for the initial isolation of marine bacteria (7, 43, 48, 49), although contradictory reports by Lloyd (24), Burke (6), and Wood (47) continued to appear. For the most part, however, by 1950 it was recognized that either aged, natural seawater or a synthetic salts mixture (26) would provide the best liquid base for the enumeration of marine bacteria. Modifications in this requirement do appear; in fact, Hidaka in 1965 (17) proposed a scheme based on salt requirements for initial isolation and subsequent growth: terrestrial (growth in distilled water and seawater), halophilic (no growth in distilled water media or 0.5 percent NaCl, but growth in one-sixth seawater and full-strength seawater), and marine (requirement for seawater, either artificial or natural). It is unfortunate that he chose the term halophilic, because today this term is generally applied to organisms requiring greater than 15 percent NaCl; but he did emphasize the extremely broad ranges of salinity requirements that are encountered in a natural population. Undoubtedly, many of the "low counts" recorded by investigators in the first part of this century were due to their failure to recognize the salt requirements of marine microbes.

Much of the concern with a proper isolation medium and the subsequent incubation conditions stems from the large discrepancy between direct microscopic counts and viable plate counts or most probable-number determinations from dilution tubes. Possible explanations for this discrepancy include large numbers of dead or injured cells, difficulties in distinguishing between detrital matter and microorganisms under direct microscopy, cell injury due to prior sample treatment, improper pH or light conditions, or nutritional deficiencies in all the enumeration media.

Even the time and temperature of sample storage have been found to influence the counts, as Tanner and Schneider (41) showed with nonmarine bacteria, while Waksman and Carey (42) demonstrated that bacteria would multiply in stored seawater. Earlier, Bertel (3) had demonstrated a diurnal cycle in the numbers of bacteria in seawater, and had also shown that the numbers decreased with depth, but that the decreased values were greatly influenced by the sampling time. Systematic investigations of these factors were undertaken by several other scientists.

Lloyd was among the first to report the specific procedural details employed in obtaining, holding, and treating the samples; portions of his paper are reprinted here (Paper 7). In addition, he studied the Clyde Sea on a diurnal and seasonal basis in order to provide greater validity to his data and considered the possible effects of sunlight on the bacterial populations in surface waters. Similarly detailed investigations were conducted by Lloyd on the sedimentary bacteria in the Clyde Sea, establishing a high standard of control and careful interpretation of data for those who followed (25).

Meanwhile, in the Soviet Union, Butkevich and Butkevich had been examining some of the same parameters while especially concentrating on the effects of light, temperature, and pH on the recovery of marine bacteria. Paper 8, an English summary and figures from their paper, shows the depth of their studies and its far-reaching implications.

In 1941, the summation of eight years of study on the nutrition and growth conditions of marine bacteria resulted in a classic paper by ZoBell, a summary of which is presented here as Paper 9. In this paper, ZoBell thoroughly demonstrated the importance of lowered levels of nutrients for the initial isolation of marine bacteria. The formulation of medium 2216 (now widely available) is presented. It contains one tenth the organic level found in nutrient broth, along with an elevated pH and level of iron. In addition, ZoBell refutes the standard practice of that time, still in use today, of adding KNO_3 to the medium by showing decreased levels of colony-forming units in media with 0.05 or 0.1 percent KNO_3. ZoBell and Grant (49) again in 1943 demonstrated the ability of marine bacteria to utilize low nutrient levels.

Despite these investigations on nutritional factors influencing the distributions of marine bacteria, Carlucci and Pramer maintained a concern with the nutritional problems of culturing marine bacteria. They reported their comparative studies on nutritional enrichments in the isolation medium in a paper that reemphasized the need for low nutrient levels for culturing these bacteria (Paper 10). This has been repeatedly verified (14, 22) and should be standard practice in any marine microbiology laboratory.

One of the most comprehensive studies of the physical effects of sample treatment on bacterial survival is contained in a series of papers by Jones and Jannasch (20). They compared the Cholodney procedure, pour plating, filtration, and direct microscopic methods of enumeration; their first paper is presented here as Paper 11. Based on this work, others have attempted less violent means of concentrating the bacterial population in offshore waters: bubble scavenging (4, 9, 44), coulter counter combined with hydrodynamic focusing (37), reverse-flow filtration (12, 19), which has been critically reviewed by Griffiths et al. (15), and the plate dilution frequency technique (16, 30).

Prior to 1960 it was standard procedure to use the pour plate technique for mixing the medium and the sample. Indeed, all the previously mentioned reports employed this technique. However, in 1960, Buck and Cleverdon reported that higher counts could be obtained by the simple expedient of using the surface spreading technique of the desired dilution (Paper 12). This is now standard practice (11), and has even been adapted for use with psychrophiles (46), in all marine microbiology laboratories.

Additional specialized techniques have been developed or proposed. Microbial growth tubes for testing already isolated strains have been suggested by Lloyd and Morris (25), and techniques for simulating natural conditions have been presented. These have included the diffusion chamber of Caldwell and Hirsch (8) and the membrane filter system of Kunicka-Goldfinger and Kunicka-Goldfinger (21). Attempts to improve direct counting techniques have also been extensive. Sheldon (36) suggested a preliminary size separation using membrane and glass fiber filters before counting. Perhaps the most promising technique is the rhodamine-labeled lysozyme procedure of McElroy and Casida (28). This has not, as yet, been widely tested with marine samples, but previously suggested specific fluorescent antibody methods (38, 39) do not really seem applicable with such a diverse population as one finds in oceans and estuaries. However, adaptations of this procedure with phage or labeled proteins hold promise.

Several other potentially useful techniques have been developed for other purposes and need to be tested under laboratory and field conditions on marine sediments and seawater. Among these newer methods are the dielectrophoretic separation of cells

(27), continuous particle electrophoresis (1), or laser-beam modification (40), and molecular sieving, which, although proposed for phytoplankton by Morris and Yentsch (29), should be adaptable to bacteria, as should ultracentrifugation, a procedure that to date has been successfully employed only with zooplankton by Bowen et al. (5). The use of infrared fluorescence photomicroscopy to detect photosynthetic bacteria has also been suggested by Pierson and Howard (33). To date, no field studies have been reported, but the method should be useful for estimating the viable photosynthetic bacteria on membrane filters, for example. In addition, scanning electron microscopy is replacing fluorescent and transmission microscopy because of the three-dimensional view it provides of the actual orientation and placement of microorganisms on particles and other solid surfaces.

Expectations have changed in recent years, and there is less surprise at the presence of bacteria in the marine environment. Indeed, there is so much more concern with the total viable microbial biomass that several interesting approaches have been proposed. One of these which has successfully been used in phytoplankton studies involves membrane filtration and the reduction of triphenyltetrazolium chloride by the *in situ* population (23). This technique deserves further testing in bacterial ecosystems. Meanwhile, others have attempted the estimation of the microbial population by measuring the incorporation of ^{32}P (45) into intracellular constituents. Although the ^{32}P method should be applicable to both autotrophic and heterotrophic populations, this technique has been largely superceded by the greater ease of using the adenosine triphosphate (ATP) content as an estimate of total viable biomass.

In Paper 13, Hamilton and Holm-Hansen advance the concept that Holm-Hansen had presented only the year before in a paper he co-authored with Booth (18) suggesting a relationship between ATP and phytoplankton numbers. Inasmuch as ATP is found in all living cells and is rapidly hydrolyzed upon cellular death, quantitation of the amount of product provides an estimate of the biomass present in the sample, assuming that the concentration of ATP per cell is reasonably constant. The relative ease of sample preparation, availability of sensitive instrumentation that can be used on board ships, and the sensitivity of the method (as few as 10^2 cells can be detected) have combined to make this an appealing procedure to many workers. Like all biomass determinations, though, it does not distinguish between bacteria, fungi, and phytoplankton. For biomass determinations based on ATP, the conversion factors, out of necessity, have been based on laboratory-grown pure cultures and plate counts or direct counts of the organisms. Consequently, relatively large correction factors are involved in relating the ATP content to biomass. However, unlike direct microscopy, ATP and other similar biomass indicators are not influenced by dead cells. Thus, widespread studies comparing plate counts, biochemical activities, and ATP biomass may eventually provide answers to the question first asked in the 1880s: how many microorganisms are really there in the oceans?

REFERENCES

1. Bayne, D. R., and Lawrence, J. M. 1972. Separating constituents of natural phytoplankton populations by continuous particle electrophoresis. *Limnol. Oceanogr. 17*:481–489.
2. Beijerinck, M. W. 1889. Le *Photobacterium luminosum,* bactérie lumineuse de la mer du nord. *Arch. Neerl. Sci. Exactes Nat. Haarlem 23*:401–415.
3. Bertel, R. 1912. Sur la distribution quantitative des bactéries planctoniques des côtes de Monaco. *Bull. Inst. Oceanogr. No. 224*:2–12.
4. Bezdek, H. F., and Carlucci, A. F. 1972. Surface concentration of marine bacteria. *Limnol. Oceanogr. 17*:566–569.
5. Bowen, R. A., St. Onge, J. M., Colton, J. B., Jr., and Price, C. A. 1972. Density-gradient centrifugation as an aid to sorting planktonic organisms. I. Gradient materials. *Mar. Biol. 14*:242–247.
6. Burke, V. 1934. The interchange of bacteria between fresh water and the sea. *J. Bacteriol. 27*:201–205.
7. Butkevich, V. S. 1932. Zur Methodik der bakteriologischen Meeresuntersuchungen und einige Angaben über die Verteilung der Bakterien im Wasser und in den Böden des Barents meeres. *Trans. Oceanogr. Inst. Moscow 2*:35–39.
8. Caldwell, D. E., and Hirsch, P. 1973. Growth of microorganisms in two-dimensional steady-state diffusion gradients. *Can. J. Microbiol. 19*: 53–58.
9. Carlucci, A. F., and Williams, P. M. 1965. Concentration of bacteria from sea water by bubble scavenging. *J. Cons. Perm. Int. Explor. Mer 30*:28–33.
10. Certes, A. 1884. Sur la culture, a l'abri des germes atmospheriques, des eaux et des sédiments rapportés par les expéditions du *Travailleur* et du *Talisman*: 1882–1883. *C. R. Hebd. Seances Acad. Sci. 98*:690–693.
11. Clark, D. S. 1971. Studies on the surface plate method of counting bacteria. *Can. J. Microbiol. 17*:943–946.
12. Dodson, A. N., and Thomas, W. H. 1964. Concentrating plankton in a gentle fashion. *Limnol. Oceanogr. 9*:455–456.
13. Fischer, B. 1894. Die Bakterien des Meeres nach den Untersuchungen der Plankton Expedition under gleichzeitiger Berücksichtigung einiger älterer und neuer Untersuchungen. *Zentralbl. Bakteriol. 15*: 657–666.

14. Gray, T. R. G. 1963. Media for the enumeration and isolation of heterotrophic salt-marsh bacteria. *J. Gen. Microbiol. 31*:483–490.
15. Griffiths, R. P., Hanus, F. J., and Morita, R. Y. 1973. Applicability of the reverse-flow filter technique to marine microbial studies. *Appl. Microbiol. 26*:687–691.
16. Harris, R. F., and Sommers, L. E. 1968. Plate-dilution frequency technique for assay of microbial ecology. *Appl. Microbiol. 16*:330–334.
17. Hidaka, T. 1965. Studies on the marine bacteria. II. On the specificity of mineral requirements of marine bacteria. *Mem. Fac. Fish. Kagoshima Univ. 14*:127–180.
18. Holm-Hansen, O., and Booth, C. R. 1966. The measurement of adenosine triphosphate in the ocean and its ecological significance. *Limnol. Oceanogr. 11*:510–519.
19. Holm-Hansen, O., Packard, T. T., and Pomeroy, L. R. 1970. Efficiency of the reverse-flow filter technique for concentration of particulate matter. *Limnol. Oceanogr. 15*:832–835.
20. Jones, G. E., and Jannasch, H. W. 1959. Aggregates of bacteria in sea water as determined by treatment with surface active agents. *Limnol. Oceanogr. 4*:269–275.
21. Kunicka-Goldfinger, W., and Kunicka-Goldfinger, W. J. H. 1972. Semicontinuous culture of bacteria on membrane filters. I. Use for the bioassay of inorganic and organic nutrients in aquatic environments. *Acta Microbiol. Pol., Ser. B, 4*:49–60.
22. Litchfield, C. D., Rake, J. B., Zindulis, J., Watanabe, R. T., and Stein, D. J. 1974. Optimization of procedures for the recovery of heterotrophic bacteria from marine sediments. *Microb. Ecol. 1*:219–233.
23. Loshakov, J. T. 1962. The determination of total bacterial numbers in water on ultrafiltration membranes with the use of triphenyltetrazolium chloride. *Hyg. Sanit. SSSR 27*:48–49.
24. Lloyd, B. 1930–1931. Muds of the Clyde Sea area. II. Bacterial content. *J. Mar. Biol. Assoc. U.K. 17*:751–765.
25. Lloyd, G. I., and Morris, E. O. 1971. An apparatus for measuring microbial growth or survival in the marine environment. *Mar. Biol. 10*:295–296.
26. Lyman, J., and Fleming, R. H. 1940. Composition of sea water. *J. Mar. Res. 3*:134–146.
27. Mason, B. D., and Townsley, P. M. 1971. Dielectrophoretic separation of living cells. *Can. J. Microbiol. 17*:879–888.
28. McElroy, L. J., and Casida, L. E., Jr. 1972. An evaluation of rhodamine-labeled lysozyme as a fluorescent stain for *in situ* soil bacteria. *Can. J. Microbiol. 18*:933–936.
29. Morris, I., and Yentsch, C. S. 1972. A new method for concentrating phytoplankton by filtration with continuous stirring. *Limnol. Oceanogr. 17*:490–493.
30. Neal, J. L., Jr. 1971. A simple method for enumeration of antibiotic-producing microorganisms in the rhizosphere. *Can. J. Microbiol. 17*:1143–1145.
31. Otto, M., and Neumann, R. O. 1904. Ueber einige bakteriologische Wasseruntersuchungen im Atlantischen Ozean. *Zentralbl. Bakteriol. 13*:481–489.

32. Pacha, R. E., and Kiehn, E. D. 1969. Characterization and relatedness of marine vibrios pathogenic to fish: physiology, serology, and epidemiology. *J. Bacteriol. 100*:1242–1247.
33. Pierson, B. K., and Howard, H. M. 1972. Detection of bacteriochlorophyll-containing micro-organisms by infrared fluorescence photomicrography. *J. Gen. Microbiol. 73*:359–363.
34. Russell, H. L. 1893. The bacterial flora of the Atlantic Ocean in the vicinity of Woods Hole, Massachusetts. *Bot. Gaz. 18*:383–395.
35. Schmidt-Nielsen, S. 1901. Beitrag zur Biologie der marinen Bakterien. *Biol. Zentralbl. 21*:65–71.
36. Sheldon, R. W. 1972. Size separation of marine seston by membrane and glass-fiber filters. *Limnol. Oceanogr. 17*:494–498.
37. Shuler, M. L., Aris, R., and Tsuchiya, H. M. 1972. Hydrodynamic focusing and electronic cell-sizing techniques. *Appl. Microbiol. 24*: 384–388.
38. Strange, R. E., and Martin, K. L. 1972. Rapid assays for the detection and determination of sparse populations of bacteria and bacteriophage T7 with radioactively labelled homologous antibodies. *J. Gen. Microbiol. 72*:127–141.
39. Strange, R. E., Powell, E. O., and Pearce, T. W. 1971. The rapid detection and determination of sparse bacterial populations with radioactively labelled homologous antibodies. *J. Gen. Microbiol. 67*: 349–357.
40. Stull, V. R. 1972. Size distribution of bacterial cells. *J. Bacteriol. 109*: 1301–1303.
41. Tanner, F. W., and Schneider, D. L. 1935. Effect of temperature of storage on bacteria in water samples. *Proc. Soc. Exp. Biol. Med. 32*: 960–965.
42. Waksman, S. A., and Carey, C. L. 1935. Decomposition of organic matter in sea water by bacteria. I. Bacterial multiplication in stored sea water. *J. Bacteriol. 29*:531–543.
43. Waksman. S. A., Reuszer, H. W., Carey, C. L., Hotchkiss, M., and Renn, C. E. 1933. Studies on the biology and chemistry of the Gulf of Maine. III. Bacteriological investigations of the sea water and marine bottoms. *Biol. Bull. 64*:183–205.
44. Wallace, G. T., Jr., Loeb, G. I., and Wilson, D. F. 1972. On the flotation of particulates in sea water by rising bubbles. *J. Geophys. Res. 77*: 5293–5301.
45. White, L. A., and MacLeod, R. A. 1971. Factors affecting phosphate uptake by *Aerobacter aerogenes* in a system relating cell numbers to ^{32}P uptake. *Appl. Microbiol. 21*:520–526.
46. Wiebe, W. J., and Hendricks, C. W. 1971. Simple, reliable cold tray for the recovery and examination of thermosensitive organisms. *Appl. Microbiol. 22*:734–735.
47. Wood, E. J. F. 1953. Heterotrophic bacteria in marine environments of Eastern Australia. *Aust. J. Mar. Fresh. Res. 4*:160–200.
48. ZoBell, C. E. 1946. *Marine Microbiology*. Chronica Botanica, Waltham, Massachusetts, 240 pp.
49. ZoBell, C. E., and Grant, C. W. 1943. Bacterial utilization of low concentrations of organic matter. *J. Bacteriol. 45*:555–564.

7

Reprinted from pp. 879, 881–900 of *J. Mar. Biol. Assoc. U.K.*, **16**(3), 879–900 (1929–1930)

Bacteria of the Clyde Sea Area: A Quantitative Investigation.

By

Blodwen Lloyd, M.Sc., Ph.D.,

From the Dept. of Bacteriology and Botany, Royal Technical College, Glasgow, and the Marine Biological Station, Millport.

With 7 Figures in the Text.

[*Editor's Note:* In the original, material precedes this excerpt.]

Experimental Technique.

Sampling Stations.

Regular monthly samples were taken at three stations which were selected as likely to be different yet characteristic areas; in addition, samples were taken at less regular intervals at other places.

The regular stations were :—

1. *Loch Striven.* This loch is notably free from steamer traffic, and the adjoining land is sparsely populated. It thus represents an area remarkably free, in view of its proximity to land, from industrial or other human contamination.
2. *Loch Long* (*Thornbank Station*). This station is moderately free from land contamination, but there are habitations along the shores of the loch, and there is a certain amount of boat traffic. Terrestrial influences are thus more marked here than in Loch Striven.
3. *Greenock.* Here the water is estuarine in character, and highly polluted with sewage, with waste from sugar refineries and with industrial effluents generally, which are emptied into the River Clyde. This area was chosen in order to determine to what extent the true water bacteria would be outnumbered by those species present as a result of contamination. A detailed account of the Clyde Sea Area is given by Mill (**19**); a map of the area and a summary are also given by Marshall and Orr (**17**).

Sampling Technique.

Vertical series of samples were taken periodically at the above-mentioned stations. It was necessary to employ an apparatus capable of collecting water at the desired depth without taking in water from other depths when being hauled up. The closing water-bottles generally employed in marine work are not suitable for the collecting of sea-water destined for bacteriological analysis. The following are the principal objections :—(i.) Such bottles are made of metal, and most metals have a marked bactericidal effect. This objection is not a cogent one if the sample is in the bottle for only a short time (**3**). (ii.) When a series of samples has to be taken with one bottle, each sample has to be siphoned into a sterile container for transport ashore, and while this operation is being carried out there is a risk of contamination. (iii.) Between the taking of successive samples it would be necessary to sterilise the bottle.

and these water-bottles are usually so unwieldy that it is laborious to accomplish this effectively.

Several bacteriological samplers have been described, notably those by Bertel (**4**), Russel (**21**), Portier and Richard (**20**), Drew (**8**, **9**), Matthews (**18**), Birge (**6**) and Sclavo-Czaplewski (**14**). The simplest type consists essentially of a stoppered bottle, the stopper being pulled out at the required depth. The chief disadvantage of the method is that the removal of the stopper requires a second line which is apt to foul the lowering wire. The second type consists of a glass container with a drawn-out sealed tip ; this tip can be broken by a messenger when the apparatus is at the required depth. Matthews has described a deep-sea bacteriological water-sampler of more complicated design than the foregoing.

For the work described in the following pages a comparatively simple apparatus was required, capable of use to a depth of 60 fathoms. Eventually a sampling apparatus similar to that described by Birge (**6**) and Wilson (**24**) was selected.

It consists of (i.) a glass sampling tube and (ii.) a metal tube-holder. The tube is an ordinary combustion tube 15 cm. $\times$ 3 cm., fitted with a one-holed rubber cork. Through this projects a glass tube bent at a right angle, having the end drawn out and sealed at about 12 cm. from the bend. The right-angle tubes are fitted into the corks, and together with the combustion tubes sterilised by steam. These are then fitted together, due precautions being taken to avoid contamination of the parts.

The tube-holder consists of a plate-sinker A (Fig. 1), with a spring clamp B to hold the sampling tube in position. At the top of the sinker is a projecting arm at the free end of which is a small brass breaking pin C. There is also a lever arm D so placed that when the sampling tube is in position the tip of the inlet tube lies immediately above the end of the lever arm and immediately under the breaking pin. When the apparatus reaches the required depth a messenger is sent down the connecting wire to operate the lever arm ; in this manner the tip of the capillary tube is broken.

The above apparatus was modified by Wilson from the sampler used by Russel (**21**) and was designed for limnological work in shallow waters down to 23 metres ; Wilson evacuated his sampling tubes, but the author found experimentally that at a depth of 10 fathoms a tube filled with air at atmospheric pressure would take a sample whose volume would be about half that of the whole tube ; at greater depths, the increased pressure caused proportionally greater filling of the tube. Most of the samples were taken at depths greater than this, and it was necessary to evacuate only those tubes destined to take surface samples.

The apparatus was sent down on a Kelvin sounding wire worked from a sounding machine recording in fathoms.

Evacuation of the surface sampling tubes was accomplished in the following manner :—sterile glass tubing was prepared as shown in Figure 2 and attached to a sterile sampling tube. This was evacuated to a pressure

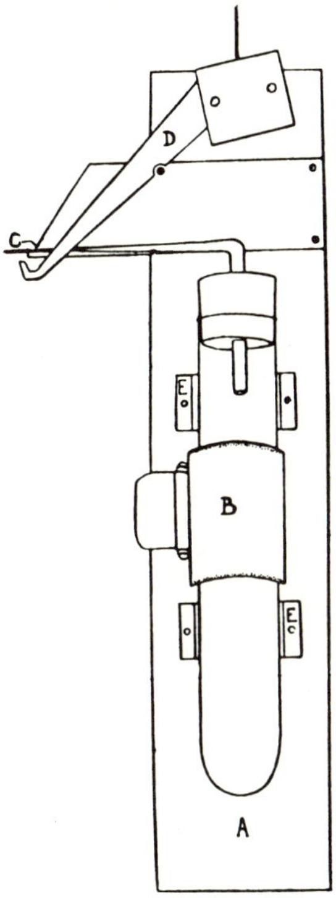

FIG. 1.—Wilson's bacteriological water-sampler, showing sampling tube in position and lever arm open. A, plate-sinker; B, spring clamp; C, breaking pin; D, lever arm; E, rubber cushion.

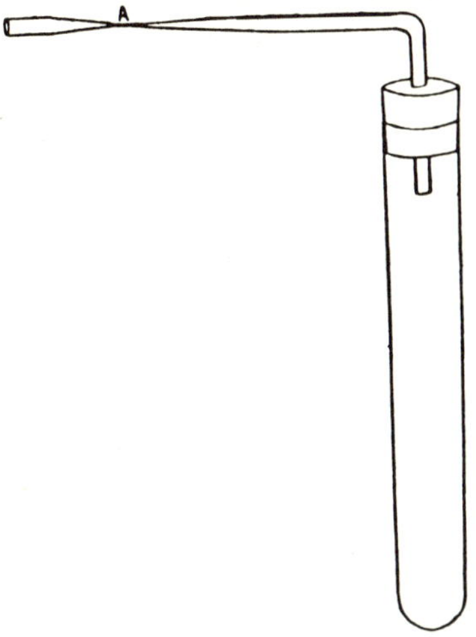

FIG. 2.—Surface sampling tube prepared for evacuation. The tube is attached to a filter pump, and sealed off at A.

of 30–60 mm. and sealed off at A. The connections at the rubber were made air-tight with paraffin wax.

The advantages of this apparatus lie chiefly in its simplicity, the extreme rapidity with which a series of samples can be taken at any one station, and the circumstance that the same tube serves both for taking the sample and for transporting it to the laboratory. The rubber cork is liable to be forced inwards as the sampler descends; if this occurs the tube breaks at the right-angle bend. In practice this can be avoided by selecting well-tapered corks, though for constant deep-water work it would be necessary to have the sampling tube and the inlet arm blown in one piece after the retort-shaped type used by Drew (**8**) for depths greater than 70 fathoms.

While the sampler is being hauled up, the tip of the tube is of course open, but there is no admixture with water from the upper layers through which it is passing, for as the apparatus is being raised the pressure in the tube is being continuously reduced and the compressed air is flowing out of the tube at the narrow orifice. An inflow of water is therefore prevented.

For the quantitative work described below, samples were taken in vertical series at intervals of 10 fathoms from the surface downwards. For transit ashore the open tips of the tubes were sealed off in the flame of a spirit lamp, and packed in sterilised cotton-wool.

Transport of Samples.

In making total counts of the bacterial content of water the following are significant factors :—

(i.) The time interval which elapses between the taking of the sample and its examination in the laboratory, because during that period the bacteria present in the sample may be actively dividing. In the course of this work, the samples at any one station were always taken at the same time of day, and dealt with in the laboratory after approximately the same time interval. In this way uniformity of treatment was ensured, and the numerical results so obtained at any one station for different dates are comparable with one another.

	Time of Sampling.	Time of Inoculating.
Loch Striven	12.30 p.m.	5.0 p.m.
Loch Long	1.0 p.m.	5.30 p.m.
Greenock	4.0 p.m.	6.0 p.m.

(ii.) The temperature at which the samples are kept during transport. It is advisable to inhibit bacterial reproduction by keeping the samples on ice until they are examined in the laboratory. It was found that when samples were transported in a padded box, the temperature of the water on arrival at the laboratory was not raised more than 2° C.; the rate of multiplication of the bacteria would thus not be appreciably increased.

Laboratory Technique.

The technique adopted for routine work followed as nearly as possible the procedure recommended by the American Society of Bacteriologists

(**1**) for the standard examination of water. The regular bacteriological examination of sea-water does not appear to have called for special attention in their schedule.

The following culture media were used :—

Standard Agar.

Tap-Water	1000 gm.
Lab-Lemco	3 gm.
Peptone (Witte's)	10 gm.
Agar	15 gm.
NaCl	5 gm.

The medium was prepared in the usual way, cleared with 10 gm. egg-albumen dissolved in 100 c.c. of water, and adjusted with NaOH to neutral, using phenolphthalein as an indicator. The agar was sterilised by autoclaving for 30 minutes at a pressure of 2 atmospheres (Giltner, **12**. p. 40).

The following modifications were tried experimentally :—

(i.) Tap water and NaCl were replaced by filtered sea-water.
(ii.) NaCl was replaced by 34 grams of evaporated sea salt.
(iii.) Tap water and Lab-Lemco were replaced by fish extract. The fish extract was prepared by slowly heating 1 kilog. of cod in 1 litre of sea-water for about 4 hours ; it was then filtered and made up to 1 litre with tap water.

These media all favoured growth, but they were not found to be markedly superior to the ordinary bacteriological media, and therefore they were not used for routine quantitative work.

Standard Gelatine. This was prepared in the same manner as the foregoing, using 150 grams of gelatine per litre instead of the agar (**12**, p. 35).

Conradi-Drigalski Agar.

(*a*)	Water	2 litres.
	Agar	60 gm.
	NaCl	10 gm.
	Nutrose	20 gm.
	Peptone	20 gm.
	Lab-Lemco	6 gm.
(*b*)	Azolitmin (2·5%)	40 c.c.
	Lactose	30 gm. in 100 c.c. water.
	Na_2CO_3 (10%)	4 c.c.
	Crystal Violet (0·1 gm. in 100 c.c.)	20 c.c.

The agar (*a*) is prepared in the usual way and sterilised. The ingredients (*b*) are sterilised separately and added to the hot agar (**12**, p. 387).

On this medium the growth of non-intestinal organisms is inhibited. The presence of colonies is presumptive evidence of fæcal contamination; the coliform lactose-fermenting species appear as red colonies, and the typhoid-dysentery group as white or blue colonies.

McConkey Agar.

Water	500 c.c.
Peptone	10 gm.
Agar	7·5 gm.
Neutral Red (1%)	1·25 c.c.
Sodium taurocholate	2·5 gm.
Lactose	5 gm.

The lactose is added to the hot agar after filtration. This medium also inhibits the growth of non-intestinal species.

Litmus-lactose-bile-salt Broth.

Tap Water	1000 gm.
Sodium taurocholate	5 gm.
Peptone	20 gm.
Lactose	10 gm.
Azolitmin (2%)	20 c.c.

The peptone and sodium taurocholate are boiled and the lactose and azolitmin added afterwards. Fermentation of the lactose is presumptive evidence of the presence of intestinal organisms (**12**, p. 382).

Of the above media, the first two were used for ordinary quantitative work, and the last three for the detection of pollution.

Plate cultures were made by adding measured volumes of the sample to the nutrient media. The inoculations were made at as low a temperature as possible, i.e. above 35° C. for agar, and above 25° C. for gelatine. At this temperature the medium is still liquid, and is not so hot as to kill the ordinary micro-organisms.

From each water sample the following cultures were made :—

(*a*) Agar : 1·0 c.c., 1·0 c.c. (duplicate), 0·5 c.c., 0·1 c.c., and a control plate, not inoculated.

(*b*) Gelatine : as for agar.

(*c*) Conradi-Drigalski agar : 1·0 c.c.

(*d*) McConkey agar : 1·0 c.c.

(*e*) Litmus-lactose-bile-salt broth : one tube culture.

Quantitative Work. Cultures (*a*) and (*b*) were incubated in dark containers at room temperature. At the fifth day after inoculation the number of colonies was counted with the naked eye; after this period of incubation the number was fairly constant.

Some departures were made from the accepted routine for the bacteriological examination of water. Firstly, the incubation temperature was some six to twelve degrees lower than that usually employed. This was done for two reasons: (i.) to encourage the growth of the true water bacteria, which appear to thrive better at a temperature lower than that which favours soil bacteria; freshwater bacteria, and presumably marine bacteria also, are extremely sensitive to high temperatures both during plating and during incubation. (ii.) to discourage the growth of any organisms present which grow best at higher temperatures, as, for example, coliform bacteria. In this way the colonies which grew on the agar and the gelatine plates were representative chiefly of the true water bacteria.

A second departure from the usual routine was that the dilution method not being necessary was not employed. It had been ascertained by preliminary tests that the number of colonies obtained from an inoculum of 1·0 c.c. could easily be counted. In the Greenock area the total number of bacteria was very high, but this was due largely to putrefactive and intestinal organisms, and since their growth was discouraged by the low incubation temperature employed, the numbers obtained on agar and gelatine plates represent the approximate number of true water bacteria and not the total number of bacteria present.

A third point of difference from ordinary routine lay in the method of recording results. The American Society of Bacteriologists (1) recommend that, in order to avoid fictitious accuracy, the numbers obtained from bacterial counts should be approximated as under :—

1– 50 recorded as enumerated.
51– 100 recorded to the nearest 5.
101– 250 recorded to the nearest 10.
251– 500 recorded to the nearest 25.
501–1000 recorded to the nearest 50.

This recommended procedure is, however, intended for use in reports based on the evidence of one sample only for any one locality. This plan was not adopted in this work, but the actual figures obtained are given, since a number of plates was made for each sample and a number of samples was taken for each place at intervals throughout a year.

Detection of Pollution. Conradi-Drigalski and McConkey plates (*c*) and (*d*) and broth tube (*d*) were incubated at 37° C. for 24 hours. Any colonies which appeared by the end of that period were presumptive coliform organisms, the red colonies being of course lactose fermenters. When the numbers so obtained were compared with the gelatine and agar counts for the same sample, it was possible to estimate the proportion of high-temperature intestinal organisms to low-temperature free-living organisms. This indicated not only whether the water sampled was polluted or not, but also the degree of pollution of the sample.

Results.

Loch Striven. This is the type area of natural sea-water, with minimal contamination despite its proximity to land. The numerical results of the plate cultures made from the samples taken each month are given in

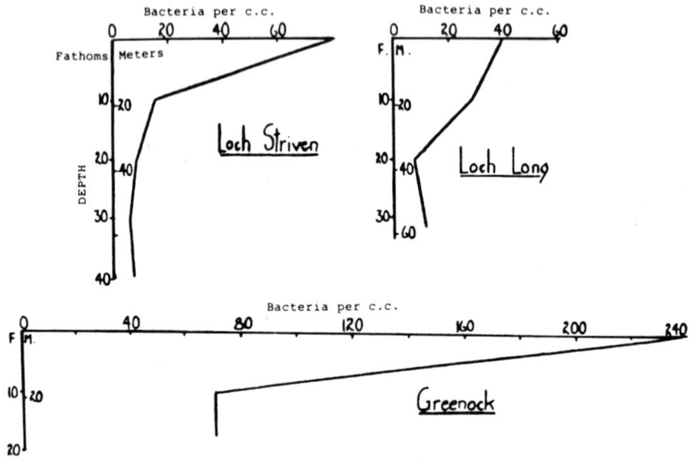

FIG. 3. Average number of bacteria per c.c. for the year, May, 1928 - April, 1929.

Table I. The average number of bacteria per c.c. is estimated for each sample from the total volume of water-sample plated out (usually 5·2 c.c.); these averages are given in the last column.

Samples were taken at the surface, and at intervals of 10 fathoms down to the bottom; the bottom sample was taken immediately above the mud, in approximately the same place, but as the sides of the loch were steep, the depth at which the bottom sample was taken varied from 37 to 40 fathoms.

The following results are noted :—

Vertical Variation. 1. The number of bacteria per c.c. is low compared with that of other stations in the Clyde Sea Area.

2. The number of bacteria is greater at the surface than at lower levels (Fig. 3).

3. At the surface, the bacterial content of the water fluctuates much more than in deep waters, where it is more nearly constant.

4. At and below 10 fathoms, the number of micro-organisms does not exceed 30 per c.c. at any season.

5. Bottom samples show a rather lower bacterial content than the water at higher levels. This is an unusual feature, which will be discussed in the following section.

Seasonal Variation. 1. The numbers of bacteria per unit volume vary only little throughout the course of a year. Any variations outside the limits of accuracy of the experimental methods adopted appear to be erratic and therefore cannot be correlated with any factor varying seasonally. Such fluctuations may be due to external factors which operate only intermittently. This is particularly applicable to the surface waters, whose composition varies so much with conditions affecting drainage from the land (**17**).

2. At the surface and below, the bacterial content is relatively high in August.

Diurnal Variation. The samples just described were taken at intervals of several weeks, so they do not show short-period variations due to such factors as tide, sunshine, and perhaps diatom increases. Accordingly, samples were taken in vertical series quarterly at 3-hour intervals over a period of 24 hours. The experimental conditions differed from those under which the other periodical samples were examined. Plating was done on the boat immediately after sampling; two agar plates were made from each sample, using 1·0 c.c. as inoculum. Gelatine plates were not employed, as the lower solidifying point of gelatine made it unsuitable for plating work carried out on a boat. Four such 24-hour series were made. The results are given in Tables IIA, IIB, IIC and IID.

Winter Series, December 19/20, 1928. The following results were observed :—

1. The general vertical distribution over the 24 hours agrees with that of the monthly samples, i.e. there is a progressive decrease in the bacterial content from the surface to the bottom and the surface samples show wider fluctuations.

2. The number of surface bacteria is unusually high : this may be correlated with the fact that the herring fishing was in progress at the time.

3. At any given depth the numbers were higher at night, with a tendency to decrease from 7 p.m. till the early hours of the morning

(Fig. 4). With increasing depth the maximum occurs at a later hour, but this may be due to chance variations.

Spring Series, March 7/8, 1929. From Table IIB the following observations were made :—

1. The number of surface bacteria is unusually low. This may have been due to the low temperature then prevailing, and to the fact that the herring fishing had ceased some weeks earlier. These possible causes are discussed in the following section.

2. There is an increase in the number of bacteria during the night hours, the highest numbers being attained at 3 a.m. in all except the bottom samples (Fig. 5).

Summer Series, July 23/24, 1929. From Table IIC the following observations were made :—

1. In this series, the number of surface bacteria is again relatively high and fluctuates very much between the 3-hour samplings.

2. The lowest surface numbers are at noon and the highest at 5 p.m.

3. Below the surface the numbers were fairly constant, but at 10 and at 20 fathoms there were increases after midnight (Fig. 6).

Autumn Series, October 24/25, 1929. These samples confirmed the findings of the three previous series. The following conclusions were drawn (Table IID) :—

1. The number of surface bacteria fluctuates very much, but on this occasion there were relatively few organisms in the surface water.

2. At lower levels there was in general an increase in numbers during the night, so that the bacterial content tends to be highest in the early morning hours (Fig. 7).

In summing up the results of these series of bacteriological analyses spread over a period of a year, it is seen that (i.) the number of surface organisms is usually relatively high, and is much subject to short-period variations, (ii.) in deeper water layers the bacterial content is very much lower and remains almost constant throughout the year, and (iii.) during the hours of darkness the numbers tend to increase slightly.

Results of Pollution Tests.

Conradi-Drigalski agar and McConkey agar were used as routine media for the detection of pollution. Both these media inhibit, or at least retard, the growth of all but intestinal bacteria, so that the presence of colonies on a plate inoculated from a water sample is presumptive evidence of the contamination of that water. Most of such colonies will appear red and will be the lactose-fermenting *B. coli* or its congeners. This can be confirmed by a lactose-broth culture made from the same

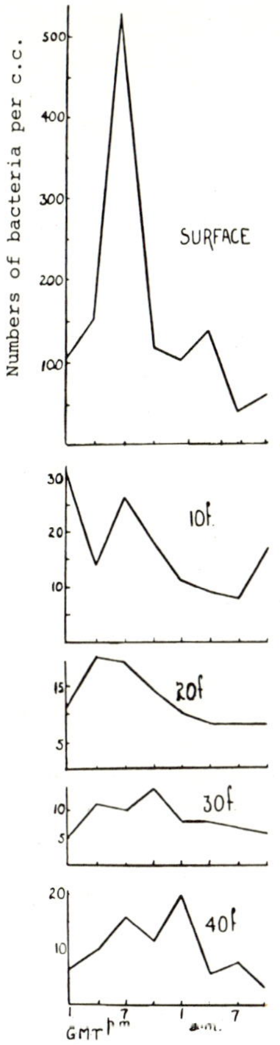

FIG. 4. Diurnal variation in the numbers of bacteria at different depths. Samples taken 19/20.xii.29. Loch Striven

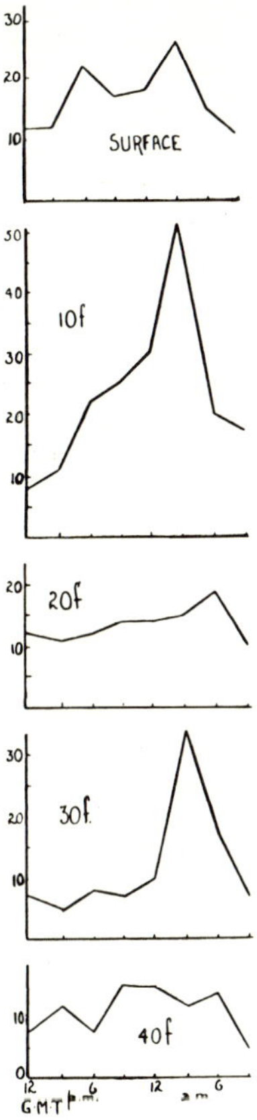

FIG. 5. Samples taken 7/8.iii.29. Loch Striven

water sample. The intestinal organisms, e.g. *B. typhosus*, Morgan's bacillus, which do not ferment lactose, are less common in polluted water ; on the above media they appear as blue-white colonies.

It was found that bacteriologically the Loch Striven waters were remarkably pure throughout the greater part of the year. During the winter months, however, when the herring fishing was in progress, the surface samples showed an increase in the total bacterial content, and this was due in part to an increase in the numbers of presumptive coliform organisms, when they formed about 20% of the total. When the herring season was over, however, the water became free from these organisms.

Loch Long. The results given in Table III show that at Loch Long the vertical distribution of bacteria is very similar to that of Loch Striven, though the bacterial content is in general higher. This is possibly to be connected with the greater accessibility of this loch to steamer traffic.

Here again there is little evidence of regular seasonal variation (Fig. 3). Table III shows that the bacterial content of the surface water varies widely and apparently erratically. The samples taken at other levels show a midsummer minimum, an autumn increase and a minimum in the months of January and February. Very few bacteria were present in the January samples ; at this time the temperature was so low that the waters at the head of the adjoining branch, Loch Goil, were frozen. This seems to bear out the opinion generally held that water bacteria are specially sensitive to changes in temperature.

Normally there are more bacteria at the surface than at other levels ; there is only one exception, namely, the May series of samples taken in Loch Long. Here a much higher bacterial content is found at 10 fathoms. For this there does not appear to be a suitable explanation, unless we have regard to the fact that the number of presumptive coliform organisms obtained on McConkey plates on that date was specially high and that 3 out of 4 lactose-broth cultures produced gas. This points to some special circumstances causing localised subsurface contamination of the water at the spot where the samples were taken.

Greenock. In this area the samples were taken at the deepest part of the river channel off Greenock, where the water is much polluted with sewage and with industrial effluents. The effect of such pollution will of course be more marked near the shore-line. Between the coast and mid-channel there is a certain amount of self-purification of the water ; nevertheless, the bacterial content of the open water is affected. The number of organisms per c.c. is considerably higher than in the lochs (see Table IV and Fig. 3) ; further, the number of colonies on McConkey plates is many times more than the number on gelatine and agar plates, so that it is not possible to estimate the number of coliforms without dilution of the

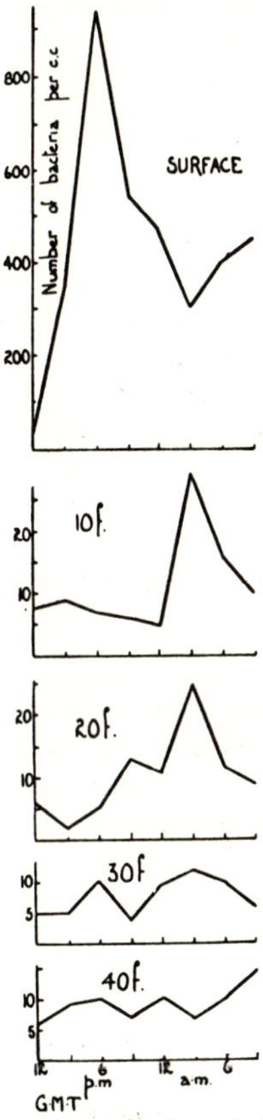

FIG. 6. Diurnal variation in the numbers of bacteria at different depths. Samples taken 23/24.vii.29. Loch Striven.

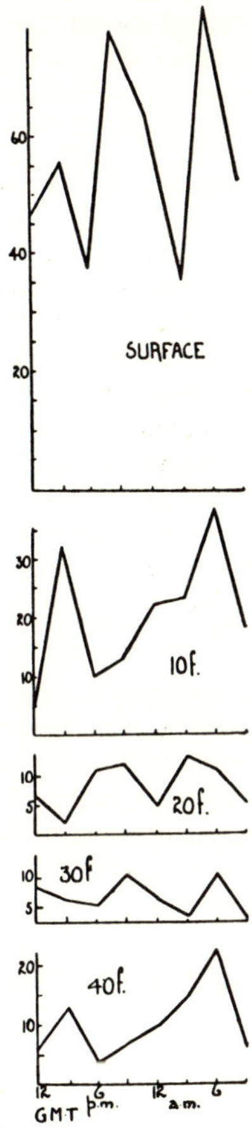

FIG. 7. Samples taken 24/25.x.29. Loch Striven.

sample before plating. In this area then a high proportion of the constituent organisms are of the intestinal type.

The surface samples again showed by far the highest numbers; at 10 fathoms it was found always that there were fewer organisms; at the bottom, however, there was an increase. In none of these was any regular seasonal quantitative variation found.

Cumbrae Deep. At the above-mentioned stations samples could not be obtained at great depths, the deepest being 40 fathoms in Loch Striven. At Cumbrae Deep, however, there is a depth of 62 fathoms. The results from one series of samples (Table V) show that the vertical distribution is similar to those of the lochs, namely, that the surface waters have a relatively higher bacterial content, that the numbers decrease with increasing depth and that they increase again near the bottom.

It is interesting to note that the number of bacteria here is higher than it is in the lochs, but Cumbrae Deep is not typical of the Clyde Sea Area, since it is used as a dumping-ground for the Glasgow sewage treated by the activated sludge process at the Shieldhall Sewage Disposal Works. Buchanan states that such activated sludge, like crude sewage, has a bacterial flora of largely intestinal species (see **13**, p. 2); this sludge therefore appears to affect the bacterial content of the sea-water into which it is discharged.

Millport. The foregoing samples were taken in open water. For purposes of comparison, samples were taken (*a*) in the intertidal zone and (*b*) in a high-water rock pool, and the bacterial content investigated. From the results given in Tables VI and VII it is seen that the numbers of bacteria are very much higher than in the open water—a fact to be correlated with the congestion of other living organisms in the strip of water bordering the land.

Factors Affecting Distribution of Bacteria.

The foregoing work deals only with the free-living bacteria; these may be differentiated by their mode of nutrition into prototrophic and metatrophic species. It is not yet definitely established to what extent the prototrophic forms, as for example, the nitrogen-fixing bacteria, are found free-living in the sea. Ordinary marine conditions are not always favourable for their growth, so their existence is problematic. Experimental proof of their activity is available, but much of the work done on the subject is contradictory.

The majority of free-living bacteria are metatrophic, deriving their organised food-stuffs directly or indirectly from other organisms. In the sea two classes may be conveniently distinguished :—

(i.) The saprophytes living on and attached to particles of an organic nature, such as decaying seaweed, dead plankton and terrigenous detritus. This group will include the putrefactive bacteria ; they will be common where suspended matter is common and will be particularly abundant where such particles will accumulate, as for instance at the sea-bottom.

(ii.) The true planktonic bacteria, which have a simple metabolic cycle, and whose source of food is the dissolved organic material—amino-acids, proteins, and carbohydrates,—present in the sea, particularly in the surface waters near the coast-line.

(*a*) *Sunlight*. Among the physical factors affecting bacteria, sunlight is well-known to have a deleterious action on the growth of micro-organisms. This is due to the bactericidal action of ultra-violet light, and is to be distinguished from the ordinary retarding effect of light on growth. If the insolation is sufficiently intense, it would be expected that this effect would be most marked at the surface. The foregoing results for the Clyde Sea Area, however, show a greater number of bacteria at the surface, even when the samples were taken on sunny days, so that the bactericidal effect of light was apparently negligible. Schmidt-Neilsen (**23**), on the other hand, records finding at Dröbak 26 bacteria per c.c. at the surface and 420 per c.c. at a depth of 25 metres, and states that this progressive increase with depth is possibly due to the influence of sunlight. Similarly Bertel (**5**) found that off the coast of Monaco the numbers increased with depth. From two series of samples he found that the numbers increased from 1 per c.c. at the surface, (*a*) to 30 at a depth of 200 metres, (*b*) to 36 at a depth of 400 metres ; these results were obtained during the months of May and June, when insolation is intense.

In the same way, the bactericidal effect of sunlight would be expected to be more marked during the summer months, and, if it were the limiting factor, the numbers of bacteria would be lower in the summer than in the winter months. For the Clyde Sea Area, the bacterial content of the water on the whole is lower in the period from June to August, but it is not possible at this stage to correlate such seasonal variations with definite variations in the intensity of light. There do not appear to be other seasonal records with which these results may be compared.

The diurnal variation in the bacterial content may be due in part to the effect of sunlight. Tables IIA, B, C, and D show that in general there is an increase during the hours of darkness. This confirms the findings of Fischer (**10**), that more bacteria are to be found at sunrise than in the afternoon. Bertel (**5**) also states that there is a night-time bacterial increase which persists even to the early hours of the morning. He bases his conclusions, however, on the evidence of three series of samples taken on different dates, so that it is possible that other factors came into play during the intervals. From his experimental work Bertel deduces that the night-time is more favourable for reproduction by bacteria ; but the fact that bacteria are more numerous at night does not prove that they necessarily reproduce more rapidly at night, since other factors such as the vertical mixing of water may cause movement of particles supporting bacteria from one water level to another.

(*b*) *Temperature.* An increase of temperature within certain limits favours the growth of micro-organisms in general. Sudden and marked changes in temperature may, however, have an unfavourable effect on bacteria, particularly on those species which do not form spores and which have not a highly resistant cell-membrane. Water bacteria are in general small non-sporing bacilli, notoriously sensitive to sudden changes in temperature (**2**).

Ordinarily, in the Clyde Sea Area the range of temperature variation is not sufficiently wide to show its effect on the bacterial content of sea-water. An examination of the results of the January samples taken in Loch Long shows the number to be very low ; these samples were taken during an extremely cold spell when the head waters of the adjoining loch were frozen.

(*c*) *Movements of Water.* In a still body of water it would be reasonable to expect a greater number of micro-organisms at the surface and at the bottom than in the intervening layers. However, any movements which cause vertical mixing from one layer to another may cause short-period variations in the bacterial content of the water at any depth. In view of this, it is remarkable that for the Clyde Sea Area the vertical progressive diminution in the number of bacteria is so constant.

Near land masses, especially off the steep western coast of Scotland, there is a tendency for vertical upwelling of water from the bottom and a surface flow away from the land. Foodstuffs are thus brought up to the surface, and become available for aerobic marine bacteria. At the same time, land drainage contributes its quota of detritus and soluble organic compounds, the less dense fresh water tending at first to distribute itself over the surface (**17**). Thus the surface waters appear to be best furnished with foodstuffs for the support of a more numerous bacterial population than is found at lower water levels. Almost all the samples taken illustrate this.

Tidal movements, similarly, may affect the bacterial content of the water. The results of the German Plankton Expedition (**10**) show that the number is higher during the ebb than during the flood tide. This influence is more marked when there is a slightly shelving coast-line with a comparatively broad intertidal zone which supports much plant and animal life, the number of bacteria being correspondingly high. In the lochs investigated, however, the steeply sloping sides afforded only a narrow intertidal zone, and the tidal variations do not appear to have much effect on the bacterial content of the water. This is best shown in Tables IIA, B, C and D, where samples were taken at 3-hour intervals over periods of 24 hours at various seasons of the year.

(*d*) *Sedimentation of Organic Particles.* Attention has already been drawn to the fact that the saprophytic bacteria are not truly planktonic, but are attached to suspended organic particles of various origins. Any factor which determines the distribution of these particles will affect also the number of saprophytic bacteria. In water where there are no currents and no vertical mixing, it would be reasonable to expect at the surface an accumulation of such organic matter as has lower density than sea-water, with a corresponding accumulation of bacterial saprophytes. Similarly, the deposits at the bottom induce a high number of bacteria. In the intervening layers, however, through which suspended matter, as for example, the "plankton rain," tends to sink slowly, the number of bacteria also decreases progressively with depth.

Almost all the vertical series of samples taken in the Clyde Sea Area show this, viz., a relatively high bacterial content at the surface. with a gradual decrease till the bottom is reached, when the numbers again rise. Probably the increase at the bottom is due to the proximity of the mud which exists at all the stations. Some work now in progress has shown that the number of bacteria in the mud itself far exceeds that of the water immediately above it.

(*e*) *Biological Factors.* Since most water bacteria are heterotrophic, they will enter into competition for foodstuffs with the simpler animal forms, and will in turn be preyed upon by protozoa. It is well known that in the soil the numbers of protozoa and bacteria bear a numerical relationship to one another, namely, that when one group is abundant, the other is scarce and vice versa. It is highly probable that a similar state of affairs exists in the sea, though experimental evidence is lacking as yet.

After a period of great activity in the development of any group of organisms, there will be a glut of dead organisms and bacteria may thereafter multiply very rapidly. After the spring diatom increase in 1929, the bacterial content of the water was high, especially at the surface. There was a comparable increase during the autumn of 1928, when the herring season was in progress in Loch Striven, but not at the other stations sampled. In Loch Striven there was a great increase in the numbers of bacteria at the surface, with no corresponding increase at the other stations.

The following table shows the average number of bacteria at the surface in Loch Striven, compared with the herring catches for the season. The latter figures are available through the courtesy of the Fishery Board for Scotland.

	1928								1929			
	May	June	July	Aug.	Sept.	Oct.	Nov.	Dec.	Jan.	Feb.	Mar.	Ap.
Quantity of herrings taken monthly (crans)	0	0	40	0	0	4,644	16,841	14,075	4,727	–	–	–
No. of bacteria taken at surface in one sample per month	43	5	–	39	–	17	136	26	44	120	12	1

Thus it appears that the surface maximum in November coincides with the maximum herring catch ; this is followed by a decrease in numbers during December, and a second increase in February.

The writer wishes to make acknowledgments to the Scottish Marine Biological Association for permitting the frequent use of their boat, the *Nautilus ;* to Mr. Elmhirst, the Superintendent of their Laboratory at Millport, for readily affording facilities for work there ; to Professor D. Ellis and to Dr. J. A. Cranston of the Royal Technical College, Glasgow, for much helpful criticism and advice ; and to Mr. R. J. Nairn for frequently deputising for the writer in the collecting of samples.

SUMMARY.

1. A bacteriological survey of the Clyde Sea Area has been made over a period of a year. Water samples have been taken monthly at three stations, and also less frequently at other places. A uniform routine technique has been adopted for studying their bacterial content.

2. *Vertical Variation.* The surface waters were found to have the highest bacterial content. With increased depth until near the bottom, there was a progressive decrease in numbers. At the bottom there was usually a slight increase.

3. *Seasonal Variation.* Throughout the year the numbers were found to be remarkably constant for all layers except the surface, with only slight evidence of rhythmic seasonal variation. At the surface, the bacterial content fluctuated widely, apparently in relation to factors which are not seasonal but irregular.

4. *Diurnal Variation.* At the surface the bacterial content is irregular throughout the day ; at lower levels, there is a slight increase during the hours of darkness, the maximum occurring in the evening hours in December, and at 3–6 a.m. in March, July and October.

5. *Purity of the Water.* The waters of both Loch Striven and Loch Long were found to be remarkably free from pollution. In Cumbrae

Deep, however, and in the estuary off Greenock, the numbers of bacteria were high, and a large proportion of these were found to be presumptive coliform organisms.

LITERATURE.

1. American Public Health Association. Standard Methods for the Examination of Water and Sewage. New York, 1923.

2. Bergey, H. A Manual of Determinative Bacteriology. Baltimore, 1923.

3. Berkeley, C. A Study of Marine Bacteria. Trans. Roy. Soc. Canada, XII, 3rd series, 1919.

4. Bertel, R. Ein einfacher Apparat zur Wasserentnahme aus beliebigen Meerestiefen für bakteriologische Untersuchungen Biol. Centralbl., XXXI, 1911.

5. ——. Sur la distribution quantitative des Bactéries planctoniques des côtes de Monaco. Bull. de l'Inst. Ocean. Monaco, No. 224, 1912.

6. Birge, E. A second report on limnological apparatus. Trans. Wisconsin Acad. Sci., XX, 1922.

7. Bohart, R. M. Bibliography of Marine Bacteria. Publ. Puget Sound Biol. Stat., Vol. V, p. 309, 1928.

8. Drew, G. H. Preliminary investigations on the marine denitrifying bacteria. Year Book, No. 10, Carnegie Inst., Washington, 1911.

9. ——. Investigations on marine bacteria at Andros Island, Bahamas. *Ibid.*, No. 11, 1912.

10. Fischer, B. Die bakterien des Meeres. Ergebnisse Plankton Expedition, Bd. IV, 1894.

11. de Giaxa. Verhalten pathogener Mikro-organismen im Meerwasser. Zeitschr. f. Hygiene, VI, 1889.

12. Giltner, W. Laboratory Manual in General Microbiology. New York, 3rd edition, 1926.

13. Harris, F. W., Cockburn, T., and Anderson, T. Biological and physical properties of activated sludge. Corporation of Glasgow, Sewage Purification Dept., 1926.

14. Janke, A. u. Zikes, H. Arbeitsmethoden der Mikrobiologie, Dresden, 1928.

15. Johnstone, J. Conditions of Life in the Sea. Camb. Univ. Press.

16. Letts. Report of the Royal Commission on Sewage Disposal, No. 3, Vol. II, 1903.

17. Marshall, S. M., and Orr, A. P. The relation of the Plankton to some chemical and physical factors in the Clyde Sea Area. Journ. Mar. Biol. Assoc., N.S., XIV, 1927.

18. Matthews, D. A deep-sea bacteriological water-bottle. Journ. Mar. Biol. Assoc., N.S., IX, 1911.

19. Mill, H. R. The Clyde Sea Area. Trans. Roy. Soc. Edin., XXXVI–XXXVIII, 1891.

20. Portier et Richards. Sur une methode de prélèvement de l'eau de mer destinée aux études bactériologiques. Bull. de l'Inst. Oceanog. Monaco, No. 97, 1907.

21. Russel, H. Untersuchungen uber in Golf von Neapel lebende Bakterien. Zeitschr. f. Hyg., XI, 1892.

22. ——. The bacterial flora of the Atlantic Ocean. Bot. Gaz., XVIII, 1903.

23. Schmidt-Nielsen, S. Beitrag zur Biologie der marinen Bakterien. Biol. Centralbl., XXI, 1901.

24. Wilson, F. Description of an apparatus for obtaining samples of water at different depths for bacteriological analysis. Journ. Bacteriol., V, 1920.

[*Editor's Note:* Material has been omitted at this point.]

8

Reprinted from *Microbiol. USSR*, **5**, 322-343 (1936) (*excerpts*)

MULTIPLICATION OF SEA BACTERIA DEPENDING ON THE COMPOSITION OF THE MEDIUM AND ON TEMPERATURE

N. V. Butkevich and *V. S. Butkevich*

[*Editor's Note:* In the original, material precedes this excerpt.]

Summary

Typical sea-water bacteria in usual broth cultures (with 3 p. c. of salt) show in the course of their development characteristic features, which fully correspond to the scheme of Buchanan.

Our experiments, similarly to those of Hess, show that the temperature optimum for the rate of multiplication of these bacteria does not coincide with the temperature optimum for their maximum accumulation. Within a certain temperature range the lowering of the temperature delays the course of the phases of development, but the maximum accumulation may reach at lower temperature the same or even a higher level, than at higher ones.

The values, obtained by means of registering the numbers of bacteria in broth cultures by plating and those obtained in membrane filters under the microscope, did not check. At different periods these differences varied not only in size but even in direction. In the logarithmic phase of growth the filter counts gave figures exceeding 3—4 times those of the plate counts, while later the first were 1,5—2 times less than the latter. These digressions under the conditions of a pure culture experiment may be explained: a) in the case of higher filter counts — by the presence of dying or dead forms, which do not give rise to colonies on agar plates, but are preserved well enough to be detected under the microscope; and b) in the case of higher plate counts — by the presence in these cultures of forms, capable of developing in plates, but not detectable under the microscope due to their too small size or to the loss of their ability to stain.

In counting bacteria in pure cultures, in spite of the deviations indicated above, a certain parallelism is, however, observed between the values obtained by plate and filter counts. In counting the bacterial population of natural reservoirs by both methods there generally is a deviation in the direction of the prevalence of the values obtained by filter counts and no definite or constant relationship between these figures and those of plate counts is revealed. This fact and the not always observed correspondence between the consumption of oxygen in sea water and the number of bacteria, found in it by plate counts, indicate, that in natural reservoirs there exist besides the microorganisms, able to grow on the usual media, also such forms, which do not grow on these media.

When the bacteria are transferred to a new medium, at first usually a decrease, or at least the lack of an increase are observed, and only later, after some time, a numeric increase is observed. This phenomenon, as well as the further development in separate parallel cultures, started and conducted under perfectly similar conditions, often does not appear to be quite uniform, and the differences between parallel cultures are the more considerable the less favourable are the conditions for the development of the bacteria. Particularly marked is this difference in cultures on synthetic media with a low content in the source of carbon and at low temperatures.

The phenomena just noted testify: 1) that the composition of the bacterial complex introduced by the inoculation is not uniform, that only a part of the organisms composing it is able to keep viable and to develop under the new conditions; and 2) that the composition of the separate bacterial complexes, transferred into the parallel cultures from the same source, varies.

Experiments with cultures on synthetic media with a low content in the sources of carbon (salts of the malic and acetic acids) have shown, that their utilization by the microorganisms proceeds up to a certain limit concentration, and that this limit concentration depends not only on the character of the source of carbon and on the microorganisms utilizing it, but also on the temperature conditions, and that it varies within very wide limits in the temperature interval, in which the growth of the bacteria is possible. Within the temperature interval tested in our experiments (25—5° C) the limit value of the source of carbon sank with the rise of temperature and rose with the sinking of the latter.

The relations, revealed by our experiments, conditions doubtless to a certain degree the phenomenon of a considerable and speedy increase of the numbers of bacteria in sea water samples kept under laboratory conditions, which has been observed and registered by us and by other workers, as well as the greatly varying accumulation of bacteria in sea water kept at various temperatures, which has received notice in some recent works.

In connection with the results of our experiments arises a number of further questions, which demand a closer investigation and concern the conditions, determining the limits of the utilization of the organic complex, found in water of natural water-reservoirs by their bacterial population. The investigation and elucidation of these conditions has not only a theoretical, but also a practical interest, as they determine to a certain degree the total production of natural reservoirs in respect of the animal world, populating them, for which the substances which are present there in a dissolved state become available mainly through the bacteria which consume them

Bacteriologic Laboratory
of the All-Union Scientific Research Institute
of Fish Economy and Oceanography.

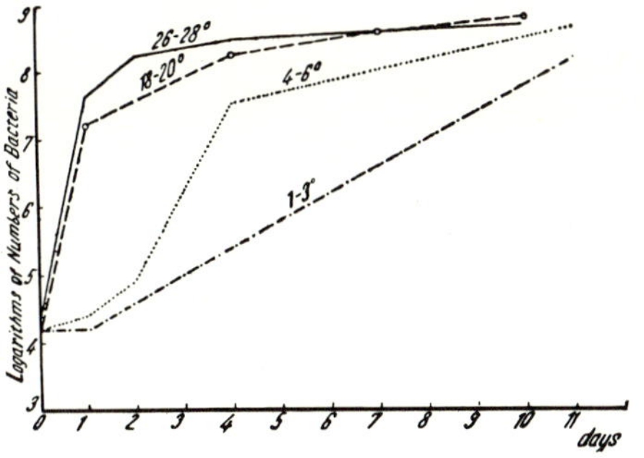

Graph. 1. (Exp. 1) Development of bacteria in broth at varying temperatures according to gelatin plate counts.

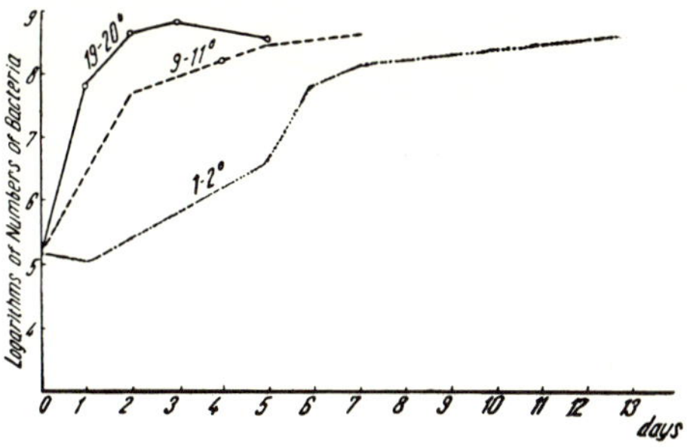

Graph. 3. (Exp. 3) Development of bacteria in broth at varying temperatures according to agar plate counts.

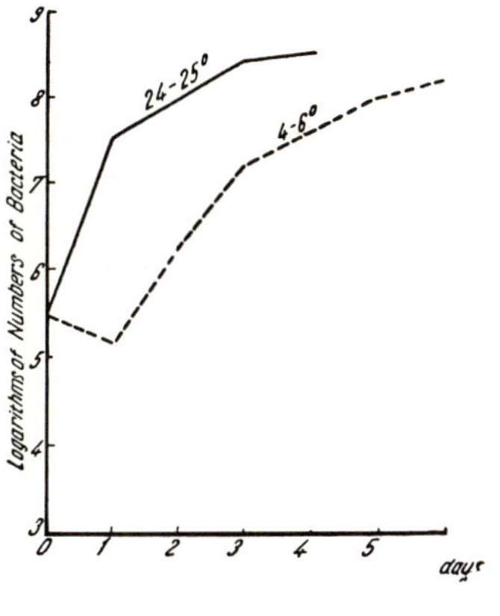

Graph. 2, (Epx. 2) Development of bacteria in broth at varying temperatures according to membrane filter counts under the microscope.

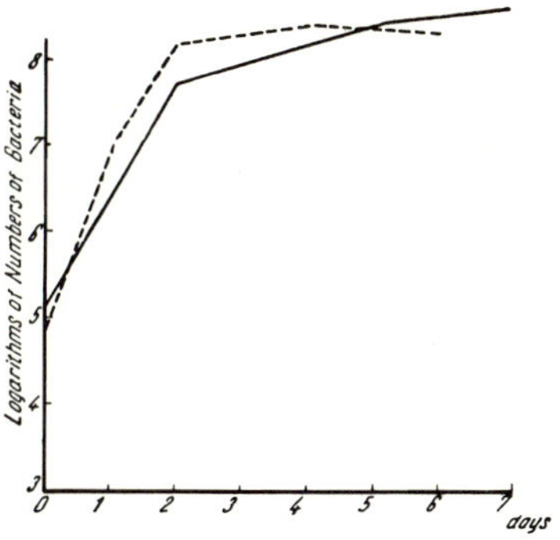

Graph. 4. (Exp. 3) Development of bacteria in broth at 9—11° C according to agar plate counts and membrane filter counts under the microscope. Filter counts — — — —. Plate counts ————.

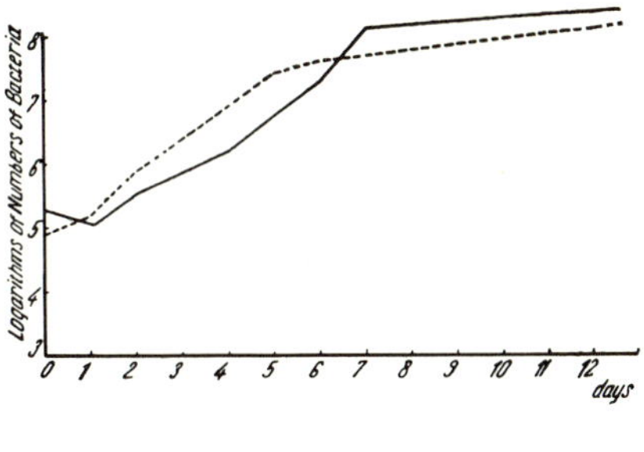

Graph. 5. (Exp. 3) Development of bacteria in broth at 1—2° C according to agar plate counts and membrane filter counts under the microccope. Filter counts — — — — — Plate counts ————

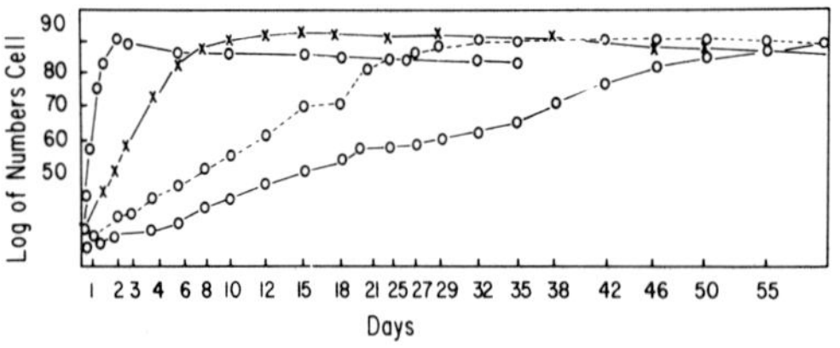

Graph 7. Development of Pseudomonas fluorescens, isolated from cod-fish slime, in a buffered broth at varying temperature (according to Hess)

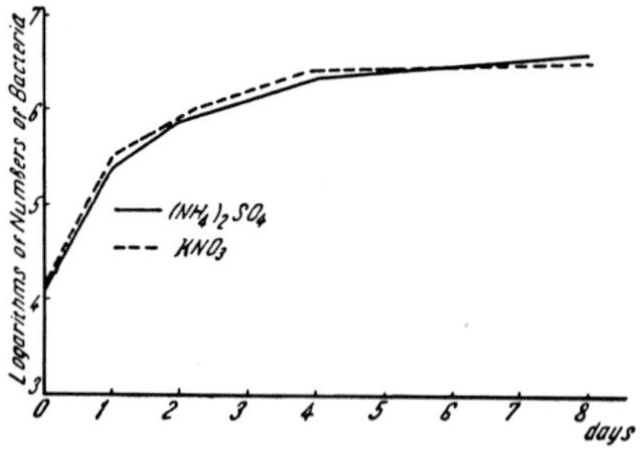

Graph. 8. (Exp. 4) Development of bacteria in a synthetic medium with 0,01 p. c. of malic acid and with $(NH_4)_2SO_4$ and KNO_3 as sources of nitrogen. According to agar plate counts.

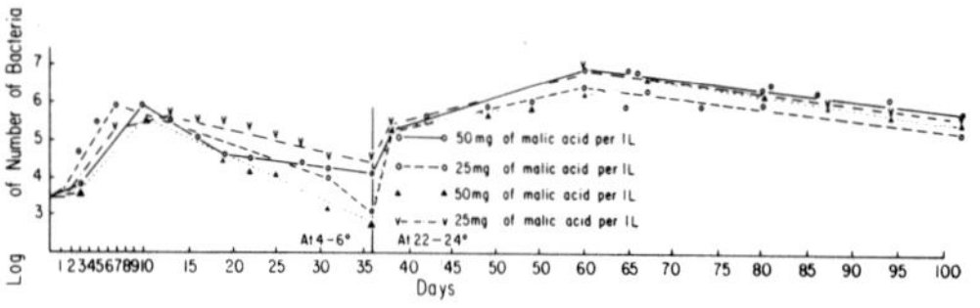

Graph 9. Development of bacteria in synthetic media with malic acid as a source of carbon and KNO_3 and $(NH_4)_2SO_4$ as sources of nitrogen, first for 36 days at 2—4°C, later at 22—24°C

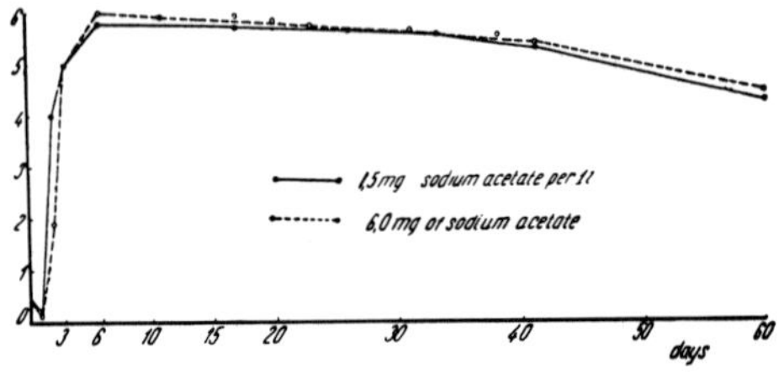

Graph. 10 (Exp. 7) Development of bacteria in synthetic media with varying amounts of acetate and with $(NH_4)_2SO_4$ as source of nitrogen at 22—25° C.

9

Reprinted from pp. 69–70 of *J. Mar. Res.*, **4**, 42–75 (1941)

STUDIES ON MARINE BACTERIA: I. THE CULTURAL REQUIREMENTS OF HETEROTROPHIC AEROBES

C. E. ZoBell

[*Editor's Note:* In the original, material precedes this excerpt.]

SUMMARY

Most of the bacteria recovered from sea water as well as bottom sediments at places remote from possibilities of terrigenous contamination have specific salt requirements which are satisfied best by natural sea water. Neither synthetic sea water nor other isotonic salt solutions are satisfactory substitutes. Merely diluting sea water with fresh water materially reduces the number of marine bacteria which will grow in it. Conversely very few terrestrial or freshwater bacteria are able to grow in nutrient sea water media upon initial isolation.

There is no evidence that the addition of nitrate to nutrient sea water media as has been recommended by others is beneficial, and concentrations exceeding 0.1 per cent potassium nitrate are inhibitory.

There are indications for the addition of 0.01 per cent dibasic potassium phosphate to peptone sea water media although it normally contains sufficient phosphate to provide for the multiplication of marine bacteria.

The enrichment of nutrient sea water media with a trace of iron increases plate counts 18 to 76 per cent, 0.01 per cent ferric citrate or ferric phosphate being almost equally beneficial. The use of the latter is recommended because it provides for the phosphate requirements as well as the iron. This concentration of iron in sea water is toxic only when the medium is acid in reaction.

Marine bacteria appear to grow best in media having a pH of 7.5 to 7.8 although sea water *in situ* is usually somewhat more alkaline than this.

Successive dilution method counts using tubes of nutrient sea water broth demonstrated the presence of 12 per cent more bacteria than plate counts on comparable nutrient sea water agar but the latter yielded much more reproducible results and require much less medium and time.

A comparative study of the merits of different solidifying agents recommends the use of 1.2 to 1.5 per cent Bacto-agar as the best for plating media. Gelatin is liquefied too rapidly by marine bacteria to be satisfactory and although silica gel prepared by a simplified method has certain advantages, it does not give results as good as agar for the demonstration of the heterotrophic bacterial population.

The use of various algae and fish infusions, carbohydrates, organic acids, nitrogen compounds and other organic nutrients in different concentrations and combinations have not proved to be as good as Bacto-peptone for the growth of marine bacteria as indicated by the number and kinds of bacteria on plates as well as the reproducibility of plate counts.

Although special media are required for certain groups of organisms and for special purposes, for maximum plate counts and for the cultivation of most aerobic heterotrophic marine bacteria Medium 2216 is recommended. It contains 0.5 per cent Bacto-peptone, 0.01 per cent ferric phosphate and 1.5 per cent Bacto-agar dissolved in aged sea water, the pH being around 7.6 after autoclave sterilization.

[*Editor's Note:* Material has been omitted at this point.]

10

Reprinted from *Proc. Soc. Exp. Biol. Med.*, **96**, 392–394 (1957)

Factors Influencing the Plate Method for Determining Abundance of Bacteria in Sea Water.*† (23487)

A. F. Carlucci and D. Pramer (Introduced by G. Litwack)
Department of Agricultural Microbiology, Rutgers State University, New Brunswick, N. J.

Total number of bacteria in sea water as determined by plate counts is influenced by numerous factors(1). In this report some effects of medium composition and plating procedure are described.

Materials and methods. Sea water was collected at Long Branch, N. J., approximately 600 feet off shore in an area subject to neither pollution nor dilution. All samples were taken at high tide, packed in ice, and transported

* This investigation was supported in part by Research Grant from Natl. Inst. of Allergy and Infectious Diseases, N.I.H., Public Health Service.

† Journal Series Paper, Department of Agricultural Microbiology, N. J. Agric. Exper. Station, Rutgers, State University, New Brunswick, N. J.

TABLE I. Influence of Composition of Nutrient Medium on Plate Counts of Bacteria from Sea Water.

Medium constituents	Medium % composition* 1	2	3	4	5	6	7
Peptone	.5	.1	.5	.5	.5		.5
Gelatin						.3	
Dextrose		.1					
Glycerol					1.3		
Yeast extract			.01				
Beef extract				.3			
Ferric ammonium citrate				.01	.01	.01	
Na_2EDTA							.015
K_2HPO_4		.005		.01	.01	.01	.01
$FePO_4$	.01		.01				
$FeCl_3$							.002
Agar	1.5	1.5	1.5	1.5	1.5	1.5	1.5
Plate count as % of medium 1 †	100	6.2	123.5	36.6	47.4	20.5	38.9

* In 80% sea water. † Mean of 8 replicates.

to the laboratory for study. The water had a salinity of 2.9-3.0% and a reaction of pH 7.9-8.1. Dilution blanks were prepared with aged sea water, sterilized by autoclaving. Appropriate dilutions of the water samples were plated, using 8 replicate plates at each dilution. Counts were made after 7-10 days of incubation at 20°C.

Results. Table I shows the extent to which composition of the nutrient medium influenced the plate counts. The results are expressed as percentages of the average plate count obtained with medium 1. Medium 1 was described by ZoBell(2) and is used frequently in studies of the abundance and distribution of bacteria in marine environments. Medium 2 was recommended by Reuszer(3) for the cultivation of marine bacteria. The remaining five media were our own formulations. Medium 3 is medium 1 supplemented with yeast extract. Media 4-7 contained different organic substrates, but were similar in that phosphate was added as K_2HPO_4 and iron was present as the citrate or ethylenediaminetetraacetate chelate. All the media were approximately pH 7.5.

Maximum plate counts were obtained on medium 3. Medium 2 yielded the lowest counts. The increased count that resulted from the addition of yeast extract to ZoBell's medium confirms Jones' report(4) that yeast extract is "stimulatory" to marine bacteria.

Although medium 3 gave maximum plate counts, it contained a precipitate ($FePO_4$), the particles of which were easily confused with the smaller subsurface colonies. The enumeration of colonies on medium 3 was facilitated greatly when the liquified agar was left undisturbed until the precipitate settled, and only the clear supernatant was used for the preparation of plates. In media with chelating agents (media 4-7) difficulties due to precipitation were absent but plate counts were low. When medium 3 was prepared using distilled water, the plate counts were only 10% of those obtained when the medium was prepared with sea water.

The plating procedure also influenced the total number of colonies that developed from sea water on medium 3. Pour plates were prepared by adding agar to Petri dishes containing 1.0 ml of various dilutions of sea water. The medium and sea water were mixed thoroughly before the agar solidified. A second series of Petri dishes was prepared by adding approximately 15 ml of medium to each plate and allowing the agar to solidify. The surface of the agar was then inoculated with 0.1 ml of various dilutions of sea water, and the inoculum was distributed over the surface of the agar by rotating and tilting each plate. Petri dish tops made of metal and containing absorbent disks were used to desiccate the agar surface partially and thus to prevent spreading of colonies.

The number of bacterial colonies that developed on pour plates was consistently greater (30-40%) than the number that developed on surface-inoculated plates. This difference may have resulted from the growth of microaerophilic or anaerobic bacteria below the surface of the agar in pour plates. Colony counts were made more easily when a surface inoculum was employed; the colonies were larger, more uniform in size, and better distributed than those on pour plates.

Summary. The nutritional requirements of marine bacteria are so varied that no one nutrient medium can suffice for the growth of all. The influence of composition of the nutrient medium and plating procedure on the plate method for determining the number of bacteria in sea water was investigated. Seven media were tested. Maximum and reproducible plate counts were obtained with a medium containing 0.5% peptone, 0.01% yeast extract, 0.01% $FePO_4$, and 1.5% agar in 80% sea water. The number of colonies that developed on pour plates was 30-40% greater than the number that developed on surface inoculated plates.

The technical assistance of Miss E. Winkler is gratefully acknowledged.

1. ZoBell, C. E., *Marine Microbiology,* 1946, Chronica Botanica Co., Waltham, Mass.

2. ———, *J. Mar. Res.,* 1941, v4, 42.

3. Reuszer, H. W., *Biol. Bull.,* 1933, v65, 480.

4. Jones, G. E., *Bacteriol. Proc.,* 1957, Detroit, Mich., p16.

Received August 19, 1957. P.S.E.B.M., 1957, v96.

11

Reprinted from *Limnol. Oceanogr.*, **4**(2), 128–139 (1959)

Bacterial Populations in Sea Water as Determined by Different Methods of Enumeration[1]

Holger W. Jannasch[2] and Galen E. Jones

Scripps Institution of Oceanography, University of California, La Jolla, California

ABSTRACT

Five different cultural and two direct microscopic methods were employed for estimating the abundance of bacteria in samples of sea water collected from both oceanic and neritic areas. The cultural methods included macrocolony counts on nutrient agar, silica gel, and membrane filters, microcolony counts on membrane filters, and the extinction dilution method. Direct microscopic counts were made of microbes on membrane filters and of microbes transferred from membrane filters to glass slides. Direct counts showed the presence of from 13 to 9,700 times as many bacteria as cultural methods. The extinction dilution method and the microcolony membrane filter method gave counts 20 and 35 times higher, respectively, than did any of the macrocolony methods. Direct microscopic counts on membrane filters were approximately 150 times higher than plate counts, and the numbers of microbes transferred from membrane filters to glass slides were approximately 2,000 times higher than plate counts. In all of the cultural methods, a peptone-yeast extract medium was used.

Differences in the abundance of microorganisms obtained by the various methods are attributed to a variety of factors: the presence of bacteria in aggregates, selective effects of the media, and the presence of inactive cells. A marked decrease in bacterial numbers was observed just below the thermocline as reflected in the macrocolony methods but not by the direct microscopic methods. A considerable population of spirilli-like forms was noticed directly under the microscope but not after cultivation.

INTRODUCTION

Studies of microorganisms in natural waters involve enumeration as a general index of activity and as a measurement of biomass. Little has been done in evaluating the intrinsic significance of the results of counting methods. Although cultural methods of enumeration are widely used, they yield only a small percentage of the microorganisms actually present (ZoBell 1946). The application of direct microscopic methods in water bacteriology led to counts 200 to 5,000 times higher than plate counts (Butkevich 1932, 1938). Radsimovsky (1930) explained this difference in bacterial numbers by the presence of autotrophic and organophobic organisms. After comparing the bacterial numbers and the oxygen demand of corresponding water samples, Butkevich and Butkevich (1936) concluded that a considerable portion of the bacteria must be present in the resting stage. However, Alfimov (1954) doubted the occurrence of many "dead" bacteria in sea water and sediments. Thus, there exists uncertainty concerning the state of bacteria in natural waters as well as the value of the various methods used for enumeration.

In the present investigation, the abundance of microorganisms in oceanic and neritic areas of the Pacific Ocean was investigated by seven methods of enumeration.

We wish to thank Dr. Claude E. ZoBell, Scripps Institution of Oceanography, University of California, La Jolla, California, for his critical review of the manuscript. This project was supported in part by a grant, E-1768, from the National Institute of Allergy and Infectious Diseases of the National Institutes of Health, Bethesda 14, Maryland.

METHODS

Sampling locations

Stations where sea water samples were collected:

1) On October 31, 1957, sea water was ob-

[1] Contribution from the Scripps Institution of Oceanography.

[2] Present address: Department of Bacteriology, University of Wisconsin, Madison, Wisconsin.

tained from an area approximately equidistant between Baja California and Guadalupe Island (29° 18′ N latitude; 116° 56′ W longitude) at depths of surface, 25, 50, 75, 100, and 200 meters.

2) On December 20, 1957, sea water was obtained from an area 16 miles off the coast of San Diego, California (32° 38′ N latitude; 117° 32′ W longitude), at depths of surface, 25, 50, 75, 100, and 200 meters.

3) On the following dates, surface sea water samples were collected off the Scripps Institution of Oceanography pier: January 3, 1958; January 17, 1958; and February 6, 1958 (a local tide pool was also sampled on this date).

Types of water samplers used:

1) Water samples obtained at sea for cultural methods of enumeration were collected in sterile J-Z bacteriological water samplers using collapsible rubber bottles (ZoBell 1946).

2) Water samples collected for direct methods were obtained in Van Dorn (1956) samplers.

3) Surface water samples obtained near shore were collected in sterile one-liter glass-stoppered bottles.

Media

Basal medium, 2216E (Oppenheimer and ZoBell 1952)—Bacto-peptone, 5 g; Bacto-yeast extract, 1 g; $FePO_4$, 0.01 g; Bacto-Agar used for pour plate method, 15 g; aged sea water (salinity adjusted to 28 g/L), 1,000 ml; pH adjusted to 7.6 to 7.8. Medium 2216, not containing yeast extract, was found superior to other media for development of maximum numbers of aerobic heterotrophic marine bacteria (ZoBell 1941). This medium was improved by the addition of yeast extract as confirmed by Carlucci and Pramer (1957). Two other media were also used in one experiment.

Succinate medium—Succinic acid, 2.0 g; NH_4NO_3, 1.0 g; K_2HPO_4, 0.1 g; aged sea water (salinity adjusted to 28 g/L), 1000.0 ml; pH adjusted with NaOH to 7.6 to 7.8.

Casein hydrolysate medium—Enzymatic casein hydrolysate (NBC), 1.0 g; NH_4NO_3, 1.0 g; K_2HPO_4, 0.1 g; aged sea water (salinity adjusted to 28 g/L), 1000.0 ml; pH adjusted to 7.6 to 7.8.

Methods of enumeration

Agar pour plate method—Samples of sea water estimated to produce between 20 and 500 colonies per plate were transferred aseptically to sterile (90-mm) plastic petri plates. In certain oceanic areas samples as large as 5 to 10 ml may be required. When working at sea the plates were supported by a weighted swinging table.

Sterile nutrient agar was transported in screw-cap prescription bottles and remelted at sea. Plates were poured with nutrient agar cooled to 40–45° C and mixed thoroughly with the sample. After solidifying on the swinging table, the plates were incubated for at least five days before counting colonies.

Extinction dilution method—Medium 2216E broth was prepared and sterilized in 9 ml amounts in test tubes. One ml portions of the water samples were inoculated into five replicates of the broth. Each of the replicates was diluted by 10-fold increments from 10^0 to 10^{-6} ml. Growth was determined by turbidity in the broth after incubation and the most probable number (MPN) of microorganisms estimated from the distribution of positive and negative tubes from the table of Hoskins (1934).

Silica gel method—The advantages of silica gel as a solidifying agent for microbiological media and some practical improvements in its preparation have been presented by Pramer (1957). In this investigation the method was modified for use with sea water.

The resin column and silica sol were prepared as described by Pramer (1957). The silicate source was $Na_2SiO_3 \cdot 9H_2O$ (3% SiO_2, w/v). The silica sol was sterilized by autoclaving at 10 lbs pressure for five minutes.

Three parts of the sterile silica sol were mixed with one part of sterile double-strength 2216E broth. Full strength aged sea water was employed as the liquid phase of the nutrient solution. This gave a final concentration of sea water of 25% (salinity of 8.5 g/L). Gelling time with this mixture was approximately six minutes. However, when triple strength artificial sea water (Lyman and Fleming 1940) was substituted for full strength aged sea water, the gel formed almost instantly. Careful adjustment of the pH to 7.5 to 7.8 with NaOH resulted in a firm silica gel.

Macrocolony membrane filter method—From 2 to 50 ml of sea water sample were passed through a sterile 47-mm type PH Millipore filter to concentrate the microorganisms. The filters were sterilized at 112°C for a few minutes and stored in sterile distilled water until used. The inoculated filters were placed on adsorbent cellulose pads soaked with 2216E nutrient solution in 60-mm petri plates. Three such plates were incubated in a 150-mm petri plate which served as a moist chamber. The time and temperature of incubation for each water sample are presented in tables of data. Macrocolonies are defined as those which are large enough to be visible to the naked eye. The filters were stained with Löffler's methylene blue in order to make very small colonies visible.

Microcolony membrane filter method—Employing methods described by Jannasch (1958), colonies developing on nutrient-impregnated membrane filters were counted under a microscope at a magnification of 430 to 970×. In the microcolony method a 25-mm Millipore filter was used to concentrate microbes in water samples, and only 1 per cent as much nutrient was added as in the macrocolony method. Decreased incubation times and low nutrient concentration avoid the possible toxicity of the latter and regulate overgrowth of rapidly growing cells.

Direct microscopic method on membrane filters—Water samples were treated with formaldehyde, and several dilutions of each sample were filtered to obtain optimal distribution of microorganisms on the filter surface (Beling 1950, Alfimov 1954, and Jannasch 1953, 1958). Sterile distilled water was added during the last stages of filtration to remove salts (Jerusalimsky 1932). Staining with methylene blue yielded better results in counting microcolonies, but a 1 per cent solution of erythrosine in 5 per cent phenol often proved superior for differentiating individual cells. Even with this staining technique, the low optical contrast of the microscopic fields occasionally made it difficult or impossible to differentiate small microbial forms from inorganic particles on the filter surface. Indistinct forms were not counted.

Mechanical cleaning of surfaces which come in contact with the water sample is required in direct methods rather than sterilization. Consequently, carefully cleaned and dried Van Dorn (1956) samplers were used for collecting large volumes of sea water for this direct method. The number of bacterial cells remaining on the inner surface of the samplers was estimated by examining the water used to rinse the samplers by the Cholodny method. The contamination was less than 0.1 per cent of the total cell count.

Cholodny method—The Cholodny (1928) method involves microscopic enumeration of microorganisms on glass slides after concentration by filtration. The Cholodny method was modified somewhat in the following experiments. Depending on the expected bacterial density, 100 to 2,000 ml of the sea water sample obtained with the Van Dorn (1956) sampler were fixed immediately with 1 per cent formaldehyde. The sample was passed through a membrane filter using the apparatus illustrated in Figure 1. By turning the two-way stopcock (E), the vacuum filtration was stopped when two to three ml were left above the filter. Salts were removed by washing several times with sterile distilled water. Care was taken to leave a few ml of solution above the filter. Two precautions in the design of the apparatus prevented deposits from sticking to the filter surface which would be difficult to remove quantitatively: (1) an ultrafine sinter-glass plate (B) with a lower filtration speed than the membrane filter and (2) a collodium loop (C) 2 mm wide, which was rotated mechanically on the filter surface during filtration (Bachman 1926). The numbers of microorganisms remaining on the membrane filters were determined by microscopic examination and served as a control. The bacterial cells remaining amounted to 1.5 per cent on the average and did not exceed 8.5 per cent of the total count. The concentrated sample was transferred quantita-

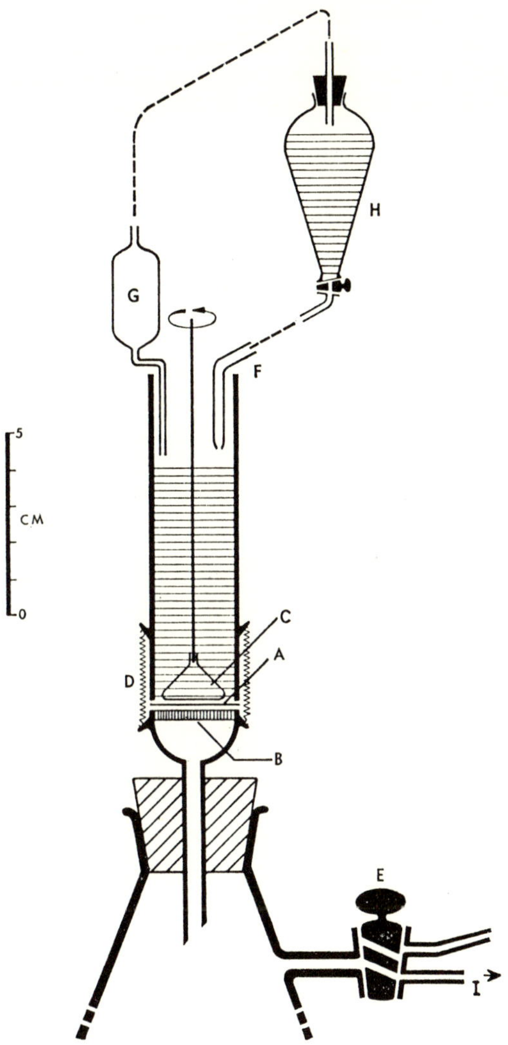

FIG. 1. Filtration apparatus for the Cholodny method. (A) membrane filter, (B) ultra-fine sinter glass plate, (C) stirring loop of collodium, (D) spring, (E) two-way stopcock, (F) sample entry, (G) level control, (H) separatory funnel (reduced in scale) containing sample, and (I) outlet connected to vacuum.

tively into small test tubes and adjusted to a certain volume (3 to 5 ml) with 5 per cent formaldehyde. Formaldehyde also was used to rinse the filter surface. With a calibrated capillary tube or micropipette, 0.01 ml of the concentrated sample was transferred to a slide, dried, and stained with a 2 per cent solution of erythrosine in 5 per cent phenol for at least five hours. The size of the dried area was measured with an ocular grid at an enlargement of 35 times. Possible contamination during 15 to 29 minutes of filtration was checked by examining filtered water. Bacterial cells introduced from the air did not influence the cell count. The per cent error of the counts obtained by this method is based on examination of 25 microscopic fields for each sample.

In the tables, per cent error is calculated by $S\bar{x}/\bar{x} \cdot 100$ where $\bar{x}$ is the mean and $S\bar{x}$ is the standard deviation of the mean.

RESULTS

The numbers of microorganisms detected by the various methods of enumeration from an oceanic area (Station 1) are presented in Table 1. The water temperatures at the various depths were recorded by a bathythermograph. The abundance of bacteria recovered from an area approaching the neritic zone (Station 2) are plotted in Figure 2. The number of bacteria recorded from the neritic waters off the Scripps Institution of Oceanography pier (Station 3) are presented in Table 2.

Evaluation of methods

As observed in Tables 1 and 2 as well as in Figure 2, the agreement among bacterial counts obtained by the agar pour plate method, silica gel method, and the macrocolonies on membrane filters was quite close. Submerged growth in the agar pour plate method which might tend to favor the development of anaerobic and microaerophilic microorganisms was no higher than that obtained by macrocolonies on membrane filters. In addition, no significant deviation was observed in samples from different depths of the sea. These results are not in agreement with those of Carlucci and Pramer (1957) who found as many as 30 to 40% more colonies arising after pouring plates than on surface-inoculated plates. These investigators have suggested the existence of a large percentage of microaerophilic and anaerobic bacteria in Atlantic ocean water.

ZoBell and Conn (1940) have shown that some marine bacteria are thermosensitive and are killed at the congealing temperature of agar. However, there is no direct evidence that the heat of the agar adversely affects the microorganisms developing on

TABLE 1. *Comparative numbers of microorganisms per ml indicated by different methods at Station 1*
Per cent error in parentheses. In all cultural methods the incubation temperature was 18 ± 1°C.

Depth m	Water temp. °C.	Plate[1] method		Serial dilution method (MPN)		Macrocolonies[1] on MF 5 Days	Microcolonies on MF 3 Days	Direct counts on MF
		5 Days	21 Days	5 Days	21 Days			
Surface	20.1	6 (10%)	11 (10%)	7	7	8 (9%)	68 (3.1%)	244 (1.8%)
25	20.1	14 (12%)	14 (12%)	5	5	14 (7%)	31 (5.4%)	262 (2.1%)
50	19.0	9 (19%)	10 (6%)	7	7	10 (12%)	30 (8.7%)	166 (6.4%)
75	15.0	1	2	0	0	6 (14%)	24 (11.2%)	147 (3.1%)
100	13.0	4 (14%)	6 (29%)	2	2	1 (15%)	29 (6.6%)	82 (11.3%)
200	10.0	3 (39%)	4 (58%)	2	2	5 (12%)	46 (4.0%)	179 (7.2%)

[1] An average of three replicates.

TABLE 2. *Comparative numbers of microorganisms per ml indicated by different methods of enumeration at Station 3*
Per cent error in parentheses. In all cultural methods the incubation temperature was 18 ± 1°C.

Area sampled	Water temp. °C.	Plate method		Silica gel method		Serial dilution method (MPN)		Macroc. on MF[4] 5 days	Microc. on MF 3 days	Direct count on MF	Cholodny method
		5 days	20 days	5 days	20 days	5 days	20 days				
Off pier	16.2	218[1] (8%)	239 (8%)	191[1] (7%)	218 (17%)	3500	3500	191 (11.2%)	1344 (6.2%)	1619 (7.4%)	25,800 (0.7%)
Off pier	16.0	37[2] (10%)	56 (11%)	17[1] (32%)	43 (14%)	220	350	61 (15%)	822 (9%)	1810 (5.3%)	41,100 (1.2%)
Off pier	15.9	25[3] (12%)	71 (5%)	—	—	64	1800	19 (13%)	108 (5.8%)	2290 (4.4%)	55,400 (3.2%)
Tide pool	16.2	75[3] (10%)	107 (9%)	—	—	700	9500	68 (17%)	136 (4.8%)	1780 (6.2%)	46,100 (2.2%)

[1] An average of 10 replicates.
[2] An average of 6 replicates.
[3] An average of 4 replicates.
[4] An average of 5 replicates.

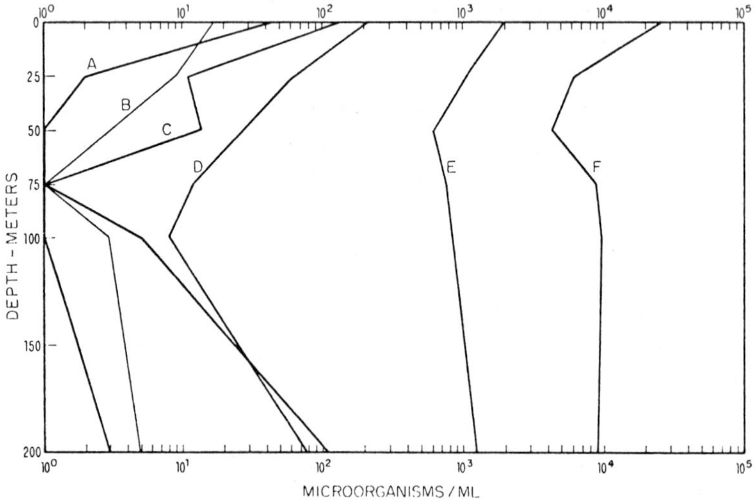

FIG. 2. Bacterial populations at depths obtained by different methods from Station 2. (A) agar pour plate method, (B) macrocolonies on membrane filters, (C) extinction dilution method, (D) microcolonies on membrane filters, (E) direct count on membrane filters, and (F) Cholodny method.

TABLE 3. *Number of microcolonies per ml as influenced by nutrients, period of incubation, and concentration of the water samples from Station 1*

Medium	Peptone-yeast extract (diluted 1:100)									Succinate (diluted 1:100)									Casein hydrolysate (diluted 1:100)								
Amount of water filtered...	2 ml			4 ml			8 ml			2 ml			4 ml			8 ml			2 ml			4 ml			8 ml		
Period of incubation (18 ± 1° C) hr	24	48	72	24	48	72	24	48	72	24	48	72	.24	48	72	24	48	72	24	48	72	24	48	72	24	48	72
Depth, meters																											
Surface.........	22	59	**68**	29	52	—	34	—	—	11	14	17	9	17	**31**	16	29	22*	22	25	34	19	41	—	16	**53**	—
25.............	20	23	31	28	**55**	—	48	54	—	12	14	20	16	17	**26**	19	21	24	21	24	36	16	28	31	24	**39**	—
50.............	14	25	30	16	36	**52**	21	44	—	0	4	7	4	13	14	7	10	**18**	0	11	29	8	19	28	6	23	**44**
75.............	6	23	24	8	37	**46**	10	36	—	0	3	9	4	5	11	6	**19**	10*	0	0	14	1	6	9	6	8	**15**
100.............	0	13	29	6	21	30	6	27	**36**	0	8	14	0	12	18	13	17	**22**	0	4	13	0	16	21	5	**26**	20*
200.............	15	27	**46**	22	39	—	23	38	—	8	14	22	2	10	**26**	4	16	19	8	12	31	6	18	26	13	**42**	—

— = overgrowth of membrane filters
* = decrease in the number of colonies due to overgrowth
Bold face type = maximum number of microcolonies per depth and medium used

TABLE 4. *Ratio of microbial counts, range of values, and per cent error of five different methods compared to plate counts computed from all data*

	Plate method	Macro-colonies on MF	Serial dilution method	Micro-colonies on MF	Direct counts on MF	Cholodny method
Mean	1	1.16	21.8	32.1	147	2100
Range	—	0.2–4.0	0.3–35	1.6–34	13–840	108–9700
Per cent error	—	17.1	19.2	12.7	8.5	20.7

agar plates as compared to those on membrane filters or silica gel.

Bacterial counts from agar pour plates were higher than those from silica gel plates (Table 2). This was true at sea as well as in near-shore samples. This effect may be attributed to specific salt requirements of marine bacteria (ZoBell 1941). In addition, the composition of unwashed agar is variable (Yaphe 1957), and micronutrients in the agar might be expected to stimulate marine bacteria (MacLeod *et al.* 1954, Jones 1957). ZoBell (1941) found about three to four per cent of the bacteria in the sea digest agar.

The most probable number (MPN) of microorganisms determined by the extinction dilution method was considerably higher than the macroscopic colony count, although the nutrient solution used in this method was identical with that used in the other cultural methods (Fig. 2 and Tables 2 and 4). This condition was not evident in oceanic water samples (Table 1). The increased counts in the majority of samples due to the liquid medium are attributed to the dispersion of bacterial aggregates and clumps on detritus as regulated by the surface tension depression of the medium. La Riviere (1955) has demonstrated that 1% peptone drops the surface tension of water by 18 dynes/cm, and 1% yeast extract by 25 dynes/cm. Surface tension depressions approaching this order of magnitude would be sufficient to disrupt bacterial aggregates causing the observed increase in the final count (Jones and Jannasch 1959). These results agree with those of Butkevich (1932) who obtained from ten to a hundred times as many bacteria by extinction dilution as by plate counts, using water samples from the Barents Sea. Butterfield (1933) reported good agreement between plate counts and the MPN of extinction dilution employing 50 replicates for the plates and 50 of each of three dilutions using *Escherichia coli.* He found 37% higher values by extinction dilution using the same experimental design when *Aerobacter aerogenes* was used as the test organism. The difference between the bacterial numbers obtained using the two test bacteria was ascribed to the greater tendency for the *Aerobacter aerogenes* cells to clump due to the presence of mucoid substances.

In the microcolony technique on membrane filters three small petri dishes were

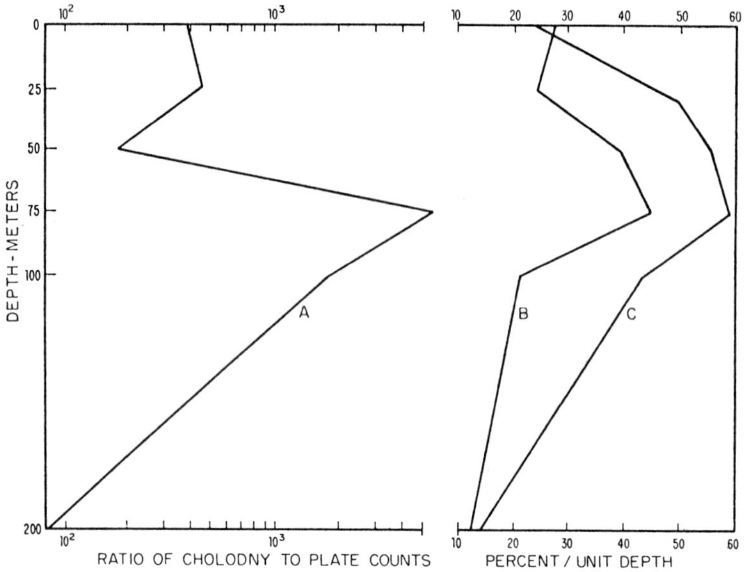

FIG. 3. Variations with the vertical distribution of microbial populations at Station 2. (A) ratio of microorganisms/ml obtained by the Cholodny method to microorganisms/ml recorded by agar pour plate method, (B) per cent of spirilli-like organisms per unit depth as determinated from the Cholodny count, and (C) per cent of aggregated microorganisms per unit depth as determined by Cholodny method.

placed in a larger one which served as a moisture chamber during incubation. Each of the small petri dishes contained three membrane filters which were divided, again, in three parts before incubation. In this way, three combinations of three different factors (nutrients, time of incubation, and concentration of sample) were compared, and 27 different values were obtained for each of the samples tested (Table 3).

The bacterial numbers obtained by the microcolony method demonstrate dependence on the incubation time. After 24 hours, groups of 2 to 20 cells developed on the membrane filter surface (Figs. 5 and 6). Single cells, presumably inactive, were not counted. Often colonies appear to have spread (Fig. 5). After approximately 48 hours in these experiments, colonies were composed of several hundred cells (Figs. 7 and 8). In many cases, colonies had begun to merge in 72 hours. Therefore, the colony counts may decrease before the membrane filter has been overgrown completely (Table 3). Sometimes dense colony centers indicate the existence of several original loci (Fig. 9).

Deviations from this general pattern of development are caused by density of bacteria on the filter surface (Table 3). In general, maximum counts appeared earlier when larger amounts of sample were used. Due to the density of bacteria in the water sample the maximum count occurred at a characteristic place in the series of nine experiments. Only the counts in the same columns are comparable. Of the three media tested, the peptone-yeast extract medium proved superior as indicated by the development of more colonies in a short period of time with early overgrowth of the filters. Similar growth ratios for these nutrients were obtained by the agar pour plate method using the same water samples. Generally, higher counts obtained by this method may result from the low nutrient level. In only a few cases were macrocolonies apparent on these filters after several weeks of incubation. Comparing the photomicrographs of Frost (1921) taken from microcolonies on the surface of agar plates with the type of growth on membrane filters, no considerable difference was noticed (Fig. 10).

Direct microscopic examination of membrane filters yielded lower numbers than

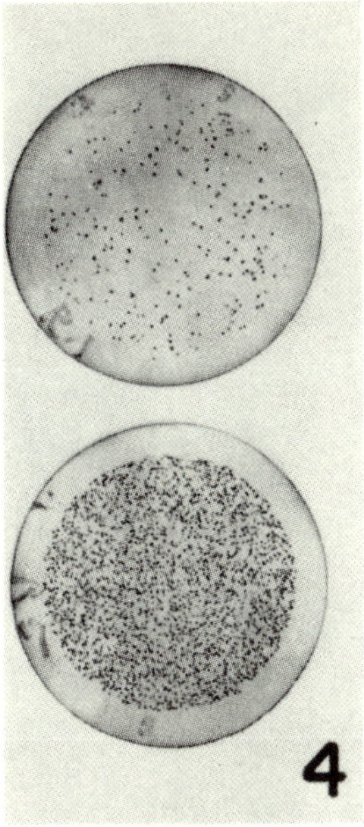

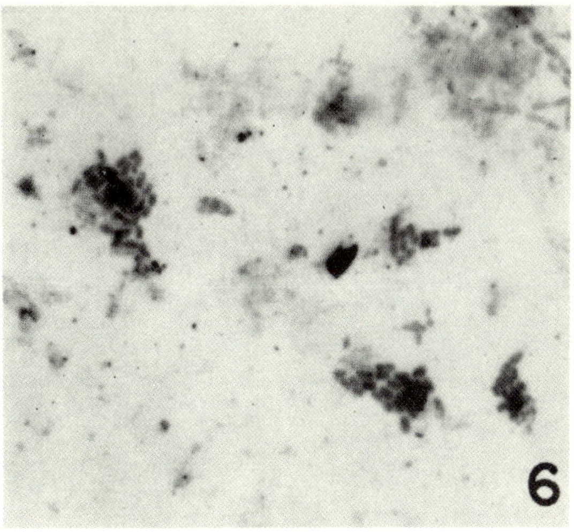

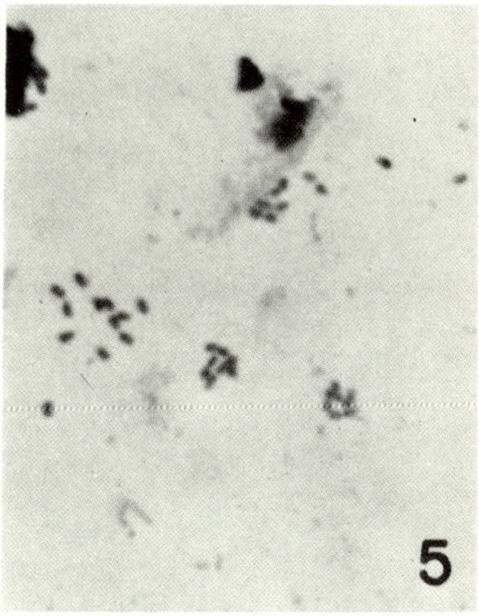

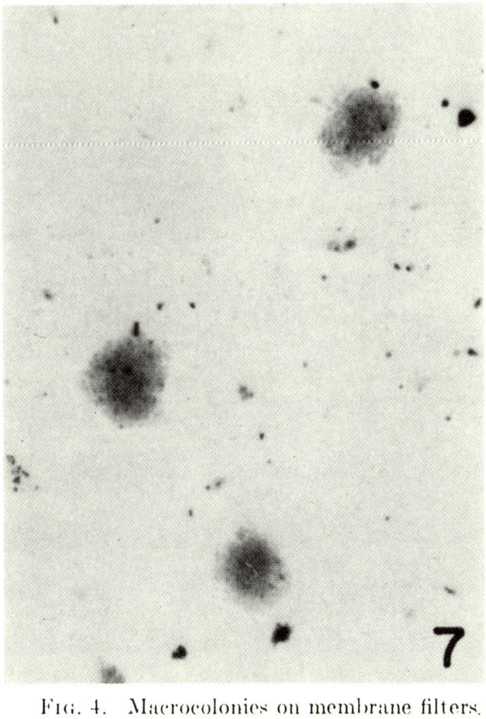

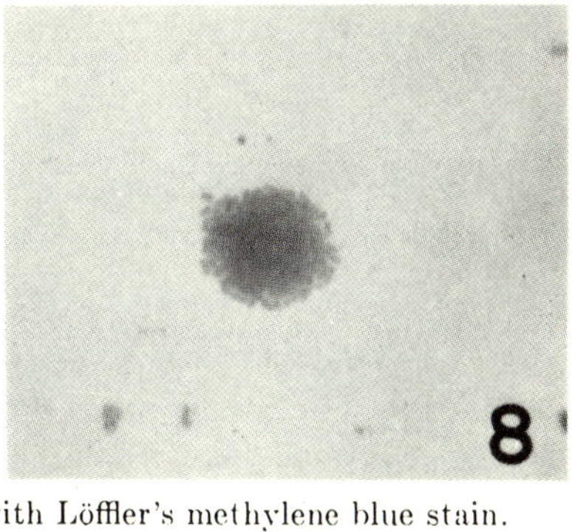

FIG. 4. Macrocolonies on membrane filters stained with Löffler's methylene blue stain.

FIGS. 5 and 6. Microcolonies on 2216E medium (diluted 1:100) after 24 hours incubation at 18 ± 1°C on membrane filters. Magnification 970×.

FIGS. 7 and 8. Colonies on 2216E medium (diluted 1:100) after 48 hours incubation at 18 ± 1°C on membrane filters. Magnification 500× and 700×, resp.

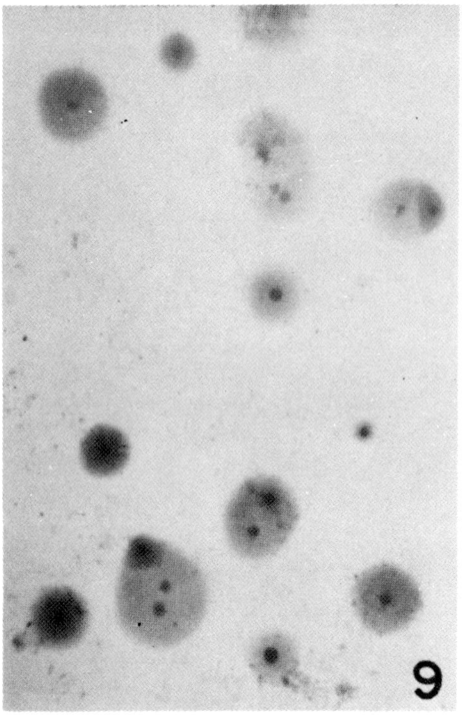

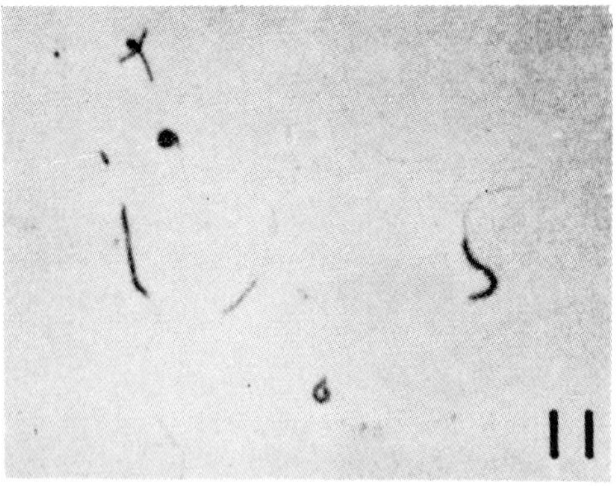

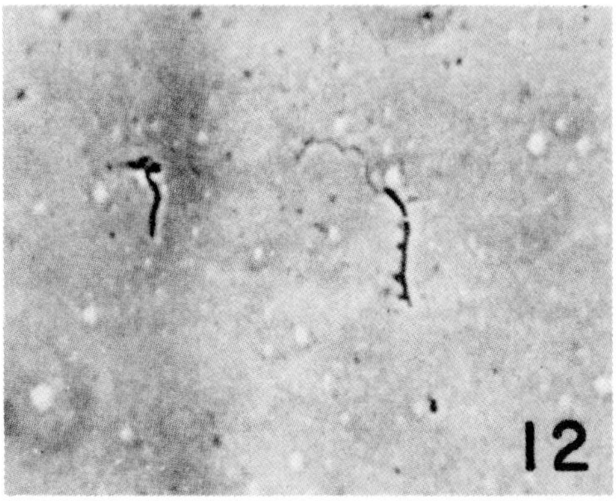

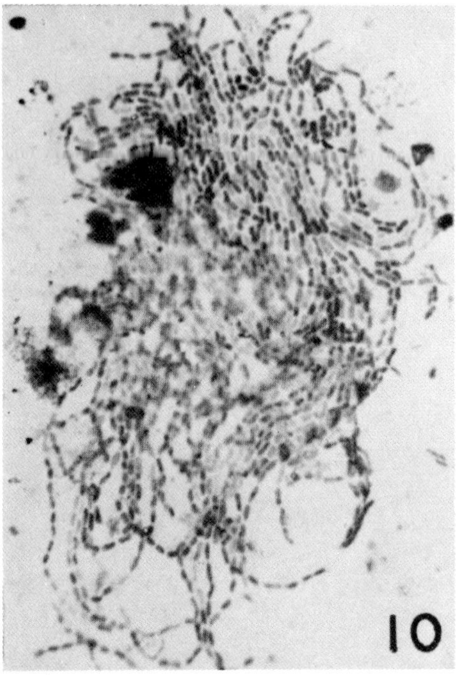

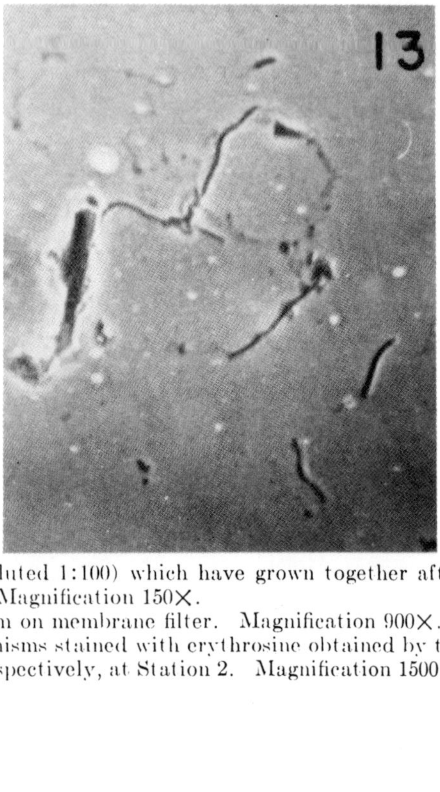

FIG. 9. Colonies on casein hydrolysate medium (diluted 1:100) which have grown together after 72-hours incubation at 18 ± 1°C on membrane filters. Magnification 150×.

FIG. 10. Typical colony of a chain-forming organism on membrane filter. Magnification 900×.

FIGS. 11–13. Spirilli-like and filamentous microorganisms stained with erythrosine obtained by the Cholodny method from samples of 75 and 200 meters, respectively, at Station 2. Magnification 1500×. Figures 12 and 13 reproduced from phase contrast.

those obtained by the Cholodny method (Table 4). Small and weakly stained cells may escape observation on the membrane filters due to the lower optical contrast of the preparation. The Cholodny method has been improved by the use of membrane filters of smaller porosity than those of Cholodny (1928, 1929) and Novobrantzev (1932), 0.3 μ as compared with 2.0 μ.

Evaluation of the vertical distribution of marine bacteria

As indicated in Table 1 and Figure 2, the results obtained by the various methods at depths down to 200 meters indicated a similar distribution of microorganisms in both vertical casts. The largest counts of bacteria occurred in surface waters, decreasing down to 75 meters, with an increase in abundance at 200 meters. Similar vertical distributions of microorganisms in the sea have been observed by Kriss and Rukina (1952), who demonstrated minimum bacterial counts at depths of 25 to 100 meters. This minimum corresponds roughly to the location of the thermocline.

A marked deviation of the ratio between direct microscopic counts and cultural method counts was observed at the various depths. The ratio of microorganisms obtained by the Cholodny method compared to the plate method at Station 2 are plotted at the various depths in Figure 3. These results indicate very few colonies developing on agar plates as opposed to relatively large numbers of cells observed directly, especially at 75 and 100 meters.

With both of the direct methods large spirilli-like organisms (4–15 μ long, 1–3 μ wide), as shown in Figures 11–13, were observed throughout the vertical casts. The long flagella (5–20 μ) were stained by erythrosine. Some of these organisms were motile in wet mounts prepared from water samples concentrated by filtration. The vertical distribution of these microorganisms in per cent of the Cholodny count is shown in Figure 3. None of these forms were observed in microscopic preparations made from colonies or dilution tubes, developing from the same water samples.

The percentage of microbes at various depths attached to detritus or observed in clumps as determined by the Cholodny method are plotted in Figure 3. The curve indicates less clumping where the plate count figures were highest (surface and 200 meters). In addition, the per cent errors of the direct counting methods (Table 1) show the lowest values in the same samples. The highest errors appear in samples of 75 and 100 meters.

DISCUSSION

Agar still appears to be the best solidifying agent for enumeration of bacteria at sea. However, silica gel has certain advantages for studying the nutritional requirements of bacteria, because it is a chemically defined substance and biologically inert. In addition, silica gel solidifies at any desired temperature within the biological range. Both of these solidifying agents are superior to gelatin, because many marine bacteria liquefy gelatin, which results in the merging of colonies before slow-growing bacteria have had time to develop visible colonies (ZoBell 1941).

Direct microscopic methods possess the advantages of revealing a more exact enumeration of the microorganisms in a sample than cultural methods regardless of their growth requirements or their physiological condition, and these results can be obtained in a very short period of time. However, there is no way to determine whether the bacteria observed are living or dead, and they cannot be used for cultural studies. Direct counts may be increased by the presence of non-proliferating, inactive, or dead cells. Karsinkin and Kusnetsov (1931) and Alfimov (1954), using erythrosine stain to differentiate living from dead protoplasm, found low percentages of dead bacteria in lake and sea water. Kusnetsov (1958) reported that dead bacteria constitute about 10 per cent of the total number of bacteria in lake water as determined by Peshkov's staining method. Strugger (1949), using acridine orange, obtained similar results in soil.

Direct microscopic methods are complicated by particulate matter in the sample simulating the appearance of microorganisms. Clumps are difficult to count. Therefore, no organism was counted unless it could be distinguished clearly from detritus. Cocciform bacteria were counted only when clearly differentiated. Due to these precautions, the actual bacterial counts are probably higher than those reported for the direct microscopic methods. The staining reaction of erythrosine was not used to distinguish between organic and inorganic material or living and dead protoplasm as suggested by Karsinkin and Kusnetsov (1931).

Ratios comparing the results obtained by cultural and direct microscopic methods for all of our experiments are presented in Table 4. The range of comparable ratios in lake water determined by Kusnetsov and Karsinkin (1931) using an evaporation technique for concentrating the sample for microscopic examination was 1:2,000–4,000. Unfortunately, this method is not applicable for sea water samples. Collins and Kipling (1957) obtained from 6 to 11,000 times as many bacteria by their direct method as by plate counts from Lake Windemere North Basin water. Salimovskaya-Rodina (1938) found up to 5,000 times higher counts in lake water by direct microscopic methods than by plate counts. According to Butkevich (1938), these ratios increase with the decrease of organic matter in the water.

Bacterial populations were reported to be directly proportional to organic matter by Novobrantzev (1932) and Chartulari and Kusnetsov (1937). According to Kriss (1953), at depths descending from the thermocline the bacterial numbers are regulated by the concentration of organic matter. As suggested by Butkevich (1938), the the converse of curve A in Figure 3 may present a rough index of the relative distribution of organic matter with depth. In this respect, it is noteworthy that our highest numbers of microorganisms were at the surface in Pacific Ocean water as were those of Lloyd (1930) in the Clyde Sea. Another possible explanation for the deviation of the ratio between direct microscopic and cultural counts at various depths is the different nutritional requirements of the populations at these depths. The spirilli-like organism is an example of a microorganism which escapes detection by cultural techniques. It is probable that other marine bacteria in these samples also did not develop into colonies with the nutrients used.

Another factor which decreases the plate counts is aggregation of bacteria as indicated by the direct methods. According to Jennison (1937) and Ziegler and Halverson (1935), the occurrence of clumps of bacteria is the main reason for the differences between direct microscopic and cultural counts of cells in cultures. A detailed study of the existence of microbial aggregates in the sea will appear shortly (Jones and Jannasch 1959).

REFERENCES

Alfimov, N. M. 1954. Comparative evaluation of methods for the determination of bacterial counts in sea water. Microbiology (Moscow), **23:** 693. (Russian).

Bachmann, H. 1926. Der Mikrofiltrierapparat von Gimesi. Z. Hydrol., **11:** 271–276.

Beling, A. 1950. Bakteriologische Untersuchungen während der Fulda-Expedition 1948. Ber. Limnol. Flussstat. Freudenthal, **2:** 4–10.

Butkevich, V. S. 1932. Zur Methodik der bakteriologischen Meeresuntersuchungen und einige Angaben über die Verteilung der Bakterien im Wasser und in den Böden des Barents Meeres. Trans. Oceanogr. Inst. Moscow, **2:** 5–39. (Russian, German summary).

———. 1938. On the bacterial populations of the Caspian and Azov Seas. Microbiology (Moscow), **7:** 1005–1021. (Russian, English summary).

Butkevich, N. V., and V. S. Butkevich. 1936. Multiplication of sea bacteria depending on the composition of the medium and on temperature. Microbiology (Moscow), **5:** 322–343. (Russian, English summary).

Butterfield, C. T. 1933. Comparison of the enumeration of bacteria by means of solid and liquid media. U. S. Public Health Repts., **48** (42): 1292–1297.

Carlucci, A. F., and D. Pramer. 1957. Factors influencing the plate method for determining abundance of bacteria in sea water. Proc. Soc. Expt. Biol. Med., **96:** 392–394.

Chartulari, E. M., and S. J. Kusnetsov. 1937. Die Ergebnisse der Gesamtzählung der Bakterien im Wasser einer Reihe von Seen des Wyschne-Volotzkij Rayons. Arb. Limnol. Sta. Kossino, **21:** 117–124. (Russian, German summary).

Cholodny, N. 1928. Contributions to the quantitative analysis of bacterial plankton. Trav. Sta. Biol. Dniepre, **3:** 157–171. (Russian, English summary).

——— 1929. Zur Methode der quantitativen Erforschung des bakteriellen Planktons. Zbl. Bakt., Abt. 2, **77:** 179-193.

COLLINS, V. G., AND C. KIPLING. 1957. The enumeration of waterborne bacteria by a new direct count method. J. Appl. Bacteriol., **20:** 257-264.

FROST, W. D. 1921. Improved technique for the micro- or little-plate method of counting bacteria in milk. J. Infectious Diseases, **28:** 176-184.

HOSKINS, J. K. 1934. Most probable numbers for evaluation of Coli-aerogenes tests by fermentation tube method. U.S. Public Health Repts., **49:** 393-405.

JANNASCH, H. W. 1953. Zur Methode der quantitativen Untersuchung von Bakterienkulturen in flüssigen Medien. Arch. Mikrobiol., **18:** 425-430.

———. 1958. Studies on planktonic bacteria by means of a direct membrane filter method. J. Gen. Microbiol., **18:** 609-620.

JENNISON, M. W. 1937. Relations between plate count and direct microscopic count of *E. coli* during the logarithmic growth period. J. Bacteriol., **33:** 461-477.

JERUSALIMSKY, N. D. 1932. Ein Versuch die Bakterienpopulation des Moskauflusses und seiner Zuflüsse nach der direkten Methode der Bakterioskopie zu untersuchen. Microbiology (Moscow), **1:** 147-175. (Russian, German summary).

JONES, G. E. 1957. The effects of organic metabolites on the development of marine bacteria. Bacteriol. Proc., pp. 16-17.

JONES, G. E., AND H. W. JANNASCH. 1959. Aggregates of bacteria in sea water as determined by treatment with surface active agents. Limnol. Oceanogr. (In press).

KARSINKIN, G. S., AND S. J. KUSNETSOV. 1931. Neue Methoden in der Limnologie. Arb. Limnol. Sta. Kossino, **13:** 47-68. (Russian, German summary).

KRISS, A. E., AND E. A. RUKINA. 1952. Biomass of microorganisms and their rates of reproduction in oceanic depths. Zhur. Obsc. Biol., **13:** 346-362. (Russian).

———, 1953. Microorganisms and biological productivity of natural waters. Perioda (Leningrad), **5:** 49-59. (Russian).

KUSNETSOV, S. J. 1958. A study of the size of bacterial populations and of organic matter formation due to photo- and chemosynthesis in water bodies of different types. Verh. Internat. Ver. Limnol., **13:** 156-169.

KUSNETSOV, S. J. AND G. S. KARSINKIN. 1931. Direct method for the quantitative study of bacteria in water and some considerations on the causes which produce a zone of oxygen-minimum. Zbl. Bakt., Abt. 2, **83:** 169-179.

LA RIVIÈRE, J. W. M. 1955. The production of surface active compounds by microorganisms and its possible significance in oil recovery. I. Some general observations on the change of surface tension in microbial cultures. Antonie Van Leeuwenhoek, **21:** 1-8.

LLOYD, B. 1930. Bacteria of the Clyde Sea area: a quantitative investigation. J. Mar. Biol. Assoc., U. K., **16:** 879-908.

LYMAN, J., AND R. H. FLEMING. 1940. Composition of sea water. J. Mar. Res., **3:** 134-146.

MACLEOD, R. A., E. ONOFREY, AND M. E. NORRIS. 1954. Nutrition and metabolism of marine bacteria. I. Survey of nutritional requirements. J. Bacteriol., **68:** 680-686.

NOVOBRANTZEV, P. V. 1932. The development of bacteria in lakes depending on the presence of easily assimilable organic matter. Microbiology (Moscow), **6:** 28-36. (Russian, English summary).

OPPENHEIMER, C. H., AND C. E. ZOBELL. 1952. The growth and viability of sixty-three species of marine bacteria as influenced by hydrostatic pressure. J. Mar. Res., **11:** 10-18.

PRAMER, D. 1957. The influence of physical and chemical factors on the preparation of silica gel media. Appl. Microbiol., **5:** 392-395.

RADSIMOVSKY, R. 1930. Vorläufige Angaben über die Dichtigkeit der bakteriellen Besiedlung einiger Gewässer. Trav. Sta. Biol. Dniepre, **5:** 385-402. (Russian, German summary).

SALIMOVSKAJA-RODINA, A. G. 1938. Concerning the vertical distribution of bacteria in the waters of lakes. Microbiology (Moscow), **7:** 789-803. (Russian, English summary).

STRUGGER, S. 1949. Fluoreszensmikroskopie und Mikrobiologie. Hannover, M. V. H. Schaper. pp. 151-173.

VAN DORN, W. G. 1956. Large volume water samplers. Trans. Am. Geophys. Union, **37:** 682-684.

YAPHE, W. 1957. The use of agarase from *Pseudomonas atlantica* in the identification of agar on marine algae (Rhodophyceae). Can. J. Microbiol., **3:** 987-994.

ZIEGLER, N. R., AND H. O. HALVORSON. 1935. Application of statistics to problems in bacteriology. IV. Experimental comparison of the dilution method, the plate count and the direct count for the determination of bacterial populations. J. Bacteriol., **29:** 609-634.

ZOBELL, C. E. 1941. Studies on marine bacteria. I. The cultural requirements of heterotrophic aerobes. J. Mar. Res., **4:** 42-75.

———. 1946. Marine microbiology: a monograph on hydrobacteriology. Chronica Botanica Co., Waltham, Mass. pp. 41-58.

ZOBELL, C. E., AND J. E. CONN. 1940. Studies on the thermal sensitivity of marine bacteria. J. Bacteriol., **40:** 223-238.

12

Reprinted from *Limnol. Oceanogr.*, **5**(1), 78–80 (1960)

THE SPREAD PLATE AS A METHOD FOR THE ENUMERATION OF MARINE BACTERIA[1, 2]

John D. Buck and Robert C. Cleverdon

Department of Bacteriology, University of Connecticut, Storrs, Connecticut

ABSTRACT

A comparison was made of agar plates spread with glass rods and poured agar plates for the enumeration of bacteria in the waters of Fisher's Island Sound, salinity 30‰. Spread plates were shown to be markedly superior. Highest counts were obtained by spreading, using rods treated with Desicote (a silicone solution) and incubating plates at 25°C, rather than at 16°C, and in air rather than in air with CO_2 content increased.

INTRODUCTION

For the enumeration of bacteria in marine waters, extensive use is made of the poured agar plate. The possibility that some organisms indigenous to waters at less than 20°C were killed by even short exposure to 45°C and might grow more promptly on the surface of an agar plate suggested a challenge of the pour plate enumeration against a spread plate technique. The variations in conditions commonly employed suggested also the comparison of the effect of incubation at different temperatures and in an atmosphere of increased CO_2.

METHODS AND MATERIALS

Water was obtained at various tidal conditions during the summer and fall of 1958, from one location at Latimer Reef, which is about 6 miles off the coast of Noank, Conneticut, and east of East Point, Fisher's Island, New York. Top samples were collected in sterile 500 ml wide-mouth reagent bottles. Samples were iced until examination at the laboratory, within 2 hours.

Preliminary studies with samples from Fisher's Island Sound showed that counts could be obtained using decimal dilutions of 10^{-1} to 10^{-4}. Dilution blanks (9 and 99 ml of sea water) were sterilized in the autoclave. Pipettes were 1.1 ml (milk) pipettes, sterilized in the oven, and stored in cans.

[1] Supported in part by Grand #E-706, National Institutes of Health.

[2] Contribution #1 from the Marine Research Laboratory, University of Connecticut, Noank, Connecticut.

The plating medium was that used at the Woods Hole Oceanographic Institute reported by ZoBell (1946) and had the following composition: sea water from the tap at the Noank Marine Research Laboratory (salinity about 30‰); peptone (Gelysate, Baltimore Biological Laboratory), 0.1%; glucose, 0.1%; K_2HPO_4, 0.005%; Bacto-agar (Digestive Ferments Company), 1.5%; final pH 7.6. It was necessary to filter through cheesecloth.

All "platings" were made in duplicate. For the pour plates, the melted agar was tempered at 45°C, then mixed with the sample or dilution. For the spread technique, about 15 ml was poured into plates prior to use. Dilutions were made so that deposition of 0.1 ml on the surface resulted in the desired dilution. Using a separate rod for each, the inoculum was spread evenly over the entire surface of the agar by a rotary twirling motion of the plate under the rod. The rods were made of 5 mm Pyrex, formed into a spreading portion with handle as follows: a 90° bend was made near the center of a 25 cm length, about 2 cm from which another 90° bend was made, thus producing a crank with parallel but opposite legs; a third "upward" bend gave a spreading portion about 8.0 cm in length. The rods were oven-sterilized and stored in cans.

Pour plates were incubated at 16°±1°C; the spread plates at 16°±1°C and 25°±1°C. The lower temperature was chosen to approximate that of the waters sampled, while the 25°C is the upper limit suggested by ZoBell (1954). For incubation in an atmosphere of increased CO_2, the plates were in-

TABLE 1. *Comparison of colony counts (× 10^2) obtained by spread and pour plates at 16°C incubated in air and in air with CO_2*

Sample No.	Date and Tide	Air		Ratios S/P	Air with CO_2		Ratio S/P	Ratios of counts	
		Spread (S)	Pour (P)		Spread	Pour		S air/S CO_2	P air/P CO_2
1	7/17 Flood	26	14	1.8	23	20	1.2	1.1	0.71
2	7/25 Ebb	53	12	4.4	41	23	1.8	1.3	0.52
3	7/31 Ebb	11	2	5.5	19	2	9.5	0.58	1.0
4	8/7 Low	140	77	1.8	86	51	1.7	1.6	1.5
5	8/14 High	11	3	3.7	7	6	1.2	1.6	0.5
6	8/21 Flood	150	64	2.4	130	99	1.3	1.2	0.65
7	8/28 High	8	2	4.0	11	3	3.7	0.73	0.67
8	9/4 Flood	35	12	2.9	41	31	1.3	0.86	0.39
9	9/10 High	8	5	1.6	5	2	2.5	1.6	2.5
10	9/19 Ebb	190	120	1.6	110	100	1.1	1.7	1.2
		Average ratios		2.9			2.5	1.2	0.96

Average ratio of spread:pour in both air and CO_2: 2.74
Average ratio of air:CO_2 with spread and pour: 1.1

cubated in a wide mouth jar sealed after lighting a candle.

After 7 days of incubation, plates were counted with the methods of the American Public Health Association (1955); counts were computed from no fewer than 4 plates (2 dilutions) and in most cases 8 plates (4 dilutions). Prolonged incubation resulted in larger colonies, not appreciably higher counts.

When spread plates appeared superior, a brief study was made to estimate the number of organisms recoverable from the rods following spreading. Several used rods were washed in 10 ml of sterile sea water, and the washings were plated by the spread method. Each rod showed a count of about 300. An appraisal was made of the practical value of Desicote, a silicone solution marketed by the Beckman Instrument Company, in order to obviate adherence of the water film entrapping bacterial cells. Treatment of the spreading rods consisted of dipping them in Desicote and shaking vigorously to remove excess. The treated rods were then sterilized as usual and used for spreading additional samples.

RESULTS AND DISCUSSION

In contrast to the findings of Carlucci and Pramer (1957), the spread plates always revealed higher counts, although the two sets of data are not strictly comparable. Table 1 shows that at 16°C, in air and in air with increased CO_2, the spread plates in all cases resulted in higher counts; the average of the ratios of counts, spread:pour, was 2.5 (range 1.1 to 9.5). Incubation of spread plates in air was in general superior to

TABLE 2. *Comparison of colony counts (× 10^2) of plates spread with Desicoted and plain rods incubated in air and in air with CO_2*

Sample No.	Date and Tide	16°C				Ratio of counts 16°C				25°C				Ratio of counts 25°C				Ratio of counts 25°C/16°C			
		Air		CO_2						Air		CO_2									
		D*	P*	D	P	Air D/P	CO_2 D/P	D Air/CO_2	P Air/CO_2	D	P	D	P	Air D/P	CO_2 D/P	D Air/CO_2	P Air/CO_2	D Air	D CO_2	P Air	P CO_2
14	10/25 Flood	13	8	7	6	1.6	1.2	1.9	1.3	20	12	12	10	1.7	1.2	1.7	1.2	1.5	1.7	1.5	1.7
15	11/1 High	12	9	8	6	1.3	1.3	1.5	1.5	18	18	7	15	1.0	0.47	2.6	1.2	1.5	0.88	2.0	2.5
16	11/8 Low	190	240	300	240	0.79	1.3	0.63	1.0	470	700	260	240	0.67	1.1	1.8	2.9	2.5	0.87	2.9	1.0
17	11/15 Ebb	46	29	9	16	1.6	0.56	5.1	1.8	48	35	14	14	1.4	1.0	3.4	2.5	1.0	1.6	1.2	0.88
18	12/6 Low	1300	1130	390	290	1.2	1.3	3.3	3.9	730	860	120	190	0.85	0.63	6.1	4.5	0.56	0.38	0.76	0.66
						1.3	1.1	2.5	1.9					1.1	0.88	3.1	2.5	1.4	1.1	1.7	1.3

Average ratio of D/P at all conditions: 1.1
Average ratio of air/CO_2 at all conditions: 2.5
Average ratio of 25°C/16°C at all conditions: 1.4
* Desicoted rod: D
* Plain rod: P

incubation in added CO_2 as shown by counts and by ratios (average ratio 1.2). With pour plates, incubation in CO_2 was superior, although with smaller differences in plate counts (average ratio 0.96).

Table 2 shows plate counts and ratios obtained with Desicoted and plain rods at two incubation temperatures both in air and in an atmosphere of increased CO_2. With cultures in air, it was observed that at both 16°C and 25°C, slightly higher counts were generally obtained by spreading with Desicoted rods (average ratios, 1.3 and 1.1); with plates incubated in CO_2 at both temperatures, the counts were generally not affected by the use of treated rods. The average ratio of all counts observed at all conditions, Desicoted:plain rods, was 1.1. Incubation in air is seen to be superior at both temperatures, whether the inoculum was spread with Desicoted or plain rods (average of all ratios 2.5). Incubation of plates at 25°C was superior to that at 16°C whether spread with treated or plain rods, in air or in air with increased CO_2 (average ratio 1.4). For other cooler or warmer waters, alternative incubation temperatures might be superior.

In comparing approximately 100 plates each of spread:pour, it was found that the reproducibility of counts obtained by the spread method was equal to that of the pour plate. There were apparently no unusual cultures obtained by this spread method as indicated by incomplete studies of several hundred strains. While spreading requires opening the plate, no excessive aerial contaminants were encountered even in our crowded laboratory.

REFERENCES

APHA, AWWA AND FSIWA. 1955. Standard methods for the examination of water, sewage, and industrial wastes. Tenth ed. APHA, Inc. New York. 522 pp.

CARLUCCI, A. F., AND PRAMER, D. 1957. Factors influencing the plate method for determining abundance of bacteria in sea water. Proc. Soc. Exptl. Biol. Med., **96:** 392–394.

ZOBELL, C. E. 1946. Marine microbiology. Chronican Botanica Co., Waltham Mass. 240 pp.

ZOBELL, C. E. 1954. Bacteriology of the sea. pp. 503–516. *In* Salle, A. J., Fundamental principles of bacteriology. McGraw-Hill Book Co., Inc. New York. 782 pp.

13

Reprinted from *Limnol. Oceanogr.*, **12**(2), 319–324 (1967)

ADENOSINE TRIPHOSPHATE CONTENT OF MARINE BACTERIA[1]

Robert D. Hamilton and Osmund Holm-Hansen
Institute of Marine Resources, University of California, La Jolla 92037

ABSTRACT

The adenosine triphosphate (ATP) content of seven selected cultures of marine bacteria, representing five genera, has been determined during growth in chemostats and during the various growth phases in batch cultures. The range of ATP content in chemostat-grown cells was 0.5–6.5 $\times 10^{-9}$ μg ATP/cell, or 0.3 to 1.1% of the cell carbon. The ATP content of senescent cells in batch cultures and in starved cells was generally about one-fifth the concentration found in logarithmically growing cells. The average content of ATP in all the bacteria examined was 0.4% of cell carbon. It is concluded that ATP data can provide a useful estimate of heterotrophic biomass in the oceans.

INTRODUCTION

Dissolved organic compounds and particulate, nonliving, organic matter account for over 99% of the total organic carbon in the oceans. Little is known about the role of this vast reservoir of material in marine community metabolism. One reason is the lack of methods for assaying the standing stock of heterotrophic microorganisms as well as their *in situ* activities.

Isolation plating techniques, extinction dilution, and direct microscopic counting are commonly used to enumerate marine bacteria. It is generally conceded that these methods produce erroneous estimates of the numbers of living bacteria in a water sample. The plating and extinction dilution techniques are selective because of the chemical composition of the media and because of inherent physical parameters such as temperature and pressure. In addition, extinction dilution may be biologically selective in that one form may easily overgrow another. Therefore, these two approaches are believed to produce low estimates of standing bacterial stocks. Direct counting, on the other hand, is believed to yield high estimates because of the difficulty in distinguishing bacteria from bacteria-sized, inert particles. Moreover, this method gives no estimate of viable cells and is therefore inapplicable to any problem involving *in situ* activities.

Recently a method was developed for the quantitative assay of adenosine triphosphate (ATP) in seawater samples (Holm-Hansen and Booth 1966). This method is based on the firefly luciferin-luciferase reaction as described by Strehler and McElroy (1957). The ATP content of a water sample may be related to its heterotrophic population if one assumes that no ATP is associated with nonliving material and if one takes samples at a depth which precludes the occurrence of autotrophic plants. Macroscopic forms may be removed from samples by filtration through a net of 35-μ mesh size.

To relate ATP concentrations in the ocean to microbial biomass, it is necessary to know the approximate ATP content of the microbial species present. As it is impossible to make a taxonomic analysis of the microbiota in each water sample, the accuracy of such biomass estimations would depend upon the inter-generic constancy of ATP concentration. In addition, such estimates would depend upon the range of ATP concentrations to be found in cells of the same species in different physiological conditions. In this paper, we report the results of an investigation into these variations in the ATP content of marine bacteria selected as representative of the genera that most commonly occur in seawater samples.

[1] This research was supported in part by U.S. Atomic Energy Commission Contract AT(11-1)-34, Project 108 and, in part, by U.S. Public Health Service Grant No. 1-RO1 ES-00035-03. We are indebted to Miss Hester Kobayashi and Miss Kay Austin for their technical assistance during the course of this investigation. We also thank Dr. J. D. H. Strickland for his comments and suggestions and Dr. R. R. Colwell for providing certain of the cultures.

MATERIALS AND METHODS

The bacteria used in these experiments were all isolated from the marine environment and required media with the ionic composition of seawater for growth.

Chromobacterium marinum is a pelagic isolate obtained by Dr. C. E. ZoBell and identified in our laboratory (Hamilton and Austin 1967). The *Pseudomonas* spp. were isolated in our laboratory, using a chemostat enrichment technique, from coastal seawater (Hamilton, Morgan, and Strickland 1966), while the *Serratia* sp. was isolated from the same environment by conventional techniques. The latter isolate was identified by Dr. R. R. Colwell, who also supplied us with the *Vibrio* sp. and an estuarine isolate, the *Micrococcus* sp. All cultures were maintained in chemostats during the experimental period. These cultures provided physiologically reproducible cells for routine ATP assay as well as inocula for batch culture experiments. Dilution rates varied slightly between isolates but were all close to 0.2/hr.

All media were prepared with artificial seawater (Lyman and Fleming 1940). The medium for the *Pseudomonas* spp. contained 1.0 mg glucose, 1.0 g $(NH_4)_2SO_4$, and 0.01 g KH_2PO_4/liter. The glucose content of batch cultures was occasionally raised to 100 mg glucose/liter when large cell yields were required. The *Vibrio* sp., *Serratia* sp., and *Micrococcus* sp. were grown on 1.0% Bacto-peptone; *C. marinum* was grown in a medium consisting of 0.5% Bacto-peptone, 0.1% Bacto-yeast extract, 0.1% Bacto-casitone, and 0.01% KNO_3.

Batch cultures were grown in approximately 1 liter of medium in Fernbach flasks on a reciprocating shaker at 20C. Samples were withdrawn aseptically, and the cell number and ATP content were determined as outlined below.

The cells of a logarithmic growth phase culture of *Pseudomonas* sp. (GL-7) were collected by centrifugation at 5C at 27,000 × g for 30 min. The cells were washed once in artificial seawater and centrifuged. The washed cells were added to 1 liter of artificial seawater containing the usual amounts of nitrogen and phosphate but with no added carbon source. They were added in sufficient numbers to approximate their concentration in the original batch culture. These starved cells were incubated in the same manner as the usual batch culture and their ATP content followed with time.

Viable cells in chemostat or batch cultures were determined by a drop-plate technique. However, in the case of the *Micrococcus* sp. direct counting procedures had to be used in addition to drop plates because the cells grew in clumps that could not be dispersed.

Cell carbon was analyzed after the method of Menzel and Vaccaro (1964) as modified by Holm-Hansen et al. (1966). Dry weight determinations were also made. Late log-phase cultures were used for both determinations. The cells were centrifuged from their culture media at 5C. They were washed once with artificial seawater, centrifuged, and resuspended in a small volume of distilled water. Aliquots were taken for the carbon analyses while other aliquots were dried (16 hr at 70C) on tared planchets and weighed on a Cahn electro-balance.

For determination of ATP, samples (0.1–1.0 ml) of the bacterial cultures were filtered through 25-mm HA Millipore® filters (pore size, 0.45 μ) as soon as possible after removal from the culture vessels. The filter was quickly placed in a small beaker and the cells killed by the addition of 3–4 ml of boiling Tris buffer (0.02 M, pH 7.75). After heating at 100C for 5 min, to allow for complete extraction of ATP, the supernatant liquid was transferred to a test tube, and the filter again extracted with 1 ml of Tris buffer. The extracts were combined, diluted to 5.0 ml with Tris buffer, and frozen at -20C until the time of analysis. In some cases the filtration was omitted, and 0.1-ml samples were pipetted directly into 5.0 ml of boiling Tris buffer. As the concentration of salts in the nutrient medium interfered with the luciferase reaction, the final volume of extract had to be at least 25 times that of the sample containing the bacterial cells. Comparable re-

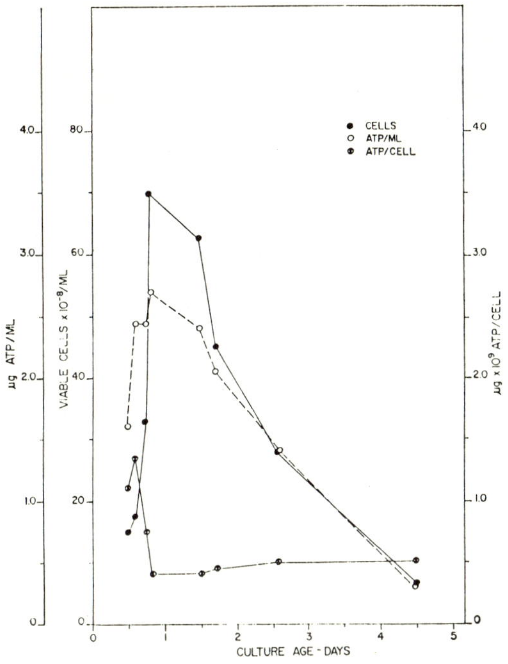

FIG. 1. Variations in ATP concentrations and viable cell numbers observed during the batch culture of *Serratia* sp.

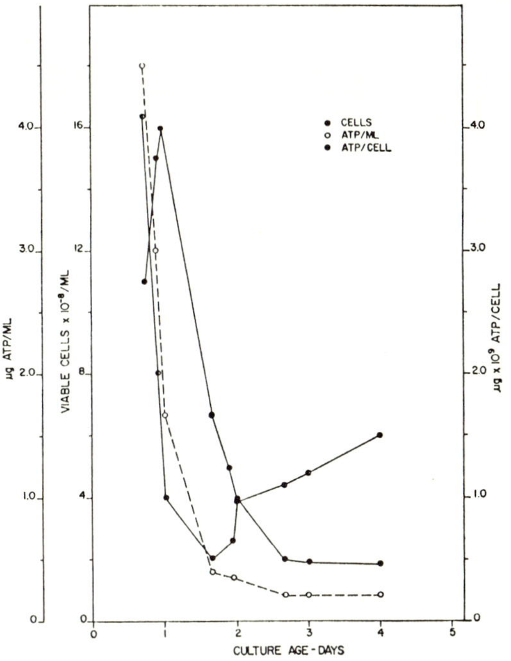

FIG. 2. Variations in ATP concentrations and viable cell numbers observed during the batch culture of *Vibrio* sp.

sults were obtained with these two methods. However, the filtration procedure was generally used to obtain larger cell samples.

The ATP in the extracts was quantitatively determined by measuring the amount of light emitted when the sample was mixed with a luciferin-luciferase preparation obtained from firefly tails. Complete details of this procedure have been described by Holm-Hansen and Booth (1966).

RESULTS

Data showing the concentration of ATP in the seven different strains of marine bacteria, when grown in a chemostat, are presented in Table 1.

Figs. 1–4 show the changes with time in the viable cell count, the concentration of ATP, and the amount of ATP per viable cell for the *Serratia* sp., *Vibrio* sp., *Micrococcus* sp., and the *Pseudomonas* sp. GL-7 when these bacteria were grown in batch culture. In all four cases, the amount of ATP per viable cell was high ($1.3–4.0 \times 10^{-9}$ μg/cell) during the period of logarithmic growth and began to decline rapidly during the late logarithmic phase or early stationary phase of growth.

It was difficult to determine the amount of ATP in each cell of the *Micrococcus* sp. because this organism formed clumps which could not be dispersed without damaging the cells. Thus, colony counts of this organism can only be used as an indication of the over-all pattern of population growth and the use of such data produced artificially high concentrations of ATP/cell. Fig. 3 shows the same ATP data coupled with cell numbers obtained by means of direct counting. Direct count data were presumed to be reliable until the stationary phase of growth was reached. At such time there would be a significant number of dead cells present that would not be recognized as dead.

After a short stationary phase almost all cultures showed a marked phase during which the viable cell count decreased drastically and the concentration of ATP in the cells decreased to a plateau (0.2–

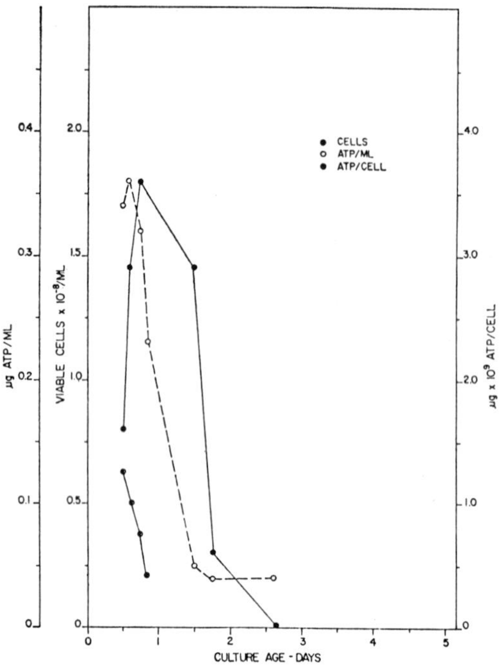

FIG. 3. Variations in ATP concentrations and viable cell numbers observed during the batch culture of *Micrococcus* sp.

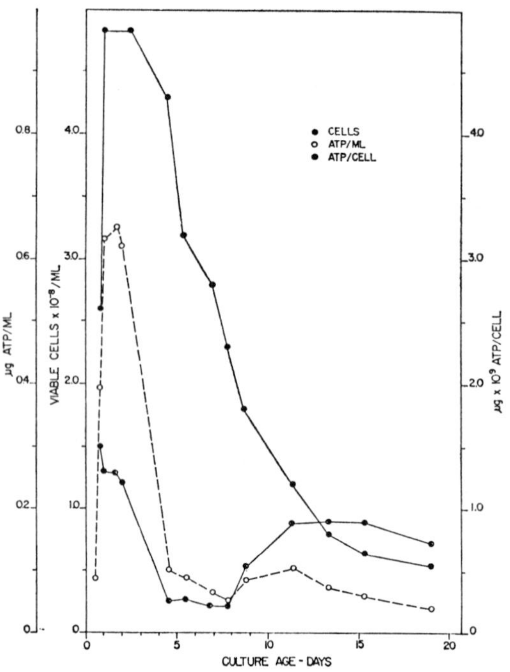

FIG. 4. Variations in ATP concentrations and viable cell numbers observed during the batch culture of *Pseudomonas* sp. (GL-7).

0.5×10^{-9} μg/cell). There was usually a slight increase in the amount of ATP per viable cell after this plateau had been reached, but this increment was most marked in *Pseudomonas* sp. GL-7.

In the case of *Pseudomonas* sp. GL-7, there was a prolonged phase during which the viable cell count fell slowly, and there was a slight increase in ATP after the eighth day when results were expressed as μg ATP/ml of suspension. The reason why *Pseudomonas* sp. behaves in this fashion is not known. As we wanted to follow the ATP content of cells starved of respiratory substrates, and not of cells which were perhaps inhibited by toxic material in senescent cultures or stimulated by metabolites produced by lysing cells, a culture of *Pseudomonas* sp. (GL-7) was removed during logarithmic growth, washed once, and then resuspended in nutrient medium devoid of any added organic substrates. The

TABLE 1. *The concentration of ATP in seven different strains of marine bacteria*

Bacterium	No. samples analyzed	μg ATP per cell (×10⁹) Range	Avg	μg carbon ×10⁷ per cell	ATP as % carbon*
Chromobacterium marinum	20	5.5–7.8	6.5	8.3	0.8
Vibrio sp.	11	3.0–4.1	3.6	6.1	0.6
Pseudomonas sp. (GL-7)	7	1.8–2.2	2.0	1.7†	1.1
Pseudomonas sp. (C-6)	8	0.7–1.3	1.0	1.7†	0.6
Pseudomonas sp. (GU-1)	1	–	1.3	1.7†	0.7
Serratia sp.	8	1.2–1.7	1.4	3.1	0.5
Micrococcus sp.	9	0.2–0.8	0.5	1.7	0.3

* The average for the seven strains was 0.7%.
† From dry weight data.

viable cell count and ATP concentration were followed with time as before (Fig. 5). Although the decline in viable cells followed the same pattern as before (Fig. 4), the ATP concentration in the suspension showed a continual decrease that paralleled the decline in number of viable cells per ml from the fifth day to the end of the experiment. The content of ATP/cell during this period was thus constant at about 0.2×10^{-9} μg/cell.

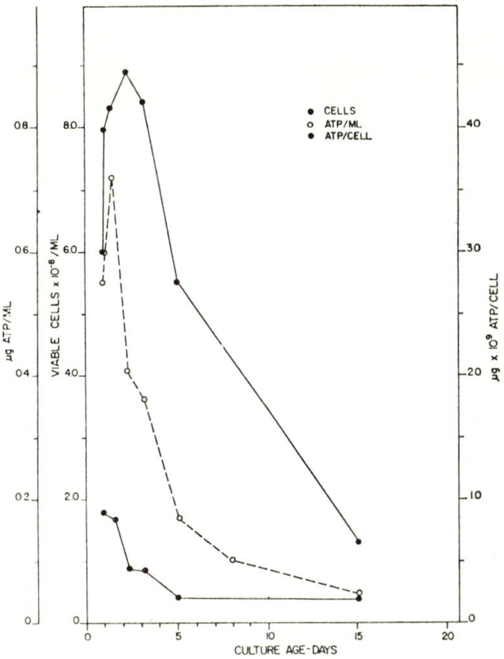

Fig. 5. Variations in ATP concentrations and viable cell numbers observed during the starvation of a washed cell suspension of *Pseudomonas* sp. (GL-7).

DISCUSSION

Previous studies (Holm-Hansen and Booth 1966) have demonstrated the applicability of the luciferin-luciferase method to the determination of ATP at various depths in the ocean. On the basis of our studies during the past two years, there are apparently no serious problems or errors inherent in the quantitative determination of extracted ATP. The extracted ATP samples may be stored in the frozen state for months without any loss in activity, and the light-measuring apparatus is extremely sensitive and reliable.

There are, however, many factors that might affect the conversion of ATP data into reliable estimates of heterotrophic biomass. The reliability and usefulness of this method for the estimation of biomass depend on the following assumptions: 1) ATP is a constituent of all living cells. There is no evidence in the literature contrary to this assumption. 2) During the filtration and killing procedures there is no significant alteration of the cellular ATP content, and the extraction procedures dissolve all ATP without significant breakdown of this compound. Laboratory studies have indicated that these conditions are fulfilled if the recommended killing and extraction procedures are rigorously followed. 3) There are no significant changes in cellular ATP from the time the water sampler closes until it reaches the deck. There is no feasible way to check the validity of this assumption. 4) ATP is not found in dead cells nor associated with nonliving, detrital material. Laboratory studies have shown the lack of any detectable ATP in cells killed by heat or various chemical methods (Holm-Hansen and Booth 1966). A batch culture of *C. marinum* contained 14.1 μg ATP/ml at 0.8 days and only 0.014 μg ATP/ml after 4.7 days. If ATP, liberated from dying cells, did adsorb onto dead cells or detrital material, such low ATP concentrations would not be found in batch cultures. It seems safe to conclude that the ATP measured in natural samples of deep ocean water is indicative of living cells and cannot be ascribed to ATP associated with detrital material. 5) Microbial cells of different species contain a relatively constant amount of ATP when expressed as a fraction of the cellular constituents. On the basis of our results, covering what we believe to be a representative selection of types of marine bacteria, the maximum variation encountered was 0.3–1.1% of the cell-carbon. 6) The ATP content of cells of any one species does not fluctuate greatly with age or with exposure to changing nutrient conditions. The greatest variation encountered was 0.5–4.0 $\times 10^{-9}$ μg ATP/cell in the *Vibrio* sp.

The chemostat-grown cultures and the senescent batch cultures probably represent the extremes of cellular condition one might expect to find in the sea. We have, therefore, averaged a representative number of observations from both types of cells and obtained a figure for the average ATP content of these bacteria. This figure is 1.5×10^{-9} μg ATP/cell. Taking an average of the cell carbon data reported in Table 1 for the five genera and making the assumption that a cell would have the same carbon content whether exponentially growing or senescent, an average of the ATP/cell data may be expressed as percentage cell carbon. The figure so obtained is 0.4%.

While the ATP/cell value might be used to convert ATP yield data into the number of heterotrophic bacteria present, it would seem more reasonable to use the average ATP concentration of the cells expressed as percentage carbon. This avoids the implication that all the heterotrophs in the ocean are bacteria.

Some investigators (*see* Strickland 1965; Fournier 1966) have reported the occurrence of microscopic, nonbacterial heterotrophs below the euphotic zone. These forms may be a significant fraction of the total heterotrophic biomass of deep ocean waters, but such forms are poorly described in terms of their distribution and ecology. It would therefore be difficult, even if cultures were available, to establish an estimate of the ATP content of a representative selection of these forms. We can say, however, that most algae investigated to date (Coombs et al. 1967*a*, 1967*b*; Holm-Hansen and Booth 1966; Syrett 1958) have had an ATP content (on a dry weight or carbon basis) in about the same range as that reported here.

REFERENCES

COOMBS, J., P. J. HALICKI, O. HOLM-HANSEN, AND B. E. VOLCANI. 1967*a*. Studies on the biochemistry and fine structure of silica shell formation in diatoms. Changes in concentration of nucleoside triphosphate in silicon-starvation synchrony of *Navicula pelliculosa* (Bréb.) Hilse. Exptl. Cell Res., (in press).

———, ———, ———, AND ———. 1967*b*. Studies on the biochemistry and fine structure of silica shell formation in diatoms. Changes in concentration of nucleoside triphosphate during synchronized division of *Cylindrotheca fusiformis* Reimann and Lewin. Exptl. Cell Res., (in press).

FOURNIER, R. O. 1966. North Atlantic deep sea fertility. Science, **153**: 1250–1252.

HAMILTON, R. D., AND K. E. AUSTIN. 1967. Physiological and cultural characteristics of *Chromobacterium marinum* sp. n. Antonie van Leeuwenhoek, J. Microbiol. Serol., (in press).

———, K. M. MORGAN, AND J. D. H. STRICKLAND. 1966. The glucose uptake kinetics of some marine bacteria. Can. J. Microbiol., **12**: 995–1003.

HOLM-HANSEN, O., AND C. R. BOOTH. 1966. The measurement of adenosine triphosphate in the ocean and its ecological significance. Limnol. Oceanog., **11**: 510–519.

———, J. COOMBS, B. E. VOLCANI, AND P. M. WILLIAMS. 1966. Quantitative micro-determination of lipid carbon in microorganisms. Anal. Biochem., (in press).

LYMAN, J., AND R. H. FLEMING. 1940. Composition of sea water. J. Marine Res., **3**: 134–149.

MENZEL, D. W., AND R. F. VACCARO. 1964. The measurement of dissolved organic and particulate carbon in seawater. Limnol. Oceanog., **9**: 138–142.

STREHLER, B. L., AND W. D. MCELROY. 1957. Assay of adenosine triphosphate, p. 871–873. *In* S. P. Colowick and N. O. Kaplan [eds.], Methods in enzymology, v. 3. Academic, New York.

STRICKLAND, J. D. H. 1965. Phytoplankton and marine primary production. Ann. Rev. Microbiol., **19**: 127–162.

SYRETT, P. 1958. Respiration rate and internal adenosine triphosphate concentration in *Chlorella*. Arch. Biochem. Biophys., **75**: 117–124.

Editor's Comments on Papers 14 Through 18

THE PRESERVATION AND IDENTIFICATION OF MARINE BACTERIA

The taxonomy of isolates and the preservation of cultures are necessary partners, for to obtain a sufficient number of cultures for taxonomic analysis, culture collections must be available. For these to exist, efficient and successful means of preserving the isolates must be developed. Five papers that have contributed to the advancement of both of these arts for marine bacteria are reprinted in this section; the following is a discussion of their historical relationships.

Expanding activity in the early twentieth century in marine microbiology resulted in a tremendous increase in the numbers of isolated strains of marine bacteria. With these increased numbers of isolates, it quickly became apparent that the medical bacteriological approaches to classification were not totally applicable to marine bacteria because (1) they had a requirement for salt, (2) they frequently had different optimal growth temperatures, and (3) they were generally unreactive to sugars, but were very reactive toward complex substances such as chitin, protein, starch, and even alginic acid and agar. Thus, initially, everyone named his own isolates, and there was no central repository or list of identified species, or even a general identification scheme for these numerous strains. Three of the key papers that have helped to organize the classification and taxonomy of the marine bacteria are presented here.

The first major attempt at the classification and taxonomic listing of marine bacteria was produced by ZoBell and Upham in 1944 and is partially reproduced here (Paper 14). It is true that previous investigators had produced lists of bacterial isolates (4, 11, 13, 22), but none of these compilations had been so detailed in presenting the biochemical characteristics, the effects on growth of various temperatures and salinities, and the morphological and cultural details.

Advances in our knowledge of bacterial nutrition and physiology brought more biochemical tests into general use. These were quickly applied to marine microbiology, and it soon became apparent that the majority of the marine isolates were gram-negative, non-spore-forming rods, which did not belong to the Enterobacteriales as described in *Bergey's Manual* prior to the eighth edition (6). Instead, most of these isolates appeared to be members of the Pseudomonadales. Gradually, as the fisheries industry recognized the importance of microbial identification, especially in relation to fish spoilage, the need for a relatively rapid and simple diagnostic key for this group of microorganisms became apparent.

The need was satisfied by Shewan and his co-workers at the Torry Research Station. The outcome of their work was two papers on the bacteriology of fish, both fresh and spoiled. These papers have found widespread acceptance today for the initial identification of marine isolates. The first of these papers is presented here as Paper 15. Not only does it provide a workable scheme with a minimum number of media for the generic identification of these bacteria, but it also introduces the concepts of numerical taxonomy to marine microbiologists.

Numerical taxonomy, Adansonian taxonomy, was suggested as the logical way to handle bacterial taxonomy by Sneath (25). The usefulness of giving equal weight to all the biochemical and cultural characteristics of an isolate was recognized by many investigators struggling with classical identification schemes in which one or two properties made the difference between genera or families (1, 6). This was especially true for marine microbiologists, and Colwell and Liston demonstrate the usefulness of the numerical approach in Paper 16.

Since that time, numerous uses have been made of numerical taxonomy to study marine isolates at the generic level, including the flavobacteria (12), the flexibacteria (8), the vibrios (10, 15), and the coryneforms (26). Furthermore, Adansonian taxonomy has been found useful in classifying isolates from various ecological niches such as seawater (20), sediments (21), algal cultures (17), and fish (24). Meanwhile, Boffi and Boffi (5) have attempted to simplify the analysis of data obtained through numerical taxonomic procedures in their analysis of Tyrrhenian Sea heterotrophic bacteria. Their procedure is especially applicable for an investigator who does not have access to a computer.

As with most techniques, numerical taxonomy has its limitations: reference strains (which have generally been identified by classical methods) must be used, a great deal of time and effort is involved, and a great many isolates need to be studied. The selection of characteristics to be tested and the interpretation of the results are still subject to investigator bias, and the phenotypic expression of a biochemical trait still does not examine the basic genetic relatedness of the organisms. Supplemental to any identification scheme, then, is the determination of the guanine plus cytosine mole ratio (18), DNA homology studies (2, 7, 9, 16, 19), and serotyping. Ancillary procedures can include electrophoresis of enzyme patterns (21) and even N-terminal amino acid patterns (23). The grouping of all these approaches has been termed *polyphasic* taxonomy, which is the approach now being used to study marine microorganisms.

For example, the budding and prosthecate bacteria (19), the nonfermentative eubacteria (1, 3), and even single species of the marine bacterium *Leucothrix mucor* (16) have come under such intensive study. By combining all available tools for bacterial identification, the similarities and diversities of the microbial flora throughout the marine environment can be studied. Eventually, perhaps, distribution patterns of genetic or physiological types may emerge that will enable us to recognize certain geographical areas by their microbial flora. Papers 14 through 16 have helped to direct us to that ultimate goal.

In the introduction to his paper describing the 60 new species of marine bacteria, ZoBell states, "The numbers are not consecutive because many of the cultures originally isolated have been lost, some have proved to be duplicates..." (27). Anyone who has worked with fresh marine isolates knows the high mortality rate of the cultures until they have been "trained" to laboratory conditions. It is frequently assumed that freeze-drying of marine isolates will be harmful to the cultures because of salt crystal formation during the desiccation process. In the first detailed study of preservation techniques (Paper 17), Floodgate and Hayes compared freeze-drying, low-temperature storage at –29°C in a glycerol-nutrient medium, and low-temperature storage at –2°C under paraffin oil. Freeze-drying was the least deleterious method employed, except for vibrios and aeromonads. After 10 years the same cultures remained viable that were found viable by Floodgate and Hayes two years after freeze-drying. However, those stored in the glycerol-nutrient medium did not survive as well (14, 24).

One possible means of preserving these more delicate forms could be the technique proposed by Choate and Alexander in Paper 18. These authors successfully freeze-dried *Spirillum atlanticum* by using 24 percent sucrose in seawater. The critical rehydration step was accomplished in liquid nitrogen, and they obtained 7.6 percent survival of the organisms. To date, however, no extensive testing of this method with other marine isolates has been reported. It seems likely, though, that a combination of the freeze-drying and the liquid nitrogen rehydration method will provide a means for the successful preservation of the frequently lost marine vibrio, aeromonads, and spirillum isolates.

REFERENCES

1. Allen, R. D., and Baumann, P. 1971. Structure and arrangement of flagella in species of the genus *Beneckea* and *Photobacterium fischeri*. *J. Bacteriol.* *107*:295–302.
2. Anderson, R. S., and Ordal, E. J. 1972. Deoxyribonucleic acid relationships among marine Vibrios. *J. Bacteriol.* *109*:696–706.
3. Baumann, L., Baumann, P., Mandel, M., and Allen, R. D. 1972. Taxonomy of aerobic marine eubacteria. *J. Bacteriol.* *110*:402–429.
4. Benecke, W. 1912. *Bau und Leben der Bakterien*. B. G. Teubner, Berlin, 650 pp.
5. Boffi, V., and Boffi, A. M. 1970. Tyrrhenian Sea heterotrophic marine bacteria: a coding scheme for classification and identification of various strains. *G. Microbiol.* *18*:1–19.
6. Breed, R. S., Murray, E. G. D., and Smith, N. R. 1957. *Bergey's Manual of Determinative Bacteriology*, 7th ed. The Williams & Wilkins Co., Baltimore, Md., 1094 pp.
7. Citarella, R. V., and Colwell, R. R. 1970. Polyphasic taxonomy of the genus *Vibrio*: polynucleotide sequence relationships among selected *Vibrio* species. *J. Bacteriol.* *104*:434–442.

8. Colwell, R. R. 1969. Numerical taxonomy of the *Flexibacteria*. *J. Gen. Microbiol.* *58*:207–215.
9. Colwell, R. R. 1970. Numerical analysis in microbial identification and classification. *Dev. Ind. Microbiol.* *11*:154–160.
10. Colwell, R. R. 1970. Polyphasic taxonomy of the genus *Vibrio*: numerical taxonomy of *Vibrio cholerae*, *Vibrio parahaemolyticus*, and related *Vibrio* species. *J. Bacteriol.* *104*:410–433.
11. Fischer, A. 1897. *Vorlesungen über Bakterien*. Fischer, Jena, 186 pp.
12. Floodgate, G. D., and Hayes, P. R. 1963. The Adansonian taxonomy of some yellow pigmented marine bacteria. *J. Gen. Microbiol.* *30*: 237–244.
13. Gibbons, N. E. 1936. Studies on salt fish. I. Bacteria associated with the reddening of salt fish. *J. Biol. Board Can.* *3*:70–76.
14. Greig, M. A., Hendrie, M. S., and Shewan, J. M. 1970. Further studies on long term preservation of marine bacteria. *J. Appl. Bacteriol.* *33*: 528–532.
15. Johnson, R. M., Katarski, M. E., and Weisrock, W. P. 1968. Correlation of taxonomic criteria for a collection of marine bacteria. *Appl. Microbiol.* *16*:708–713.
16. Kelly, M. T., and Brock, T. D. 1969. Molecular heterogeneity of isolates of the marine bacterium *Leucothrix mucor*. *J. Bacteriol.* *100*:14–21.
17. Litchfield, C. D., Colwell, R. R., and Prescott, J. M. 1969. Numerical taxonomy of heterotrophic bacteria growing in association with continuous-culture *Chlorella sorokiniana*. *Appl. Microbiol.* *18*:1044–1049.
18. Marmur, J., and Doty, P. 1962. Determination of the base composition of deoxyribonucleic acid from its thermal denaturation temperature. *J. Mol. Biol.* *5*:109–118.
19. Moore, R. L., and Hirsch, P. 1972. Deoxyribonucleic acid base sequence homologies of some budding and prosthecate bacteria. *J. Bacteriol.* *110*:256–261.
20. Pfister, R. M., and Burkholder, P. R. 1965. Numerical taxonomy of some bacteria isolated from Antarctic and tropical seawaters. *J. Bacteriol.* *90*:863–872.
21. Quigley, M. M., and Colwell, R. R. 1968. Properties of bacteria isolated from deep-sea sediments. *J. Bacteriol.* *95*:211–220.
22. Russell, H. L. 1892. Untersuchungen über im Golf von Neapel lebende Bacterien. *Z. Hyg. Infectionskr.* *11*:165–206.
23. Sarimo, S. S., and Pine, M. J. 1969. Taxonomic comparison of the amino termini of microbial cell proteins. *J. Bacteriol.* *98*:368–374.
24. Shewan, J. M., Hobbs, G., and Hodgkiss, W. 1960. The *Pseudomonas* and *Achromobacter* groups of bacteria in the spoilage of marine white fish. *J. Appl. Bacteriol.* *23*:463–468.
25. Sneath, P. H. A. 1957. Some thoughts on bacterial classification. *J. Gen. Microbiol.* *17*:184–200.
26. Vanderzant, C., Judkins, P. W., Nickelson, R., and Fitzhugh, H. A., Jr. 1972. Numerical taxonomy of coryneform bacteria isolated from pond-reared shrimp (*Penaeus aztecus*) and pond water. *Appl. Microbiol.* *23*:38–45.
27. ZoBell, C. E. and Upham, H. C. 1944. A list of marine bacteria including descriptions of sixty new species. *Bull. Scripps Inst. Oceanogr. Univ. Calif.* *5*:239–292.

14

Reprinted from pp. 239–240, 246–247, 251–253, 280–281 of *Bull. Scripps Inst. Oceanogr. Univ. Calif.*, **5**(2), 239–292 (1944)

A LIST OF MARINE BACTERIA INCLUDING DESCRIPTIONS OF SIXTY NEW SPECIES

BY

CLAUDE E. ZOBELL AND HARVEY C. UPHAM

INTRODUCTION

THE WORK OF Lloyd (1930), Bavendamm (1932), Benecke (1933), Waksman (1934*a*), ZoBell (1942*a*), and others has shown that bacteria are quite widely distributed in the sea where they influence chemical, geological, and biological conditions. Although they are more abundant in the topmost one or two hundred meters of water, bacteria have been found in the ocean at all depths sampled (ZoBell, 1938). Thousands to millions of living bacteria per gram of mud are found on the sea floor (Rittenberg, 1939). Some of the factors which influence the distribution of bacteria in the sea have been discussed by Reuszer (1933), ZoBell and Feltham (1934), Waksman (1934*b*), ZoBell and Anderson (1936), ZoBell (1944), and others.

Marine bacteria mineralize organic matter (Waksman, Carey, and Reuszer, 1933), oxidize ammonia to nitrate, transform sulfur compounds, influence the phosphorus cycle (Renn, 1937), and otherwise affect the chemical composition of sea water and bottom deposits. They produce plant nutrients, and the bacteria themselves may serve as an important source of food for certain animals (ZoBell and Feltham, 1938). There are several ways in which they affect the pH of sea water or sedimentary materials. Collectively the bacteria of the sea probably consume more oxygen than all other organisms combined, thereby influencing the distribution of oxygen in the sea. Likewise they are important geological agents, which influence the diagenesis of sedimentary materials in many ways (ZoBell, 1942*b*). They may be instrumental in petroleum genesis (ZoBell, 1943*b*).

So far as is known, very few marine bacteria are pathogenic for the inhabitants of the sea, and none are pathogenic for man. After inoculating several white rats with cultures of different bacteria found in the sea, Russell (1893) declared that "in no case was there any evidence that would lead one to think that any of the species tested possessed any pathogenic peculiarities." However, there are bacteria which cause the deterioration of fish nets, ropes, cork floats, and wooden structures submerged in the sea. Also they promote the fouling of ships' bottoms, water conduits, and other submerged surfaces (ZoBell, 1939*b*, 1943*a*).

Marine bacteria are very troublesome in the spoilage of food products from the sea (Hunter, 1922; Griffiths, 1937; Wood, 1940; Snow and Beard, 1939), particularly since many of them are active at refrigeration temperatures (Bedford, 1933*a;* Kiser and Beckwith, 1942) and in the presence of high salt concentrations (Gibbons, 1937). The importance of bacteria as an ecological factor for plants as well as for animals has been studied by ZoBell and Feltham (1942).

Although the factors which influence the distribution and activities of marine bacteria have been extensively studied, very little is known concerning the characteristics of the organisms themselves. For want of adequate descriptions they are commonly referred to merely as "marine bacteria" or by "strain" numbers which may be assigned differently by various workers. The few marine bacteria which have been studied sufficiently to warrant the application of generic and species names have, for the most part, been described in obscure publications which are not generally available. In certain cases the same species names has been applied to totally unrelated organisms because the later workers had no way of knowing that the name had already been used.

We have made an effort in this paper to catalog previously described species of marine bacteria, and to describe several new species which occur commonly in the sea. The latter have been selected from several hundred species of bacteria which have been isolated from sea water, bottom deposits, marine animals, submerged surfaces, and other marine materials, by bacteriologists at the Scripps Institution of Oceanography during the last twelve years. Approximately 140 species have been extensively studied. The commonest of these which appear to be important chemical, geological, or biological agents are herewith described and named.

Of the 1335 species of bacteria listed in Bergey's (1939) Manual of Determinative Bacteriology only 86 were isolated from the sea. This is a commentary on the fact that the study of marine bacteria has been sadly neglected. There are probably just as many or more different species of bacteria in the sea, including bottom deposits and marine organisms, as there are on the land, and future studies may reveal that marine bacteria are of as great importance as land-dwelling bacteria. Actually the surface area of the sea is nearly four times as great as the land surface, and the volume of the ocean habitable by bacteria is hundreds of times greater than that of land. Exclusive of insects, four-fifths of all animal species known to man dwell in the sea.

[*Editor's Note:* Material has been omitted at this point.]

TERRIGENOUS BACTERIA IN THE SEA

It is not surprising that *Bacillus subtilis, Bacillus mesentericus, Escherichia coli,* and other common terrigenous bacteria have been found in brackish or coastal waters. Large numbers of bacteria are carried from the land into the sea by wind, water, migrating animals, and other agencies, and the gradual transition from fresh to salt water at the mouths of large rivers would seem to afford an opportunity for terrigenous bacteria to become adapted gradually to the marine environment. Moreover, terrigenous bacteria are carried all over the world by continental air masses (ZoBell, 1942*c*) and many of them must be precipitated into the ocean. Finally, laboratory experiments (Burke and Baird, 1931; Doudoroff, 1940) indicate that many fresh-water or terrigenous bacteria can be adapted to tolerate salt concentrations which are hypertonic to sea water. However, in spite of this a priori evidence, very few terrigenous bacteria are found in the sea at places which are remote from possibilities of terrigenous contamination.

From data given by Warren and Rawn (1938), we estimate that sewage outfalls on the west coast of North America daily discharge enough coliform bacteria to give nearly ten per ml. for the entire North Pacific Ocean, yet coliform bacteria are not found in the ocean except within a few miles of sewage outfalls. In reviewing the literature on the subject Griffiths (1937) points out that *Escherichia coli* is not a normal inhabitant of the intestines of marine fish although it may be found in the intestinal contents of fish which have fed in polluted waters. *Escherichia coli* and even *Eberthella typhosa* may survive in mussels, oysters, or other marine animals for several weeks (Dodgson, 1928), but these bacteria and allied species apparently do not remain viable very long in natural sea water (Browne, 1917).

According to de Giaxa (1889), intestinal bacteria lose their viability very soon after reaching sea water. In the Gulf of Naples he found 100,000 bacteria per ml. 50 meters from the sewage effluent, 26,000 at a distance of 350 meters and less than 100 per ml. 3000 meters from the effluent. Similar observations have been made on Los Angeles ocean outfalls (Knowlton, 1929). De Giaxa (1889) observed that *Vibrio comma* remained alive only three or four days in sterilized sea water and perished much sooner in unsterilized sea water. *Eberthella typhosa* and *Staphylococcus* species were even less resistant. *Bacillus anthracis* tolerated raw sea water very poorly. Carpenter *et al.* (1938) found that sea water kills 80 per cent of the organisms in sewage in half an hour.

Korinek (1926) observed that although fresh-water bacteria might grow in sea water enriched with meat extract and peptone, there was very little growth in sterilized sea water to which marine algae were added as a source of nutriment. Typical marine bacteria grew very well in sea water to which marine algae were added. Only hardy varieties of fresh-water bacteria like *Bacillus subtilis, Escherichia coli,* and *Pseudomonas aeruginosa,* for example, grew at all in the natural sea-water medium, nor was he able to acclimatize them to sea water. From these experiments Korinek (1927) concluded that even if certain fresh-water bacteria can tolerate sea water, they would soon perish if carried out to sea.

Korinek (1928) found that mixtures of *Bacillus mesentericus, Bacterium fluorescens, Escherichia coli,* and *Serratia marcescens* could not successfully compete with marine bacteria in sea water, leading him to conclude that fresh-water bacteria are incapable of prolonged survival in the sea. The failure of various workers to find significant numbers of the common fresh-water or terrigenous bacteria in the sea at places which are remote from possibilities of terrigenous contamination substantiates Korinek's conclusions.

[*Editor's Note:* Material has been omitted at this point.]

TEMPERATURE REQUIREMENTS

The temperature of the sea water and bottom deposits from which the bacteria were isolated ranges from 2° to 20° C. There was no evidence to indicate that any of the bacteria recovered from great depths where the temperature is perpetually colder than 5° C. were injured by exposure to temperatures as high as 25° C. However, approximately 25 per cent of the bacteria recovered from the sea floor were killed in ten minutes at 30° C., and 80 per cent of them were killed in ten minutes at 40° C. (ZoBell and Conn, 1940). These thermal death points are considerably lower than those of most terrigenous or fresh-water bacteria.

Some thermo-sensitive species are killed by the plating temperature of agar, 42° C., and when the media are poured at 45° C. the plate count of marine materials may be reduced 15 to 20 per cent. If the media are poured at 50° C., as may occur when working under the trying conditions on a small rolling boat at sea, all of the thermo-sensitive species representing 30 to 35 per cent of the total flora may be killed.

Although they are by no means psychrophilic, all of the pure cultures of marine bacteria grew in the refrigerator at 0° to 4° C. Some of them multiplied and were otherwise physiologically active at temperatures as low as –11° C. Since ice begins to form in sea water of average salinity at around –1.9° C., in such experiments it is necessary to depress the freezing point by the addition of sodium chloride, urea, glycerol, or other solutes. According to Hess (1934), Bedford (1933*b*), ZoBell (1934), Kiser and Beckwith (1942), and others, it is a common property of marine bacteria to grow at sub-zero temperatures. However, the optimum temperature for rapid multiplication and physiological activity is around 20° C. for most marine bacteria which have been studied.

The pure cultures described below have been maintained in a refrigerator at 0° to 4° C. for from one to twelve years. We have noted that refrigeration temperatures favor the constancy of species characteristics. Prior to being used to inoculate differential media, the stock cultures were permitted to incubate at 20° C. for a day or two. The differential media were incubated at 20° C.

CHARACTERISTICS OF MARINE BACTERIA

Besides the specific salinity requirements of marine bacteria, they are not morphologically or physiologically distinctive, although there are statistical differences between marine and terrigenous bacteria. For example, there are very few autochthonous marine species which ferment any of the common carbohydrates with the formation of detectable gas. In fact, acid is only slowly produced from simple sugars, and neither acid nor gas is produced from most carbohydrates.

Typical of water bacteria, more than half of those occurring in the sea are chromogenic. Of the 60 cultures described below, 19 are yellow, 5 are brown, 5 are pink or salmon colored, 4 are orange, and 1 is red. Other pigments, including indigo, black, and green, have been observed. In making a survey of the relative abundance of pigment producers in the sea with no regard for species, ZoBell and Feltham (1934) noted that 31.3 per cent of the colonies from marine materials were yellow, 15.2 per cent orange, 9.9 per cent brown, 7.4 per cent fluorescent, and 5.4 per cent red or pink. Many marine bacteria tend to lose their ability to produce pigment during prolonged laboratory cultivation. A low incubation temperature and meat infusion media are conducive to pigment formation. The junior author has found octopus infusion media prepared with sea water to be particularly conducive to pigment formation by marine bacteria.

The production of an acid reaction in glucose broth indicated that only 46 of the 60 cultures attacked this sugar, but in a medium containing glucose as the only source of carbon, it was found that all of the cultures multiplied and consumed oxygen. The failure of bacteria to produce an acid reaction in carbohydrate media may be due to a lack of fermentative ability, or it may be because the carbohydrate is quantitatively converted into bacterial protoplasm, carbon dioxide, and water without the accumulation of organic acids. The carbon dioxide is converted into carbonate as fast as it is produced and thus does not appear as a gas. In a well-buffered solution carbon dioxide has little effect on the pH. Working with dilute solutions, ZoBell and Grant (1943) found that 30 to 35 per cent of the glucose added to sea water was converted into bacterial protoplasm and the rest was quantitatively oxidized to carbon dioxide and water.

In view of the sanitary significance of the coliform bacteria, it is noteworthy that lactose fermenters are not indigenous to the sea nor do they live in the intestinal tracts of marine fish (Gibbons, 1934; Griffiths, 1937; Wood, 1940). Although laboratory experiments show that coliform bacteria can be acclimatized to grow in sea water, they occur in the sea only near sources of terrigenous contamination.

As a class, marine bacteria are very actively proteolytic, rapidly attacking most kinds of proteins. Half of the cultures hydrolyze casein, 47 of the 60 cultures liquefy gelatin, and all of them liberate ammonia from peptone. However, only 2 of them form detectable quantities of indol from tryptophane broth, a property which is more common among terrigenous bacteria.

Seven of the new species described below digest agar although they tend to lose this ability after prolonged incubation in the laboratory. Several species of marine agar digesters have been described by Angst (1929), Waksman and Bavendamm (1931), Waksman, Carey, and Allen (1934), Stanier (1941), and others (see page 242). We have encountered some of these previously described agar digesters on nutrient agar plates from sea water and marine mud. Such plates usually show the presence of agar digesters. Because they have already received much attention, we have made no special effort to isolate new species of agar digesters.

Nearly all of the bacteria isolated from sea water or marine mud prove to be facultative aerobes. Of the 60 cultures which were isolated under aerobic conditions only 5 are obligate aerobes, although most of them grow better in the presence of free ogygen than in its absence. The facultative aerobes tend to lose their ability to grow anaerobically after prolonged cultivation in the laboratory under aerobic conditions. While this paper is not concerned with anaerobes, many have been isolated from marine mud and other materials. Very few of them are obligate anaerobes.

Most of the bacteria which we isolated from plates proved to be asporogenous, although numerous species of sporogenous *Bacillus* and *Clostridium* occur in the sea. Using selective methods, Newton (1924) isolated 80 different pure cultures of spore-formers from sea water and the alimentary tract of marine animals.

Gram-negative bacteria predominate in the sea. Many of the cultures are highly pleomorphic. Most of the bacteria are small rods, being either straight or helicoidal. Cocci are encountered only infrequently in the sea. On the average, marine bacteria are smaller than the bacteria found in milk, sewage, fresh water, or soil. Typical of marine bacteria 49 of the 60 cultures are actively motile, including four flagellated cocci. Motile cocci are very rarely reported (Bergey *et al.*, 1939).

Many of the bacteria found in sea water are sessile or periphytic (ZoBell, 1936*a*), growing preferentially or exclusively attached to solid surfaces. The sessile habit of marine bacteria is most pronounced when they are growing in very dilute nutrient solutions such as sea water to which nothing has been added. Unpolluted sea water generally contains about 5 mgm. of organic matter per liter (Krogh, 1931), much of which is fairly refractory to bacterial attack (Waksman and Carey, 1935). In such a solution solid surfaces tend to adsorb organic matter, thereby enhancing bacterial activity on such surfaces (ZoBell, 1943*a*). Solid surfaces also retard the diffusion of exoenzymes and hydrolyzates away from the cell, thus promoting the assimilation of nutrients which must be hydrolyzed extracellularly prior to ingestion. Most sessile bacteria appear to attach themselves tenaciously to solid surfaces by exuding a mucilaginous holdfast. A few have stalks (Henrici and Johnson, 1935). Some of the sessile bacteria grow on the walls of the culture receptacle without clouding the medium itself.

[*Editor's Note:* Material has been omitted at this point.]

SUMMARY

Bacteria appear to be widely distributed in the sea where they live in the water and bottom deposits or associated with plants, animals, and marine particulate detritus. The bacteria of the sea play an important role in chemical, geological, and biological processes. Some are known to be of economic importance.

Although the study of marine bacteria has received comparatively little attention, some of the first bacteria described were found in the sea. Ehrenberg described the marine-dwelling *Spirochaeta plicatilis* in 1838, and Cohn described *Beggiatoa mirabilis* in 1865. Eighteen species of marine photogenic or luminescent bacteria were named by 1891 and several more since then.

Large numbers of bacteria have been found in marine fish and shellfish but only a few species are described in the available literature. Some of them are active in the presence of high concentrations of salt and at refrigeration temperatures, thereby presenting a problem in the preservation of sea foods.

Nine different species of halophilic bacteria have been isolated from the Dead Sea, a body of water which contains from 25 to 30 per cent salt.

Being very conspicuous on nutrient agar plates, 23 species of marine agar digesters have been described, and many more have been observed.

Several of the 17 named species of marine denitrifiers, some of which are believed to promote the precipitation of calcium carbonate, may prove to be synonymous. *Desulfovibrio aestuarii* is the only known marine species which reduces sulfate, although many others may exist.

Species of *Nitrosomonas, Nitrobacter, Azotobacter,* and nitrogen-fixing *Clostridium* have been found in the sea but it has not been established that they are indigenous.

Sulfur bacteria, which appear to live equally well in fresh or salt water, occur abundantly in brackish and other coastal waters and associated with marine vegetation. Only a few species of iron bacteria have been described.

Most of the known species of *Saprospira* and *Cristospira* and some species of *Spirochaeta* live exclusively in the sea.

There are no marine bacteria known to be pathogenic for man. Species which parasitize marine plants and animals are probably quite common although very few species have been investigated.

Many terrigenous or fresh-water bacteria are found in the sea in brackish or coastal water. They are rarely found at places which are remote from possibilities of contamination from land. Most of the bacteria occurring in unpolluted sea water appear to be specifically marine.

We are aware of no infallible method of distinguishing marine bacteria from those of soil, fresh water, or other habitats. If bacteria are found in the sea at places which are remote from possibilities of terrigenous contamination and if upon initial isolation they require sea water for their growth, the bacteria are regarded as marine species.

As a class marine bacteria are only mildly halophilic. They prefer sea water or isotonic salt solutions of similar composition. They do not tolerate increases or decreases in the salt concentration very well. Only a few marine bacteria grow in media containing 12 per cent sodium chloride although halophilic bacteria from other environments flourish in media containing 18 to 30 per cent sodium chloride.

Though coming from an environment where the temperature is generally lower (2° to 18° C.), most marine bacteria grow best at around 20° C. Some are killed in 10 minutes at 30° C., and thermo-sensitive species may be destroyed by being subjected to the plating temperature of agar (42° C.).

Using nutrient sea-water media, 140 new and previously undescribed species of bacteria have been isolated from sea water, bottom deposits, and other marine materials during the last twelve years by bacteriologists at the Scripps Institution. Sixty of the species which appear to be the most important in marine transformations are described and named. These include 18 species of *Pseudomonas,* 11 *Vibrio,* 8 *Bacillus,* 6 *Flavobacterium,* 6 *Micrococcus,* 4 *Achromobacter,* 3 *Bacterium,* 2 *Actinomyces,* 1 *Serratia,* and 1 *Sarcina.*

Small, Gram-negative, motile, straight or helicoidal rods predominate. Pleomorphism is common. Most marine bacteria are facultative aerobes. They are generally actively proteolytic and weakly saccharolytic. Although most marine bacteria can assimilate the simple carbohydrates, little acid and no detectable gas is produced as a rule.

Since it is likely that many described species of marine bacteria have probably been overlooked in this initial attempt to bring them all together under one cover, readers are invited to direct the attention of the authors to other species which should be included in our list. Likewise we shall appreciate receiving information on the characteristics and synonymy of named species of marine bacteria.

REFERENCES

Angst, E. C. 1929. Some new agar-digesting bacteria. Publ. Puget Sound Biol. Sta. *7*:49–63.

Bavendamm, W. 1932. Die mikrobiologische Kalkfallung in der tropischen See. Arch. f. Mikrobiol., Vol. 3, pp. 206–276.

Bedford, R. H. 1933. The discoloration of halibut by marine chromogenic bacteria at 0°C. Can. Biol. Bd. Contrib. *7*:425–430.

Benecke, W. 1933. Bakteriologie des Meeres. Abderhalden's Handb. der biol. Arbeitsmethoden, Abt. IX, Lfg. 404, pp. 717–872.

Browne, W. W. 1917. The presence of the *B. coli* and *B. welchii* groups in the intestinal tract of fish (*Stenomus chrysops*). J. Bacteriol. *2*:417–422.

Burke, V., and Baird, L. A. 1931. Fate of fresh water bacteria in the sea. *J. Bacteriol. 21*: 287–298.

Carpenter, L. V., Setter, L. R., and Weinberg, M. 1938. Chloramine treatment of sea water. Amer. J. Pub. Health *28*:929–934.

de Giaxa, Professor. 1889. Ueber das Verhalten einiger pathogener Mikroorganismen in Meerwasser. Zeitschr. f. Hyg. 6:162–225.

Dodgson, R. W. 1928. Report on mussel purification. His Majesty's Stationery Office, London, 498 pp.

Doudoroff, M. 1940. Experiments on the adaptation of *Escherichia coli* to sodium chloride. J. Gen. Physiol. *23*:585–611.

Gibbons, N. E. 1934. Lactose-fermenting bacteria from the intestinal contents of marine fish. Can. Biol. Bd. Contrib., N.S. *8*:291–300.

Gibbons, N. E. 1937. Studies on salt fish. I. Bacteria associated with the reddening of salt fish. J. Biol. Bd. Can. *3*:70–76.

Griffiths, F. P. 1937. A review of the bacteriology of fresh marine-fishery products. Food Res. *2*:121–134.

Henrici, A. T., and Johnson, D. E. 1935. Studies of freshwater bacteria. II. Stalked bacteria, a new order of Schizomycetes. J. Bacteriol. *30*:61–93.

Hunter, A. C. 1922. The sources and characteristics of the bacteria in decomposing salmon. J. Bacteriol. *7*:85–109.

Kiser, J. S., and Beckwith, T. D. 1942. Effect of fast-freezing upon bacterial flora of mackerel. Food Res. *7*:255–259.

Knowlton, W. T. 1929. *B. coli* surveys, Los Angeles ocean outfalls. Calif. Sew. Works Jour. *2*:150–152.

Korinek, J. 1926. Ueber Susswasserbakterien im Meere. Centralbl. f. Bakt., II Abt. *66*:500–505.

Korinek, J. 1927. Ein Beitrag zur Mikrobiologie des Meeres. Centralbl. f. Bakt., II Abt. *71*:73–79.

Korinek, J. 1928. Ueber die Zeesetzungsprocesse der organischen Substanz im Meere. Biochem. Zeitschr. *192*:230–237.

Krogh, A. 1931. Dissolved substances as food of aquatic organism. Rapports et Proces-Verbaux des Reunions *75*:7–36.

Lloyd, B. 1930. Bacteria of the Clyde Sea area: a quantitative investigation. J. Mar. Biol. Ass. *16*:879–907.

Newton, D. 1924. Marine spore forming bacteria. Can. Biol. Bd. Contrib., N.S. *1*:377–400.

Renn, C. E. 1937. Bacteria and the phosphorus cycle in the sea. Biol. Bull. *72*:190–195.

Reuszer, H. W. 1933. Marine bacteria and their role in the cycle of life in the sea. III. Distribution of bacteria in the ocean waters and muds about Cape Cod. Biol. Bull. *65*: 480–497.

Rittenberg, S. C. 1939. Investigations on the microbiology of marine air. J. Mar. Res. *2*: 208–217.

Russell, H. L. 1893. The bacterial flora of the Atlantic Ocean in the vicinity of Woods Hole, Mass. Bot. Gaz. *18*:383–395, 411–417, 439–447.

Snow, J. E., and Beard, P. J. 1939. Studies on bacterial flora of north Pacific salmon. Fod. Res. *4*:563–585.

Stanier, R. Y. 1941. Studies on marine agar-digesting bacteria. J. Bacteriol. *42*:527–559.
Waksman, S. A. 1934a. The role of bacteria in the cycle of life in the sea. Sci. Month. *38*: 35–49.
Waksman, S. A. 1934b. The distribution and conditions of existence of bacteria in the sea. Ecol. Monogr. *4*:523–529.
Waksman, S. A., and Bavendamm, W. 1931. On the decomposition of agar-agar by an aerobic bacterium. J. Bacteriol. *22*:91–102.
Waksman, S. A., and Carey, C. L. 1935. Decomposition of organic matter in sea water by bacteria. I. Bacterial multiplication in stored sea water. J. Bacteriol. *29*:531–543.
Warren, A. K., and Rawn, A. M. 1938. Disposal of sewage into the Pacific Ocean. New York, Modern Sewage Disposal, Federation of Sewage Works Assoc., pp. 202–208.
Waksman, S. A., Carey, C. L., and Allen, M. C. 1934. Bacteria decomposing alginic acid. J. Bacteriol *28*:213–220.
Waksman, S. A., Carey, C. L., and Reuszer, H. W. 1933. Marine bacteria and their role in the cycle of life in the sea. I. Decomposition of marine plant and animal residues by bacteria. Biol. Bull. *65*:57–79.
Waksman, S. A., Hotchkiss, M., and Carey, C. L. 1933. Bacteria concerned in the cycle of nitrogen in the sea. Biol. Bull. *55*:137–167.
Wood, E. J. F. 1940. Studies on the marketing of fresh fish in eastern Australia. II. The bacteriology of spoilage of marine fish. Melbourne, Australia Counc. for Sci. & Indust. Res. *100*:1–92.
ZoBell, C. E. 1936. Periphytic habits of some marine bacteria. Proc. Soc. Exp. Biol. Med. *35*:270–273.
ZoBell, C. E. 1938. Studies on the bacterial flora of marine bottom sediments. J. Sed. Petrol. *8*:10–18.
ZoBell, C. E. 1939. The biological approach to the preparation of anti-fouling paints. Proc. Scientific Section of the National Paint, Varnish and Lacquer Assoc., Circular 588, pp. 149–163.
ZoBell, C. E. 1942a. Bacteria of the marine world. Sci. Month. *55*:320–330.
ZoBell, C. E. 1942b. Changes produced by microorganisms in sediments after deposition. J. Sed. Petrol. *12*:127–136.
ZoBell, C. E. 1942c. Microorganisms in marine air. Aerobiology, Amer. Assoc. Advancement Science, Symposium Series No. 17, pp. 55–68.
ZoBell, C. E. 1943a. The effect of solid surfaces upon bacterial activity. J. Bacteriol. *46*: 39–56.
ZoBell, C. E. 1943b. Influence of bacterial activity on source sediments. Oil Weekly *109*: 15–26.
ZoBell, C. E. 1944. *Marine Microbiology—A Monograph on Hydrobacteriology*. Waltham, Mass., Chronic Botanica Company, pp. 240.
ZoBell, C. E., and Anderson, D. Q. 1936. Vertical distribution of bacteria in marine sediments. Bull. Amer. Assoc. Petrol. Geol. *20*:258–269.
ZoBell, C. E., and Feltham, C. B. 1934. Preliminary studies on the distribution and characteristics of marine bacteria. Bull. Scripps Inst. Oceanogr. Tech. Ser. *3*:279–296.
ZoBell, C. E., and Feltham, C. B. 1938. Bacteria as food for certain marine invertebrates. J. Mar. Res. *1*:312–327.
ZoBell, C. E., and Feltham, C. B. 1942. The bacterial flora of a marine mud flat as an ecological factor. Ecology *23*:69–78.
ZoBell, C. E., and Grant, C. W. 1943. Bacterial utilization of low concentrations of organic matter. J. Bacteriol. *45*:555–564.

15

Reprinted from *J. Appl. Bacteriol.*, **23**(3), 379–390 (1960)

(Symposium on *Pseudomonas* and *Achromobacter*: Paper II)

A DETERMINATIVE SCHEME FOR THE IDENTIFICATION OF CERTAIN GENERA OF GRAM-NEGATIVE BACTERIA, WITH SPECIAL REFERENCE TO THE PSEUDOMONADACEAE

By J. M. SHEWAN, G. HOBBS and W. HODGKISS
Torry Research Station, Aberdeen

CONTENTS

1. Introduction

The identification and classification of bacteria isolated from marine environments, including fresh and spoiling fish, have been subjects of study in our laboratory for many years and we have found that the majority of our isolates are Gram-negative asporogenous rods. Wherever these bacteria are encountered, as in studies concerned with aquatic environments generally, soils, foods, and plant and animal diseases, investigators are generally agreed that classification presents considerable difficulty. Widely divergent views on taxonomy are reflected in the various proposed systems of classification (Ingram & Shewan, 1960), and resolution of the difficulty has been delayed by the lack of readily applicable discriminatory tests. However, over the past decade the work of identification has been greatly eased by the development of new tests as well as by improvements in existing apparatus. e.g. phase contrast and electron microscopes.

Precise identification of organisms is of course not simply of academic interest; it often has great practical significance. In our own field, for example, it is important to know whether the *Pseudomonas* spp. causing fillet spoilage are present on the fish in the sea, in the ice used to chill the fish, or are picked up from the filleting benches. Studies of the changes in the bacterial flora of fish during storage for up to about 16 days in ice involve the handling of several hundreds of micro-organisms, and this led us to seek speedier methods of identifying large numbers of isolates. In the first instance differentiation at the generic level was the aim. Our early results (Shewan *et al.* 1954) suggested some broad differentiations of groups corresponding roughly to the genera *Pseudomonas*, *Achromobacter* and *Vibrio* and the family Bacteriaceae of Bergey's *Manual of Determinative Bacteriology*, 7th ed. (Breed, Murray & Smith, 1957), and from the experience gained since then the scheme presented here and now in use in our laboratory was developed.

2. The Determinative Scheme

The medium used for isolation and cultivation of the bacteria investigated was 'Oxoid' Lab-Lemco nutrient agar made up either with tap water or with 75% of aged sea water as required (ZoBell, 1946). The original isolate was plated to obtain a pure culture and then put through the tests outlined in Table 1. The colony appearance, the Gram reaction, the morphology and the behaviour in the oxidase test (Kovacs, 1956) were obtained from the agar plate culture, and observations of motility and morphology (under phase contrast) were done on a Lab-Lemco broth culture. This latter culture was also used to surface seed an agar plate for the

Table 1. *A summary of the tests used in the determinative scheme*

Medium	Observation
Nutrient agar	Colony appearance Gram stain Morphology Oxidase test (Kovacs, 1956) Antibiotic and O/129 sensitivity tests (Shewan *et al.* 1954)
Nutrient broth	Motility, morphology
Medium of King *et al.* (1954) Medium of Paton (1959)	Diffusible fluorescent pigments detected under U.V. light
Medium of Paton (1959)	2-ketogluconic acid formation
Medium of Hugh & Leifson (1953)	Dissimilation of carbohydrates
Nutrient agar+30% skim milk	Pigment production by *Flavobacterium* spp.
Nutrient agar slope	Flagella stain (Casares-Gil)

sensitivity tests described by Shewan *et al.* (1954) and to inoculate two tubes of the medium of Hugh & Leifson (1953), one of which was incubated anaerobically under paraffin. The production of diffusible fluorescent pigment was determined using either the medium of King *et al.* (1954) or that of Paton (1959). When growth occurred in the latter medium, the production of 2-ketogluconic acid was recorded using an aniline oxalate or aniline phthalate reagent. A sheet of chromatography paper (Whatman no. 1) was impregnated with a saturated solution of recrystallized aniline oxalate or phthalate (Hough *et al.* 1950) and the culture heavily spotted on the paper, which was then heated in an oven at 105° for 2–3 min. A red spot indicated the presence of 2-ketogluconic acid. Occasionally the gluconate (as also glucose) appeared to be dissimilated beyond the 2-ketogluconic acid stage and this gave rise to yellow colours with the aniline reagent (Floodgate, pers. comm.). For the detection of intracellular pigments (e.g. those of flavobacteria) a nutrient agar incorporating 30% of skim milk is used. In motile cultures the type of flagellation was determined by the use of a modified Casares-Gil flagella stain (Manual, 1957) or by electron microscopy. Most of our isolates were psychrophiles and the tests were normally carried out at 20°; but they are equally applicable to mesophiles.

On the basis of the above tests the broad groupings outlined in Table 2 were obtained. It is important to stress, however, that the scheme was devised primarily for the organisms from the special environment concerned, mainly fresh and spoiling marine fish, although we have tested the scheme with many strains isolated from different environments by other workers. Thus the only members of the Enterobacteriaceae we normally encountered were 'paracolons' and *Escherichia coli.* We

Table 2. *A broad grouping of organisms on the results of the tests in Table* 1

Motile rods		Nonmotile rods	
Oxidase positive, polar flagella	Oxidase negative, peritrichous flagella	Nonpigmented colonies Short stout rods, often coccoid	Pigmented colonies (yellow, greenish-yellow, orange)
Pseudomonas, Xanthomonas, Aeromonas, Vibrio	Paracolons, *E. coli,* etc.	*Achromobacter, Alcaligenes*	*Flavobacterium, Cytophaga*

have thus had no experience with, for instance, the *Shigella* group, and it is possible that they would give rise to difficulties in our scheme. But in our experience the oxidase test of Kovacs (1956) is invaluable for the differentiation of the Enterobacteriaceae which we encounter from members of the Pseudomonadaceae, and this seems to be true for the Enterobacteriaceae generally, according to Ewing & Johnson (1960). The Enterobacteriaceae are of course being deliberately excluded from our present discussion, as are other Gram-negative rods such as the members of the genera *Azotobacter, Acetobacter* and *Cellulomonas*, which are not normally isolated except by the use of special media.

3. Discussion of the Groups

(a) *The polar-flagellate, oxidase positive group*

This comprises members of the genera *Pseudomonas, Aeromonas* and *Vibrio*. Some members of the genus *Xanthomonas* also fit in here, but at present our experience is too limited for any general conclusions to be drawn.

By means of the medium of Hugh & Leifson (1953) these organisms can be differentiated into those which give (1) an oxidative reaction (*Pseudomonas* spp.); (2) no reaction (*Pseudomonas* spp.); (3) an alkaline reaction (*Pseudomonas* spp.) and (4) a fermentative reaction (*Aeromonas* and *Vibrio* spp.) (Table 3).

(i) Pseudomonas *spp.* Since the oxidative *Pseudomonas* spp. can be further subdivided on the basis of the presence or absence of fluorescence under ultraviolet light, we therefore postulate four groups of *Pseudomonas* spp. Group I includes those, such as *Pseudomonas fluorescens* and *Ps. aeruginosa*, which are oxidative and produce a diffusible fluorescent pigment. Group II strains attack glucose oxidatively, but in our hands never produce a diffusible fluorescent pigment. Group III strains produce no diffusible pigment and have no action on glucose, but produce an alkaline reaction in Hugh & Leifson's medium, and group IV strains produce no change at all.

Table 3. *A grouping of the Gram-negative asporogenous rods, polar-flagellate, oxidase positive and not sensitive to 2·5 i.u. of penicillin, on the results of four other tests*

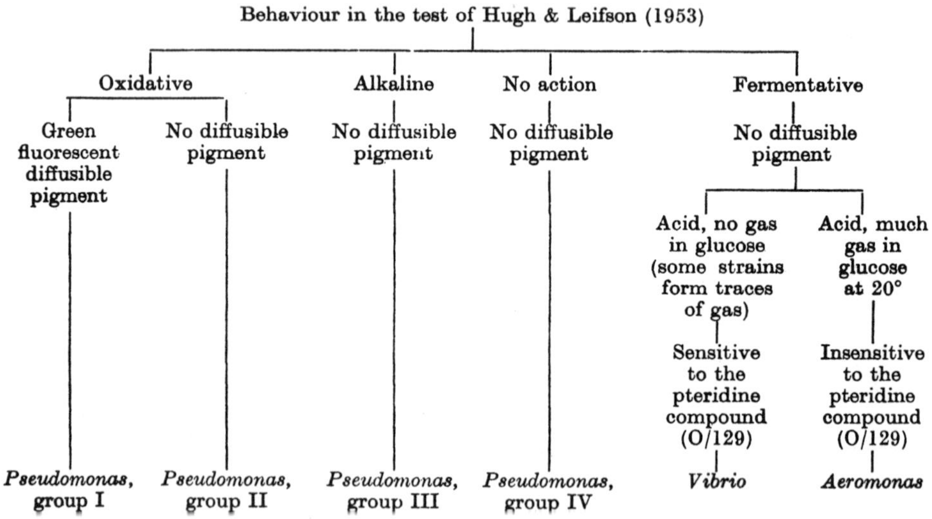

Further subdivision of groups I and II may be possible, since both contain strains which appear to effect the further breakdown of 2-ketogluconic acid.

Recently Sneath (1957*a*) has suggested that the most logical way of classifying bacteria is to compare the overall similarities of the organisms, applying the Adansonian principle of giving equal weight to each character. Stewart (1958), at the Torry Research Station, has recently studied in detail about 60 organisms isolated from marine sources by various workers, and on the basis of the determinative scheme outlined above classified them as *Pseudomonas* spp. The overall similarities of about 80 tests for 55 of these organisms have been calculated by G. D. Floodgate of the Torry Research Station, using the formula of Sneath (1957*b*). The results (Fig. 1) show that although there is some overlapping three groups of *Pseudomonas* can be differentiated, and these correspond generally to the organisms of groups I, II, and III and IV together. Of the recently described strains of *Pseudomonas* available to us, *Pseudomonas aureofaciens* (Kluyver, 1956) fits into group III, *Ps. fragi* (from the American Type Culture Collection) into group II and *Ps. rubescens* (Pivnick, 1955) into group IV.

(ii) Aeromonas *and* Vibrio *spp.* These are differentiated (Table 3) from the *Pseudomonas* species by their fermentative action upon glucose. *Aeromonas* spp. are characterized by the production of both acid and gas from glucose (at 20°) whereas *Vibrio* spp. ferment glucose anaerogenically. Moreover, only the latter are sensitive to 2:4-diamino-6:7-di-*iso*propylpteridine (O/129) (Shewan *et al.* 1954).

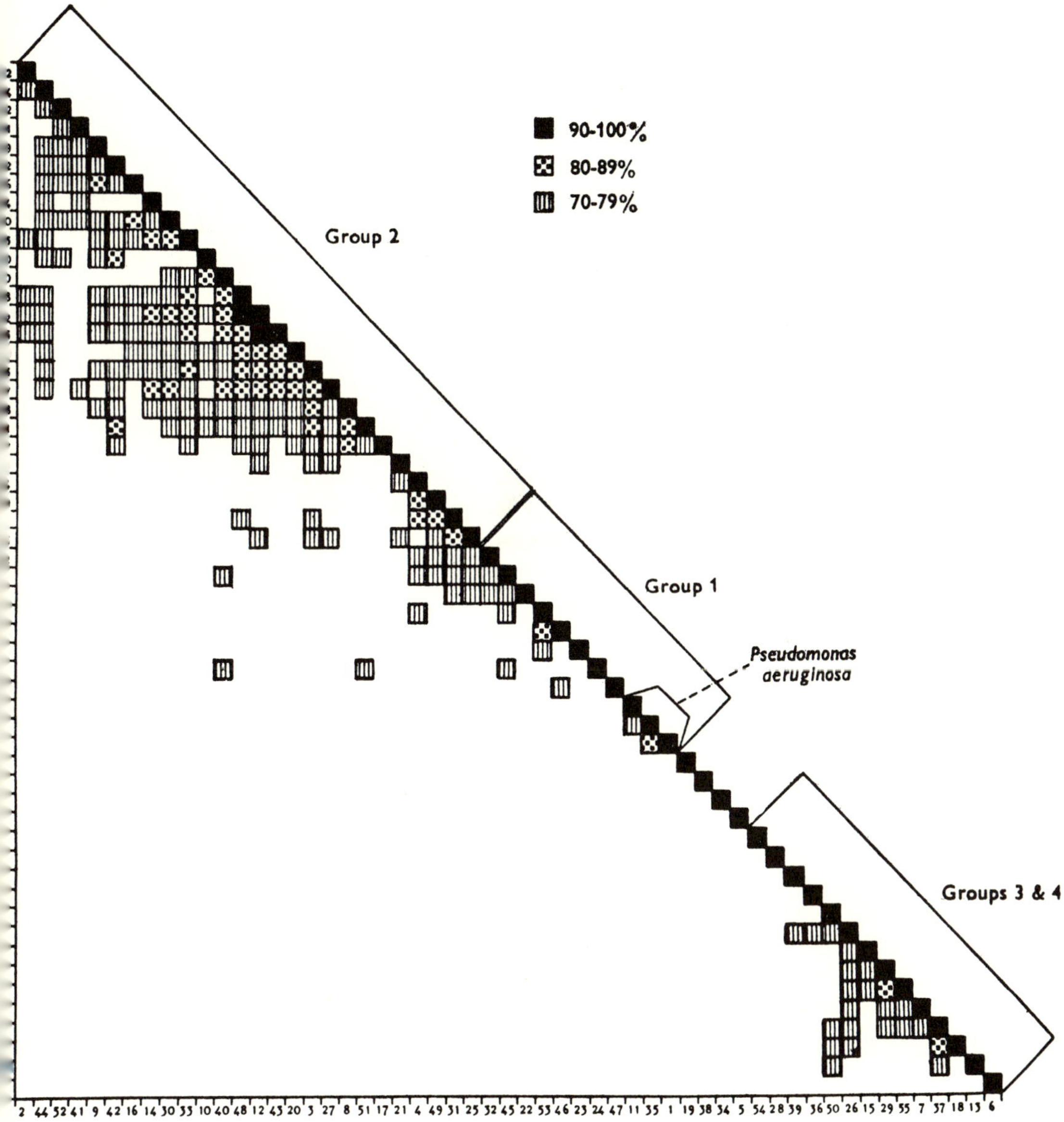

Fig. 1. A diagrammatic representation of the similarity values (% S values of Sneath, 1954) among 55 strains of *Pseudomonas*, obtained by the use of 80 features.

Stevenson (1959) has questioned the validity of the genus *Aeromonas* and has suggested that all its members could be regarded as nonchromogenic species of the genus *Serratia*. Several authors, Stevenson points out, have stated that flagellation in the genera *Aeromonas* and *Serratia* may vary according to the conditions of culture, and he believed that his *Aeromonas margarita* was in reality a nonchromogenic strain of *S. marcescens*. '*Aeromonas margarita*', however, is peritrichous (Plate 1, a) and this along with its biochemical characteristics indicates that it is a typical member of the Enterobacteriaceae. Members of the genus *Kluyvera*, defined as consisting of polar-flagellate Gram-negative rods which attack glucose fermentatively (Asai *et al.* 1956), and the so-called polar-flagellate member of the Enterobacteriaceae (Cook, 1958) (Plate 1, b), were also found to be peritrichous.

Two well recognized organisms, *Pseudomonas formicans* (Crawford, 1954) and *Bacterium salmonicida*, both regarded by many workers (Griffin, 1953; Snieszko, 1953) as *Aeromonas* spp., appear to occupy anomalous positions in our suggested scheme. *Ps. formicans* is a Gram-negative, oxidase positive, polar-flagellate asporogenous rod, which is fermentative in Hugh & Leifson's medium, forms no gas from glucose and yet is insensitive to the pteridine compound (O/129). In many supposedly key characters, e.g. formation of 2:3-butanediol and production of gas from glucose, it would appear to be excluded from the genus *Aeromonas*, and from its morphology and its insensitivity to the pteridine compound from *Vibrio* also. *Bacterium salmonicida* is also a Gram-negative rod, oxidase positive, forming acid and gas from glucose and fermentative in Hugh & Leifson's medium, but is nonmotile. Moreover, unlike *Aeromonas* spp. it is Voges-Proskauer negative.

Hugh & Leifson (1953) suggested that only fermentative organisms be included in the genus *Vibrio*. *V. percolans* and *V. cuneatus*, they pointed out, were not related either morphologically or physiologically to *V. comma* (Leifson, 1960). We also would exclude *V. cyclosites*, *V. neocistes* (Hugo & Rogers, 1958), *V. jamaicensis* (Caselitz, 1955), and many others listed as vibrios in some Culture Collections, on morphological and physiological grounds. In our experience most of these organisms are *Pseudomonas* spp. of the nonpigmented groups. *V. jamaicensis*, as Caselitz (1958) later pointed out, is an *Aeromonas* sp.

A new genus, *Lophomonas*, has been suggested for some organisms formerly known as *V. alcaligenes* (Galarneault & Leifson, 1956); but from the data presented these fit into our suggested group IV of *Pseudomonas*. On the other hand, certain organisms formerly described as *Pseudomonas* spp. should on morphological and physiological grounds be placed in the genus *Vibrio*, for example *Ps. ichthyodermis* (Hodgkiss & Shewan, 1950) and *Achromobacter ichthyodermis* (Wells & ZoBell, 1934; ZoBell & Wells, 1934).

V. comma, fresh water vibrios of the paracholera group, *V. foetus*, some organisms pathogenic for fish (Rucker, 1959; Smith, 1961) and *V. anguillarum* (Bagge & Bagge, 1956), all fit into the *Vibrio* group of our determinative scheme.

Previously it had always been difficult to differentiate the saprophytic vibrios and the vibrios causing disease in poikilothermic animals from other Gram-negative, polar-flagellate rods. This was usually done on the basis of the curvature of the cell, a quite unsatisfactory criterion. The fermentative character of their attack on carbohydrates, together with their sensitivity to the pteridine compound O/129, are thus valuable additional criteria. It could be added that in addition to the curvature of the cell, *V. comma* was characterized by the pleomorphism of the cells and in particular by the presence of round or coccoid bodies (Hammerl, 1906; Henrici, 1925). Quite recently Hallock (1959) has claimed that this is a distinguishing feature of the *Vibrio* group as a whole and more attention might be given to this character than hitherto. Coccoid bodies have been described in certain luminous bacteria (Johnson & Gray, 1949), which on our criteria have been classified as *Vibrio* spp. (Spencer, 1955), and they are a regular feature of the 'gut-group' vibrios of Liston (1955) (Plate 1, c and d).

(iii) Spirillum *spp.* Members of the genus *Spirillum* have been infrequently encountered in our field. It is considered that morphology alone, including the bipolar tufts of flagella (Williams, 1960), is sufficiently characteristic for identification. Other important features of members of this genus are sensitivity to penicillin, but not to the pteridine compound O/129, and no action upon glucose in Hugh & Leifson's medium.

In the 6th edition of Bergey's *Manual* (Breed, Murray & Hitchens, 1948) the *Vibrio* spp. were included in the Tribe Spirillae of the Family Pseudomonadaceae, whereas in the 7th edition (Breed, Murray & Smith, 1957) they form part of the separate Family Spirillaceae. In our determinative scheme, however, the *Vibrio* spp. appear to be more closely allied to the *Pseudomonas* spp. than to the *Spirillum* spp.

(b) *The nonmotile nonpigmented group*

In Table 2 the *Achromobacter* spp. are recorded as being nonmotile and nonpigmented, despite the fact that the type species, *Achromobacter liquefaciens*, is defined in the 7th edition of Bergey's *Manual* (Breed, Murray & Smith, 1957) as being peritrichous. However, owing to the absence of the original culture from any known collection and to the paucity of data concerning it in Bergey's *Manual*, it cannot be identified with any degree of certainty. Moreover, over the past ten years, in which many thousands of isolates have been screened at the Torry Research Station, we have never once encountered an organism which is peritrichous and yet possesses the typical morphology given in the original description (Eisenberg, 1891 and Frankland & Frankland, 1894). Breed was well aware of the difficulty arising from the absence of the type strain and in a private communication in 1947 stated that he believed an organism, corresponding to the original strain, had at last been isolated. However, no description of such a culture has yet appeared in the literature. The organism recently described as *Achr. liquefaciens* (Youatt, 1954; Skerman *et al.* 1958) has not been available to us, nor have the mesophilic *Achromobacter* spp., including one designated *Achr. liquefaciens*, of Rush (1947). Buttiaux & Gagnon (1959) have also recently concluded that members of the genus *Achromobacter* are nonmotile, since they never once encountered a peritrichous organism corresponding to *Achr. liquefaciens*, despite the fact that they were working with organisms isolated from very widely differing environments. A motile organism with peritrichous flagella, isolated by Kimata & Kawai (1953) and originally named *Achr. histamineum*, is fermentative in its action on glucose, producing acid and gas, and has recently been assigned by Kimata *et al.* (1958) to the genus *Proteus*.

We would define the *Achromobacter* group as being composed of nonmotile, non-pigmented, short, stout or coccoid rods, occurring singly, in pairs or short chains, forming grey or off-white slightly opaque colonies on agar, sensitive to penicillin, and usually biochemically inactive. In Hugh & Leifson's medium most strains produce an alkaline reaction, a few no reaction, and some strains are oxidative. In general these organisms appear to correspond to the genus *Acinetobacter* proposed by Brisou & Prévot (1954).

Achromobacter spp. are distinguished from *Pseudomonas* spp. primarily on motility and morphology, and they are usually less active biochemically. In addition most strains are sensitive to penicillin, whereas the majority of *Pseudomonas* strains are not.

(c) *The nonmotile pigmented group*

Although the genera *Flavobacterium* and *Cytophaga* are not being considered in this Symposium, for the sake of completeness they merit some mention in relation to the determinative scheme outlined in Table 2.

All the flavobacteria and many of the cytophagas are characterized by the formation on agar of yellow to orange translucent colonies. Pigmentation is most easily detected on a nutrient agar containing 30% of skim milk. This not only appears to enhance pigmentation but makes its detection easier.

Morphologically the cells of *Flavobacterium* spp. are generally longer and more slender than those of *Achromobacter* spp. They occasionally form filaments in fluid media, and the cells are usually less refractile under phase contrast. Although in Table 2 *Flavobacterium* spp. are designated nonmotile, a few motile strains have been encountered.

Cytophaga spp. generally are defined in Bergey's *Manual*, 7th ed. (Breed, Murray & Smith, 1957) as thin rods, often with pointed ends, which show flexing and gliding motility; hence on agar plates the colonies have a characteristic diffuse margin, spreading outwards and into the agar. Recent experience at the Torry Research Station (Hayes, 1960) has shown that the differentiation of *Flavobacterium* spp. from the yellow to orange *Cytophaga* spp. is by no means as clear cut as the definitions in Bergey's *Manual* imply, and more work will have to be done before the position can be clarified.

4. The Effectiveness of the Scheme

The determinative scheme outlined above and set out in detail with regard to the major groups in Tables 4 and 5, follows closely the system of Bergey's *Manual* (Breed, Murray & Smith, 1957) in that the genera all have the definitive characters given there. It has now been in use at the Torry Research Station for a number of years, although naturally refinements and adjustments are constantly taking place. During this time several workers have examined thousands of isolates, mainly from marine sources, and the classification of most of these cultures at the generic level has been accomplished. In addition we have examined more than 200 named strains from Type Culture Collections and individual workers; these have included *Pseudomonas*, *Achromobacter*, *Vibrio*, *Aeromonas*, *Xanthomonas* and *Flavobacterium* spp. In almost every instance we have been able to identify them unequivocally by means of our scheme, although in some cases this classification has differed from that originally given. In addition our scheme has been in use for some time in other laboratories, and other workers to whom the scheme has been demonstrated have published data (e.g. Browne & Weidemann, 1958) which show its general usefulness.

Table 4. *An outline of the determinative tests applied to certain Gram-negative asporogenous rod-shaped organisms*

Motility	Morphology	Colony appearance*	Penicillin†	Behaviour in Hugh & Leifson's medium‡	Kovacs' oxidase test‡	Growth at 37°	Genus, species or type
–	Short stout rods 0·8 × 1·0–1·5 μ and coccal forms	Grey to greyish-white, slightly opaque	++ or +	Alkaline or oxidative	N.A.	–	*Achromobacter Alcaligenes*
–	Slender rods, occasionally filamentous	Yellow to orange	–	N.A.	N.A.	–	*Flavobacterium*
+	Slender rods	Yellow-orange	–	N.A.	N.A.	–	*Flavobacterium*
+	Slender straight or curved rods	Translucent, colourless, diffusible fluorescent pigment ±	–	Oxidative	+	– or +	*Pseudomonas*
+	Slender, straight or curved rods	Translucent, colourless, occasionally pigmented	–	Alkaline or no action	+	–	*Pseudomonas*
+	Slender rods, straight or curved	Translucent, colourless	–	Fermentative, gas abundant	+	+	*Aeromonas*
+	Slender, straight or curved rods	Translucent, colourless, occasionally yellow pigmented	–	Fermentative with no gas or (very seldom) traces	+	–	*Vibrio* (marine origin)
+	Straight rods	Translucent, colourless	–	Fermentative, gas abundant	–	+	Paracolon *E. coli*

* ±, may or may not be present. † Sensitivity to 2·5 i.u.; ++, very sensitive; +, sensitive; –, insensitive; ‡ N.A., test not applied.

We need hardly stress that the scheme as it now stands is a determinative one and does not claim to express any phylogenetic relationships between groups. We are well aware of its present deficiencies and only future work will show whether these can be removed without altering its general pattern.

Table 5. *An outline of the behaviour of certain members of the family Pseudomonadaceae, and of the genus* Spirillum, *in certain determinative tests*

Genus or group	Behaviour in Hugh & Leifson's carbohydrate medium	Sensitivity* to Penicillin	Terramycin	Comp. O/129
Pseudomonas group I, green fluorescent pigment formed	Oxidative, acid only	−	−	−
Pseudomonas group II,† no fluorescent pigment	Oxidative, acid only	−	+	−
Pseudomonas group III,‡ no fluorescent pigment	Alkali formed aerobically	−	+	−
Pseudomonas group IV, no fluorescent pigment	No action aerobically or anaerobically	−	+	−
Aeromonas (Kluyver & van Niel)	Fermentative, acid with abundant gas	−	+	−
Vibrio	Fermentative, but acid only	−	+	+
'Gut-group' vibrios (Liston) and related luminous types (Spencer)	Fermentative, acid with little gas	−	+	+
Spirillum	No action	+	+	−

* +, sensitive; −, insensitive. Penicillin, 2·5 i.u.; Terramycin, 10 mg.
† *Xanthomonas* spp. give the same reactions. ‡ See text.

The work described in this paper was carried out as part of the programme of the Department of Scientific and Industrial Research.

5. References

Asai, T., Okumura, S. & Tsunoda, T. (1956). On a new genus, *Kluyvera*. *Proc. imp. Acad. Japan* **32**, 488.

Bagge, J. & Bagge, O. (1956). Vibrio anguillarum som årsag til ulcussygdom hos torsk (Gadus callarias, Linné). *Nord. VetMed.* **8**, 481.

Breed, R. S., Murray, E. G. D. & Hitchens, A. P. (1948). Bergey's *Manual of Determinative Bacteriology*, 6th ed. London: Ballière, Tindall & Cox.

Breed, R. S., Murray, E. G. D. & Smith, N. R. (1957). Bergey's *Manual of Determinative Bacteriology*, 7th ed. London: Baillière, Tindall & Cox.

Brisou, J. & Prévot, A. R. (1954). Études de systématique bactérienne. X. Révision des espèces réunies dans le genre *Achromobacter*. *Ann. Inst. Pasteur* **86**, 722.

Brown, A. D. & Weidemann, J. F. (1958). The taxonomy of psychrophilic meat-spoilage bacteria: a reassessment. *J. appl. Bact.* **21**, 11.

Buttiaux, R. & Gagnon, P. (1959). Au sujet de la classification des *Pseudomonas* et des *Achromobacter*. *Ann. Inst. Pasteur, Lille* **10**, 121.

Caselitz, F. H. (1955). Ein neues Bakterium der Gattung: *Vibrio* Müller-*Vibrio jamaicensis*. *Z. Tropenmed u. Parasit.* **6**, 52.

Caselitz, F. H. (1958). Grundsatzliche Erwagungen uber den *Vibrio jamaicensis*. *Zbl. Bakt.* (Abt. 1 Orig.) **173**, 238.

COOK, K. A. (1958). A polarly flagellated member of the Enterobacteriaceae. *J. appl. Bact.* **21**, 1.

CRAWFORD, I. P. (1954). A new fermentative pseudomonad, *Pseudomonas formicans* n.sp. *J. Bact.* **68**, 734.

EISENBERG, J. (1891) *Bakteriologische Diagnostik.* Leipzig.

EWING, W. H. & JOHNSON, J. G. (1960). The differentiation of *Aeromonas* and C27 cultures from *Enterobacteriaceae. Int. Bull. bact. Nomen. Taxon.* **10**, 223.

FRANKLAND, P. & FRANKLAND, G. C. (1894). *Microorganisms in Water.* London: Longmans Green & Co.

GALARNEAULT, T. P. & LEIFSON, E. (1956). Taxonomy of *Lophomonas* n.gen. *Canad. J. Microbiol.* **2**, 102.

GRIFFIN, P. J. (1953). The nature of bacteria pathogenic to fish. *Trans. Amer. Fish. Soc.* **83**, 241.

HALLOCK, F. A. (1959). The coccoid stage of vibrios. *Trans. Amer. micr. Soc.* **78**, 231, 273.

HAMMERL, H. (1906). Studien uber die morphologie des *Vibrio cholera asiatica. Zbl. Bakt.* (Abt. 1 Orig.) **43**, 384.

HAYES, P. R. (1960). A study of certain types of pigmented marine bacteria with particular reference to the *Flavobacterium* and *Cytophaga* species. Thesis, University of Birmingham.

HENRICI, A. T. (1925). A statistical study of the form and growth of the cholera vibrio. *J. infect. Dis.* **37**, 75.

HODGKISS, W. & SHEWAN, J. M. (1950). *Pseudomonas* infection in a plaice. *J. Path. Bact.* **62**, 655.

HOUGH, L., JONES, J. K. N. & WADMAN, W. H. (1950). Quantitative analysis of mixtures of sugars by the method of partition chromatography. Part V. Improved methods for the separation and detection of the sugars and their methylated derivatives on the paper chromatogram. *J. chem. Soc.* 1702.

HUGH, R. & LEIFSON, E. (1953). The taxonomic significance of fermentative versus oxidative metabolism of carbohydrates by various Gram-negative bacteria. *J. Bact.* **66**, 24.

HUGO, W. B. & ROGERS, M. K. (1958). The nutrition and some biochemical studies of *Vibrio cyclosites* and *Vibrio neocistes. J. appl. Bact.* **21**, 20.

INGRAM, M. & SHEWAN, J. M. (1960). Introductory reflections on the *Pseudomonas-Achromobacter* group. *J. appl. Bact.* **23**, 373.

JOHNSON, F. H. & GRAY, D. H. (1949). Nuclei and large bodies of luminous bacteria in relation to salt concentration, osmotic pressure, temperature, and urethane. *J. Bact.* **58**, 675.

KIMATA, M. & KAWAI, A. (1953). A new species of bacterium which produces large amounts of histamine on fish meats, found in spoiled fresh fish. *Mem. Res. Inst. Fd Sci. Kyoto Univ.* no. 6, pp. 1-2.

KIMATA, M., KAWAI, A. & AKAMATSU, M. (1958). Classification and identification of the bacteria having an activity which can produce a large amount of histamine. *Mem. Res. Inst. Fd Sci. Kyoto Univ.* no. 14, pp. 33-42.

KING, E. O., WARD, M. K. & RANEY, D. E. (1954). Two simple media for the demonstration of pyocyanin and fluorescin. *J. Lab. clin. Med.* **54**, 301.

KLUYVER, A. J. (1956). *Pseudomonas aureofaciens* nov. spec. and its pigments. *J. Bact.* **72**, 406.

KOVACS, N. (1956). Identification of *Pseudomonas pyocyanea* by the oxidase reaction. *Nature, Lond.*, **178**, 703.

LEIFSON, E. (1960). *Atlas of Bacterial Flagellation.* London: Academic Press.

LISTON, J. (1955). A quantitative and qualitative study of the bacterial flora of skate and lemon sole trawled in the North Sea. Thesis, University of Aberdeen.

MANUAL (1957). *Manual of Microbiological Methods.* New York: McGraw-Hill.

PATON, A. M. (1959). Enhancement of pigment production by *Psuedomonas. Nature, Lond.*, **184**, 1254.

PIVNICK, H. (1955). *Pseudomonas rubescens*, a new species from soluble oil emulsions. *J. Bact.* **70**, 1.

PRÉVOT, A.-R. (1948). *Manuel de Classification et de Détermination des Bactéries Anaérobies*, 2nd ed. Paris: Masson et Cie.

RUCKER, R. R. (1959). *Vibrio* infections among marine and fresh-water fish. *Progr. Fish Cult.* **21**, 22.

RUSH, J. M. (1947). A taxonomic study of the mesophilic *Achromobacter.* Thesis, Purdue University.

SHEWAN, J. M., HODGKISS, W. & LISTON, J. (1954). A method for the rapid differentiation of certain non-pathogenic asporogenous bacilli. *Nature, Lond.*. **173**, 208.

Skerman, V. B. D., Carey, B. J. & MacRae, I. C. (1958). The influence of oxygen on the reduction of nitrite by washed suspensions of adapted cells of *Achromobacter liquefaciens*. *Canad. J. Microbiol.* **4**, 243.

Smith, I. W. (1961). A disease of finnock due to *Vibrio anguillarum*. *J. gen. Microbiol.* **24**, 247.

Sneath, P. H. A. (1957*a*). Some thoughts on bacterial classification. *J. gen. Microbiol.* **17**, 184.

Sneath, P. H. A. (1957*b*). The application of computers to taxonomy. *J. gen. Microbiol.* **17**, 201.

Snieszko, S. F. (1953). Therapy of bacterial fish diseases. *Trans. Amer. Fish. Soc.* **83**, 313.

Spencer, R. (1955). The taxonomy of certain luminous bacteria. *J. gen. Microbiol.* **13**, 111.

Stevenson, J. P. (1959). A note on the genus *Aeromonas*. *J. gen. Microbiol.* **21**, 366.

Stewart, D. J. (1958). A study of *Pseudomonas*-like organisms isolated from freshly caught fish in the North Sea. Thesis, University of Durham.

Wells, N. A. & ZoBell, C. E. (1934). *Achromobacter ichthyodermis* n.sp., the etiological agent of an infectious dermatitis of certain marine fish. *Proc. nat. Acad. Sci., Wash.* **20**, 123.

Williams, M. A. (1960). Flagellation in six species of *Spirillum*—A correction. *Int. Bull. bact. Nomen. Taxon.* **10**, 193.

Youatt, J. B. (1954). Denitrification of nitrite by species of *Achromobacter*. *Nature, Lond.*, **173**, 826.

ZoBell, C. E. & Wells, N. A. (1934). An infectious dermatitis of certain marine fishes. *J. infect. Dis.* **55**, 299.

ZoBell, C. E. (1946). *Marine Microbiology*. Waltham, Mass.: Chronica Botanica Co.

Explanation of Plate

(a) *Serratia margarita* (Stevenson, 1959), culture SL 10a. Electron micrograph, Au-Pd shadowed, of cells from a 24 hr agar slope culture at 20°, ×13,300.

(b) Culture NCTC 9960 (Cook, 1958). Cells from a 24 hr agar slope culture at 20°, modified Casares-Gil flagella stain, ×1,330.

(c) 'Gut-group' vibrios (Liston, 1955), showing curved rods and the characteristic round bodies. Phase contrast, ×800.

(d) 'Gut-group' vibrios (Liston, 1955), showing rounded bodies in culture from sea water agar. Phase contrast, ×800.

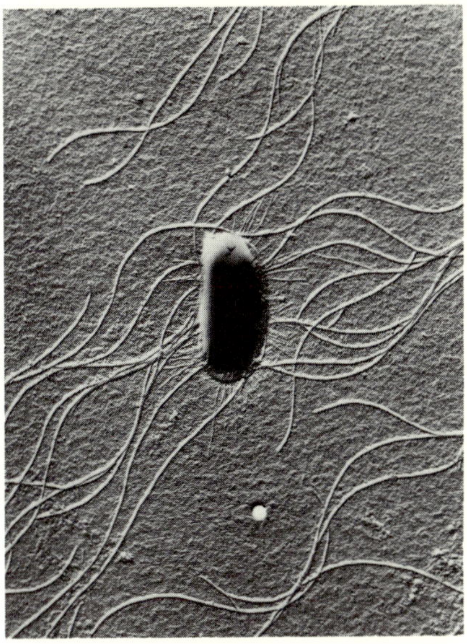

(*a*)

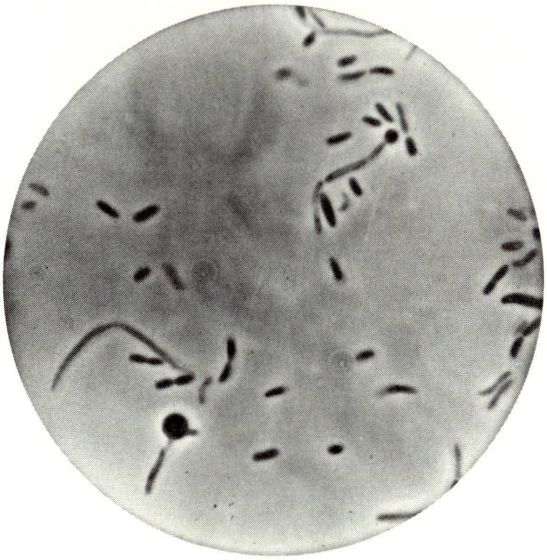

(*c*)

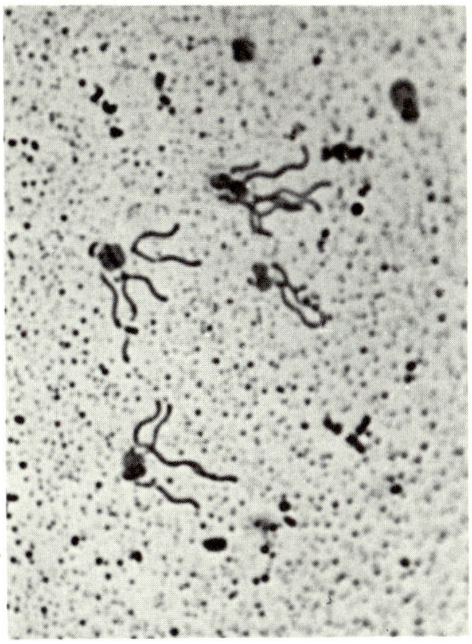

(*b*)

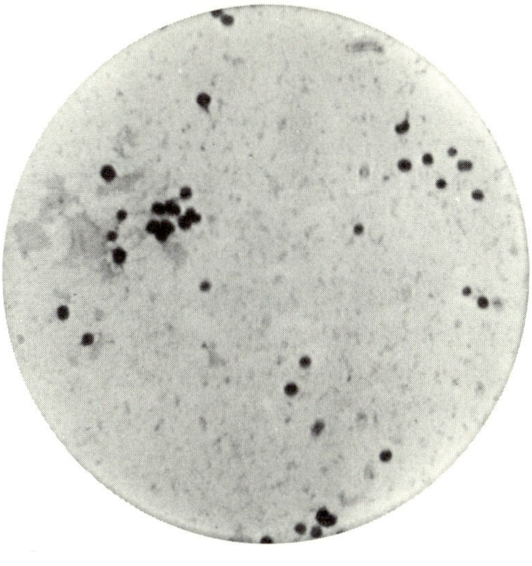

(*d*)

Shewan, J. M., Hobbs, G. & Hodgkiss, W. — A Determinative Scheme for the Identification of Certain Genera of Gram-Negative Bacteria, with Special Reference to the Pseudomonadaceae. Plate 1

16

Reprinted from *J. Bacteriol.*, **82**, 1–14 (1961)

TAXONOMIC RELATIONSHIPS AMONG THE PSEUDOMONADS[1]

R. R. COLWELL AND J. LISTON

College of Fisheries, University of Washington, Seattle, Washington

Received for publication October 11, 1960

ABSTRACT

COLWELL, R. R. (University of Washington, Seattle), AND J. LISTON. Taxonomic relationships among the pseudomonads. J. Bacteriol. **82:** 1–14. 1961.—An electronic computer technique, utilizing the Adansonian principle that every feature should have equal weight, was applied in an effort to derive a taxonomy of the Pseudomonas-Achromobacter group of gram-negative, asporogenous, rodlike bacteria. The validity of the general method was tested by an analysis of 40 well defined strains, principally derived from type culture collections and representative of different genera and families of the *Pseudomonadales* and the *Eubacteriales*. The analysis clearly separated groups which are recognized to be taxonomically distinct.

Aerogenic Aeromonas were most similar to the *Enterobacteriaceae* and taxonomically distinct from the anaerogenic *Aeromonas formicans* which was more similar to the Pseudomonas group. Oxidative *Vibrio* species grouped with Pseudomonas and probably should be renamed Pseudomonas. Fermentative *Vibrio* species showed affinities with *A. formicans*.

Fifty-four *Pseudomonas* strains out of 58 tested in an analysis of 80 bacterial cultures, clustered into four large groups: group 1, a marine group within which appeared a psychrophilic, nonproteolytic species subgroup related to *Vibrio beijerinckii* and a subgeneric subgroup which included *Pseudomonas elongata*; group 2, a subgeneric mesophilic group including *Vibrio percolans* and associated with *Pseudomonas fragi*; group 3, a fluorescent pigment-producing group, including three subgroups, a psychrophilic aeruginosa-like species group, a mesophilic *Pseudomonas aeruginosa* species group, and a heterogeneous subgeneric group containing *Vibrio cuneatus* and related to *Pseudomonas fluorescens* and *Pseudomonas pavonacea*; group 4, a subgeneric nonpigment-producing, mesophilic group, which included *Vibrio tyrogenus*, *Pseudomonas ovalis*, and *Pseudomonas denitrificans*.

The sharing of common characteristics by bacteria has provided the basis for most bacterial classifications. In practice, arbitrarily selected key characters are used to define taxa, more weight being given to certain properties of bacteria and perhaps less to others of equal value. Thus, properties of prime importance in defining one particular group of organisms may be of little significance in defining another. This method is well adapted for use in determinative classifications and has yielded many good bacterial catalogs, but few taxonomies. This point has been discussed by Sneath (1957*a*, *b*) in his presentation of the Adansonian method of classification as applied to bacteria. In this method, all characteristics are given equal weight and division into taxa is based on the correlated features. The quantitative approach to problems of taxonomy has also been taken by Sokol and Michener (1957) in relation to the classification of bees of the family *Megachilidae*, using a matrix of species-×-species correlation coefficients in a weighted variable group method. Since all features of the organisms to be classified must be considered, many data have to be handled. This is done using an electronic computer, and Sneath (1957*b*) has established a simple mathematical procedure, using a numerical index S for similarity of each pair of organisms examined. The derivation of S is shown in Fig. 1. S values, when tabulated, indicate a clustering of organisms into groups according to their mutual similarities. Intragroup and intergroup S values may be calculated and used to define taxonomic relationships among the groups. In a survey of a wide taxonomic range of bacteria, Sneath and Cowan (1958) observed such groupings as: gram-positive organisms; predominantly gram-negative organisms; *Corynebacterium diphtheriae*; acid-fast bacilli; and spirilla. In a more narrow application of the Adansonian classification, Hill (1959) subjected

[1] Contribution no. 89, College of Fisheries, University of Washington, Seattle.

$$S = \frac{N_s}{N_s + N_d}$$

Where N_s = number of positive features shared; N_d = number of features positive in one strain and negative in the other.

FIG. 1. *Formula showing the derivation of the S value.*

staphylococci to analysis and demonstrated two major taxonomic branches within the group: *Staphylococcus* and *Micrococcus.*

In the present study, summarized, in part, elsewhere (Liston and Colwell, 1960), the Sneath technique has been applied to the taxonomy of pseudomonad-like organisms isolated from marine and other sources. The present classification of the pseudomonads is satisfactory neither from the determinative nor the taxonomic point of view. Indeed, much confusion seems to exist as to the precise definition of the genus *Pseudomonas.* Recent publications by Buttiaux and Gagnon (1958/1959) and Rhodes (1959) have underscored this point, one which has been recognized for a considerable time. Moreover, the taxonomic position of groups which are probably similar to the genus *Pseudomonas,* such as *Aeromonas,* fermentative pseudomonads and even the long established genus *Vibrio,* is not firmly fixed. Psychrophilic pseudomonads, which are of major importance in the marine field (Liston, 1956), and in the field of food bacteriology (Ayers, 1960), have proved particularly difficult to fit into the present schema of pseudomonad classification. The essentially objective method of Sneath appeared to provide a convenient procedure for a complete reassessment of the taxonomic position of many of these groups.

MATERIALS AND METHODS

The organisms used in this study, together with their sources and presumed species name, are shown in Tables 1 and 2. Each strain was obtained in pure culture. Organisms from marine sources were maintained on the medium described by MacLeod, Onofrey, and Norris (1954) but other organisms were maintained on a medium consisting of nutrient agar (Difco) plus 0.5% yeast extract. Each pure culture was subjected to examinations and tests as outlined in Table 3. Morphology was determined by observation of 24- to 48-hr cultures, using the phase contrast microscope and the Gram stain. Colonial characteristics were determined from 3-day cultures grown on MacLeod's solid medium. Growth characteristics in liquid media were determined from cultures grown in 1% peptone in aged seawater (ZoBell, 1946), except in a few cases where growth was not obtained in this medium and trypticase soy broth (Difco) was subsequently used. Motility was determined by direct observation of a 24- to 48-hr liquid culture, the medium being 0.5% NaCl in 1% peptone water or 1% peptone in aged seawater, depending on the organism and its requirements. The methods of Casares-Gil (*Manual of Microbiological Methods,* 1957), Leifson (1951), and Novel (1939) for staining flagella were tested. The Casares-Gil

TABLE 1. *American Type Culture Collection organisms used*

Organism no. (ATCC)	Genus and species name	Organism no. (ATCC)	Genus and species name
8689	*Pseudomonas aeruginosa*	11471	*Vibrio* sp.
8707	*P. aeruginosa*	13137	*Aeromonas formicans*
10145	*P. aeruginosa*	11163	*Aeromonas punctata*
7700	*P. aeruginosa*	4335	*Achromobacter parvulus*
11251	*P. fluorescens*	7744	*Photobacterium fischeri*
10144	*P. elongata*	11040	*Photobacterium phosphoreum*
8209	*P. ovalis*	11947	*Flavobacterium aquatile*
12133	*P. denitrificans*	9682	*Corynebacterium poinsettiae*
4973	*P. fragi*	7469	*Lactobacillus casei*
951	*P. pavonacea*	8750	*Alcaligenes faecalis*
6972	*Vibrio cuneatus*	211	*Aerobacter aerogenes*
7085	*Vibrio tyrogenus*	4157	*Escherichia coli*
7708	*Vibrio metschnikovii*	4669	*Proteus vulgaris*
8461	*Vibrio* sp.	398	*Micrococcus luteus*

TABLE 2. *Organisms, isolated in our laboratory or from sources indicated, used*

Organism no.	Name	Source
WC-8	*Micrococcus* sp.	*Gadus macrocephalus* skin
WC-12	Brown-pigmented *Pseudomonas* sp.	*Sebastodes melanops* skin
WC-15	*Pseudomonas* sp.	
WC-20	*Flavobacterium* sp.	*Parophrys vetulus* skin
WC-23	*Bacillus* sp.	*Gadus macrocephalus* flank lesion
WC-44	*Pseudomonas* sp.	*Ophiodon elongatus* skin
WC-47	Green-pigmented *Pseudomonas* sp.	*O. elongatus* gill
WC-49	*Pseudomonas* sp.	
WC-53	Green-pigmented *Pseudomonas* sp.	*Pleuronichthys* sp. skin
PS-02-i	*Achromobacter* sp.	*Sebastodes caurinus* gill
PS-05-ii	*Vibrio* sp.	*O. elongatus* gill
PS-08-iii-A	*Flavobacterium* sp.	*S. melanops* gill
PS-016-v	*Aerobacter* sp.	*Anoplopoma fimbria* skin
PS-019-i	*Achromobacter* sp.	*G. macrocephalus* skin
PS-019-iii	*Vibrio* sp.	
PS-023-iii	*Vibrio* sp.	*Damalichthys vacca* gill
PS-203	Green-pigmented *Pseudomonas* sp.	*Squalus suckleyi* gut
PS-213	Agar-digesting *Pseudomonas* sp.	*G. macrocephalus* gill
PS-216	*Vibrio* sp.	
PS-218	Green-pigmented *Pseudomonas* sp.	*S. suckleyi* gut
PS-230	Agar-digesting *Pseudomonas* sp.	*P. vetulus* skin
PS-235	Green-pigmented *Pseudomonas* sp.	*P. vetulus* gill
PS-236	*Pseudomonas* sp.	*P. vetulus* gut
PS-241	Green-pigmented *Pseudomonas*	
PS-242	*Pseudomonas* sp.	*Merluccius productus* skin
PS-247	*Pseudomonas* sp.	*M. productus* gill
PS-302	*Achromobacter* sp.	*G. macrocephalus* skin
PS-305	*Achromobacter* sp.	*G. macrocephalus* gill
PS-314	Green-pigmented *Pseudomonas* sp. (agar digesting)	*P. vetulus* gill
PS-316	*Pseudomonas* sp.	
PS-325	*Achromobacter*	*G. macrocephalus* skin
E-67C	*Aerobacter* sp.	*Aprion virescens* gut
E-70A	*Aerobacter* sp.	*Epinephalus merra* gut
E-71a, E-71b, E-71d-1, E-71d-2, E-71e	*Pseudomonas* sp.	*E. merra* skin
E-74b	*Pseudomonas* sp.	*Acanthurus* sp. skin
E-75c	*Pseudomonas* sp.	*Acanthurus* sp. gill
E-76a-2, E-76a-3	*Pseudomonas* sp.	*Holothuria atra* skin
E-76b	*Bacillus* sp.	
E-81, E-82	Green-pigmented *Pseudomonas* sp.	*Lambis lambis* foot
E-86a	*Alcaligenes* sp.	*Coenobita perlatus* carapace
E-88	Green-pigmented *Pseudomonas* sp.	*C. perlatus* gill
E-89a, E-89b	Green-pigmented *Pseudomonas* sp.	Algae "leaf"
E-91a-1, E-91a-2, E-91c-1, E-91-c-2	*Pseudomonas* sp.	*Caranx* sp. skin
E-92	*Vibrio* sp.	*Caranx* sp. gill
E-93A, E-93B	*Pseudomonas* sp.	*Siganidae* sp. gut
G-1	Green-pigmented *Pseudomonas* sp.	South African Hake (isolated by D. Georgala, University of Capetown, South Africa)
G-24	*Pseudomonas* sp.	
S-21	*Pseudomonas* sp.	Fillet of Sole (isolated by L. Farber, University of California Medical Center, San Francisco)
S-27, S-34	Green-pigmented *Pseudomonas* sp.	

TABLE 2—*Continued*

Organism no.	Name	Source
HC-11, HC19	*Pseudomonas* sp.	*Crassostrea gigas* body fluid (obtained from A. K. Sparks, College of Fisheries, University of Washington)
HC-15	Green-pigmented *Pseudomonas* sp.	
WB-28	Yellow *Pseudomonas* sp.	
JL-A	*Achromobacter* sp.	Torry Research Station, Aberdeen, Scotland
JL-224, JL-226	Brown-pigmented *Pseudomonas* sp.	
R-2	Irregular R-2 (*E. coli* type)	Frozen seafood sample
RL-1	*Aeromonas* sp.	
UW-2	*Escherichia coli* strain I (K-12)	Department of Microbiology, School of Medicine, University of Washington
UW-4	*Bacillus subtilis*	
UW-6	*Corynebacterium hoffmanii*	
FWS-2	*Aeromonas hydrophila*	R. Rucker, Fish and Wildlife Service, Seattle

Key: WC = Washington Coast; PS = Puget Sound; E = Eniwetok Atoll; HC = Hood Canal, Washington; WB = Willapa Bay, Washington.

method was found to be most satisfactory and was adopted as the routine procedure.

Growth characteristics at 0, 25, and 37 C were measured in liquid nutrient media. The requirement for seawater was tested by streak plating on agar containing seawater and on a duplicate distilled water-nutrient agar plate. Sensitivity to antibacterial agents was determined by disc tests (Difco penicillin discs) or "ditch" tests, in the case of 0/129 pteridin compound (Shewan, Hodgkiss, and Liston, 1954), on MacLeod agar plates. Except for the following, all biochemical tests were carried out as described in the *Manual of Microbiological Methods* (1957): the Hugh and Leifson test for anaerobic fermentation of sugar was performed according to the procedure described by these authors (Hugh and Leifson, 1953), the oxidase test as described by Gaby and Free (1958), the Kovacs oxidase test by Kovacs (1956), and the trimethylamine oxide reduction test by Wood and Baird (1943). All cultures were tested immediately after isolation and final purification. All the American Type Culture Collection (ATCC) cultures were retested after 6 months, as were most of the other organisms used in this study. Incubation of inoculated test media, except where otherwise stated, was at 25 C, aerobically, for 28 days. Detailed recordings were made after 1, 2, 7, 14, and 28 days.

The coded characteristics listed in Table 3, representing 134 "states" or characteristics, were punched onto IBM cards. The presence of a characteristic was scored 1 and the absence of a characteristic, 0.

The equation for calculating the *S* value was programmed for the IBM 650 magnetic drum computer. From the machine calculated data, the table of *S* values was printed. The cards were machine sorted according to decreasing *S* value. Using these data, the groups were determined manually. The calculations of the inter- and intragroup mean *S* values were machine programmed and calculated for the groups derived by the preceding step. (The detailed program is available and may be obtained from the authors.)

RESULTS

S values for 40 well defined strains, principally derived from type culture collections, grouped by the machine sorting procedure, are shown in Fig. 2. Representatives of the genus *Pseudomonas* constitute a block lying at the left of the diagram and clearly separated from the other organisms. Strains of *Aerobacter* and of *Escherichia coli* cluster with *Aeromonas* strains in the center of the diagram in what possibly represents the *Enterobacteriaceae*. Corresponding closely with the *Enterobacteriaceae* are three strains representative of the genus *Bacillus*. One *Vibrio*, one *Aeromonas*, and various unrelated species, fail to group and lie in an undifferentiated series of columns to the right of the graph.

According to the criteria proposed by Haynes

TABLE 3. *Coded characteristics used in the IBM 650 computer program*

Morphology:
Rods
Curved rods
Ovals (spheres)
Filaments
Singles
Pairs
Chains
Short (0.2–0.6 μ)
Medium (0.6–1.2 μ)
Slender (0.2–0.6 μ)
Stout (0.6–1.0 μ)
Rounded end
Motile
Polar flagellation
Gram-negative
Gram-variable
Culture:
Small colony (< 2 mm)
Medium colony (2–5 mm)
Entire edge
Convex
White
Off-white
Gray
Green pigment (diffusible)
Opaque
Translucent
Heavy turbidity
Moderate turbidity
Slight turbidity
Even turbidity
Granular sediment
Ring
Pellicle
Fluorescent
Physiology:
Growth at 0 C
Growth at 25 C (RT)
Growth at 37 C
Seawater required
Catalase positive
Penicillinase present
0/129 insensitive

Biochemistry:
Agar digested
Gelatin liquefied
Litmus milk peptonized
Litmus milk surface peptonized
Litmus milk acid
Litmus milk alkaline
Litmus milk reduced
Litmus milk acid → alkaline
Litmus milk alkaline → acid
Ammonia produced in peptone water
Voges-Proskauer positive
Methyl red positive
Indole positive
Growth in Koser's citrate
Urease production on Christiansen's urea agar slants
Cytochrome oxidase positive (Gaby)
Oxidase positive (Kovacs)
Glucose positive (Hugh and Leifson, 1953)
Glucose acid
Glucose gas
Maltose acid
Maltose gas
Sucrose acid
Sucrose gas
Lactose acid
Lactose gas
Galactose acid
Galactose gas
Mannitol acid
Mannitol gas
Starch hydrolyzed
Gluconate oxidized
Nitrates reduced to nitrites
Nitrites reduced
Trimethylamine oxide reduced to trimethylamine
Hydrogen sulfide produced from cysteine

and Burkholder in *Bergey's Manual of Determinative Bacteriology* (Breed, Murray, and Smith, 1957), the genus *Pseudomonas* may be considered complete in this diagram at *Vibrio tyrogenus*. This indicates a possible lower generic level of 60% which is close to the level of similarity between the *E. coli-A. aerogenes* group and the genus *Bacillus*. The four strains of *Pseudomonas aeruginosa* show a lowest limit of similarity between 75 and 80%. The mean *S* value of this small species group was actually 85%.

Taking into account the intergroup *S* values, it is possible to rearrange the information according to the format of Fig. 3. The rearrangement indicates strongly the clustering of the pseudomonads and the separate clustering of *Aerobacter*,

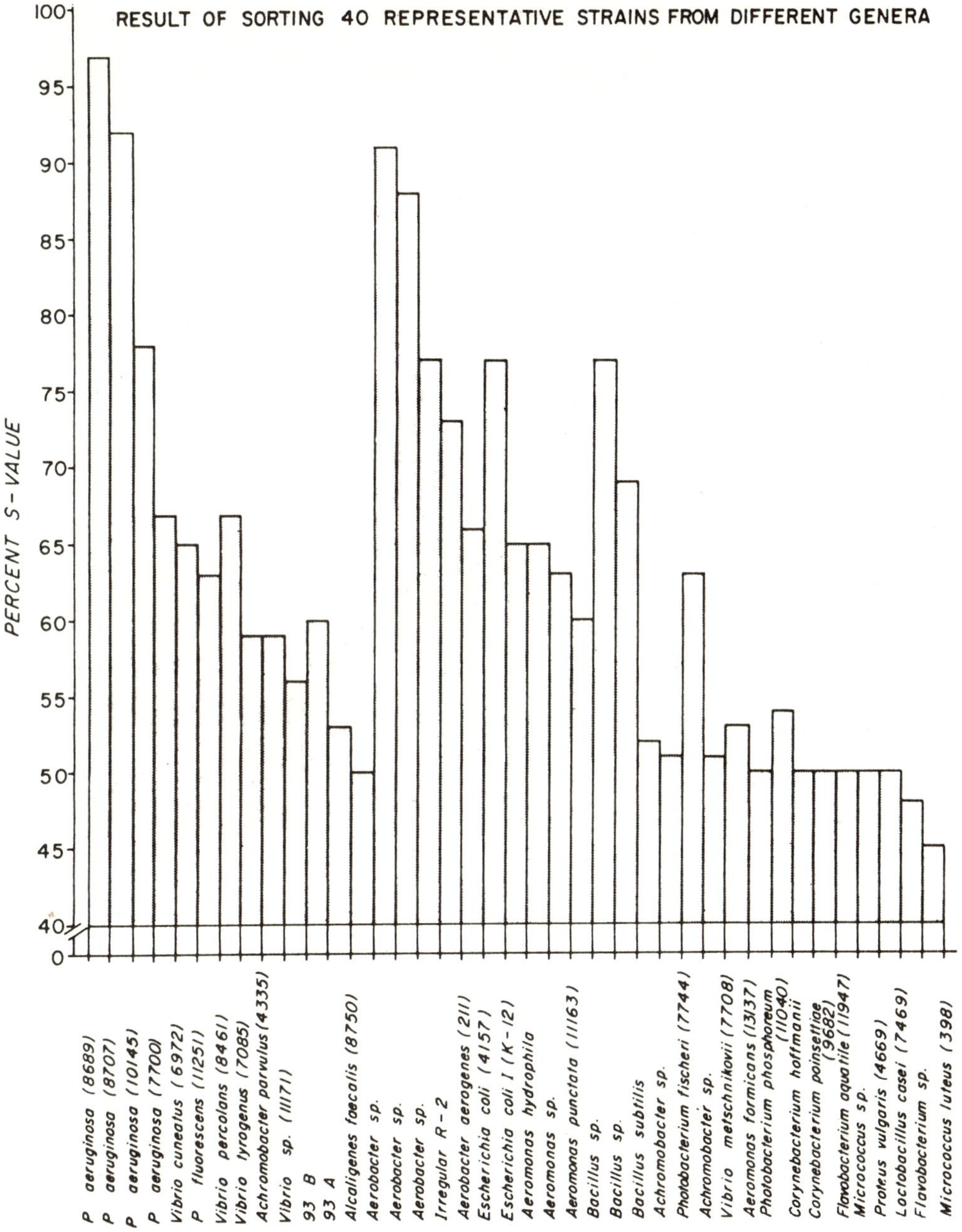

FIG. 2. *Result of the machine sorting procedure*

Escherichia, and *Aeromonas* strains. This illustrates perhaps more clearly the levels at which the various subgroups and groups fuse. *Vibrio percolans* (8461), *V. tyrogenus* (7085), and *V. cuneatus* (6972) fall within the pseudomonas group. *Vibrio* sp. (11171) lies outside the pseudomonas group next to *Achromobacter parvulus* (4335). *V. metschnikovii* (7708) also falls outside the pseudomonas group and lies next to *Aeromonas formicans* (13137).

The intragroup *S* value for all 40 members of the *Eubacteriales* is at approximately 38 to 39%.

The results obtained by sorting 80 pseudomonads, 20 of which were ATCC strains (including two strains of the type species *Pseudomonas aeruginosa* (Schroeter-Migula) and the remainder

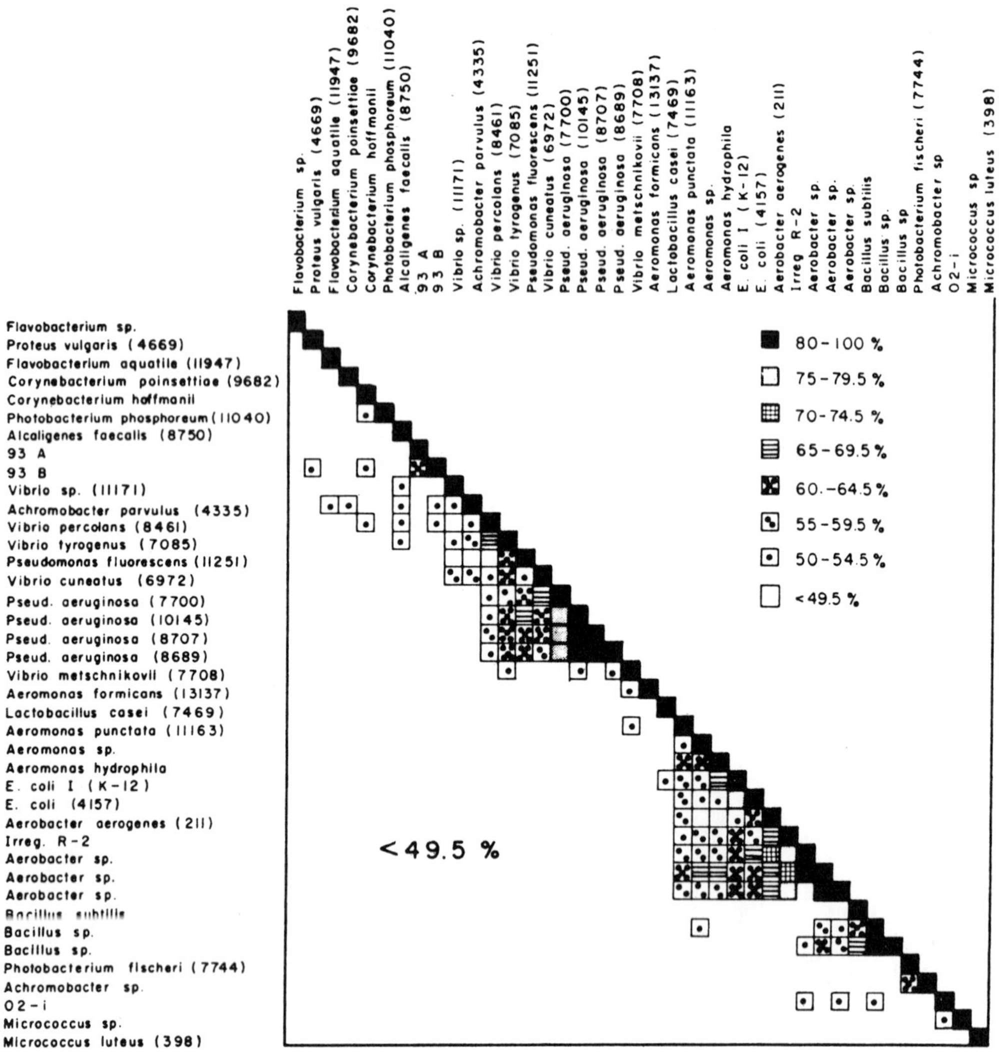

FIG. 3. *Full table of S values obtained from the analysis of 40 representative strains from different genera*

fresh isolates, are presented in Fig. 4. According to the 60% lower genus level, the genus *Pseudomonas* might be considered to terminate on the diagram with *Pseudomonas pavonacea* (951). This is in general accord with the characteristics of the organisms excluded from this group, i.e. Alcaligenes, Flavobacterium, etc. The picture presented by these data is of a somewhat homogeneous group. However, a few subgeneric groups other than simple pairs are apparent. There is a clustering of a number of pigmented strains around the *P. aeruginosa* type cultures. Another group with some pigmented strains occurs at the extreme left of the diagram and a group of seawater-requiring strains toward the right. From this sorting arrangement, with the exception of *P. aeruginosa*, there is no apparent clustering of freshly isolated organisms around defined ATCC strains. The strain groups fuse together in a very limited range of $S = 70\%$.

To delineate the multidimensional relationships represented, the full S-value table was rearranged as shown in Fig. 5, on the basis of mean S values between groups of strains. Four groupings of organisms are immediately apparent from this arrangement: PS247 to E71d$_2$ (group 1); *P. fragi* (4973) to E71a (group 2); G24 to PS236 (group 3); E91C$_2$ to *P. denitrificans*

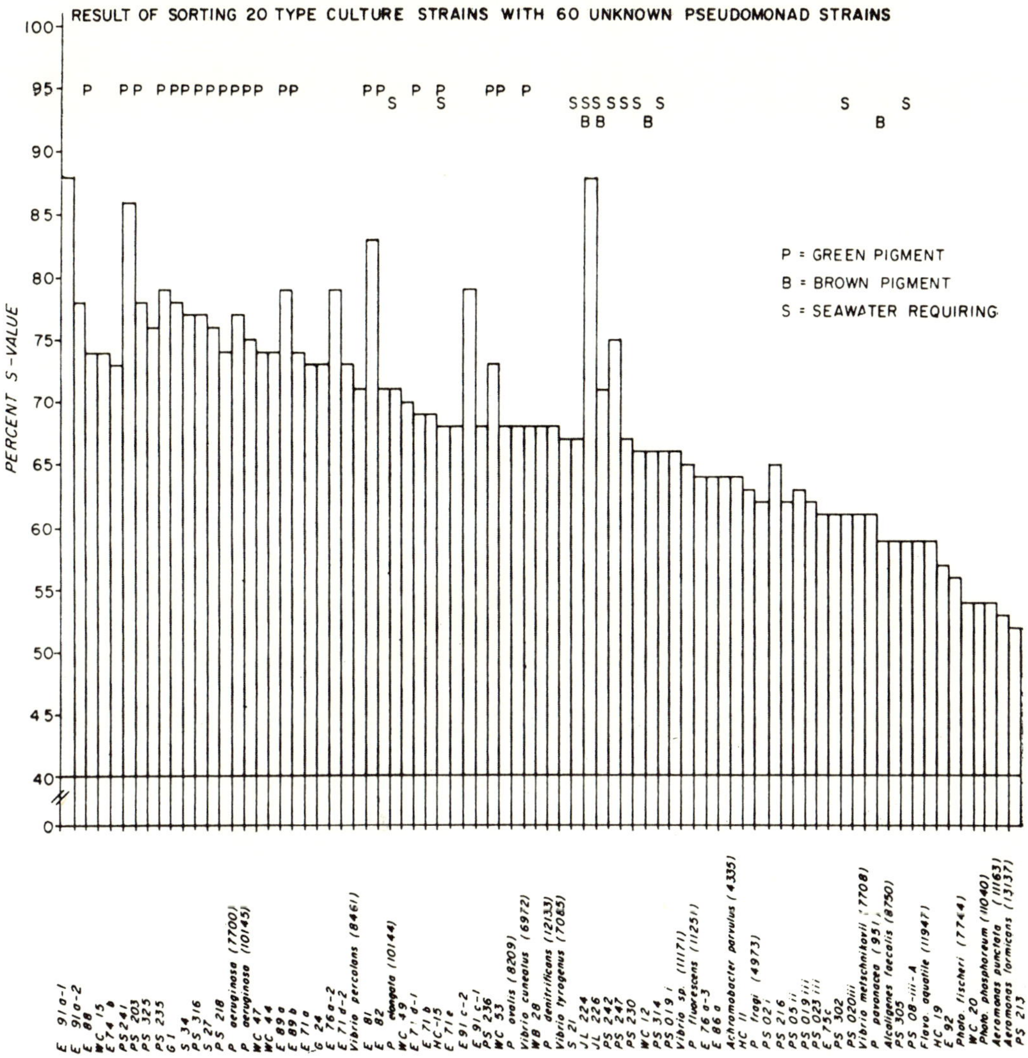

FIG. 4. *Result of the machine sorting procedure for the pseudomonads*

(12133) (group 4). Group 3 may be subdivided into three subgroups: G24 to PS218 (group 3a); *P. aeruginosa* (7700) to E89b (group 3b); WC12 to PS236 (group 3c). The intra- and intergroup *S* values are shown in Table 4.

Group 1, which has an intragroup value of $S = 61\%$ is composed of organisms which require seawater for growth or whose growth is enhanced by seawater. Although the group is somewhat heterogeneous in its properties, certain characteristics are generally present among its members. Morphologically they are usually small, slender, straight, motile rods with occasional coccal bacilli and pleomorphic forms and may produce brown soluble or yellowish nonsoluble pigments that are not usually fluorescent. They are mainly psychrophilic but a few forms may be eurythermic (able to grow at 0 C and 37 C). Some members of the group are able to digest agar. Most strains do not reduce nitrate but do liquefy gelatin. Reaction in litmus milk may be no change, acid, or reduction; with a few strains additionally causing peptonization. Ammonia is produced by all strains from peptone water. Trimethylamine oxide is reduced by most strains but H_2S is not produced and urease is usually not present. Indole is not produced. Catalase is always present and both the Kovacs oxidase test and the Gaby

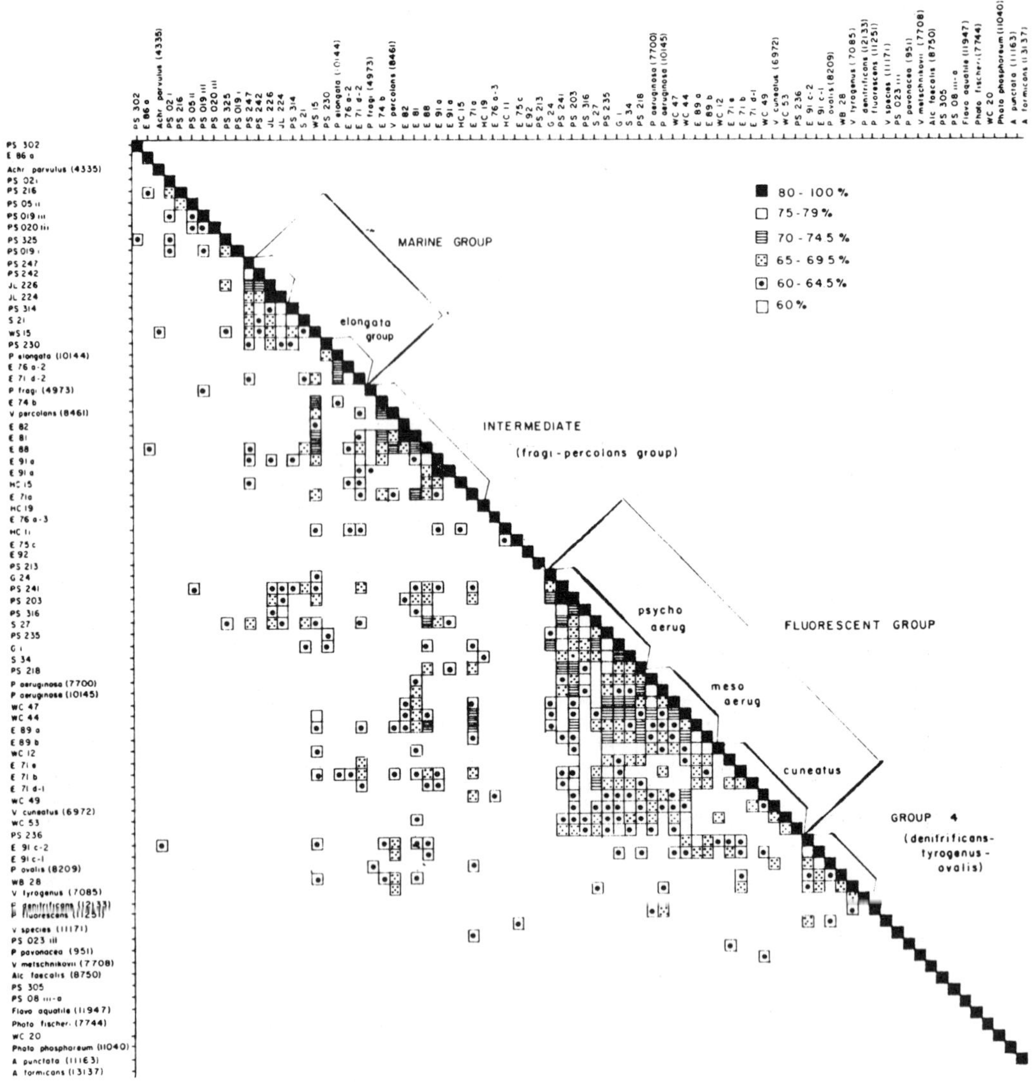

FIG. 5. *Full table of S values obtained from the analysis of 80 bacterial cultures*

cytochrome oxidase test are positive. Gluconate is not oxidized to the 2-keto gluconate and there is usually no acid produced in the Hugh and Leifson test (using glucose) or in peptone water sugars, except occasionally transiently. Starch may be hydrolyzed, particularly by agar-digesting strains. The Voges-Proskauer and methyl red tests are consistently negative and citrate is usually not used as sole carbon source. All members of the group are resistant to penicillin.

There are two subgroups within the marine group, 1a with an intragroup S value of 73% and 1b, with $S = 65\%$. Organisms of 1a liquefy gelatin, reduce trimethylamine oxide, and may produce brownish diffusible pigment, whereas members of 1b are nonpigmented, do not liquefy gelatin, usually fail to reduce trimethylamine oxide, and may digest agar.

Group 2, which has an intragroup S value of 64%, is composed of slender, straight, or, occasionally, slightly curved rods and, rarely, coccal bacillus forms. They may produce a yellow or yellow-green diffusible fluorescent pigment and are mainly mesophilic although a few strains are eurythermic. Nitrate and nitrite

TABLE 4. *Intra- and intergroup S values*

	Marine group *Vibrio beijerinckii* subgroup 1a	Marine group *Pseudomonas elongata* subgroup 1b	Marine group 1	Group 2	Psychrophilic aeruginosa subgroup 3a	Mesophilic aeruginosa subgroup 3b	*Vibrio cuneatus* subgroup 3c	Group 3	Group 4
1a	73								
1b	58	65							
1	67	63	61						
2	52	55	55	64					
3a	53	54	55	55	69				
3b	47	52	51	56	65	69			
3c	49	51	51	54	60	61	57		
3	50	53	52	55	66	66	61	63	
4	50	49	50	57	51	56	56	54	63

are always reduced, gelatin is usually liquefied, and litmus milk is usually acidified and may be peptonized or less frequently reduced. Ammonia is produced from peptone by all strains. Trimethylamine oxide is not reduced, H_2S may be produced, and urease may be present. Indole is not produced. Catalase is always present and the Kovacs test and the Gaby test are always positive. About half the strains tested oxidize gluconate to the 2-keto form. Acid is usually produced oxidatively from glucose and occasionally slight acid may be produced in galactose but not in the other sugars. Starch is not hydrolyzed. Voges-Proskauer and methyl red tests are negative and citrate may or may not be used as a sole carbon source. All members of the group are insensitive to penicillin.

Group 3, which has an intragroup *S* value of 63%, is composed of slender straight rods with occasional curved forms, most of which produce diffuse fluorescent pigments.

Subgroup 3a which has an intragroup *S* value of 69%, is mainly psychrophilic. Nitrate may not be reduced or (usually) reduced to nitrite or (rarely) completely to gaseous products; all strains liquefy gelatin and all bring about acid peptonization of litmus milk (one exception produced slight alkaline conditions or partial reduction of the milk). Ammonia is produced by all strains from peptone. Trimethylamine oxide may or may not be reduced, H_2S is not produced, but urease is usually present. Indole is not produced. Catalase is always present and Kovacs and Gaby's tests are always positive. Gluconate is oxidized to the 2-keto form by all strains. Acid is produced oxidatively in glucose and acid may be produced in galactose, sucrose, and mannitol, but not in lactose or maltose. Starch is not hydrolyzed. Methyl red and Voges-Proskauer tests are negative but citrate is used by all strains as sole carbon source. All members of the group are insensitive to penicillin.

Subgroup 3b, which has an intragroup *S* value of 69%, is mesophilic or eurythermic. All members reduce nitrate and nitrite, liquefy gelatin, and bring about an acid peptonization of litmus milk. Ammonia is produced from peptone. Trimethylamine oxide is rarely reduced and H_2S rarely produced, but most strains possess urease. Indole is not produced. Catalase is present and the Kovacs and Gaby tests are positive. Gluconate is consistently oxidized to the 2-keto form. Acid may be produced oxidatively from glucose and galactose but not from other sugars tested. Starch is not hydrolyzed. Methyl red and Voges-Proskauer tests are negative but citrate is utilized as sole carbon source by most strains. All members of the group are insensitive to penicillin.

Subgroup 3c is somewhat heterogeneous having an intragroup *S* value of 57%. It consists of strains which are eurythermic, mesophilic, and, in two strains, stenothermic (i.e., do not grow at 0 C or at 37 C but grow well at about 25 C). Nitrate is either not reduced or is reduced through nitrite to gaseous products, gelatin may or may not be liquefied, and litmus milk is peptonized (with one exception) usually with preliminary acidification. Ammonia is produced from peptone. Trimethylamine oxide is not reduced, H_2S may or may not be produced, and urease may or may not be present. Indole is not produced. Catalase is present and Kovacs and Gaby tests are always positive. Gluconate may or may not be oxidized to the 2-keto form. Most strains produce acid oxidatively in glucose and a few produce transient slight acidity in galactose and maltose. Starch is not hydrolyzed. Methyl red and Voges-Proskauer tests are negative but citrate is commonly utilized as sole carbon source. All members of the group are insensitive to penicillin.

Group 4, which has an intragroup value of $S = 63\%$, consists of organisms which are short, slender, straight or slightly curved, motile rods which do not produce fluorescent pigments, although one strain has a yellow colony. The group is principally mesophilic but one strain is eurythermic. All strains, except *P. ovalis*, reduce nitrate and nitrite. All liquefy gelatin and produce alkaline conditions in litmus milk. Ammonia is produced from peptone water. Trimethylamine oxide is never reduced but a few strains produce some H_2S. Urease was present in half the strains tested but indole was not produced. Catalase is always present and the Kovacs and Gaby tests are always positive. Gluconate is not oxidized to the 2-keto form but acid is often produced oxidatively from glucose. Some strains produce acid slowly in galactose but not in other sugars. Starch is usually not hydrolyzed. Voges-Proskauer and methyl red tests are negative and citrate is usually not utilized as sole carbon source. All members of the group are insensitive to penicillin.

DISCUSSION

The computer technique, with the underlying Adansonian principle that every feature should have equal weight, was adopted in an effort to derive a systematic arrangement of the general group of gram-negative, asporogenous rodlike organisms of the Pseudomonas-Achromobacter-Flavobacterium group. The results of the analysis of 40 well defined strains of bacteria from a number of genera provide a measure of the validity of the computer method. Groups generally recognized to be taxonomically distinct form separate, well defined clusters.

A possible genus level may be set at $S = 60\%$ but this may represent a family level, since it appears to apply to the *Enterobacteriaceae*. Cheeseman and Berridge (1959), in their electronic computer analysis of the chromatographic patterns of 90 heterofermentative lactobacilli, did not obtain sufficient clustering to justify division into species, but it is interesting to note that the "generic" level apparent from their results was about $S = 60\%$. The possibility of a lower species level of $S = 70$ to 75% is represented by the results for the four strains of *P. aeruginosa*. However, more work will obviously be required to determine the validity of these potentially useful numerical indexes of taxonomic level.

Three points of particular interest arise from the data shown in Fig. 3. First, Pseudomonas is clearly distinct from the other defined genera. Secondly, *Aeromonas* strains, other than *A. formicans*, are more closely associated with the *Enterobacteriaceae* than with *Pseudomonas* species. This is in accord with the findings of Stanier (1953) who confirmed that *A. hydrophila* carried out a butanediol fermentation similar to that of *Aerobacter aerogenes*. Our results would suggest the grouping of the aerogenic butanediol fermentation types of Aeromonas with the *Enterobacteriaceae*. Thirdly, the organisms listed as Vibrio are dispersed among the various species of *Pseudomonas*. This is apparent in Fig. 3 and more clearly in Fig. 5. We have found, as did Hugh and Leifson (1953), that *V. percolans* and *V. cuneatus* have all the characteristics of an oxidative pseudomonad but none that could define them as *Vibrio*. Indeed, *V. cuneatus* produces a green soluble pigment. *V. tyrogenus* has similarly fallen into the oxidative pseudomonad group. *Vibrio* sp. (11171) appears to be more closely related to *Pseudomonas* than to *Vibrio*. *V. metschnikovii*, which forms a small subgroup with *A. formicans* (Fig. 3), although at a rather low level of similarity, shows some fermentative tendency but little curvature of the cells. Thus, it appears that the characterization of these named cultures as Vibrio is incorrect or that the definition of the genus *Vibrio* requires emendation.

Asai, Okumura, and Tsunoda (1956) have proposed the genus name *Kluyvera* for polarly flagellated organisms with a coliform fermentation. While we realize that too few strains were used in our analysis for definite conclusions to be drawn, our results do tend to indicate that the genus name *Vibrio* might usefully be applied to a group of organisms similar to *A. formicans* and *V. metschnikovii*, which lie taxonomically intermediate between the *Pseudomonadaceae* and the *Enterobacteriaceae*. The descriptions of *Vibrio comma* in the literature (in particular, Hugh and Leifson (1953) and Collier, Campbell, and Fitzgerald (1950)) indicate that this organism

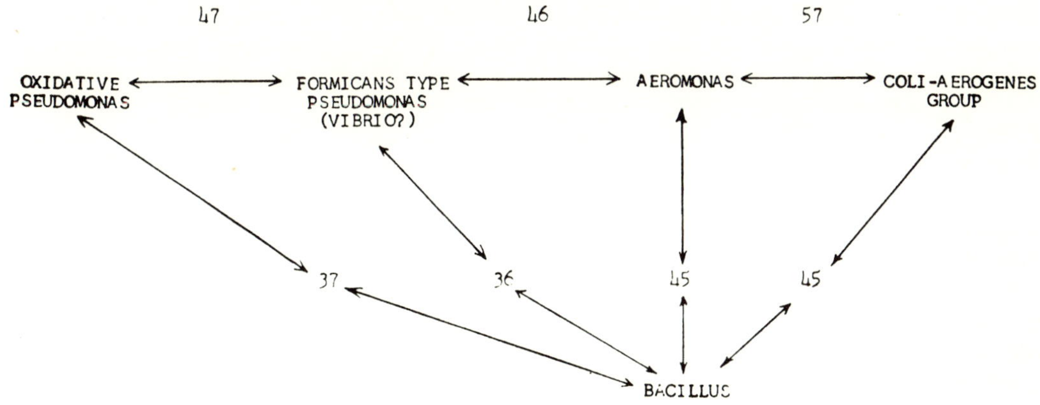

FIG. 6. *Taxonomic relationships between pseudomonads and the coli-aerogenes group. Diagrammatic representation of inter- and intragroup S values indicating "natural relationships."*

would fit well into such a grouping. A tentative schema of relationships in accordance with this suggestion is shown in Fig. 6. The highest intergroup relationships, machine calculated from the complete *S*-value data, are included in this figure.

The separation of the fermentative anaerogenic *A. formicans-Vibrio* group and the fermentative aerogenic Aeromonas from the oxidative Pseudomonas is emphasized by the low maximal intergroup *S* values.

Moreover, the intergroup relationship between the aerogenic Aeromonas and Bacillus is identical with that of Bacillus and the coli-aerogenes group and greater than that of either the oxidative Pseudomonas or the anaerogenic fermentative group with Bacillus.

From our results, a general description of the genus *Pseudomonas* (Migula, 1894) might be as follows: "gram-negative, nonendosporeforming, curved, slightly curved, or straight, short, slender, lanceolate rods demonstrating rapid corkscrew motility in peptone water culture by virtue of polar flagella, nonfermentative, occasionally utilizing dextrose, maltose, sucrose, or mannitol but only oxidatively; indole, methyl red, and Voges-Proskauer negative; insensitive to penicillin and pteridin 0/129 compound; catalase and Kovacs' oxidase positive and producing free ammonia in peptone water."

This description is in accord with that of Rhodes (1959) with the exception of the methyl red test, for which she used a different medium. The size range observed by Rhodes for Pseudomonas, 1.0 to 3.0 μ by 0.5 to 0.7 μ, is similar to our findings. However, our observations on flagellation in this group correspond with those of Buttiaux and Gagnon (1958/1959); i.e., that in a single colony or a single culture, cells are present possessing polar flagella of different types: monotrichous, amphitrichous, and lophotrichous.

The question of speciation within the genus *Pseudomonas* is difficult. We have been unable to identify most of our freshly isolated strains with any of the named strains listed by Haynes and Burkholder in Bergey's Manual (Breed et al., 1957) and in other works of reference. Most of our pseudomonad strains clustered into four main groups.

The seawater requiring organisms in our collection formed a group which contained the single seawater-requiring type species *P. elongata.* This marine group is similar in its general characteristics to the group growing only on media containing seawater, or artificial seawater, described by Tyler, Bielling, and Pratt (1960).

The subgroup 1a, which probably constitutes a single species, contains no named strains but appears most similar to *V. beijerinckii* described by Stanier (1941). Subgroup 1b contains *P. elongata* but may consist of more than one species.

Organisms described in the literature which correspond in their general characteristics with group 1, include *P. gelatica*, *P. calciprecipitans*, and *P. nigrifaciens* (Haynes and Burkholder, 1957), and *V. granii* and *P. irridescens* (Stanier, 1941).

The group described as group 2 above, contains *V. percolans. P. fragi* appears to be closely associated with this mesophilic group which is intermediate in its biochemical activity between the rather inactive marine group and the highly active fluorescent group 3. No other named species of pseudomonad corresponded to this group. *V. percolans* seems to fall more correctly into the genus *Pseudomonas* than into *Vibrio*.

Group 3, the largest group, contains two well defined, closely related subgroups 3a, 3b (intergroup, $S = 65\%$), and one rather diverse subgroup, 3c. Subgroups 3a and 3b shared the general characteristics of the type species, *P. aeruginosa*. However, 3a is composed essentially of psychrophilic organisms, able to grow at 0 C, and is a little less active proteolytically than 3b but slightly more active saccharolytically and more likely to be able to utilize citrate as sole carbon source. None of the named strains used in the analysis fell into group 3a. Among the organisms listed in the literature only *P. chlororaphis* appeared to fit the general group description. However, from the characteristics described by Rhodes (1959) it appears likely that some of the strains studied by her fall into this group, which we have called the psychrophilic aeruginosa group. The homogeneity of this group is such that it probably represents a single species.

The subgroup 3b, which is mainly mesophilic, includes the two type strains of *P. aeruginosa*, has the same level of homogeneity as 3a ($S = 69\%$), and probably represents a single species.

Subgroup 3c includes *V. cuneatus* which, as we have already indicated, appears to be a green pigment-producing pseudomonad. The group is heterogeneous having an intragroup S value of 57%. The group contains organisms which are mesophilic, eurythermic, psychrophilic, and even stenothermic, and the only consistent physiological and biochemical properties are those noted above as characteristic for the genus. The ATCC culture of *P. fluorescens* (11251), which failed to produce pigment, appears to be associated with this group. Examination of published descriptions of *P. putida*, *P. geniculata*, and *P. pavonacea* indicated that these species might fall within this group, although the ATCC strain of *P. pavonacea* did not enter this cluster in our analysis.

Group 4, which includes *P. ovalis*, *P. denitrificans*, and *V. tyrogenus*, shows relationships with the mesophilic group 2 and subgroup 3b and 3c. Group 4 obviously represents a clustering of related species. A number of nonpigmented species described in Bergey's Manual (Breed et al., 1957), including *P. mira*, and the hydrocarbon-utilizing species appear to resemble the group, as do also some of the plant pathogenic species. However, no consistent species subgroups within group 4 were revealed by our analysis.

Named species of *Achromobacter*, *Alcaligenes*, *Flavobacterium*, *Photobacterium*, and *Aeromonas* fell completely outside of the genus *Pseudomonas* in the 80-strain analysis. *V. metschnikovii* showed only a very slight relationship. Thus the results of the 80-strain analysis corresponded with those of the 40-species analysis used to demonstrate generic and suprageneric groupings. Of the unnamed, freshly isolated strains which failed to group, five are thought to be *Achromobacter*, two *Flavobacterium*, one *Alcaligenes*, four *Vibrio* sp., and one a gut-group *Vibrio* (Liston, 1955). The remaining six are probably pseudomonads which showed insufficient correlation with other strains to be put into one of the four groups.

The success of the Sneath method in putting some 54 strains of pseudomonads out of 60 into four subgeneric groups indicates the potential usefulness of this procedure for bringing about a satisfactory classification of organisms of this type. Many of the species of *Pseudomonas* in the literature are ill defined and consequently it is extremely difficult to identify strains freshly isolated from natural environments. This has resulted in a proliferation of new species names of doubtful validity which has served to compound the confusion rather than to diminish it. The separation of pseudomonads into subgeneric groups which have taxonomic significance since they are based on natural similarities (as explained by Sneath (1957a)) would facilitate the identification of new organisms and postpone the need for creation of new species names until the desirability of the present crop can be evaluated and an objective definition of a bacterial species devised. The wider range of variation of properties which is permissible in a subgeneric group would circumvent some of the difficulties arising in the assignment of organisms to species as a result of the rigid (if often incomplete) definition of the latter. The computer method shows promise of providing a numerical index of taxonomic level in terms of S value. The development of such indexes will require considerable further work, both with large groups of organisms to establish generic and larger groupings and with individual strains to establish, so far as possible, the spectrum of properties which may be encompassed by a species. Such work is now proceeding in our laboratory.

ACKNOWLEDGMENTS

The authors are indebted to the following for providing cultures: D. L. Georgala, L. Farber, J. M. Shewan, R. Rucker, A. K. Sparks, H. Raj, and the Department of Microbiology, University of Washington. The authors also wish to acknowledge the invaluable assistance of G. Constabaris in the initial programming for the electronic computer and R. Baxter in the laboratory testing of cultures.

The research was supported by a grant (E-2417) from the National Institutes of Health.

LITERATURE CITED

Asai, T., S. Okumura, and T. Tsunoda. 1956 On a new genus, *Kluyvera*. Proc. Japan Acad. **32**:488–493.

Ayers, J. C. 1960. Temperature relationships and some other characteristics of the microbial flora developing on refrigerated beef. Food Research **25**:1–18.

Breed, R. S., E. D. G. Murray, and N. R. Smith. 1957. Bergey's manual of determinative bacteriology. 7th ed. The Williams & Wilkins Co., Baltimore. 1094 p.

Buttiaux, R., and P. Gagnon. 1958/1959. Au sujet de la classification des *Pseudomonas* et des *Achromobacter*. Ann. inst. Pasteur (Lille) **10**:121–149.

Cheeseman, G. C., and N. J. Berridge. 1959. The differentiation of bacterial species by paper chromatography. VII. The use of electronic computation for the objective assessment of chromatographic results. J. Appl. Bacteriol. **22**:307–316.

Collier, H. O. J., N. R. Campbell, and M. E. H. Fitzgerald. 1950. Vibriostatic activity in certain series of pteridines. Nature **165**:1004–1005.

Gaby, W. L., and E. Free. 1958. Differential diagnosis of *Pseudomonas*-like microorganisms in the clinical laboratory. J. Bacteriol. **76**:442–444.

Hill, L. R. 1959. The Adansonian classification of the staphylococci. J. Gen. Microbiol. **20**:277–283.

Hugh, R., and E. Leifson. 1953. The taxonomic significance of fermentative versus oxidative metabolism of carbohydrates by various gram-negative bacteria. J. Bacteriol. **66**:24–26.

Kovacs, N. 1956. Identification of *Pseudomonas pyocyanea* by the oxidase reaction. Nature **178**:703.

Leifson, E. 1951. Staining, shape and arrangement of bacterial flagella. J. Bacteriol. **62**: 377–389.

Liston, J. 1955. A group of luminous and non-luminous bacteria from the intestine of flatfish. J. Gen. Microbiol. **12**:i.

Liston, J. 1956. The occurrence and distribution of bacterial types on flatfish. J. Gen Microbiol. **16**:205–216.

Liston, J., and R. R. Colwell. 1960. Taxonomic relationships among the pseudomonads. Bacteriol. Proc. **1960**:78–79.

MacLeod, R. A., E. Onofrey, and M E. Norris. 1954. Nutrition and metabolism of marine bacteria. I. Survey of nutritional requirements. J. Bacteriol. **68**:680–686.

Manual of microbiological methods. 1957. Society of American Bacteriologists. McGraw-Hill Book Co., Inc., New York. p. 315.

Migula, W. 1894. Über ein neues System der Bakterien. Arb. Bakteriol. Inst. Karlsruhe **1**:235.

Novel, E. 1939. Une technique facile et rapide de mise en évidence des cils bactériens. Ann. inst. Pasteur, **63**:302–311.

Rhodes, M. E. 1959. The characterization of *Pseudomonas fluorescens*. J. Gen. Microbiol. **21**:221–263.

Shewan, J. M., W. Hodgkiss, and J. Liston. 1954. A method for the rapid differentiation of certain non-pathogenic, asporogenous bacilli. Nature **173**:208.

Sneath, P. H. A. 1957*a*. Some thoughts on bacterial classification. J. Gen. Microbiol. **17**: 184–200.

Sneath, P. H. A. 1957*b*. The application of computers to taxonomy. J. Gen. Microbiol. **17**: 201–226.

Sneath, P. H. A., and S. T. Cowan. 1958. An electro-taxonomic survey of bacteria. J. Gen. Microbiol. **19**:551–565.

Sokol, R. R., and C. D. Michener. 1957. A quantitative approach to classification. Evolution **10**:130–162.

Stanier, R. Y. 1941. Studies on marine agar-digesting bacteria. J. Bacteriol. **42**:527–559.

Stanier, R. Y. 1953. A note on the taxonomy of *Proteus hydrophilus*. J. Bacteriol. **46**:213.

Tyler, M. E., M. C. Bielling, and D. B. Pratt. 1960. Mineral requirements and other characters of selected marine bacteria. J. Gen. Microbiol. **23**:153–161.

Wood, A. J., and E. A. Baird. 1943. Reduction of trimethylamine oxide by bacteria. J. Fisheries Research Board Can. **6**:194–201.

ZoBell, C. E. 1946. Marine microbiology. Chronica Botanica Co., Waltham, Mass. 58 p.

17

Reprinted from *J. Appl. Bacteriol.*, **24**(1), 87–93 (1961)

THE PRESERVATION OF MARINE BACTERIA

By G. D. FLOODGATE and P. R. HAYES
Torry Research Station, Aberdeen

SUMMARY: Forty-five marine bacteria and five nonmarine strains, representing 10 genera, were preserved by freeze drying, by storage in glycerol-nutrient broth at −29° and under paraffin oil at 1°. Checks on viability and viable counts were made over a two year storage period. At the end of this time 100% of the strains had been preserved by freeze drying, 86% by storage in glycerol-nutrient broth at low temperature and 72% under paraffin oil. The greatest loss of viability occurred among the vibrios and spirilla whichever of the three methods was used, but coryneforms, achromobacters and micrococci were successfully preserved by all these methods.

Since most of the work on the preservation of bacteria has been concerned with nonmarine organisms it was necessary, when setting up the National Collection of Marine Bacteria (NCMB), to find out if bacteria from the sea can be preserved by the same techniques. It was decided therefore to study the suitability for this purpose of freeze drying, of freezing in glycerol-nutrient broth and of storage under paraffin oil, and to follow, as far as was practical, the fate of the organisms during a two year storage period. Moreover, such a study would make it possible to compare the effectiveness of the methods which would be valuable in itself, because although freeze drying and other means of preserving bacteria have been in general use for some time there have been few contributions to the literature in which comparisons have been made between the various preservation techniques. It was also decided to enhance the usefulness of the study by making it as quantitative as possible, for in spite of some valuable papers, such as those of Proom & Hemmons (1949), Fry & Greaves (1951) and Hörter (1958, 1960), much is still unknown of the changes in bacterial count which occur during the preservation process or during subsequent storage.

Methods and Materials

Cultures. Forty-five marine bacteria representing 10 genera were used in the investigation. These organisms had been originally isolated by several different workers from marine fish, sea water, sea mud and other marine environments in the Arctic Ocean, the North Sea, the eastern Pacific Ocean and the Gulf of Mexico. Five nonmarine organisms, *Pseudomonas aeruginosa* (NCIB 6570 and 950), *Ps. oleovorans* (NCIB 6576), *Ps. formicans* (Crawford, 1954) and *Aeromonas hydrophila* (NCTC 7810) were included, to enable comparisons to be made between marine and nonmarine strains and as a control on the techniques. Most of the marine isolates were characterized only as far as the genus. The achromobacters, aeromonads, pseudomonads and vibrios were identified using the techniques of Shewan, Hobbs &

Hodgkiss (1960). Five organisms, NCMB 1, 2, 7, 24 and 25, were luminous and were originally placed in the genus *Photobacterium*, but have been reclassified as *Aeromonas* spp. or *Vibrio* spp. by Spencer (1955). Although all the marine organisms were isolated on sea water media, it was found that after a few transfers most of them grew very well on tap water nutrient agar. This medium was used for growing all such adapted strains. However, 17 cultures (see Table 1) could not be so adapted and were grown on sea water nutrient agar.

Preservation methods

Freeze drying. The cells were usually harvested after incubation for 3 or 4 days at their optimum temperature, that is towards the end of their logarithmic growth phase, as there is some evidence that older cultures survive freeze drying better than young ones (Fry & Greaves, 1951; Fry, 1954; Toyokawa & Hollander, 1956). A thick suspension of the cells was then made in 'Mist desiccans' (nutrient broth 1 part and horse serum 3 parts, containing 7·5% (w/v) of glucose), and 0·1 ml amounts pipetted into ampoules which had been previously prepared by being soaked overnight in 5% HCl, rinsed thoroughly, dried and sterilized. Each ampoule also contained a piece of filter paper on which the identification of the organism and the date had been stamped, using 'ENM' ink (Gutteridge Sampson Ltd., 155, Farringdon Road, London, E.C.1) which is known to be harmless to bacteria. Freeze drying was carried out in an LC5 centrifugal freeze dryer (Edwards High Vacuum Ltd., Crawley, Sussex). Primary drying was carried on for 1–2 hr, and secondary drying for 16–20 hr. The ampoules were sealed off at a pressure of 0·012–0·028 mm Hg, as measured by a Pirani gauge, and stored at room temperature.

Storage in glycerol-nutrient broth. Several workers (Hollander & Nell, 1954; Howard, 1956; Fox & Hotchkiss, 1957; Quadling, 1960) have shown that the presence of glycerol in a medium reduces the number of bacteria which are killed when the suspension is frozen. In this study cultures which had been growing for 3 or 4 days on tap water nutrient agar or sea water nutrient agar were preserved, using the glycerol-nutrient broth method of Howard (1956). The cells were harvested, washed with normal saline or sea water, suspended in nutrient broth to which 15% (v/v) of sterile glycerol had been added, and stored at −29°.

Storage under mineral paraffin oil. Mineral paraffin oil was first employed for preserving cultures by Lumière & Chevrotier (1914) and has been successfully used by many workers since then. Recently this method has been re-examined by Hartsell (1953, 1956) and Rhodes (1956, 1957). For the present investigation the organisms were grown on short, deep slopes for 2 or 3 days until there was abundant growth. This was then covered with a layer of sterile paraffin oil (B.P. quality), care being taken to cover the slope completely. The cultures were stored at 1°.

Viability checks

Freeze drying method. Using the Miles & Misra (1938) technique, the numbers of bacteria were determined in the 'Mist desiccans' suspension shortly before drying and in the dried material immediately after sealing. Further counts were made after storage for 1, 3, 6, 12 and 24 months.

It is known that the method and speed of resuscitation may have a profound effect on the recovery of freeze dried cultures (Wasserman & Hopkins, 1957; Leach & Scott, 1959). The method of resuscitation used here was to add 0·5 ml of tap water nutrient broth or sea water nutrient broth to the contents of an ampoule and to resuspend the freeze dried material as rapidly as possible.

Glycerol-nutrient broth and paraffin methods. The viability of the organisms preserved by the glycerol-nutrient broth and paraffin methods was checked by streaking one loopful on a tap water nutrient agar or sea water nutrient agar plate after 1, 3, 6, 12 and 24 months. In the case of the glycerol-nutrient broth method a fresh tube, i.e. one that had not previously been thawed, was taken for each check; in the case of the paraffin method, all the samples were taken from the same slope.

Results and Discussion

Freeze dried cultures

The bacterial counts on the freeze dried material are shown in Table 1: they are expressed as the logarithm to base 10 of the number of organisms/ml of 'Mist desiccans'. There was a much greater variation in the bacterial counts than could be accounted for by the inherent statistical error of the Miles & Misra method, and this was believed to be due to the excessive clumping that occurred when the organisms, particularly the *Achromobacter* spp., were mixed with the 'Mist desiccans'. This effect was so marked in the case of NCMB 138 that the organism formed a precipitate and settled at the bottom of the ampoule. In other cases the effect was less marked, but could be observed by mixing the organism with 'Mist desiccans' on a microscope slide. These aggregations were not easily dispersed; even continuous flushing into and out of a Pasteur pipette for 20 min failed to break up all the clumps in resuscitated cultures. The cause of this excessive clumping is not known, but it has been observed occasionally particularly among certain slime producing genera such as *Azotobacter* (MacKenzie, pers. comm.). These aggregations were also reminiscent of those formed by rough strains of coli-aerogenes bacteria in low salt concentrations (Mitchell, 1951), although the colonies appeared to be smooth. The effect would be explained if it could be shown that marine bacteria have a strongly hydrophobic surface, or that other cohesive forces which exist between the cells came into play when the bacteria were mixed with 'Mist desiccans'. Further work on the action of electrolytes or colloids in various concentrations on these cohesive forces would be valuable. Undoubtedly the clumping accounted for some unexpectedly low counts, particularly in the case of NCMB 131 and 138. Moreover, the degree of clumping made a statistical evaluation of the results impracticable. In spite of these difficulties, however, certain conclusions may legitimately be drawn from the results. First, although in some cases there was a considerable reduction in count, freeze drying successfully preserved marine as well as nonmarine organisms. That the requirement of some of the organisms for salt for growth is not an important factor in their susceptibility to freeze drying is shown by the fact that several of the organisms requiring sea water showed no marked loss of viability when freeze dried whereas others were killed off to a considerable extent. Within the limits imposed by the

Table 1. *The influence of freeze drying in 'Mist desiccans' and storage at room temperature on the survival of various bacteria*

Identification	Culture collection* and number	Log$_{10}$ (colony count/ml of 'Mist desiccans') Before drying	After drying and storage for (months) 0	1	3	6	12	24
Achromobacter spp.	NCMB 9†	8·3	8·0	7·9	7·5	7·3	7·6	7·0
	NCMB 20†	8·3	6·3	5·6	5·7	6·8	5·5	6·9
	NCMB 26	9·5	9·4	8·5	8·0	9·4	8·3	8·5
	NCMB 27	7·8	6·4	6·1	8·3	8·0	7·8	8·0
	NCMB 28	8·8	8·0	8·7	8·0	8·9	8·0	8·8
	NCMB 29	8·9	8·8	9·0	9·0	9·4	8·7	7·4
	NCMB 131	9·7	6·7	6·7	8·0	6·7	9·2	9·0
	NCMB 132	10·0	8·5	8·5	6·2	9·0	9·4	8·0
	NCMB 135	9·6	10·0	9·8	10·0	9·7	9·7	9·0
	NCMB 138	10·5	9·2	7·9	9·7	9·9	7·0	8·8
Aeromonas harveyi	NCMB 2†	9·1	7·0	6·5	6·2	5·9	6·5	6·4
Aer. sepiae	NCMB 24†	10·3	7·8	7·1	7·6	7·8	6·0	5·3
Aer. hydrophila	NCTC 7810	10·8	9·9	8·5	7·0	7·8	7·3	7·0
Corynebacterium erythrogenes	NCMB 5	9·8	9·7	9·7	9·8	9·8	9·7	9·5
Corynebacterium sp.	NCMB 8	10·1	9·6	9·8	10·1	9·8	10·2	10·3
	NCMB 12	9·8	9·8	9·6	9·7	9·5	9·7	9·4
	NCMB 16†	9·3	9·3	9·3	9·3	9·8	9·0	9·0
	NCMB 31	10·4	10·7	9·8	9·7	10·0	9·9	9·9
	NCMB 32	10·5	9·6	10·0	9·7	9·4	9·3	9·3
	NCMB 33	10·7	10·2	10·0	9·8	9·9	10·5	10·6
	NCMB 34	10·2	9·5	9·5	9·7	9·2	9·8	9·3
	NCMB 35	10·6	9·5	9·5	9·6	9·3	9·8	10·0
	NCMB 39	11·4	11·3	10·5	10·6	10·4	10·4	10·2
Flavobacterium spp.	NCMB 21	8·9	8·6	8·5	8·7	9·3	9·2	9·2
	NCMB 22†	6·8	6·4	5·8	6·7	6·8	6·0	6·3
Micrococcus spp.	NCMB 13	10·0	9·9	9·6	9·7	9·8	9·8	10·0
	NCMB 14	10·6	10·1	10·5	11·0	10·0	11·0	10·0
	NCMB 15	9·7	9·7	9·6	9·7	9·8	9·3	9·3
Myxobacterium sp.	NCMB 11†	9·5	9·3	9·5	8·2	7·7	8·0	7·4
Pseudomonas aeruginosa	NCIB 6750	9·3	8·0	8·5	7·8	7·7	8·2	8·3
	NCIB 950	10·3	10·0	10·0	9·4	9·3	9·5	9·3
Ps. oleovorans	NCIB 6576	9·3	8·3	8·0	7·7	7·8	8·0	7·2
Pseudomonas spp.	NCMB 17†	7·3	5·5	5·0	5·0	5·6	5·3	5·4
	NCMB 18†	9·4	9·3	9·0	8·8	9·1	8·7	8·8
	NCMB 19†	9·4	8·3	8·3	7·0	8·6	6·3	8·0
Ps. formicans	NCMB 23	9·5	9·0	8·6	7·9	7·3	7·3	6·0
Pseudomonas spp.	NCMB 125	9·3	8·6	8·5	8·7	8·0	8·8	8·4
	NCMB 126	5·7	5·4	5·7	5·7	6·7	6·4	6·3
	NCMB 127	9·8	8·7	9·0	9·7	9·0	9·6	8·2
	NCMB 128	10·4	9·9	9·5	9·6	9·3	9·9	8·8
	NCMB 133	10·3	9·4	9·7	9·5	9·7	10·0	9·3
	NCMB 136	9·7	9·3	9·2	9·2	9·8	9·7	8·6
Serratia marinorubra	NCMB 4	10·1	9·3	8·9	9·4	9·2	9·5	9·3
Spirillum spp.	NCMB 38†	10·3	8·2	8·2	8·1	8·7	8·7	8·4
	NCMB 40†	10·4	6·7	7·6	6·3	7·6	6·3	3·5
Vibrio splendidum	NCMB 1†	7·0	4·5	4·4	4·3	3·8	4·3	4·7
V. anguillarum	NCMB 6†	9·2	6·3	5·0	5·9	6·0	6·0	5·4
V. phosphoreum	NCMB 7†	8·3	4·9	4·0	5·3	5·3	5·2	4·5
V. pierantonii	NCMB 25†	9·8	4·0	4·2	5·3	5·5	4·8	4·3
Vibrio sp.	NCMB 36†	10·2	7·5	7·7	7·5	7·6	7·3	5·5

* NCTC, National Collection of Type Cultures; NCIB, National Collection of Industrial Bacteria; NCMB, National Collection of Marine Bacteria. † Require sea water for growth.

number of strains used, it is clear that the corynebacteria, flavobacteria and micrococci were resistant to freeze drying, the count dropping on an average by less than one third. The pseudomonads and achromobacters showed a greater strain to strain variation and in general the decrease was by a factor approaching one tenth. The greatest losses during freeze drying occurred among the spirilla and vibrios and, to a lesser degree, among the aeromonads. The high susceptibility of *Vibrio* spp. to freeze drying has been reported several times (e.g. Fry, 1954) and it may well be a feature of this genus.

Generally speaking the strains lost little or none of their viability during storage, although there were clear exceptions such as NCMB 23, 36, and 40. Cultures which markedly decreased in numbers during freeze drying did not necessarily die off to any extent during storage, whereas some organisms, which withstood freeze drying

Table 2. *The survival of bacteria after storage in glycerol-nutrient broth at –29° (a) and under mineral paraffin oil at 1° (b)*

Identification	Culture collection† and number	Viability* after storage for (months)									
		1		3		6		12		24	
		(a)	(b)	(a)	(b)	(a)	(b)	(a)	(b)	(a)	(b)
Achromobacter spp.	NCMB 9	+	++	+	++	+	++	+	++	+	++
	NCMB 20	++	+	++	+	++	+	++	+	++	+
	NCMB 26	++	++	++	++	++	++	++	+	+	+
	NCMB 27	++	++	++	++	++	++	+	+	+	+
	NCMB 28, NCMB 29, NCMB 131	++	++	++	++	++	++	++	++	++	++
	NCMB 132, NCMB 135, NCMB 138	++	++	++	++	++	++	++	++	++	+
Aeromonas harveyi	NCMB 2	+	++	+	+	+	–	–	–	–	–
Aer. sepiae	NCMB 24	++	++	++	+	+	–	+	–	+	–
Aer. hydrophila	NCTC 7810	++	++	++	++	++	++	+	+	+	–
Corynebacterium erythrogenes; *Corynebacterium* spp.	NCMB 5; NCMB 8, NCMB 12, NCMB 16, NCMB 31, NCMB 32, NCMB 33, NCMB 34, NCMB 35, NCMB 39	++	++	++	++	++	++	++	++	++	++
Flavobacterium spp.	NCMB 21	++	++	++	++	++	++	++	++	++	++
	NCMB 22	++	++	++	++	++	++	+	+	+	–
Micrococcus spp.	NCMB 13, NCMB 14, NCMB 15	++	++	++	++	++	++	++	| |	| |	| |
Myxobacterium sp.	NCMB 11	++	++	++	+	++	+	++	–	++	–
Pseudomonas aeruginosa	NCIB 6750	++	++	+	+	+	–	+	–	+	–
	NCIB 950	++	++	++	++	++	++	++	+	++	–
Ps. oleovorans	NCIB 6576	++	++	++	++	++	++	++	+	+	+
Pseudomonas spp.	NCMB 17	+	++	+	++	+	++	–	++	–	++
	NCMB 18, NCMB 19	++	++	++	++	++	++	++	++	+	–
Ps. formicans	NCMB 23	++	++	++	++	++	++	+	–	+	–
Pseudomonas spp.	NCMB 125, NCMB 126, NCMB 127	++	++	++	++	++	++	++	++	++	++
	NCMB 128, NCMB 133, NCMB 136	++	++	++	++	++	++	++	++	++	+
Serratia marinorubra	NCMB 4	++	++	++	++	++	++	++	++	++	++
Spirillum spp.	NCMB 38	++	++	++	++	++	++	++	++	–	+
	NCMB 40	++	++	+	+	+	–	+	–	–	–
Vibrio splendidum	NCMB 1	+	++	+	+	–	–	–	–	–	–
V. anguillarum	NCMB 6	++	++	++	++	++	++	++	++	++	++
V. phosphorerum	NCMB 7	++	++	+	++	+	++	–	+	–	+
V. pierantonii	NCMB 25	++	++	++	–	++	–	+	–	–	–
Vibrio sp.	NCMB 36	++	–	++	–	++	–	+	–	+	–

* Growth after subculture: ++, abundant to good; +, moderate to slight; –, none.

† NCTC, National Collection of Type Cultures; NCIB, National Collection of Industrial Bacteria; NCMB, National Collection of Marine Bacteria.

well, tended to die off during the 2 year storage period. Occasionally an organism appeared to increase in numbers during storage, as for example NCMB 27 and 126 and, to a less marked extent, NCMB 33 and 35. This rise may have been partly due to the experimental error within the counting method, but it is interesting to notice that Maister, Pfeifer, Bogart & Heger (1958) found similar increases in viable count of *Serratia marcescens* after storage of freeze dried pellets. They suggested that the increase may be due to the recovery of enzymic systems during storage or to the breaking up of clumps. Such a possibility should not be ruled out here.

Stored cultures

The viability of cultures preserved by the glycerol-nutrient broth method and under paraffin oil is shown in Table 2; 86% of the cultures preserved by the former method and 72% of those preserved by the latter method were viable after 2 years. It is interesting to note that in both cases, as in the freeze drying experiment, the vibrios, aeromonads and spirilla were the most easily killed of all the groups examined. In fact, both the most difficult and the easiest organisms to preserve were generally the same whichever process of preservation was used. Again there appeared to be no relationship between the requirement of sea water for growth and the effectiveness of preservation by either the glycerol-nutrient broth method or the paraffin method; marine organisms behaved in the same way as nonmarine ones. One unexpected result was the loss of the *Ps. aeruginosa* strains under paraffin oil. Hartsell (1953) was able to maintain this species for more than 9 years under paraffin; he was also able to preserve *V. comma* for more than 6 years. Gordon & Smith (1947), on the contrary, found the mineral oil method unsuitable for some strains of pseudomonads and vibrios. This variation in results may be due to storage methods; Hartsell (1953) stored his cultures at 25°, whereas in this investigation the storage temperature was 1°.

Conclusion

It may be concluded from these results that marine bacteria can be preserved by well established techniques. From the point of view of Culture Collections each method has advantages and disadvantages. The advantages of freeze drying are that the cells remain viable for a long time and that ampoules of freeze dried bacteria are light and easy to despatch; the disadvantage is that the process is slow and tedious. Both the mineral oil and glycerol-nutrient broth methods are rapid and cheap, but on the data presented here not as successful as freeze drying for preserving marine bacteria for a long time.

This work was financed by the Development Commission and was carried out by Research Fellows of the Department of Bacteriology, University of Aberdeen.

REFERENCES

CRAWFORD, I. P. (1954). A new fermentative pseudomonad, *Pseudomonas formicans* n. sp. *J. Bact.* **68**, 734.

FOX, M. S. & HOTCHKISS, R. D. (1957). Initiation of bacterial transformation. *Nature, Lond.*, **179**, 1322.

FRY, R. M. (1954). The preservation of bacteria. In *Biological Applications of Freezing and Drying*. New York: Academic Press Inc.

FRY, R. M. & GREAVES, R. I. N. (1951). The survival of bacteria during and after drying. *J. Hyg., Camb.*, **49**, 220.

GORDON, R. E. & SMITH, N. R. (1947). The preservation of certain micro-organisms under paraffin oil. *J. Bact.* **53**, 669.

HARTSELL, S. E. (1953). The preservation of bacterial cultures under paraffin oil. *Appl. Microbiol.* **1**, 36.

HARTSELL, S. E. (1956). Maintenance of cultures under paraffin oil. *Appl. Microbiol.* **4**, 350.

HOLLANDER, D. H. & NELL, E. E. (1954). Improved preservation of *Treponema pallidum* and other bacteria by freezing with glycerol. *Appl. Microbiol.* **2**, 164.

HÖRTER, R. (1958). Überlebensraten von Bakterien nach der Hochvakuumgefriertrocknung. *Zbl. Bakt.* (Abt. 1) **171**, 526.

HÖRTER, R. (1960). Überlebensraten von gefriergetrockneten Bakterien nach zweijahriger Lagerung. *Zbl. Bakt.* (Abt. 1) **178**, 364.

HOWARD, D. H. (1956). The preservation of bacteria by freezing in glycerol broth. *J. Bact.* **71**, 625.

LEACH, R. H. & SCOTT, W. J. (1959). The influence of rehydration on the viability of dried micro-organisms. *J. gen. Microbiol.* **21**, 295.

LUMIÈRE, A. & CHEVROTIER, J. (1914). Sur la vitalité des cultures des gonocoques. *C.R. Acad. Sci., Paris*, **158**, 1820.

MAISTER, H. G., PFEIFER, V. F., BOGART, W. M. & HEGER, E. N. (1958). Survival during storage of *Serratia marcescens* dried by continuous vacuum sublimation. *Appl. Microbiol.* **6**, 413.

MILES, A. A. & MISRA, S. S. (1938). The estimation of the bactericidal power of the blood. *J. Hyg., Camb.*, **38**, 732.

MITCHELL, P. (1951). Physical factors affecting growth and death. In *Bacterial Physiology*. New York: Academic Press Inc.

PROOM, H. & HEMMONS, L. M. (1949). The drying and preservation of bacterial cultures. *J. gen. Microbiol.* **3**, 7.

QUADLING, C. (1960). Preservation of *Xanthomonas* by freezing in glycerol broth. *Canad. J. Microbiol.* **6**, 475.

RHODES, M. E. (1956). The preservation of pseudomonads in mineral oil. *J. appl. Bact.* **19**, iii.

RHODES, M. E. (1957). The preservation of *Pseudomonas* under mineral oil. *J. appl. Bact.* **20**, 108.

SHEWAN, J. M., HOBBS, G. & HODGKISS, W. (1960). A determinative scheme for the identification of certain genera of Gram-negative bacteria, with special reference to the Pseudomonadaceae. *J. appl. Bact.* **23**, 379.

SPENCER, R. (1955). The taxonomy of certain luminous bacteria. *J. gen. Microbiol.* **13**, 111.

TOYOKAWA, K. & HOLLANDER, D. H. (1956). Variation in sensitivity of *Escherichia coli* to freezing damage during growth cycle. *Proc. Soc. exp. Biol., N.Y.*, **92**, 499.

WASSERMAN, A. E. & HOPKINS, W. J. (1957). Studies in the recovery of viable cells of freeze dried *Serratia marcescens*. *Appl. Microbiol.* **5**, 295.

18

Reprinted from *Cryobiology*, 3(5), 419–422 (1967)

THE EFFECT OF THE REHYDRATION TEMPERATURE AND REHYDRATION MEDIUM ON THE VIABILITY OF FREEZE-DRIED *SPIRILLUM ATLANTICUM**

R. V. CHOATE AND M. T. ALEXANDER

American Type Culture Collection, Rockville, Maryland

Freeze-drying is a method of importance for maintaining stock bacterial cultures over long periods of time. It is known that most bacteria can be successfully freeze-dried and have been found to be viable after years of storage.[11]

Although much work has been done in the area of freeze-drying,[3, 6] most of the emphasis has been placed on the methods and suspending media used for freeze-drying. There is little information available regarding variations caused by the conditions of rehydrating freeze-dried organisms. Pedersen[9] reported that reconstitution of freeze-dried *Cryptococcus terricolus* in a malt extract solution showed a significant increase in percentage recovery compared with rehydration in distilled water on the appropriate growth medium. Leach, Ohye, and Scott[4] studied the effects of rehydration on freeze-dried *Vibrio metchnikovii* and *Mycoplasma mycoides*. They found that the volume of the rehydrating fluid, the length of time the reconstituted cells were in contact with the rehydrating fluid before dilution, and the nature of the rehydrating fluid itself had a marked effect on the viable cell count. Leach and Scott[5] also indicated that the requirements for reconstitution may vary greatly in different organisms.

While conducting studies on the preservation of marine bacteria, Floodgate and Hayes[2] found that plate counts after freeze-drying were substantially more variable than was expected using the Miles and Misra technique,[7] a fact which they attributed to cell clumping. W. A. Clark (personal communication) reported that by varying the salt or sugar concentration in the rehydrating solution he obtained variations in the recovery of freeze-dried osmotically sensitive phages. Record, Taylor, and Miller[10] using *Escherichia coli* improved the percentage survival after drying by closely regulating the reconstitution of the dried material. They attributed this increase in percentage survival in part to an effect of osmotic pressure during rehydration. In light of the above, we felt it desirable to study the effects of variation in rehydration procedures.

Dehydration during freeze-drying takes place gradually and at low temperatures. To correlate rehydration more closely with these phases of dehydration, a procedure of "low temperature rehydration" was adopted. Monk and McCaffery[8] developed a method of low temperature rehydration, by which ice was quickly sublimed onto the dried cell suspension, then warmed to 25°C so that the cells became wet. This paper deals with an investigation of the effects of rehydration temperature and medium on the viability of freeze-dried *Spirillum atlanticum*, ATCC 12753.

Materials and Methods

S. atlanticum Williams and Rittenberg, ATCC 12753,[1] a salt water bacterium sensitive to freeze-drying, was used for these experiments. Cultures of the organism were propagated on the following Seven Seas marine medium: Seven Seas marine mix (Utility Chemical Company, Paterson, N. J.), 37.9 g; yeast extract (Difco), 3.0 g; proteose peptone No. 3 (Difco), 10.0 g; distilled water, 1.0 liter.

Standard cultures consisted of 5-ml quantities of broth per test tube, with a 0.1-ml inoculum from a 24-hr culture, with incubation at 26°C for 48 hrs. The suspensions for freeze-drying were prepared by centrifuging the culture for 15 min at approximately 4000 rpm (safety head centrifuge, Clay-Adams, Inc., New York), discarding the supernatant, and resuspending the cells in the appropriate freeze-dry medium at one-fifth the original volume.

The freeze-dry media tested were: 1) 20%

Received December 6, 1966.

* Supported by National Science Foundation Grant GB5756.

TABLE 1

The effect of various suspending media on the recovery of freeze-dried *S. atlanticum* with Seven Seas medium as a rehydration and dilution medium at 25°C. Control cell count 1.0×10^9 per ml.

Suspending Medium	Recovery	
		Survival
	cells/ml	%
Skim milk (Difco), 20%	3.0×10^2	0.00003
Sucrose, 24%	4.0×10^6	0.4
Seven Seas	6.0×10^6	0.6
Seven Seas sucrose	1.2×10^7	1.2

skim milk, w/v (Difco); 2) 24% sucrose, w/v, in distilled water; 3) Seven Seas marine medium; and 4) Seven Seas sucrose medium (12% sucrose and 0.5 concentration Seven Seas medium, final volume).

The method of freeze-drying was one routinely employed for bacteria and fungi at the ATCC laboratories. Aliquots of 0.1 ml were dispensed in tear-drop-type ampules (1-ml tear-drop ampule, catalog No. 4020, Bellco Glass, Inc., Vineland, N. J.) and frozen while being agitated in a Dry Ice-ethylene glycol bath. In this manner, a large drying surface with a rapid drying rate was achieved. The freeze-drying was done on a single stage manifold-type apparatus using an 18-hr drying cycle. Once a vacuum of 50 μ of Hg was attained (McLeod Gage, Stokes Machine Company, Philadelphia), the bath was allowed to warm to room temperature, a process requiring approximately 2 hrs. After the drying cycle was completed the ampules were sealed under the original vacuum with a gas-air burner.

Upon completion of the drying process, two vials of each sample were rehydrated with the designated medium. The rehydration media used were: 1) 0.85% NaCl, w/v, in distilled water; 2) 6%, 12%, 24%, 36%, and 48% sucrose, w/v, in distilled water; 3) Seven Seas marine medium (see above); and 4) Seven Seas sucrose medium (see above). Rehydration was carried out at room temperature (25°C) and liquid nitrogen temperature (−196°C). In the case of room temperature rehydration, the ampules were opened and 1.0 ml of the appropriate rehydration medium was added to each dry sample. The dilutions were made with the rehydration medium.

Low temperature rehydration was carried out as follows. The ampules were opened at room temperature and the bulb portion of the ampule containing the culture was immersed in liquid nitrogen (−196°C) for 5 min. Then 1.0 ml of the rehydration fluid at room temperature was pipetted rapidly into the ampule while it was still in the liquid nitrogen. As the rehydrating fluid was added, care was taken to prevent loss by spurting of the sample material. After the sample was allowed to remain frozen for 1 min, it was thawed rapidly in a water bath at 37°C. As soon as the sample reached 25°C, it was diluted and plated with the appropriate rehydration medium.

Recovery was based on colony counts of *S. atlanticum* and was expressed as viable cells per milliliter. Seven Seas marine medium with 2.3% agar was used for plating. Controls were plated in duplicate with physiological saline and broth as diluents. For each determination at least two ampules were titered and each determination was repeated at least once. Data given represent the average of all the determinations.

Results

The recoveries of freeze-dried specimens of *S. atlanticum* suspended in various media and rehydrated with Seven Seas medium are presented in Table 1. The data show that 20% skim milk (Difco), a medium found in this laboratory to be generally satisfactory for freeze-drying most microorganisms, was not effective for preserving *S. atlanticum*. Of the media tested, the Seven Seas sucrose medium gave the highest recovery in terms of viable cell count. In view of these results, all of the subsequent experiments were conducted using material freeze-dried in this medium.

Results of experiments using various concentrations of sucrose for rehydration and dilution are presented in Figure 1. The results in Figure 1 indicate that a higher recovery was obtained as the concentration of sucrose was increased up to 24%. Any further increase in the sucrose concentration resulted in the gradual decrease in recovery.

The effects of varying the temperature of rehydration are shown in Table 2. A definite increase in recovery was effected by "low temperature rehydration." Even physiological saline showed slightly better recovery with this method. However, the best recovery was obtained with the Seven Seas sucrose medium.

Discussion

Several factors are involved in the survival of microorganisms after freeze-drying. Of the most important is rehydration, during which there are many changes in the concentration of solutes. These changes not only affect the osmotic equilibrium of the cells but may also affect specific cellular constituents or systems of the organism.

Sucrose was chosen as a rehydration and dilution medium primarily because of its known protective effect in freeze-drying and its osmotic properties. It was thought that sucrose might regulate osmotically the rehydration of freeze-dried materials to give higher recovery. The osmotic pressures of the 24% sucrose solution (995 milliosmols) (Advanced osmometer, model 31LA, Advanced Instruments, Inc., Newton Highlands, Mass.), the Seven Seas medium (1034 milliosmols), and the Seven Seas sucrose medium (986 milliosmols) are all in the same range. Our experience indicates that the rehydration medium with an osmotic pressure in this range is necessary for optimum recovery. Efforts to increase cell recovery by rehydration in media of higher osmotic pressure were unsuccessful even though diminishing concentrations of sucrose were used to approach the concentrations of the plating medium. Rehydration and dilution in physiological saline (265 milliosmols) resulted in cell recovery corresponding to values for sucrose solutions between 6 and 12% (197 to 424 milliosmols). The correlation of one solution to another in terms of milliosmols suggests that the increased recovery using sucrose in the freeze-drying and rehydration medium was attributed to its capabilities as an osmotic buffer. As previously mentioned, Record et al.[10] first discussed the possible effects of osmotic pressures during the rehydration of freeze-dried *E. coli.*

The purpose of our rehydration technique was to attempt to replace water inside the cell in much the same way it was removed during freeze-drying. Liquid nitrogen temperature was chosen as the temperature at which the medium was added because 1 ml of rehydrating medium could be added to the freeze-dried specimen with a gross temperature increase to no higher than 0°C. It is apparent from the data that the recovery of viable organisms from freeze-dried material

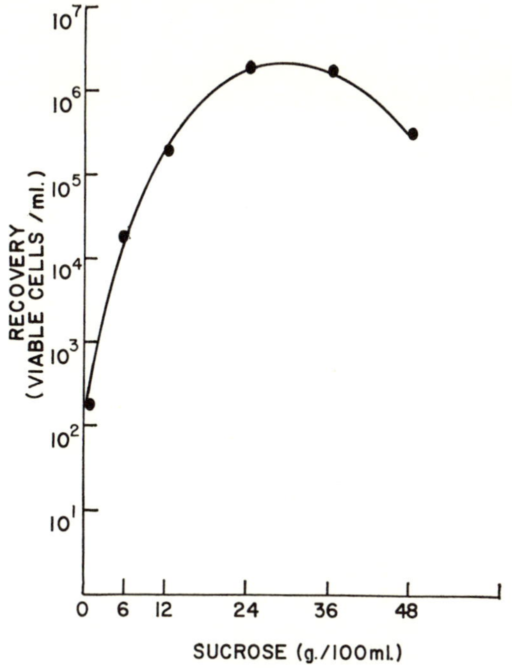

FIG. 1. Effects of rehydration and dilution with various concentrations of sucrose in distilled water at 25°C on the survival of freeze-dried *Spirillum atlanticum.* Control cell count 1.0 × 10^9 per ml.

TABLE 2

The effect of various rehydration media, added at different temperatures, on recovery of *S. atlanticum* freeze-dried in Seven Seas sucrose medium. Control cell count 1.0 × 10^9 per ml.

Rehydration and Dilution Medium	Recovery			
	25°C		−196°C	
		Survival		Survival
	cells/ml	%	*cells/ml*	%
Physiological saline.	2.5 × 10^5	0.03	9.0 × 10^5	0.09
Sucrose, 24%......	4.8 × 10^6	0.5	1.3 × 10^7	1.3
Seven Seas..	6.0 × 10^6	0.6	2.2 × 10^7	2.2
Seven Seas sucrose	1.2 × 10^7	1.2	7.6 × 10^7	7.6

was increased by adding the rehydrating medium at a very low temperature and thawing rapidly. This method of "low temperature rehydration" is in reality a rehydration occurring at an unspecified temperature. We cannot specifically isolate the area in which the beneficial effect is obtained.

During the rehydration procedure, before the addition of the rehydration medium, some gas was consensed in the vial from the atmosphere, causing a great deal of turbulence. However, this did not interefere with the consistency of our results over a considerable number of experiments. Recovery did not seem to be affected by thawing either to 25°C or 37°C.

The data showed that with the various rehydration media used recoveries were beneficially affected by the "low temperature rehydration" procedure. In all cases there was a recovery increase over rehydration at 25°C.

These data further indicate that osmotic effects, as well as temperature, play an important part in the recovery of freeze-dried material. More emphasis should be placed on rehydration effects and their relationship to the concentrated solutes of freeze-dry medium.

Summary

Recovery of freeze-dried cultures of *S. atlanticum* Williams and Rittenberg, ATCC 12753, suspended in Seven Seas sucrose protective medium, were greatly influenced by varying the conditions of rehydration. Effects of the rehydration medium were determined using different concentrations of sucrose in water and Seven Seas medium. Higher recovery was obtained as the concentration of sucrose was increased up to 24%. Rehydration carried out by adding the medium to freeze-dried cultures held at —196°C and then thawing rapidly resulted in better recovery than rehydration carried out at room temperature.

References

1. American Type Culture Collection. Catalog of cultures, 7th ed. 1964.
2. Floodgate, G. D., and Hayes, P. R. The preservation of marine bacteria. J. Appl. Bact., *24:* 87–93, 1961.
3. Fry, R. M. Freezing and drying of bacteria. In Cryobiology, pp. 665–696. Academic Press, Inc., London, 1966.
4. Leach, R. H., Ohye, D. F., and Scott, W. J. The death of micro-organisms during drying in relation to solute concentration and drying temperature. J. Gen. Microbiol., *21:* 658–665, 1959.
5. Leach, R. H., and Scott, W. J. The influence of rehydration on the viability of dried micro-organisms. J. Gen. Microbiol., *21:* 295–307, 1959.
6. Meryman, H. T. Freeze-drying. In Cryobiology, pp. 609–663. Academic Press, Inc., London, 1966.
7. Miles, A. A., and Misra, S. S. The estimation of the bactericidal power of the blood. J. Hyg. (Camb.), *38:* 732, 1938.
8. Monk, G. W., and McCaffery, P. A. Effect of sorbed water on the death rate of washed *Serratia marcescens*. J. Bact., *73:* 85–88, 1957.
9. Pedersen, T. A. Factors affecting viable cell counts of freeze-dried *Cryptococcus terricolus* cells. Antonie van Leeuwenhoek, *31:* 232–240, 1965.
10. Record, B. R., Taylor, R., and Miller, D. S. The survival of *Escherichia coli* on drying and rehydration. J. Gen. Microbiol., *28:* 585–598, 1962.
11. Weiss, F. A. Maintenance and preservation of cultures. In Manual of microbiological methods by the Society of American Bacteriologists, pp. 99–119. McGraw-Hill Book Company, Inc., New York, 1957.

The authors thank Mr. Hormoz Adle for his technical assistance.

Part III

THE STRUCTURAL AND PHYSIOLOGICAL RESPONSES OF MARINE BACTERIA TO THEIR SALINE ENVIRONMENT

Editor's Comments on Papers 19 Through 21

NUTRITIONAL NEEDS OF MARINE BACTERIA

For almost 80 years the marine microbiologist has been trying to answer the very basic question: what is a marine bacterium? What properties, requirements, and structural or chemical differences exist that can help to explain why the natural habitats for certain organisms are seawater and oceanic sediments and why will they not grow elsewhere? This theme has run throughout most of the previously mentioned papers, but in this section it comprises the main subject of the papers selected. These papers have each advanced new concepts about the uniqueness of marine bacteria: physiological processes requiring specific ions, ability to grow and survive under extreme pressures, or inability to survive at temperatures above 15 to 20°C, or the necessity for high ionic levels for the maintenance of cellular integrity. The intensification of interest in the biochemistry of marine bacteria and their physiological responses to their unique environment was evidenced by the publication in 1974 of *Effect of the Ocean Environment on Microbial Activities* (5).

Although the need for seawater for the growth of marine bacteria on initial isolation was generally recognized quite early, the overwhelming question was why. Was there something unusual about the biochemistry of these organisms that required salt or was it simply an osmotic effect? Two major schools of thought developed, and it was not until the quantitative work of MacLeod and Onofrey (17–19) that solid scientific evidence could be presented which showed that generally a need for specific ions existed. Occasionally, however, there were organisms that were truly marine but for which the apparent salt requirement could be satisfied by having an equivalent osmotic pressure. In other words, both explanations were right and it depended upon the organism, not the scientist!

Few papers have had such an impact on this field as did Paper 19, for it created the pattern for much of the subsequent research on salinity effects, nutritional patterns, and even general chemical composition. Following the lead of Richter (Na^+ requirements) (25), Bukatsch (K^+) (3), and Johnson and Harvey (osmotic effects) (9), all of whom studied luminescent marine bacteria, MacLeod and Onofrey deomonstrated that the requirement of marine bacteria for seawater was not just for the sodium ion or a total ionic effect, but that different bacteria had different requirements, which could be either for the chloride ion (19), sodium or potassium ions (14, 17, 20), or sulfate (15, 17). Several of these marine isolates could not be "trained" or "adapted" to grow on freshwater media, and freshwater isolates had been previously shown not to require such elevated concentrations of either Na^+ or K^+ (20). In addition, replacement by Li^+, Cs^+, or Rb^+ was ineffective or even toxic to the marine isolates (18).

MacLeod and his co-workers quickly followed up this lead with studies on the oxidation of propionic acid (29) and succinate and alanine (30), and showed that the oxidative process was cation dependent (14). The glyoxalate cycle could be induced in B16 only when it was grown in a low-nutrient medium in which acetate was the sole carbon source (16). They also isolated a flavobacterium that required amino acids, biotin, thiamine, and nucleotides, as well as the usual anions, for growth (13). However, the marine pseudomonad B16 was found to be inhibited by some of the same amino acids (31), so no general conclusions about amino acid requirements among marine bacteria could be reached.

About this time, Payne, in Paper 20, demonstrated the inducible nature of the enzymes for uronic acid oxidation in another pseudomonad. The dependence of enzyme induction on Na^+ and the requirement of the cells for K^+ for oxidation led him to conclude that these ionic functions constituted at least one explanation for the "marine" nature of some bacteria. Here was experimental proof of a specific function for the higher levels of sodium and potassium ions required by marine bacteria but not by terregenous forms.

Further evidence of the salt requirement of marine bacteria was provided by Merkel et al. (22) in their studies on oxygen uptake with casamino acids, glucose, galactose, or glycine present. Meanwhile, Isenberg and Lavine (8) had demonstrated another function for cations in that the production of pigment by *Flavobacterium piscicida (Pseudomonas piscicida)* was dependent on the calcium ion concentration. Furthermore, Massarini and Cazzulo (21) have shown the activation by KCl of the citrate synthase system in a marine pseudomonad. It is not clear, as of now, whether the observed effect is due to the K^+, Cl^-, or osmotic factors. However, Takacs et al. (26) have succeeded in demonstrating that the intracellular level of K^+ is about double that of the external environment, so it would seem likely that additional enzymatic studies will implicate K^+ further as either an essential ion or a stabilizing–activating factor for cellular metabolism.

A summary of the extremely active preceding 11 years was published by MacLeod in 1965 (Paper 21). Further studies have subsequently shown that Na^+ can be required for the synthesis of a permease (24) as well as its operation (7), for the maintenance of cellular integrity (6, 27), and for general macromolecular synthesis (33). In addition, Unemoto et al. (32) studied the distribution of phosphohydrolases in *Vibrio alginolyticus* and noted that the 3′-nucleotidase and 5′-nucleotidase in this organism required anions for maximal activity. Cations are also necessary for the proper functioning of the cell envelope nucleotidases (28) and for the stabilization of isocitrate dehydrogenase (15).

Although several authors have commented on the lytic properties of marine isolates, and have noted that many enzymes become inactivated when tested in distilled-water-based systems, Pratt (23) recently pointed out that it is highly unlikely that the organisms in nature would ever be exposed to such conditions, except in the intertidal or estuarine environment.

Investigation of the comparative biochemistry of marine bacteria vis-à-vis terregenous forms is essential, however, if we are to understand the selective advantage that the marine strains have in the sea. During the last 10 to 12 years, there have been a few reports on other physiological properties of these marine organisms, such as Ayers's studies (1) on the requirement of two marine flavobacteria for vitamin B_{12} and resulting exogenous respiration. Electron-transport systems and energy-producing mechanisms have received some attention. Recently, Weston and Knowles (34) reported the characteristics of the cytochrome system in *Beneckea (Pseudomonas) natriegens*. The cytochrome *c* has a high redox potential and it actively binds carbon monoxide. It is interesting to note, in conjunction with this study, that, in 1971, Junge et al. (10) had demonstrated the production of both carbon monoxide and hydrogen by bacteria from ocean waters. Meanwhile, Linton et al. (11) have found a second CO-binding cytochrome with an absorption maximum at 418 nm, which is induced when the oxygen tension drops. Because of the inductive nature of this cytochrome, and its increased level of resistance to cyanide and carbon monoxide respiration, the authors postulate that different terminal oxidases may be available to the bacterium in response to differing oxygen tensions. Thus the possibility exists of an ecological impact resulting directly from the occurrence of the cytochrome systems in marine pseudomonads.

Interest has also been expressed by a few workers in comparing the control mechanism of enzyme synthesis between marine and terrestrial bacteria: aspartokinases (2), 3-deoxy-D-arabinoheptulosomate 7-phosphate synthetase (4), and an extracellular endopeptidase (12). Unfortunately, none of these investigators has examined the specific ionic effects on the general phenomena they observed, that is, the effect Na^+ or other ions might exert toward obviating or accentuating these controls. Here, then, is a virtually untapped area for research on marine bacteria—the control of metabolic systems and protein synthesis as influenced by specific ions in conjunction with "normal" control mechanisms. Are feedback control or catabolite repression operational if certain ions are either missing or present at reduced levels? Only a tentative start has been made toward answering these types of questions.

REFERENCES

1. Ayers, W. A. 1962. The influence of cobamides on the endogenous and exogenous respiration of a marine bacterium. *Can. J. Microbiol. 8*:861–867.
2. Baumann, L., and Baumann, P. 1973. Regulation of aspartokinase activity in the genus *Beneckea* and marine, luminous bacteria. *Arch. Mikrobiol. 90*:171–188.
3. Bukatsch, F. 1936. Uber den Einfuess von Salzen auf die Lichtentwicklung von Bakterien. *Sitzungsber. Oesterr. Akad. Wiss, Math.-Naturwiss. Kl., Abt. I, 145*:259–277.
4. Chludzinski, A. M., Salter, D. S., and Nasser, De. L. 1972. Feedback regulation of 3-deoxy-D-arabino-heptulosonate 7-phosphate synthetase from a marine bacterium, *Vibrio* MB22. *J. Bacteriol. 109*:1162–1169.

5. Colwell, R. R., and Morita, R. Y. 1974. *Effect of the Ocean Environment on Microbial Activities.* University Park Press, Baltimore, Md., 587 pp.
6. DeVoe, I. W., and Oginsky, E. L. 1969. Antagonistic effect of monovalent cations in maintenance of cellular integrity of a marine bacterium. *J. Bacteriol. 98*:1355–1367.
7. Gow, J. A., and MacLeod, R. A. 1974. Growth of a marine pseudomonad at suboptimal Na^+ concentrations. *Proc. 70th Ordinary Meet., Soc. Gen. Microbiol., Apr. 8–10, Lond.*, p. 2.
8. Isenberg, H. D., and Lavine, L. S. 1963. The production of pigment by a marine bacterium in response to calcium. *Ann. N.Y. Acad. Sci. 109*: 46–48.
9. Johnson, F. H., and Harvey, E. N. 1937. The osmotic and surface properties of marine luminous bacteria. *J. Cell. Comp. Physiol.* 9:363–380.
10. Junge, C., Seiler, W., Schmidt, U., Bock, R., Greese, K. D., Radler, F., and Rüger, H. J. 1971. Kohlenmonoxid- und Wasserstoffproduktion mariner mikroorganismen im Nährmedium mit synthetischen Seewasser. *Naturwissenschaften 59*:514–515.
11. Linton, J. D., Harrison, D. E. F., and Bull, A. T. 1974. Effects of dissolved oxygen tensions on the growth and respiration of the facultative anaerobic marine bacterium *Beneckea natriegens. Proc. 70th Ordinary Meet., Soc. Gen. Microbiol., Apr. 8–10, Lond.*, p. 13.
12. Litchfield, C. D., and Prescott, J. M. 1970. Regulation of proteolytic enzyme production by *Aeromonas proteolytica*. I. Extracellular endopeptidase. *Can. J. Microbiol. 16*:17–22.
13. MacLeod, R. A., Hogenkamp, H., and Onofrey, E. 1958. Nutrition and metabolism of marine bacteria. VII. Growth response of a marine flavobacterium to surface active agents and nucleotides. *J. Bacteriol. 75*:460–466.
14. MacLeod, R. A., and Hori, A. 1960. Nutrition and metabolism of marine bacteria. VIII. Tricarboxylic acid cycle enzymes in a marine bacterium and their response to inorganic salts. *J. Bacteriol. 80*:464–471.
15. MacLeod, R. A., Hori, A., and Fox, S. M. 1960. Nutrition and metabolism of marine bacteria. IX. Ion requirements for obtaining and stabilizing isocitric dehydrogenase from a marine bacterium. *Can. J. Biochem. Physiol. 38*:693–701.
16. MacLeod, R. A., Hori, A., and Fox, S. M. 1960. Nutrition and metabolism of marine bacteria. X. The glyoxylate cycle in a marine bacterium. *Can. J. Microbiol. 6*:639–644.
17. MacLeod, R. A., and Onofrey, E. 1956. Nutrition and metabolism of marine bacteria. II. Observations on the relation of sea water to the growth of marine bacteria. *J. Bacteriol. 71*:661–667.
18. MacLeod, R. A., and Onofrey, E. 1957. Nutition and metabolism of marine bacteria. III. The relation of sodium and potassium to growth. *J. Cell. Comp. Physiol. 50*:389–401.
19. MacLeod, R. A., and Onofrey, E. 1957. Nutrition and metabolism of marine bacteria. VI. Quantitative requirements for halides, magnesium, calcium, and iron. *Can. J. Microbiol. 3*:753–759.
20. MacLeod, R. A., Onofrey, E., and Norris, M. E. 1954. Nutrition and metabolism of marine bacteria. I. Survey of nutritional requirements. *J. Bacteriol. 68*:680–686.

21. Massarini, E., and Cazzulo, J. J. 1974. Activation of citrate synthase from a marine pseudomonad by adenosine monophosphate and potassium chloride. *FEBS Lett. 39*:252–254.
22. Merkel, J. R., Carlucci, A. F., and Pramer, D. 1957. Respiratory characteristics of marine bacteria. *Nature 180*:1489–1490.
23. Pratt, D. 1974. Salt requirements for growth and function of marine bacteria, pp. 3–15. In: R. R. Colwell and R. Y. Morita (eds.), *Effect of the Ocean Environment on Microbial Activities.* University Park Press, Baltimore, Md., 587 pp.
24. Rhodes, M. E., and Payne, W. J. 1967. Influence of Na^+ on synthesis of a substrate entry mechanism in a marine bacterium. *Proc. Soc. Exp. Biol. Med. 124*:953–955.
25. Richter, O. 1928. Natrium ein notwendiges Nährelement für eine marine mikroärophile Leuchtbakterie. *Denkschr. Akad. Wiss. Wein, Math.-Naturwiss. Kl. 101*:261–294.
26. Takacs, F. P., Matula, T. I., and MacLeod, R. A. 1964. Nutrition and metabolism of marine bacteia. XIII. Intracellular concentrations of sodium and potassium ions in a marine pseudomonad. *J. Bacteriol. 87*:510–518.
27. Thompson, J., Costerton, J. W., and MacLeod, R. A. 1970. K^+-dependent deplasmolysis of a marine pseudomonad plasmolyzed in a hypotonic solution. *J. Bacteriol. 102*:843–854.
28. Thompson, J., Green, M. L., and Happold, F. C. 1969. Cation-activated nucleotidase in cell envelopes of a marine bacterium. *J. Bacteriol.* 99: 834–841.
29. Tomlinson, N., and MacLeod, R. A. 1955. The oxidation of propionic acid by a marine bacterium. *Biochim. Biophys. Acta 18*:570–571.
30. Tomlinson, N., and MacLeod, R. A. 1957. Nutrition and metabolism of marine bacteria. IV. The participation of Na^+, K^+, and Mg^{++} salts in the oxidation of exogenous substrates by a marine bacterium. *Can. J. Microbiol. 3*:627–638.
31. Tomlinson, N., and MacLeod, R. A. 1957. Nutrition and metabolism of marine bacteria. V. The inhibition of growth of a marine bacterium by amino acids and the development of resistant strains. *Arch. Biochem. Biophys. 70*:477–490.
32. Unemoto, T., Hayashi, M., Kozuka, Y., and Hayashi, M. 1974. Localizations and salt modifications of phosphohydrolases in slightly halophilic *Vibrio alginolyticus*, pp. 46–71. In: R. R. Colwell and R. Y. Morita (eds.), *Effect of the Ocean Environment on Microbial Activities.* University Park Press, Baltimore, Md., 587 pp.
33. Webb, C. D., and Payne, W. J. 1971. Influence of Na^+ on synthesis of macromolecules by a marine bacterium. *Appl. Microbiol. 21*:1080–1088.
34. Weston, J. A., and Knowles, C. J. 1973. A soluble CO-binding C-type cytochrome from the marine bacterium *Beneckea natriegens. Biochim. Biophys. Acta 305*:11–18.

19

Reprinted from *J. Bacteriol.*, **71**(6), 661–667 (1956)

NUTRITION AND METABOLISM OF MARINE BACTERIA

II. Observations on the Relation of Sea Water to the Growth of Marine Bacteria[1]

ROBERT A. MacLEOD and E. ONOFREY

The Fisheries Research Board of Canada, Pacific Fisheries Experimental Station, Vancouver, B.C.

Received for publication October 21, 1955

Marine bacteria have been arbitrarily defined as bacteria from the sea which on initial isolation require for growth a medium containing sea water as the diluent (ZoBell and Upham, 1944). The function of sea water in media for the growth of these organisms has never been clearly established. It has been stated, however, that neither isotonic salt solution nor artificial sea water is as good as natural sea water for the cultivation of recently isolated marine bacteria (ZoBell, 1946).

The nutritional requirements of a number of marine bacteria have been determined recently (MacLeod *et al.*, 1954). In the course of this study it was found that a mixture of inorganic salts approximating the composition of sea water could be used in place of natural sea water in a chemically defined medium suitable for the growth of these organisms. A limited amount of information regarding a requirement of the bacteria for specific inorganic ions was also obtained.

The present report presents the results of a more detailed study of the relation of the constituents of sea water to the growth of marine bacteria.

METHODS

Cultures. The sources of the organisms and the general methods used in their study have been described (MacLeod *et al.*, 1954). Six of the organisms used in the previous study were selected for this work. The organisms used have been tentatively identified as B9, a flavobacterium; B10 and B20, corynebacteria; B16 and B26, mycoplana; and B30, a pseudomonad. Their organic growth requirements are known and were reported in the previous communication. The organisms all satisfy the criteria established for classifying them as marine bacteria, that is, on initial isolation they grew in a complex medium prepared with sea water, but not with fresh water.

Inocula. The inoculum medium used has been described previously (MacLeod *et al.*, 1954). In this study cells were washed by centrifuging, removing the supernatant liquid and resuspending them in 1.097 M glycerol solution. This washing operation was repeated twice and one drop of the final suspension used to inoculate each flask. The concentration of glycerol in the wash solution was equal to the total molar ion concentration of sea water.

Basal medium.[2] This medium was the simplest capable of supporting the growth of all six organisms. All solutions were prepared using water demineralized by passage through a bed of ion exchange resins (Bantam demineralizer, Barnstead Still & Sterilizer Co., Boston, Mass.). The inorganic salts used were of reagent grade quality and were employed after recrystallization from demineralized water. The three amino acids were recrystallized from a solution adjusted to pH 2 to 3 with HCl. Glucose was purified by consecutive passage through "amberlite IRC-50" (H form) and "amberlite IR-4B" (OH form) ion exchange resin columns. Glycerol was distilled under vacuum in an all-glass apparatus before use. The NH_4OH and HCl solutions for neutralizing the medium were prepared by dissolving NH_3 and HCl gases, respectively, in demineralized water.

Assay procedure. The procedures used in this study differed from those employed previously (MacLeod *et al.*, 1954) in only one important respect. In this investigation the growth response in a particular assay was measured at frequent

[1] A preliminary report of these findings was made at the Ninth Science Conference, B.C. Academy of Science, University of British Columbia, April 28 and 29, 1955.

[2] Composition of the basal medium (in mg per 10 ml of final medium): Glucose, 30 mg; DL-α-alanine, 24 mg; DL-aspartic acid, 9.6 mg; L-glutamic acid, 12 mg; $(NH_4)_2SO_4$, 8.7 mg; $(NH_4)_2HPO_4$, 1.3 mg.

TABLE 1

Comparison of the ability of natural sea water, artificial sea water and fresh water, when each is acting as the diluent in a chemically defined medium, to promote the growth of six marine bacteria

Diluent	Strength	Fe^{++} Added	Organism Tested: B9, Incubation time (hr) 71	B9 144	B10 71	B10 144	B16 71	B16 144	B20 71	B20 144	B26 71	B26 144	B30 71	B30 144
		µg/10 ml	Per cent incident light transmitted†											
Natural sea water	Full*	0	78	75	100	86	72	50	16	—	93	83	100	85
Natural sea water	Full	50	19	21	90	21	28	30	2	—	87	29	98	80
Natural sea water	Half	0	66	53	58	36	73	59	36	—	68	51	58	35
Natural sea water	Half	50	19	20	5	24	30	50	27	—	19	20	29	49
Artificial sea water	Full	0	54	50	97	70	77	52	56	—	92	82	93	80
Artificial sea water	Full	50	25	23	100	85	56	31	11	—	89	71	92	67
Artificial sea water	Half	0	72	68	34	71	82	56	30	—	64	56	65	48
Artificial sea water	Half	50	19	21	12	50	32	62	28	—	16	22	21	49
Distilled water	—	—	100	100	100	100	100	100	100	—	100	100	100	100

* Salinity: 35.4 g total salts per kg.

† Evelyn colorimeter readings, 660 mµ filter, uninoculated blank = 100.

intervals during incubation instead of after a fixed period of incubation. The variation in procedure was made possible by the use of 50 ml Erlenmeyer flasks, to each of which a matched 18 by 150 mm pyrex culture tube had been attached. By tilting the contents of the flask into the side tube at appropriate intervals, and inserting the side tube into a colorimeter, measurements of the change in turbidity of a culture with time could be made without danger of contamination (Sherman and Grant, 1951). Unless otherwise indicated, the results reported for each organism in each experiment were obtained after a period of incubation which resulted in the achievement of maximum growth by the organism. Certain of the organisms, however, autolyzed very rapidly after growth in the tubes had ceased. If growth of the latter organisms was slower in some tubes than in others, often some cultures would be autolyzing with a consequently decreasing turbidity while others were still growing. In these cases, results in the particular experiment were reported after a period of incubation which represented the best possible compromise between those cultures which were still growing and those which were autolyzing.

Salt solutions. The natural sea water used in these experiments was collected in the vicinity of Departure Bay, Vancouver Island. We are indebted to the Pacific Biological Station for making the sample available to us. Despite the fact that the sea water collection was made as far from land as possible, the salinity was below that considered average for the open sea. The sea water used was therefore evaporated to the desired salinity (35 to 40 g of total salts per kg) before use.

For the composition of the artificial sea water used in these experiments, see Lyman and Fleming (1940).[3]

RESULTS

The comparative ability of natural sea water, artificial sea water and fresh water when each is acting as a diluent in a chemically defined medium to promote the growth of the six marine bacteria investigated is shown in table 1. In this table, the results are reported after two periods of incubation except in the case of B20, which had reached maximum growth in all tubes in 71 hours. In certain cases it will be noted that the turbidity of a culture was less after 144 hours than after 71. This loss in turbidity on incubation was due to autolysis. The results show that growth of the organisms occurred in media prepared with

[3] Composition of artificial sea water used: NaCl, 23.476 g/kg; Na_2SO_4, 3.917 g/kg; $NaHCO_3$, 0.192 g/kg; KCl, 0.664 g/kg; KBr, 0.096 g/kg; $MgCl_2$, 4.981 g/kg; $CaCl_2$, 1.102 g/kg; $SrCl_2$, 0.024 g/kg; H_3BO_3, 0.026 g/kg.

both natural and artificial sea water, but not with fresh water. Relatively poor growth of all organisms except B20 was obtained in the media supporting growth unless a supplement of iron was added. The table also shows that for all of the organisms except B20 both the rate and extent of their growth was increased by using half rather than full strength sea water, irrespectively of whether the sea water used was natural or artificial. When both natural and artificial sea water were supplemented with iron and tested at half strength, there was no significant difference in the ability of the two diluents to promote growth of any of the organisms. In separate experiments it was also noted that natural and artificial sea water could be used interchangeably for growth of the marine bacteria in the complex medium used in their isolation. It would thus appear that the observed dependence of the marine bacteria investigated here on the presence of sea water in the medium on initial isolation was due to the ability of sea water to supply the inorganic environment required by the organisms for growth. There is no evidence that sea water was needed to supply unknown organic factors necessary or stimulatory for the growth of these marine bacteria.

When various groups of salts were omitted in turn from artificial sea water, the growth responses obtained with the various organisms are shown in table 2. It is evident that removing either Na^+ or K^+ salts completely prevented growth in all cases. When Mg^{++}, Ca^{++} and Sr^{++} salts were omitted, four organisms failed to grow and two grew only slightly. The removal of H_3BO_3 did not significantly affect the growth of any of the organisms. Since removing various salts, particularly the Na^+ and Mg^{++} salts, would render the solution hypotonic, the osmotic pressure of the solution was restored by adding a level of glycerol equal to the molar concentration of the total number of ions removed. In no case did the addition of glycerol permit growth of the organisms in the absence of the salts. It was established in separate experiments that none of the organisms could utilize glycerol as a source of carbon in the medium, and that cells harvested from a medium containing glycerol were unable to oxidize glycerol in a Warburg respirometer.

Since each of the omissions shown in table 2 involved more than one salt, it was of interest to know whether more than one salt in each group was required for growth by the organisms. Three sodium salts, NaCl, Na_2SO_4, and $NaHCO_3$, are used in the preparation of the artificial sea water.[3] NaCl and Na_2SO_4 were each added separately to a medium deficient in Na^+ salts at a level which provided Na^+ at the concentration normally made available by the mixture. $NaHCO_3$ could not be tested at this level because of the alkalinity of the resulting solution. The results (table 3) show that either salt was capable of satisfying the Na^+ salt requirements of all of the organisms if the period of incubation was

TABLE 2

Growth response of marine bacteria in the presence of glycerol when various salts are removed from the artificial sea water used as a diluent in the medium

Composition of Artificial Sea Water*	Glycerol Added†	Organism Tested					
		B9	B10	B16	B20	B26	B30
		Per cent incident light transmitted‡					
Complete	−	19	22	26	32	15	15
No Na^+ salts	−	100	100	100	100	100	100
No Na^+ salts	+	100	100	100	100	100	100
No K^+ salts	−	100	100	100	100	100	100
No K^+ salts	+	100	100	100	100	100	100
No Mg^{++}, Ca^{++} or Sr^{++} salts	−	100	100	100	100	91	81
No Mg^{++}, Ca^{++} or Sr^{++} salts	+	100	100	100	100	91	86
No H_3BO_3	−	19	13	24	28	15	17

* Supplemented with Fe^{++} 50 μg per flask.

† The molar concentration of glycerol added in each case was equal to the molar concentration of the total number of ions removed.

‡ See table 1. Incubation times: B10 and B20, 40 hr; B9, B16 and B30, 48 hr; B26, 72 hr.

TABLE 3

The ability of single salts to replace groups of salts for the growth of marine bacteria in artificial sea water medium

Composition of Artificial Sea Water*	Additions	Organism Tested: B9	B10	B16	B20	B26	B30
		Per cent incident light transmitted†					
Complete	—	19	6	24	26	17	17
No Na^+ salts	—	100	100	100	100	100	100
No Na^+ salts	NaCl (2.29 mM)	18	6	27	7	16	19
No Na^+ salts	Na_2SO_4 (1.145 mM)	20	16	28	100§	16	16
No Na^+ salts	KCl (2.29 mM)	100	100	100	100	100	100
No K^+ salts	—	100	100	99	97	100	97
No K^+ salts	KCl (48.6 μmoles)	18	5	23	19	16	17
No K^+ salts	KBr (48.6 μmoles)	20	4	23	26	16	17
No K^+ salts	NaCl (48.6 μmoles)	100	100	100	100	100	100
No Mg^{++}, Ca^{++}, or Sr^{++}	—	100	99	98	99	95	91
No Mg^{++}	—	100	99	96	99	68	83
No Ca^{++}	—	96	95‡	21	14	16	17
No Sr^{++}	—	19	16	23	11	16	17

* See table 2.
† See table 1. Incubation times: B20, 40 hr; B10, 48 hr; B16 and B30, 66 hr; B9 and B26, 90 hr.
‡ Growth after 90 hr, 32.
§ Growth after 120 hr, 28.

sufficiently long. Maximum growth was achieved much more rapidly in some cases, however, with NaCl than with Na_2SO_4. This was particularly noticeable in the case of B20, where maximum growth in the presence of Na_2SO_4 was not reached until after an incubation period of 120 hours. A level of KCl equal to that of the NaCl added did not permit growth of the organisms. These findings indicate that lack of growth in the absence of Na^+ salts was due primarily to a deficiency of Na^+ in the medium.

Two K^+ salts are present in artificial sea water, KCl and KBr. Growth resulted when the total K^+ provided by these two salts was made available by either one, but not when the K^+ salts were replaced by equimolar concentrations of NaCl (table 3). These results indicate that, in addition to a requirement for Na^+, all of the organisms tested need K^+.

When Mg^{++}, Ca^{++}, and Sr^{++} salts were omitted from the medium, little or no growth of any of the organisms was obtained. A deficiency of Mg^{++} alone prevented growth of four of the organisms and greatly restricted the growth of the other two. Without added Ca^{++}, B9 would not grow, while B10 grew only after a relatively long incubation period. A deficiency of Sr^{++} had no significant effect on the growth of any of the organisms. One can conclude, then, that at least four and probably all of the organisms tested have an absolute requirement for Mg^{++} for growth, while one needs Ca^{++} as well.

A salt solution was prepared[4] which provided those ions in sea water which this study has shown to be required for the growth of the six marine bacteria. Na^+, K^+, Mg^{++}, and Ca^{++} were each added to the medium at the level at which they are present in one-half concentrated artificial sea water. Growth of the bacteria in a medium prepared with this salt solution was compared with growth when half strength artificial sea water (supplemented with Fe^{++}) was used as the diluent. The results (table 4) reveal that growth in the medium prepared with the salt solution was in every case equal to or better than growth in the medium prepared with artificial sea water. A supplement of yeast extract ash was also added to tubes containing the salt solution to determine whether further growth could be obtained by supplying traces of other inorganic ions. In the case of only one organism, B16, was any ap-

[4] Composition of salt solution containing ions found to be required for growth by the marine bacteria studied: NaCl, 12.705 g/L; KCl, 0.72 g/L; $FeSO_4(NH_4)_2SO_4$, 0.0254 g/L; $MgCl_2$, 2.49 g/L; $CaCl_2$, 0.551 g/L.

TABLE 4

Comparison of the growth response of marine bacteria in media prepared with artificial sea water and with salts providing ions known to be required for growth

Diluent	Organisms Tested					
	B9	B10	B16	B20	B26	B30
	Per cent incident light transmitted*					
Artificial sea H_2O (half strength + Fe^{++} (50 μg per 10 ml)	19	12	32	32	43	18
Salt solution	20	16	31	29	16	16
Salt solution + 1 ml yeast extract ash	17	23	23	40	15	15

* See table 1. Incubation times: B20, B26, and B30, 66 hr; B9, B10, and B16, 71 hr.

preciable stimulation obtained when yeast extract ash was added.

Since phosphorus is intimately involved in the metabolism of all living forms so far examined, and as sulfur amino acids are normal constituents of proteins, one would expect that marine bacteria would require sources of both phosphorus and sulfur in the medium for growth. To test the need for a source of phosphate, a phosphate deficient medium was prepared by replacing the $(NH_4)_2HPO_4$ used to introduce phosphate into the medium with an equivalent amount of $(NH_4)_2SO_4$. None of the organisms grew to more than a very limited extent in this medium unless $(NH_4)_2HPO_4$ was added back (table 5). The addition of the same amount of NH_4^+ introduced as $(NH_4)_2SO_4$ did not permit growth, indicating that the response to $(NH_4)_2HPO_4$ was due to the phosphate it supplied. When the sulfate salts in the medium were replaced by chloride, two of the organisms, B20 and B30, grew to an appreciable extent. Even in the case of these organisms, however, the addition of K_2SO_4 to the SO_4^{--} deficient medium greatly improved growth. The results indicate that four of the organisms definitely require SO_4^{--} for growth, and two others probably require it also, although in amounts capable of being partially satisfied by contaminating traces present in the medium.

To determine if chloride was required for growth by any of the organisms, a medium was prepared using a salt mixture in which all of the chloride salts with the exception of Ca^{++}, which was added as the nitrate, were replaced by their corresponding sulfates. The yeast extract nutrient broth medium used to grow the inoculum was prepared, using this same chloride deficient salt mixture. The growth response of the organisms in the chloride deficient basal medium is recorded in table 5. In the case of only one organism, B9, was no growth obtained in this medium unless NaCl was added. For another, B20, the maximum growth achieved in the absence of chloride was significantly less than was obtained in its presence. For the other or-

TABLE 5

Some anion requirements of marine bacteria

Medium*	Additions Per 10 Ml	Organisms Tested					
		B9	B10	B16	B20	B26	B30
		Per cent incident light transmitted†					
PO_4^{---} deficient	None	98	100	100	88	95	96
	$(NH_4)_2HPO_4$ 11.3 μmoles	18	10	26	10	15	14
	$(NH_4)_2SO_4$ 11.3 μmoles	97	99	100	88	98	94
SO_4^{--} deficient	None	100	100	100	57	97	66
	K_2SO_4 66.0 μmoles	19	10	22	12	14	15
	KCl 132.0 μmoles	99	100	100	98	92	87
Cl^- deficient	None	100	40	26	69	17	29
	NaCl replacing Na_2SO_4	16	34	23	30	15	22
	2.89 mM NaCl	18	23	21	15	30	71

* The salt solution of footnote 4 was used in this medium with modifications as described in the text.
† See table 1. Incubation times: For the PO_4^{---} and SO_4^{--} experiments: B20, 24 hr; B9, B10, B16, B30, 48 hr; B26, 72 hr. For the Cl^- experiment: B20 and B30, 48 hr; B10, B16, and B26, 72 hr; B9, 96 hr.

ganisms studied, growth was either not affected or only delayed by the absence of added chloride. It was necessary to establish whether or not growth of B9 and B20 on the addition of NaCl to the Cl^- deficient medium was due to the introduction of Cl^- or to an increase in either the Na^+ concentration or the total ionic strength of the solution. Since adding further Na_2SO_4 to the medium at a level equivalent to the added NaCl would have introduced a toxic level of SO_4^{--}, Na_2SO_4 present in the basal medium was replaced by a level of NaCl providing the same amount of Na^+. Although this did not provide two solutions of exactly the same ionic strength, it did provide another medium containing Cl^- ion in which neither an increased ionic strength nor additional Na^+ could be the factor determining growth. Again, maximum growth of B9 and B20 occurred in the medium containing the chloride ion. It is thus evident that B9 requires Cl^- for growth. It would appear also that B20 has an absolute requirement for Cl^- but that the amounts needed are less than are required by B9, and can be partially provided for by contaminating traces of chloride present in the medium.

ZoBell has reported that, after laboratory cultivation on sea-water media for periods of 2 to 12 years, 56 of 60 species of marine bacteria developed the ability to grow on fresh water media (ZoBell, 1946). It would appear from this observation that marine bacteria lose comparatively readily the one characteristic which has been used to distinguish them from land forms. Since it would be of interest to know what changes in specific ion requirements would be represented by such a marked change in salt requirements, an attempt was made to train two of the marine bacteria used in this study to grow in a medium prepared with fresh water. ZoBell's observations were made using a complex laboratory medium which would be expected to supply trace elements in amounts sufficient for the growth of most terrestrial bacteria. In the training studies carried out here a medium of similar complexity was used, i.e., yeast extract, 0.5 per cent; nutrient broth, 0.8 per cent. If full strength sea water was used as the diluent in this medium, the sea water concentration was considered to be 100 per cent. Lower sea water concentrations were obtained by diluting the diluent appropriately with fresh water. Initially, the minimum concentrations of sea water in the medium supporting just visible growth of the two organisms studied were 9 and 12 per cent, respectively. After serially subculturing the organisms into media containing progressively lower concentrations of sea water, cultures were eventually obtained which could produce just visible growth at sea water concentrations of, in one case, 3.25 per cent, and in the other, 3.2 per cent. The organisms, when grown at these low concentrations, however, autolyzed very rapidly and failed to survive more than 1 or 2 subcultures into media of the same sea water concentrations. Repeated attempts to obtain growth at still lower sea water concentrations were unsuccessful.

DISCUSSION

Marine bacteria have been differentiated from land forms on the basis of their need for sea water rather than fresh water for growth in a complex laboratory medium on initial isolation. The results of this investigation reveal that, at least for the marine bacteria studied here, the requirement for sea water in the medium is based on the capacity of sea water to supply the kinds and amounts of inorganic ions required for growth by the organisms. In addition it has been observed that none of the organisms investigated here has gained the capacity to grow in a complex medium prepared with fresh water, even after 2½ years of cultivation on laboratory media. An attempt to develop such a capacity in two of the organisms by the application of training techniques was unsuccessful.

There are two unusual features of the qualitative mineral requirements of the marine bacteria revealed by this study. One is the fact that all of the organisms investigated require Na^+ for growth and two of them need Cl^-. Only one organism from a terrestrial source, the red halophile *Pseudomonas salinaria*, has been reported to require Na^+ for growth (Brown and Gibbons, 1955), although the presence of Na^+ in the medium improved the response of *Lactobacillus arabinosus* to pantothenic acid (Sirny *et al.*, 1954) and shortened the lag phase in the growth of *Clostridium perfringens* (Shankar and Bard, 1952). True marine bacteria may well prove to be distinguishable from land forms present as contaminants in sea water not by having a requirement for sea water but rather by having a readily detectable need for Na^+ in the medium for growth.

The only other report of a Cl^- requiiement for the growth of bacteria is that for the halophile

P. salinaria (Brown and Gibbons, 1955). That Cl^- can be involved in the metabolism of microorganisms in capacities other than those concerned with growth, however, is apparent from the fact that two antibiotics produced by microorganisms—chlortetracycline (Broschard *et al.*, 1949) and "chloromycetin" (Ehrlich *et al.*, 1947)—contain chlorine.

Sea water does not contain sufficient iron to permit the marine bacteria investigated to grow either at their maximum rate or to the fullest extent. This, of course, applies only under the cultural conditions used in the laboratory. In their natural environment, the majority of marine bacteria are believed to be attached to the surfaces of larger organisms and to particles suspended in the sea. These surfaces, by adsorption, would be able to concentrate nutrients sufficiently to enable better growth of the marine bacteria to take place than would occur if the bacteria lived free in the sea (ZoBell, 1946).

SUMMARY

The need of six marine bacteria for sea water in a chemically defined medium has been shown to be due to the ability of sea water to supply the inorganic ions required for growth by the organisms. No evidence was obtained that sea water was capable of supplying unknown organic factors either required by or stimulatory for the growth of these organisms.

A supplement of iron in the sea water medium increased both the rate and the extent of growth of all of the organisms tested. None of the organisms grew unless Na^+ and K^+ were added to the medium. Additions of Mg^{++} were required by four of the organisms and were stimulatory for the growth of the other two. Ca^{++} was required by one organism and stimulated the early growth of another. None of the organisms grew significantly without PO_4^{---} in the medium while the absence of SO_4^{--} prevented the growth of four organisms and reduced the amount of growth of the other two. One organism required Cl^-, the growth of another was limited by its absence, while the remainder either were unaffected by its absence or needed Cl^- for optimum rate of growth. Attempts to train two of the marine bacteria studied to grow in a complex medium prepared with fresh water instead of artificial sea water were unsuccessful.

REFERENCES

Broschard, R. W., Dornbush, A. C., Gordon, S., Hutchings, B. L., Kohler, A. R., Krupka, G. Kushner, S., Lefemine, D. V. and Pidacks, C. 1949 Aureomycin, a new antibiotic. Science, **109**, 199–200.

Brown, H. J. and Gibbons, N. E. 1955 The effect of magnesium, potassium and iron on the growth and morphology of red halophilic bacteria. Can. J. Microbiol., **1**, 486–493

Ehrlich, J., Bartz, Q. R., Smith, R. M. and Joslyn, D. A. 1947 Chloromycetin, a new antibiotic from a soil actinomycete. Science, **106**, 417.

Lyman, J. and Fleming, R. H. 1940 Composition of sea water. J. Marine Research (Sears Foundation), **3**, 134–146.

MacLeod, R. A., Onofrey, E. and Norris, M. E. 1954 Nutrition and metabolism of marine bacteria. I. Survey of nutritional requirements. J. Bacteriol., **68**, 680–686.

Shankar, K. and Bard, R. C. 1952 The effect of metallic ions on the growth and morphology of *Clostridium perfringens*. J. Bacteriol., **63**, 279–290.

Sherman, F. G. and Grant, B. M. 1951 A culture flask for the estimation of the growth of micro-organisms by optical methods. J. Lab. and Clin. Med., **37**, 325–326.

Sirny, R. J., Braekkan, O. R., Klungsøyr, M. and Elvehjem, C. A. 1954 Effects of potassium and sodium in microbiological assay media. J. Bacteriol., **68**, 103–109.

ZoBell, C. E. and Upham, H. C. 1944 A list of marine bacteria including descriptions of sixty new species. Bull. Scripps Inst. Oceanog., Univ. Calif., **5**, 239–292.

ZoBell, C. E. 1946 *Marine microbiology*. Chronica Botanica Co., Waltham, Mass.

20

Reprinted from *J. Bacteriol.*, **76**(3), 301–307 (1958)

STUDIES ON BACTERIAL UTILIZATION OF URONIC ACIDS

III. Induction of Oxidative Enzymes in a Marine Isolate[1]

WILLIAM J. PAYNE

Department of Bacteriology, University of Georgia, Athens, Georgia

Received for publication May 5, 1958

MacLeod *et al.* (1954) noted that a number of strains of bacteria from the marine environment which can not grow on fresh water media are dependent upon Na^+, K^+, and Mg^{++}, whereas several similar isolates capable of growth on fresh water media have no Na^+ dependence. A subsequent investigation of the effect of these metal ions on the viability in storage and the activity of oxidative systems of a Na^+ dependent strain (Tomlinson and MacLeod, 1957) showed a definite requirement for the three ions for storage and for Na^+ and K^+ for the oxidation of succinate and alanine.

In this laboratory a marine bacterium capable of forming induced enzymes for the oxidation of uronic acids has been isolated from glucuronate enriched marsh mud from Sapelo Island, Georgia. This paper represents a study of the effects of substrate, sea salt, and Na, K, and Mg salts on the induction of the oxidative system.

MATERIALS AND METHODS

Test organism. The physiological characteristics of the microorganism M11 were determined by culturing in differential media prepared with sea water as diluent. These media were sterilized by filtration. Concentrates of media in which agar was required were filtered and added to tempered suspensions of sterile sea water agar. Stock cultures were maintained on sea water nutrient agar slants and all cultures were incubated at 30 C unless otherwise indicated.

Electron photomicrographs of the bacteria were kindly prepared by Dr. N. M. McClung of this Department. Young cells from an agar slant were used.

Respirometry. Bacteria were cultured in sea water nutrient broth, broth supplemented with 0.25 per cent galacturonate, or minimal salts and 0.25 per cent glucuronate in sea water on a rotary shaker at 30 C for 18 hr and harvested by centrifugation. The cells were twice washed with sea water or 0.052 M $MgCl_2$ and suspended in concentrations which gave 5 per cent transmittance in the various suspending media as indicated in the figures. Washing in distilled water inactivated the cells, but 0.052 M $MgCl_2$ as a washing solution prevented cytolysis (Tomlinson and MacLeod, 1957). The oxygen uptake of 2.0 ml of each of the suspensions was measured at 30 C using conventional manometric techniques. The cells were equilibrated for 15 min before the substrate was added. Substrate was 1.0 ml of 0.02 M sodium glucuronate or galacturonate and all systems were buffered at pH 7 with 0.04 M mono- and di-basic sodium phosphates.

Experiments were done as well with cells washed with solutions of $MgCl_2$ and preincubated with 0.02 M glucuronate for 2 hr in solutions of each of the chlorides on a rotary shaker at 30 C. The other salts were added from one or two side arms of the Warburg vessels in concentrations which gave the desired molarity in a volume of 3.0 ml. Further description is given in figure 6.

RESULTS

Characterization of the bacterium. Electron photomicrographs in figure 1, which show the effects of the prolonged suspension of the cells in distilled water during preparation for exposure in the microscope, establish the identity of the bacterium M11 as a pseudomonad by the revelation of polar flagellation. Burkholder and Bornside (1957) obtained morphologically similar isolates from the Sapelo marsh.

Inspection of the results in table 1 indicate that the organism does not have the characteristics of any of the members of the Pseudomonadales described in *Bergey's Manual of Determinative Bacteriology* (1957). Cultures have been

[1] Supported by a grant from the National Science Foundation (G-3323) and Equipment Loan Contract Nonr 2337(00) with the Office of Naval Research.

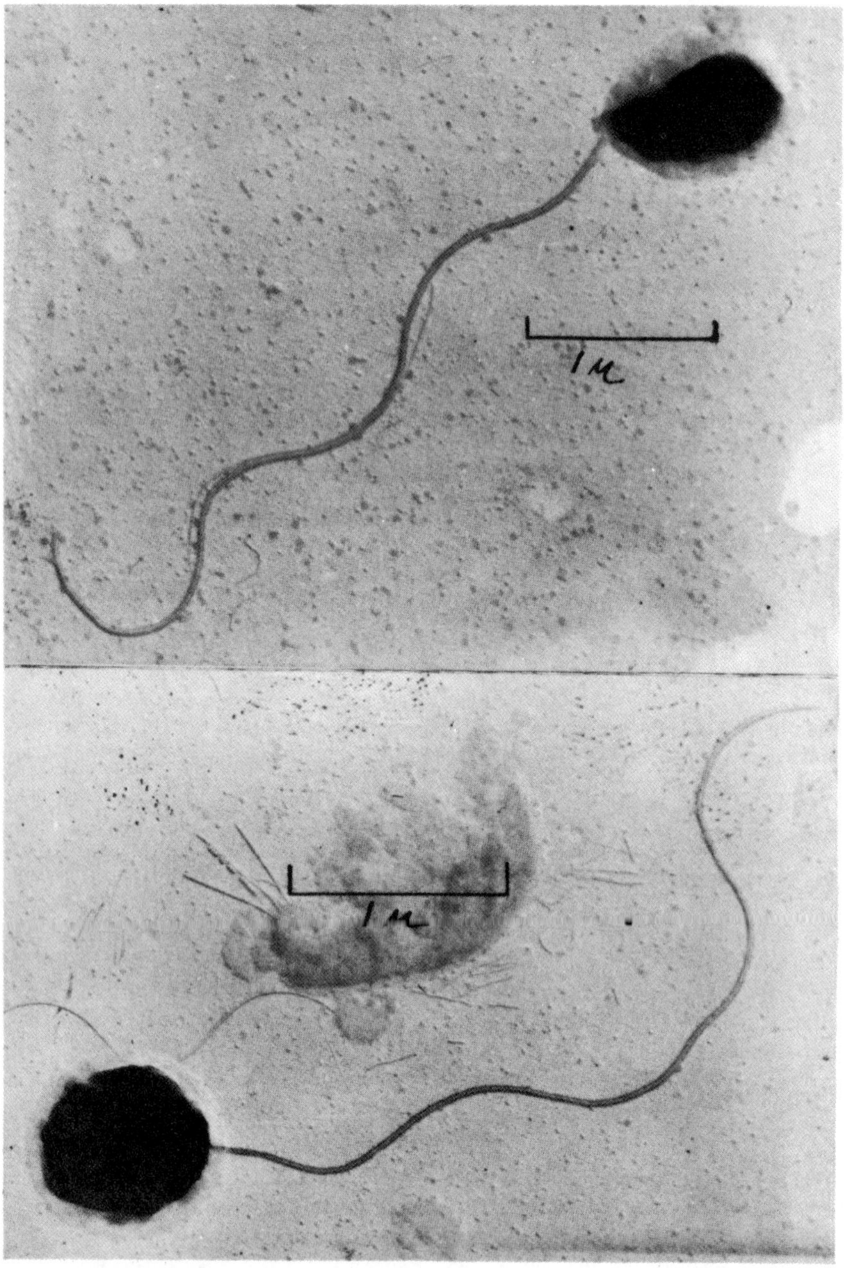

Figure 1. Electron photomicrographs of marine bacterium M11 demonstrating polar flagellation.

submitted to Dr. Einar Leifson of Loyola University of Chicago for investigation of the classification of the isolate.

Effects of substrate and sea salt on induction. The curves in figure 2 indicate a response to glucuronate, by cells grown on galacturonate, which is different than any of those previously reported (Cohen, 1949; Payne and Carlson, 1957). Oxidation of glucuronate during the first hour appears to have been accomplished by nonspecific, galacturonate induced enzymes. Subsequently, more active enzymes were induced in response to glucuronate, and the rate of oxidation of the substrate increased about eightfold.

TABLE 1

Some characteristics of marine bacterium M11

Morphology	Fermentation	Other Metabolic Activity
Colonial:	*Acid produced with:*	Nitrite produced from nitrate.
Circular, smooth, entire, convex, translucent, nonpigmented colonies on agar plates.	L-Arabinose	Indole produced.
	Glucose	Methyl red test positive.
	Fructose	Starch hydrolyzed.
	Galactose	Salicin hydrolyzed.
Abundant, spreading, glistening, viscid, white growth in agar strokes with medium unchanged.	Mannose (±)	Inulin utilized.
	Sucrose	
	Maltose	Alginic acid not hydrolyzed.
	Glycerol	Gelatin not hydrolyzed.
	Mannitol	Hydrogen sulfide not produced.
Cellular:		Urea not hydrolyzed.
Short nonsporogenous, gram-negative rods with single polar flagella.	*Acid not produced with:*	Voges-Proskauer negative.
	D-Arabinose	
	Xylose	*Effect of temperature on growth:*
Average length, 2.5 μ.	Lactose	4 C − 25 C ++
Average diameter, 0.3 μ.	Adonitol	15 C + 30 C +++
	Gas not produced.	20 C + 37 C +
	Acid, no curd in litmus milk, litmus reduced.	Facultative anaerobe.

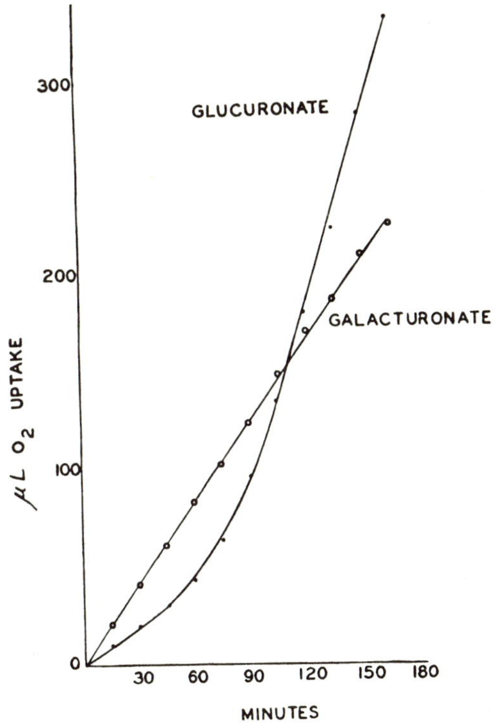

Figure 2. Oxidation of galacturonate and glucuronate by marine bacterium M11 grown on galacturonate broth, washed with sea water and suspended in 3 per cent sea salt water.

Galacturonate was not oxidized by cells grown on broth or glucuronate.

Time required for induction of glucuronate oxidation in resting suspensions of cells grown in sea water broth and washed with sea water was sharply increased by decreasing concentrations of sea salt in the suspending medium (figure 3). The maximum rate of oxidation obtainable varied directly with the concentration of sea salt, and no activity was observed in the absence of salt. To determine if enzymes were induced but inactive in the salt-free system, a concentrated solution of sea salt was tipped into a suspension of cells in distilled water after an incubation period of 45 min. Cells in sea salt water in a simultaneous system were quite active at that time, but induction apparently did not begin in the absence of salts (figure 4). Suspension in distilled water for this period did not inactivate the cells, for introduction of salts permitted induction and oxidation to occur.

Effect of specific salts on induction. Washing in 0.052 M $MgCl_2$ yielded cells which could be induced to good activity in neutralized sea water. Suspension in solutions of the major salts of sea water in various combinations enabled the cells to become active in varying degrees (figure 5). Curves *F* and *E* show that $MgCl_2$, in the absence of Na and K salts, did not support induction of

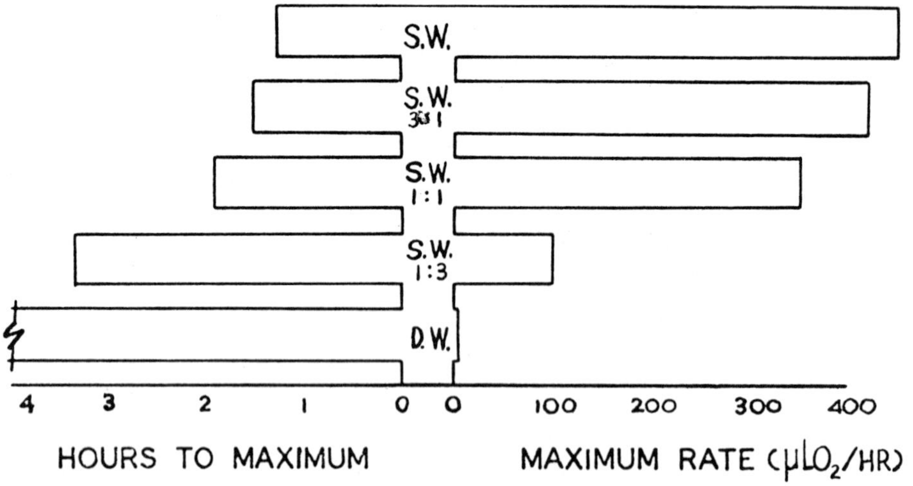

Figure 3. Effect of decreasing concentrations of sea salt on rate of induction and oxidation of glucuronate by marine bacterium M11 grown on broth and washed with sea water.

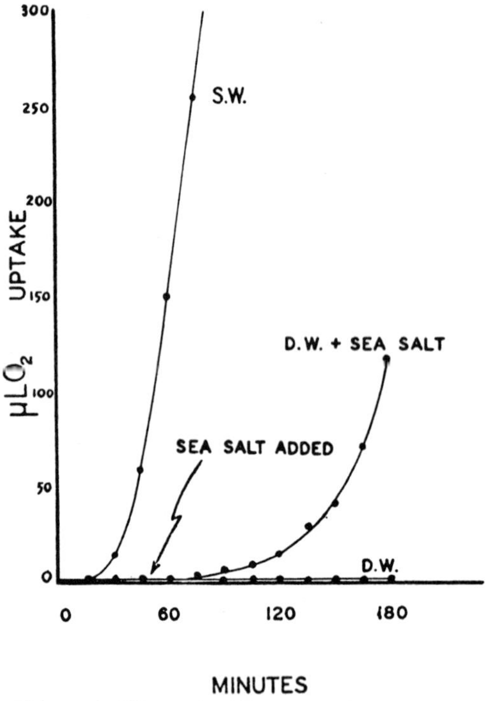

Figure 4. Effect of adding sea salt to a resting suspension of marine bacterium M11 on induction of glucuronate oxidizing enzymes. Side arms in test vessel contained 0.5 ml of 0.01 M sodium glucuronate and 0.5 ml of 18 per cent sea salt.

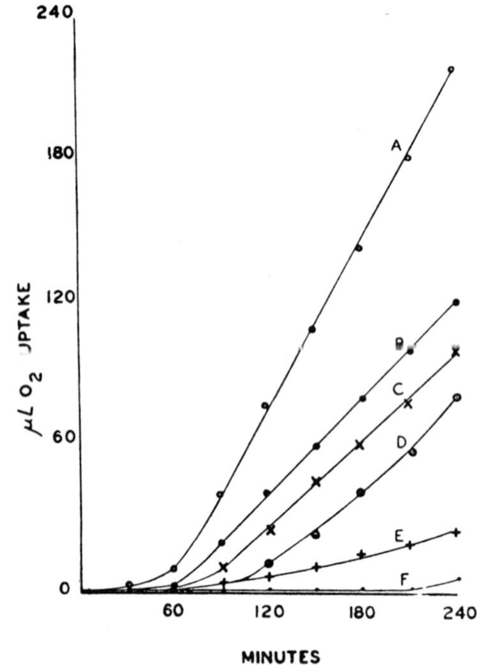

Figure 5. Influence of various combinations of NaCl, KCl, and $MgCl_2$ in the suspending media on the induction of glucuronate oxidizing enzymes in marine bacterium M11. *A*, Neutralized sea water. *B*, NaCl, 0.46 M; KCl, 0.01 M; $MgCl_2$, 0.0245 M; $MgSO_4$, 0.0275 M. *C*, NaCl, 0.46 M; KCl, 0.01 M; $MgCl_2$, 0.052 M. *D*, NaCl, 0.46 M; KCl, 0.01 M. *E*, NaCl, 0.46 M or KCl, 0.01 M (or NaCl and $MgCl_2$ or KCl and $MgCl_2$). *F*, $MgCl_2$, 0.052 M. Endogenous respiration was greatest in sea water and least in the suspensions containing single salts.

activity nor supplement the ability of Na or K salts, in the absence of one or the other, to facilitate induction. Furthermore, curve *E* indicates neither NaCl or KCl alone permits a significant degree of activity. A combination of NaCl and

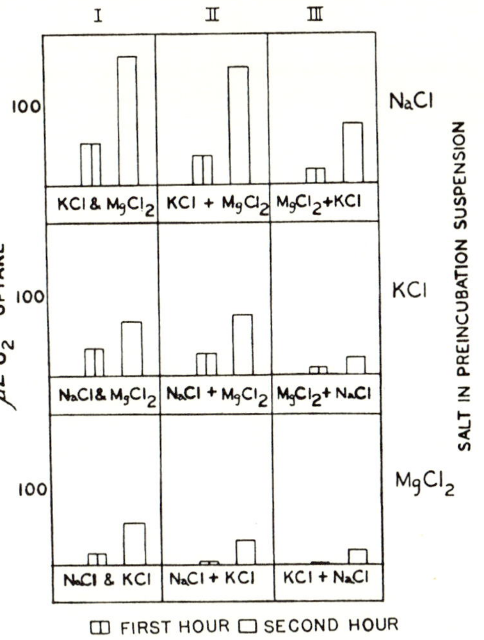

Figure 6. Response to the addition of the other salts of suspensions of marine bacterium M11 preincubated with sodium glucuronate in solutions of 0.46 M NaCl, 0.01 M KCl or 0.052 M $MgCl_2$. In the experiments described in column *I* both salts were added simultaneously, and in those in columns *II* and *III* the second salt was added 30 min after the first. Endogenous respiration was 40 to 50 μL in 2 hr in each system.

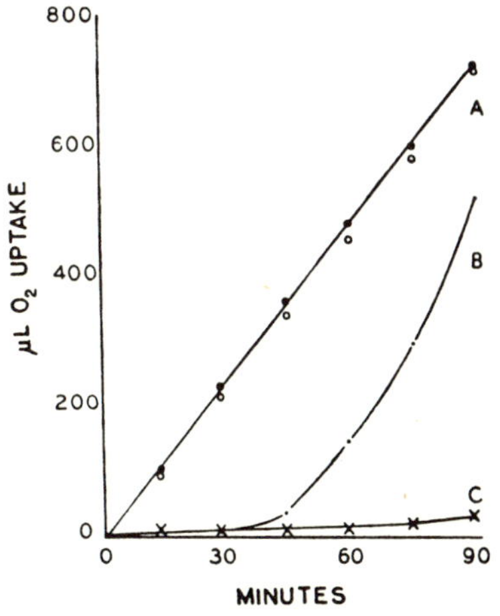

Figure 7. Effect of suspending medium on oxidation of glucuronate by suspensions of marine bacterium M11 grown on glucuronate. *A*, KCl, 0.01 M or KCl and NaCl, 0.46 M. *B*, No salts. *C*, NaCl, 0.46 M.

KCl, however, enabled the cells to be induced to nearly one-third of the activity demonstrable in sea water (curve *D*). This effect was not duplicated by increasing the concentration of KCl to 0.47 M nor by using 0.46 M erythritol in place of NaCl as a presumably inert agent for increasing osmotic pressure. Sucrose, which has been used as an osmotic agent in other studies (Tomlinson and MacLeod, 1957), was rapidly oxidized by these cells. A combination of the three metallic chlorides restored the cells to approximately one-half the activity observed in sea water (curve *C*), and the addition of $MgSO_4$ (with appropriate adjustment in the concentration of $MgCl_2$ to maintain the molarity of Mg^{++}) indicated that $SO_4^{=}$ enhances the rate of induction and activity slightly (curve *B*).

Results obtained by measuring the activity of cells preincubated in solutions of a single salt followed by the addition of the other two, or of one then the other, were revealing with respect to the effect of Na^+. Data in figure 6 indicate that preincubation in NaCl yielded cells with good activity. These bacteria were induced, for oxidation was linear from the time KCl was tipped in, and the presence of KCl served to increase the rate of subsequent induction and activity. KCl did not replace the effect of NaCl in preincubation, and $MgCl_2$ alone seemed not to be involved in induction and increased activity beyond its role in preventing cytolysis.

Cells induced to glucuronate in sea water culture seemed nearly indifferent to NaCl but oxidized the substrate rapidly when suspended in solutions containing KCl (figure 7).

DISCUSSION

The results of this and other studies of uronic acid utilization (Cohen, 1949; Payne, 1956; Payne and Carlson, 1957) reveal that glucuronic and galacturonic acids vary from species to species in their ability to induce enzymes which use either acid as a substrate. *Escherichia coli*, *Erwinia carotovora*, *Serratia marcescens*, and isolate M11 are induced by one or the other acid to oxidize its isomer, more or less rapidly than the inducer; but *Shigella flexneri* strain 2a cultured on galacturonate is inactive with glucuronate.

The essentiality of Na and K salts in the metabolism of whole cells of isolate M11 demonstrated in this study is in agreement with the results of Tomlinson and MacLeod (1957) with another marine bacterium. In addition, the range of influence of the salts in bacterial metabolism has now been shown to include induction of oxidative enzymes. No indication is given of the precise biochemical role of the inorganic ions in these studies. However, it is significant to note that cells preincubated in solutions of NaCl and substrate were induced and became active with the addition of KCl and that those incubated in a solution of both NaCl and KCl were induced to the extent of one-third of the activity demonstrable in sea water. This suggests that NaCl (or probably Na^+) has an indispensable role in the elaboration of the cell's permease system (Cohen and Monod, 1957) and the oxidase system, whereas KCl (or K^+) is involved in the activation of oxidation.

Experiments with bacteria induced in culture to glucuronate support this conclusion, for suspension of the cells in NaCl at sea water concentrations did not permit oxidation. Suspension in a marine level solution of KCl or of KCl and NaCl provided conditions for rapid oxidation. The requirement for a minimal concentration of Na^+ for the oxidation of substrate established by MacLeod *et al.* (1958) is met by the substrate and the dilute buffer in all these experiments. Cells suspended in 0.04 M buffer without added salts show that oxidation in the KCl system was not attributable to enzymes released by cytolysis. Suspensions in the more dilute solution where leakage should be expected did not begin oxidizing at a linear rate for an hour after substrate was tipped in. Activity in the KCl system was linear from the outset.

Considering these results and those of MacLeod's group, it seems likely that many of the bacteria which can grow either in the presence or absence of sea water might be considered to have adapted permeases indifferent to Na^+. On the other hand, strictly terrestrial bacteria might be thought to have evolved further to sensitivity to Na^+.

ACKNOWLEDGMENT

The technical assistance of Mr. Morris Cody and Mrs. Joy Eller is gratefully acknowledged.

SUMMARY

Resting cells of a marine psuedomonad, induced in culture to galacturonate utilization and washed with sea water, oxidized glucuronate initially at 50 per cent of the rate on galacturonate. After an incubation of 2 hr, the rate of oxidation of glucuronate increased to approximately three times that of galacturonate. Cells grown on glucuronate did not oxidize galacturonate.

Induction of glucuronate oxidation in resting cells grown in sea water nutrient broth, washed with sea water, and suspended in sea salt water was inhibited by dilution of the sea salt water. Addition of sea salt to cells incubated with substrate in distilled water enabled induction to occur. Washing with 0.052 M $MgCl_2$ provided cells which were inducible to significant activity in neutralized sea water and, to lesser degrees, in solutions of Na, K, and Mg salts.

NaCl and KCl were found to be indispensable to the cells for induction and activity. Supplementing solutions of these salts with $MgCl_2$ and $MgSO_4$ increased the rates of induction and oxidation. In experiments in which cells were preincubated with substrate in solutions of single salts, the dependence of induction on Na^+ and of oxidation on K^+ was demonstrated. Oxidation of glucuronate by resting cells grown on that substance was linear from the time the substrate was added only in systems containing K^+.

REFERENCES

BREED, R. S., MURRAY, E. F. D., AND SMITH, N. R. 1957 *Bergey's manual of determinative bacteriology*, 7th ed. The Williams & Wilkins Co., Baltimore.

BURKHOLDER, P. R. AND BORNSIDE, G. H. 1957 Decomposition of marsh grass by aerobic marine bacteria. Bull. Torrey Botan. Club, **84**, 366–383.

COHEN, G. N. AND MONOD, J. 1957 Bacterial permeases. Bacteriol. Revs., **21**, 169–194.

COHEN, S. S. 1949 Adaptive enzyme formation in the study of uronic acid utilization by the K-12 strain of *Escherichia coli*. J. Biol. Chem., **177**, 607–619.

MACLEOD, R. A., CLARIDGE, C. A., HORI, A., AND MURRAY, J. F. 1958 A possible role of Na^+ in the metabolism of a marine bacterium. Federation Proc., **17**, 267.

MACLEOD, R. A., ONOFREY, E., AND NORRIS,

M. E. 1954 Nutrition and metabolism of marine bacteria. I. Survey of nutritional requirements. J. Bacteriol., **68**, 680–686.

Payne, W. J. 1956 Studies on bacterial utilization of uronic acids. I. *Serratia marcescens*. J. Bacteriol., **72**, 834–838.

Payne, W. J. and Carlson, A. B. 1957 Studies on bacterial utilization of uronic acids. II. Growth response and oxidative activity of various species. J. Bacteriol., **74**, 502–506.

Tomlinson, N. and MacLeod, R. A. 1957 Nutrition and metabolism of marine bacteria. IV. The participation of Na, K and Mg salts in the oxidation of exogenous substrates by a marine bacterium. Can. J. Microbiol., **3**, 627–638.

21

Reprinted from *Bacteriol. Rev.*, **29**(1), 9–23 (1965)

The Question of the Existence of Specific Marine Bacteria[1]

ROBERT A. MACLEOD

Department of Bacteriology, Macdonald College of McGill University, and Marine Sciences Centre, McGill University, Montreal, Canada

INTRODUCTION

Much information has accumulated over the past 60 years on the nutrition and metabolism of bacteria from nonmarine sources. In contrast, little comparable information is available regarding bacteria from the sea. This may be due, at least in part, to the fact that there has been considerable doubt as to whether or not there actually are specific marine bacteria. Representatives of most of the well-defined bacterial genera found growing on land and in freshwater have been isolated from seawater and marine muds. If no differences exist between bacteria in the sea and their counterparts on land except superficial ones readily lost by training, there would be little purpose in studying the nutrition and metabolism of the same genera of bacteria in more than one habitat. In fact, it has been stated that the central problem of marine microbiology is the question of the existence of specific marine bacteria, and, until this problem is settled, work on marine bacteria, apart from studies on gross transformations of matter, would have very little point (76).

Most of the experiments undertaken in attempting to settle the question have dealt with the temperature range and halophilic nature of bacteria from the sea. Marine bacteria were found to be generally more psychrophilic in character than terrestrial species and to prefer seawater or 3% NaCl to freshwater in the medium for growth. Evidence was presented, however, to indicate that these physiological properties were unstable. Baars (2), for instance, reported success in interconverting, by training procedures, three varieties of sulfate-reducing bacteria which, on the basis of temperature range, salt range, and habitat, had been regarded as separate species. ZoBell and Rittenberg (92) found that chitinoclastic bacteria from the sea, after prolonged laboratory cultivation or acclimatization procedures, developed the ability to grow in freshwater media. This change was accompanied by a widening of the temperature range for growth.

Reports on the stability of the halophilic character of marine bacteria have been particularly

[1] Issued as Macdonald College Journal Series Number 520.

confusing. Korinek (36) stated that after cultivation for 1 year on laboratory media original differences in salinity requirements between freshwater and marine bacteria were not eliminated. Stanier (76) reported failure to train marine agar-digesting bacteria to grow at appreciably lowered seawater or salt concentrations. Littlewood and Postgate (40), studying strains of *Desulfovibrio desulfuricans* of both freshwater and saltwater origin, found a complete gradation of behavior towards NaCl within the genus *Desulfovibrio*, ranging from very salt-sensitive to salt-requiring types. The strain requiring NaCl could not be adapted to grow in media lacking added NaCl. ZoBell and Michener (86), however, observed that 9 of 12 cultures requiring seawater in the medium on initial isolation grew in the same medium prepared with freshwater after the cultures had been held 5 months without transfer. Paradoxically, attempts to train the original cultures to grow at lower seawater concentrations met with only limited success. ZoBell reported subsequently that 56 of 60 species of marine bacteria had developed a capacity to grow in freshwater media (89). Observations such as these led ZoBell and Upham (88) to define marine bacteria as being bacteria from the sea which on initial isolation required seawater in the medium for growth.

Reports of the growth of marine bacteria in media prepared without seawater or NaCl were all based on observations made with complex laboratory media such as nutrient broth, fish broth, or Trypticase. This type of medium could be expected to be contaminated with inorganic ions, a factor which might conceivably have a bearing on the conflicting reports on the stability and specificity of the salt requirements of marine bacteria. The importance of inorganic contaminants for the growth of marine bacteria was first clearly demonstrated by Richter (70) in 1928, but his observations were generally ignored. By using low concentrations of peptone and taking very special precautions to avoid the introduction of inorganic contaminants, he was able to show that a marine luminous bacterium had a specific requirement for Na^+ for growth and luminescence.

Recently, studies of pure cultures of marine bacteria growing in chemically defined media have been conducted for the purpose of critically evaluating the role of the various components of the media in the growth and metabolism of the cells. From the observations made, as well as from investigations with washed-cell suspensions, cell-free extracts, and particulate components of the organisms, new insight has been gained into the relation of marine bacterial cells to their environment. These studies and the bearing of the findings on the question of the existence of specific marine bacteria are the subject of the present review.

ORGANIC REQUIREMENTS

The highest plate counts on seawater and marine materials are obtained when the plating medium contains a complex carbon and nitrogen source such as peptone. Those bacteria of marine origin which grow on such a medium prepared with seawater but not on the same medium prepared with freshwater on initial isolation have been defined as marine bacteria and their characteristics, unless otherwise indicated, are the subject of the present review. The bacteria isolated from such media are heterotrophic, 95% are gram-negative rod forms, and most are motile (89). The first attempts to replace the complex carbon and nitrogen source with chemically defined components in media for the growth of these organisms were made with marine luminous bacteria. Mudrak (58) showed that 10 strains of luminous bacteria isolated from various marine fish grew and luminesced well in nutrient solutions in which peptone was replaced by asparagine or aspartic acid. Bukatsch (12) found that several amino acids, such as glutamic acid, serine, alanine, and leucine, could replace peptone in a medium containing glycerol for the growth of some marine luminous bacteria isolated from herring. Ostroff and Henry (60) studied the capacity of 15 aerobic bacteria of marine origin to utilize various classes of nitrogen-containing compounds as sources of carbon and nitrogen in a simple medium containing 3% NaCl. The different bacteria grew luxuriantly on amino acids which, as a class of compounds, were the best sole source of nitrogen and carbon. Alanine, aspartic acid, and glutamic acid, tested separately, permitted growth of the largest numbers of different organisms. Doudoroff (22) found that four species of *Photobacterium* were able to develop in inorganic media with simple organic compounds as sole carbon source and NH_4Cl as a nitrogen source. In contrast, most strains of *P. phosphoreum* did not develop readily in the basal medium with a single carbon source but required the further addition of methionine. MacLeod et al. (43) investigated the organic nutritional requirements of 33 bacteria of marine origin; 19 were found to have relatively simple nutritional requirements, in that any one of several organic compounds could act as a source of carbon and energy with an inorganic ammonium salt present as a source of nitrogen. More organisms could use succinate as sole carbon source than could utilize glucose. The one carbon and energy source acceptable to all the organisms tested was a complex mixture of 18 amino acids. Seven of the organisms grew only in the presence of the amino acids. For the latter bacteria, the complex mixture could be replaced by glutamic acid, preferably in combination with alanine and aspartic acid. Burkholder and Bornside (13) showed that a number of marine isolates from the coast of Georgia, which were able to decompose marsh grass, did not possess specific requirements for single amino acids but grew better on multiple mixtures. Among the combinations of pure amino acids studied, a mixture of alanine, aspartic acid, and glutamic acid was reported to yield very good growth.

MacLeod et al. (43) found that several marine bacteria required the addition of vitamins to the medium for growth. Requirements for biotin, for biotin and thiamine, and for biotin, thiamine, and niacin were demonstrated in the case of three organisms. Surface-active agents stimulated the growth of two others. One organism, a *Flavobacterium*, required six amino acids, biotin, thiamine, a combination of three nucleosides, and a surface-active agent in the medium to promote appreciable growth in the absence of yeast extract (47).

Burkholder (14) reported on the general growth requirements of 1,748 aerobic heterotrophic bacteria isolated from marine muds. He was able to grow 75% of these on media of known chemical composition. Biotin and thiamine were the vitamins most frequently required for growth. Cobalamin and nicotinic acid stood next, and pantothenate and riboflavine requirements occurred infrequently.

Studies so far indicate that the marine bacteria which grow on complex media possess a wide range of organic nutritional requirements, from the relatively simple to the very complex. The plankton in seawater and its residues in marine mud could be expected to serve as a source of the nutrients required by these organisms. Although these bacteria appear to have a characteristic preference for amino acids as a carbon, nitrogen, and energy source, there is nothing that could be considered unique about their organic nutritional requirements.

INORGANIC REQUIREMENTS

The need for seawater or NaCl in the medium for growth has long been considered to reflect a requirement of marine bacteria for a medium in which salts maintained a suitable osmotic pressure. This conclusion stemmed from observations that marine luminous bacteria lysed when suspended in seawater too greatly diluted with distilled water (31, 32, 33). The first indication that there might be a specific function for the ions of seawater in the growth of marine bacteria was provided by Richter (70), who showed that a marine luminous bacterium had a specific requirement for Na^+. This report was confirmed, and the observation was extended to 10 additional strains of marine luminous bacteria by Mudrak (58), who used a chemically defined medium for the growth of his organisms. Bukatsch (12), using a defined medium, showed that luminous bacteria of marine origin also required K^+. Dianova and Voroshilova (21), employing a fish broth medium, found that Na^+ salts were required for the growth of a number of marine isolates and could not be replaced by equimolar concentrations of K^+ salts. MacLeod and Onofrey (44) found that six marine isolates grew relatively poorly when either natural or artificial seawater was the diluent in a chemically defined medium unless a supplement of an iron salt was added. In addition, both the rate and extent of growth were increased when half-strength, rather than full-strength, seawater was used. When both natural and artificial seawater were supplemented with iron and tested at half strength, there was no significant difference in the capacity of the two diluents to promote the growth of the organisms. When the need for the various ions in artificial seawater was examined, all of the organisms tested could be shown to require Na^+, K^+, Mg^{++}, $PO_4^{=}$ and $SO_4^{=}$ for growth. Several of the organisms also required Ca^{++} and some Cl^-.

Requirement for Na^+

Specificity of the requirement. Since the possession of a requirement for Na^+ for growth distinguishes marine bacteria from most nonmarine species, the characteristics of the Na^+ requirement are of special interest.

When the quantitative requirements of three marine bacteria for Na^+ were determined, MacLeod and Onofrey (46) found that the maximal rate and extent of growth was achieved with 0.2 to 0.3 M Na^+, which is about one-half of the Na^+ concentration in seawater. Below this level, the rate and extent of growth were roughly proportional to the amount of Na^+ added. After a sufficiently long incubation period, growth occurred at almost one-tenth of the optimal concentration of Na^+, but never in the absence of the ion. Li^+, Rb^+, and Cs^+ showed no capacity to replace Na^+ for the growth of the organisms. K^+ exhibited a very slight sparing action at suboptimal concentrations of Na^+, an effect which disappeared on longer incubation. Sucrose had about the same limited capacity to spare the Na^+ requirement. These findings indicated that the requirement for Na^+ of the organisms examined was highly specific and that Na salts had little if any osmotic function. Two of the organisms examined in this study have been recognized more recently to be pseudomonads, and the third was a *Cytophaga* species (52).

Payne (61) studied the Na^+ requirement of a glucuronate-oxidizing marine pseudomonad and also concluded that the effects of salts could not be explained by their osmotic action. Pratt and Austin (66), on the other hand, found that a number of salts and sucrose could greatly reduce but not eliminate the requirement of a marine *Vibrio* for Na^+. They concluded, in the case of this organism and three others examined, that a considerable proportion of the salt requirement was needed to satisfy the osmotic demands of the organisms. In an extension of these studies, Pratt (67) reported that, with seawater samples plated on a Trypticase medium containing a low concentration of added NaCl, increases in counts were obtained when the medium was supplemented with either KCl or sucrose. The counts were not so high as those obtained when the medium was made equiosmolar with respect to NaCl. He concluded that approximately half the bacteria in the samples would grow in media in which a substantial replacement of NaCl by sucrose or KCl had been made. It would thus appear that different marine bacteria differ in the extent to which nonspecific solutes can replace Na^+ for growth. In all cases examined in detail, however, it has been shown that bacteria of marine origin requiring seawater in the medium for growth have an irreplaceable minimal requirement for Na^+. Tyler et al. (80) studied 96 isolates of marine bacteria from Atlantic coastal waters off Florida and found all to require Na^+.

Most marine bacteria which have been examined have Na^+ requirements which are readily detectable, because the amounts needed for optimal growth are 0.2 to 0.3 M. Such organisms require the addition of Na salts or seawater even to the complex laboratory media commonly used for their isolation, though such media are usually contaminated with appreciable amounts of Na^+. Two organisms of marine origin were isolated, however, which grew optimally in complex media prepared with freshwater. Such organisms would not have been classified as marine bacteria according to earlier criteria (88). When grown on chemically defined media, however, their requirements for Na^+ became apparent (43). Quantitative estimations of their requirements revealed that one needed 0.02 M Na^+ and the other 0.005 M Na^+ for optimal growth (MacLeod and Onofrey, *unpublished data*). By comparison, the nutrient broth-yeast extract isolation medium prepared with distilled water contained 0.03 M Na^+, an amount clearly sufficient to permit optimal growth of both organisms.

Stability of the requirement. It was of interest to determine whether the requirement of marine bacteria for Na^+ is as readily lost as the requirement for seawater had been reported to be. By plating heavy suspensions of marine bacteria on Trypticase medium prepared without added Na^+, Pratt and Waddell (63) obtained a few colonies which they concluded were mutants of marine bacteria no longer requiring Na^+ for growth. MacLeod and Onofrey (53) trained a marine pseudomonad to grow on Trypticase medium prepared without added Na^+ salts by streaking cultures serially onto the surface of plates of the medium containing progressively lower concentrations of Na^+. A flame photometric analysis of the Trypticase medium without added salts revealed a concentration of 0.028 M Na^+ present as a contaminant. When the adapted culture was tested in a chemically defined medium containing less than 6.5×10^{-5} M Na^+, the organism was found still to require Na^+ for growth. The adapted culture grew only a little more quickly and at a slightly lower Na^+ concentration in the chemically defined medium than the parent culture. All attempts to train the organism to grow in the chemically defined medium in the absence of added Na^+ failed. This organism, which had been trained to grow in a complex medium without added Na^+ salts had apparently developed a capacity to grow well at the concentrations of Na^+ present as a contaminant in the complex medium, so long as other components of the complex medium were present. The possible significance of this finding in relation to the reported ability of some marine bacterial cultures to lose their requirement for seawater remains to be established.

When a marine pseudomonad was exposed to ultraviolet irradiation, a limited number of what appeared to be mutants were obtained which grew in the chemically defined medium in the absence of added Na^+ (53). The extent and rate of growth of the mutants was still enhanced by added Na^+, but this response could be eliminated by training. The difficulty experienced in getting any appreciable number of mutants lacking a Na^+ requirement by irradiation of a Na^+-dependent culture is a further indication of the stability of the Na^+ requirement of these organisms.

Uniqueness of the requirement. The evidence which has accumulated suggests that bacteria from the sea which require seawater in the medium for growth on isolation possess a stable, highly specific, and in most cases readily detectable requirement for Na^+ for growth. To what extent is this a characteristic unique for marine bacteria? Halophilic bacteria, including representatives of the extreme halophiles, have been isolated from freshwater sources and soil. Both extreme and moderate halophiles have been reported to have specific requirements for Na^+ [*see* Larsen (39) for a review]. Among nonhalophilic species, two strains of *Rhodopseudomonas spheroides* and one strain of *R. palustris* were found to require Na^+ when grown in a chemically defined medium (75). The original source of these isolates was unknown. A strain of *Bacteroides succinogenes*, a cellulolytic organism isolated from the rumen of a steer, also has been shown to require Na^+ for growth (10). Goldman and co-workers made the interesting observation that a number of strains of lactic acid bacteria isolated from meat-curing brines developed a requirement for NaCl at elevated temperatures (28). Tests indicated that neither the Na^+ nor the Cl^- could be replaced by other ions. These are the only well-documented cases so far reported of bacteria from nonmarine sources requiring Na^+ specifically for growth. Sakazaki et al. (71), however, reported that they have confirmed an observation made by Nakagawa (cited as a personal communication) that various halophilic organisms can be found in the feces of guinea pigs, rats, and monkeys. These organisms required 3% salt for growth. Although a requirement for Na^+ has not been established in this case, it is evident that a specific need for Na^+ is not a characteristic unique for bacteria of marine origin and may well prove to be more widespread than was previously imagined.

Among those organisms needing Na^+ for growth, there is a wide range in quantitative requirements. In the case of extreme halophiles, growth ceased when the Na^+ concentration fell below 1.5 M, even in the presence of large amounts of K^+ or Mg^{++}, and for maximal growth under these circumstances 2.5 M Na^+ was required (5). The moderate halophile *Vibrio costicolus* had a nonspecific requirement for about 0.4 M salt in the medium but a specific requirement for only 0.017 M Na^+ (18). The strains of *Rhodopseudomonas* studied by Sistrom (75) required a maximum of 0.002 M Na^+ for growth. The marine bacteria so far examined have been found to have optimal requirements for Na^+ ranging from 0.005 to 0.2 M, depending on the species.

Function of Na^+. Washed-cell suspensions of two marine pseudomonads were shown to require Na^+ as well as K^+ for the oxidation of exogenous substrates (79, 48, 62). In the case of these organisms, neither related ions nor sucrose showed any significant capacity to reduce the requirement for Na^+ for oxidation. In this respect, the responses to the ions for substrate oxidation were similar to those for growth. When one of the organisms was examined in more detail, the amounts of Na^+ required for oxidation were found to vary, depending on the substrate being oxidized (48). To obtain maximal rate of oxidation of acetate, butyrate, propionate, or an oxidizable sugar, 0.05 M Na^+ was required; for malate, citrate, and succinate, 0.15 to 0.20 M Na^+ was necessary. All the enzymes of the tricarboxylic cycle were found to be present in cell-free extracts of the organism. When each of the enzymes was tested for its response to inorganic ions, the acetate-activating enzyme and malic dehydrogenase were found to require K^+, aconitase and isocitric dehydrogenase required media of appropriate ionic strength (0.3 to 0.4 μ) for optimal activity, and the remainder functioned better in the absence of added salts than in their presence. None of the enzymes, however, could be shown to require Na^+ specifically (48, 49).

Washed cells of a marine *Vibrio* species were found to require both Na^+ and K^+ for the production of indole from tryptophan (64). In the case of this organism, however, the presence of sucrose in the suspending medium reduced the Na^+ requirement for indole production from 0.3 to 0.05 M. A similar sparing action of sucrose on the Na^+ requirement has been observed with this organism during growth (66). Cell-free extracts of the *Vibrio* required K^+ and pyridoxal phosphate for indole production. Added NaCl was not required, and concentrations of NaCl giving optimal activity with intact cells partially inhibited the activity of cell-free extracts.

Payne noted that induction and activity of enzymes for catabolizing glucuronate in the marine isolate *Pseudomonas natriegens* were specifically affected by the presence of Na^+ and K^+. The role of K^+ appeared to be restricted to influencing the activity and not the induction of enzymes. The requirement for Na^+, however, seemed to be coupled to the induction of a mechanism for the uptake of glucuronate (61, 62). In subsequent experiments, the induction of resting cells of the same organisms and other marine isolates to the oxidation of L-arabinose, mannitol, and lactose was found to be dependent on the presence of Na^+ (69).

A role for Na^+ in the induction of penetration mechanisms or in the formation of adaptive enzymes would fail to account for the requirement for Na^+ observed when compounds were oxidized by pathways employing constitutive permeases and enzymes. Under these circumstances, since whole cells required Na^+ for the metabolism of substrates whereas intracellular enzymes appeared not to require the ion, it seemed likely that Na^+ might be involved in the transport of substrates into the cell. To test this possibility, it was necessary to dissociate the uptake of substrates from their subsequent metabolism. This was accomplished by using nonmetabolizable analogues of metabolizable substrates. Drapeau and MacLeod (23) found that, when washed cells of a marine pseudomonad were incubated with C^{14}-α-aminoisobutyric acid, this analogue of the naturally occurring amino acids accumulated inside the cells but could not be metabolized. The uptake required the presence of Na^+ in the suspending medium. Since uptake took place without a lag period from an incubation mixture containing chloramphenicol, the possibility that the accumulation was due to the preliminary induction of a penetration mechanism was rendered very unlikely. K^+, Rb^+, NH_4^+, Li^+, and sucrose could not substitute for Na^+ in the transport process. Sulfate and chloride salts providing the same level of Na^+ were equally effective. The uptake process was an active one, because the substrate was concentrated in the cells to a level some 3,000 times that in the medium. The uptake was stimulated by the presence of an oxidizable substrate (in these experiments, galactose). Since galactose required less Na^+ for its maximal rate of oxidation than was needed for the optimal rate of uptake of the amino acid analogue, there was clearly a role for Na^+ in the uptake process which was separate from any other possible role of Na^+ in oxidative metabolism. Because D-fucose, a nonmetabolizable analogue of galactose, also required Na^+ for uptake, it seemed likely that the requirement for Na^+ for galactose oxidation also represented a requirement for transport. The uptake of α-aminoisobutyric acid by cells of the marine luminous bacterium *Achromobacter* (*Photobacterium*) *fischeri* has also been shown to be a Na^+-dependent process (Drapeau and MacLeod, *unpublished data*). These results support the conclusion that the primary function of Na^+ in marine bacteria may be to permit the transport of substrates into the cell. Previously observed differences in the quantitative requirements for Na^+ for the oxidation of various substrates by cells of a marine bacterium (48) can now be accounted for if one assumes a number of different permeases in the cell membrane with quantitatively different requirements for Na^+ for activation. Whether or not there are Na^+-dependent transport mechanisms in Na^+-requiring bacteria of nonmarine origin has yet to be determined.

Response to Halides

Of six marine bacteria examined, three were found to have an absolute requirement for halide ions for growth, and three reached maximal growth more quickly if halide was present in the medium (45). Chloride and bromide could be used interchangeably on a mole for mole basis in these experiments. Iodide was toxic. The amounts of halide required and the effects of the anion on rate and extent of growth corresponded closely to the response to Na^+, suggesting that the function of the two ions might be closely related in the metabolism of those organisms requiring both ions.

Some moderate and extreme halophiles have specific requirements for chloride for growth, whereas others do not [*see* Larsen (39) for a review]. In all cases but one so far reported, halide requirements for growth among bacteria have been detected only in bacteria which also need Na^+ specifically, though organisms requiring Na^+ do not always need halide. The exception is a strain of *D. desulfuricans* which failed to grow without the addition of NaCl to the medium. The Na^+ but not the Cl^- could be replaced by other ions (40).

Requirement for Mg^{++}

When grown in a chemically defined medium, marine bacteria have been found to require 4 to 8 mM Mg^{++} for maximal rate and extent of growth (45). This requirement is high compared with that of most terrestrial species examined. A level of 0.02 mM Mg^{++} was established as the requirement of a strain of *Escherichia coli* (85) and 0.08 mM was needed by *Bacillus subtilis* (26). Wiebe and Liston (83), noting the high Mg^{++} requirement of classical marine bacterial types, suggested that this might be a useful criterion of the marine origin of a bacterium. In the case of the marine bacteria examined by MacLeod and Onofrey (45), however, a marked interaction between Mg^{++} and Ca^{++} was noted. For one organism, the presence of 2.5 mM Ca^{++} in the medium reduced the requirement for Mg^{++} from 8 mM to 0.04 mM but did not eliminate the need for the ion. Higher levels of Ca^{++}, on the other hand, increased the requirement for Mg^{++}. For other organisms, both Mg^{++} and Ca^{++} were required for growth and, in these, the quantitative requirements for one of the ions was much affected by the level of the other in the medium. Both Mg^{++} and Ca^{++}, therefore, appear to play an important role in the nutrition of marine bacteria, and the requirements for these two ions taken together are somewhat higher than those of most terrestrial species. The requirements of marine bacteria for divalent ions are low, however, compared with the levels required by the extreme halophiles. For the latter organisms, 100 to 500 mM concentrations of Mg^{++} are necessary for optimal growth and for the maintenance of normal morphology in media containing all the other ions necessary for growth at their optimal concentrations (5).

Salt Tolerance

It is commonly assumed that marine bacteria, since they live in the sea, must be salt-tolerant organisms. Seawater, however, contains only 0.45 M Na^+, 0.05 M Mg^{++}, 0.01 M K^+, and 0.01 M Ca^{++}, plus traces of other ions. The Na^+ level, expressed as NaCl, is about 2.6%. Three marine bacteria investigated by MacLeod and Onofrey (46) were inhibited by the presence of 0.8 M Na (4.7% NaCl) in the medium. Of 15 marine bacteria examined by Tyler et al. (80), all grew at 0.8 M (4.7%) NaCl, 9 grew at 1.4 M (8.2%) NaCl, and none grew at 2.6 M (15.2%) NaCl. In contrast, many terrestrial species, among them organisms not classed as halophiles, can tolerate much higher concentrations of salt than the marine bacteria studied. Larsen (39) stated that, among bacteria commonly found to be agents of food spoilage, aerobic sporeformers grow at 15 to 20% NaCl and many micrococci tolerate 25% NaCl. Gram-negative rods of terrestrial origin are generally completely inhibited by NaCl concentrations between 5 and 10%, and thus have a sensitivity to salt similar to that of the marine bacteria examined.

Lytic Susceptibility

Characteristics of the lytic phenomenon. Harvey (31) observed in 1915 that marine luminous bacteria failed to luminesce when the seawater in which they were suspended was too greatly diluted with distilled water. He ascribed the effect to cytolysis through lowered osmotic pressure, because light production was maintained when seawater was replaced by a 1 M sucrose solution. Hill (32) concluded that luminous bacteria are cytolyzed by water, hypotonic nonpenetrating solutions, and penetrating solutions of all concentrations. A penetrating solution in Hill's study was one which failed to prevent lysis of cells suspended in it. In a study of 96 isolates of marine bacteria (all nonluminous gram-negative rod forms), Tyler et al. (80) observed that in the majority of cases suspensions of cells of the organisms were susceptible to a loss of optical density in distilled water. MacLeod and Matula (52) found that five marine bacteria differed considerably in lytic susceptibility. Two lysed immediately and completely when suspended in less than 0.15 M NaCl, but suspensions of the other three still contained many whole cells at 0.025 M NaCl.

Pratt and Riley (65) and MacLeod and Matula (52) noted differences in the capacities of different salts to prevent lysis of marine bacteria. For a number of different isolates NaCl and LiCl were found to be more effective than KCl or NH_4Cl in preventing lysis. The same salts had the same relative capacity to prevent lysis in the case of the moderate halophile *Vibrio costicolus* (19) and the extreme halophile *Halobacterium cutirubrum* (1), suggesting that the mechanism of lysis may be basically the same in all of the organisms examined.

Divalent cations were found to be much more effective than monovalent cations in preventing lysis of marine bacteria (51). The order of effectiveness of the divalent cations appeared to be similar to that of their capacity to form chelate complexes. The Mg^{++} and Ca^{++} concentrations in seawater would have been sufficient to prevent the lysis of all but one of the marine bacteria examined, without the assistance of Na^+ salts.

The nature of the anion was found to be important in preventing lysis of marine bacteria, particularly on long incubation of suspensions of the cells (52). For four of five organisms examined, sulfate salts stabilized the cell suspensions better than did chlorides. For the fifth organism, the reverse was true.

As little as 5×10^{-4} M spermine was found to suppress lysis of the marine luminous bacterium *Achromobacter fischeri* (41).

Mechanism of lysis. The wide variation in the concentration of the different solutes required to prevent lysis made it seem extremely unlikely that all the solutes exerted their effects through osmotic action. Proof that NaCl does not prevent lysis in this way in the case of one marine pseudomonad was obtained by measuring the intracellular Na^+ and Cl^- concentrations at various levels of extracellular NaCl (78). At all levels of Na^+ in the medium, the intracellular and extracellular Na^+ concentrations within the limits of experimental error were the same. Intracellular and extracellular Cl^- concentrations were the same at the one level of Cl^- examined. Since, so far as NaCl was concerned, no gradient was maintained between the inside and outside of the cell, NaCl could not prevent lysis of the cells through osmotic action.

Brown (6, 7) prepared cell walls of a marine pseudomonad by mechanical disintegration of the cells followed by washing with distilled water. Suspensions of the cell walls, when incubated in a dilute phosphate buffer (0.05 M), showed a decrease in absorbancy with time. This decrease was prevented by increasing the buffer concentration, by heating the cell walls, or by the addition of spermine. When cell walls were incubated under conditions permitting a decrease in absorbancy of their suspensions, a dialyzable fraction and a nondialyzable fraction were released. An acid hydrolysate of the nondialyzable fraction was shown to contain hexosamine, muramic acid, and the normal amino acids of protein hydrolysates. Both diaminopimelic acid and glucose, constituents of the cell-wall residue, were absent from the nondialyzable fraction. The dialyzable fraction contained a number of peptides. The latter observation suggested to Brown that the breakdown of the crude cell wall is caused by a lytic enzyme in the cell wall, and not merely by spontaneous disintegration under appropriate physicochemical conditions. Comparison of the effect of cations on the cell-wall autolytic system and on tryptic digestion of the cell envelope suggested to Brown (8) that the simplest and most probable explanation of the effects of ionic strength and particularly di- and multivalent cations was that they operate through their influence on the conformation of membrane proteins. Proteolytic autolysis was considered to be a direct consequence of such changes.

Buckmire and MacLeod (11) did not favor this hypothesis, because the lysis of whole cells is such a rapid process that it seemed unlikely that it could be due to the action of an enzyme. Cell envelopes of a marine pseudomonad were prepared by mechanical disintegration of the cells in 0.5 M NaCl, a concentration of salt able to prevent lysis. The envelopes were washed free from cytoplasmic material with 0.5 M NaCl. This was a departure from the procedure of Brown, who washed his cell envelopes in distilled water. It was felt that this might well lead to the loss of components important in the maintenance of cell-envelope structure. When a suspension of the cell envelopes in 0.5 M NaCl was added to distilled water, a soluble nondialyzable material was found to be present in the supernatant solution. Both the nondialyzable fraction and the cell-envelope residue after acid hydrolysis contained glucosamine, muramic acid, 15 amino acids (including diaminopimelic acid), and four unidentified ninhydrin-positive compounds. It appeared from visual inspection of the paper chromatograms that not only were the same compounds present in both fractions but that they were present in the same relative proportions. When cell envelopes suspended in 0.5 M NaCl were heated at 100 C for 15 min, they still released the nondialyzable hexosamine-containing fraction on suspension in distilled water. When the suspension of walls in 0.5 M NaCl was autoclaved at 121 C for 10 min, a considerable release of hexosamine-containing material occurred. This could be largely prevented, however, by raising the NaCl concentration to 5 M. The effect of heat and salt concentration on the release of the hexosamine-containing fraction is exactly analogous to the effects of heat and salts on the denaturation of a polyanion, and is explainable in terms of polyelectrolyte theory (37).

The finding that a fraction is released from the cell envelopes into distilled water, which has apparently the same composition as the residual cell envelope, suggested that the cell envelope is made up of a series of units. The effect of heat and salt concentration could best be explained if one assumes that the units are held together by cross-linkages between polyanions on adjacent units. The units would be able to come close enough together to form a continuous wall only if the negative charges on the polyanions were screened by the cations of a salt. The effects of salts in maintaining the integrity of the envelopes could thus be explained satisfactorily on the basis of polyelectrolyte theory. Conditions which prevent denaturation of a polyelectrolyte maintain the structure of the envelope. This did not eliminate the possibility that an enzyme was involved, because enzymes are polyelectrolytes. An explanation of lysis based on spontaneous disintegration of the envelopes under appropriate physicochemical conditions, however, was more compatible with the observations than an explanation involving enzymes.

It would thus appear that the cell envelopes of the marine bacteria examined are maintained intact by salts in somewhat the same way as the envelopes of the more extreme halophiles. Abram and Gibbons (1) suggested that the cell walls of halobacteria are held together by hydrogen bonds, Coulomb forces, or "salt" linkages, and that in the presence of NaCl the electrostatic forces are screened so that the bonds hold the organism in a rod shape. Brown (9) concluded that the effects of salt concentration, bivalent cations, and *p*H on the disaggregation of cell envelopes of *H. halobium* are all consistent with a mechanism which operates principally through exposure on the membrane of a net negative charge.

In the case of organisms which lyse in distilled water, then, there is direct evidence through studies with isolated cell envelopes that inorganic ions are directly involved in holding the cell wall together. Evidence has been obtained which suggests that a somewhat similar situation may prevail in some terrestrial species. Repaske (68) reported that a number of gram-negative bacteria could be induced to lyse in the presence of lysozyme if the incubation mixture contained the metal-binding agent ethylenediaminetetraacetic acid (EDTA). Carson and Eagon (17) found that EDTA alone was capable of lysing a suspension of *P. aeruginosa*, producing large cell-wall fragments which could then be further digested by lysozyme. It is tempting to speculate that these fragments are units of the cell envelope which, in the intact cells, are held together by metal ion bridges. Thus, there may be more in common in the structures of the cell envelopes of marine and terrestrial pseudomonad species than was suspected previously.

Singularity of lytic susceptibility. Among gram-negative bacteria, there is a spectrum of susceptibility to lysis ranging from organisms which require high salt concentration to prevent disruption of the cells to those which maintain their integrity in distilled water. The bacteria which are most susceptible to lysis are the extreme halophiles, the halobacteria, which lyse below 2.0 M NaCl (1). Next come the moderate halophiles, of which *V. costicolus* is an example. This organism lyses at NaCl concentrations ranging from 0.25 to 1.0 M, depending upon the salt concentration of the growth medium (19). At the lower end of the spectrum come the marine bacteria. Some species lyse when the NaCl concentration drops below 0.15 to 0.2 M. In the case of others, only part of the population lyses in distilled water (52). Organisms of terrestrial origin are ordinarily considered not to be susceptible to lysis. However, two nonmarine species, *Pasteurella tularensis* and *Neisseria perflava* are markedly affected by a lowered solute concentration, as indicated by leakage of cell material, decay of respiratory ability, and decline of viability on brief exposure to distilled water (41, 42). Furthermore, the capacity of solutes such as Mg^{++} to maintain the respiratory activity of cells of *Azotobacter* (29), an organism otherwise stable in water suspension, may represent a further ramification of a basically similar phenomenon. It is evident, therefore, that a clear-cut distinction between marine and nonmarine species of bacteria cannot be made on the basis of lytic susceptibility alone.

METABOLIC PATHWAYS IN MARINE BACTERIA

Very little information is available regarding the intermediary metabolism of marine bacteria. All the enzymes of the tricarboxylic acid cycle have been found to be present in cell-free extracts of a marine pseudomonad (48, 49). Enzymes of the glyoxylate by-pass were also detected (50). Isocitrate lyase was demonstrated in extracts of *Agarbacterium alginicum* (84). Enzymes of both the glycolytic pathway and the hexose monophosphate pathway were demonstrated to be present in extracts of glucose-grown cells of the marine pseudomonad, *P. natriegens* (24). Data from radiorespirometric experiments indicated that approximately 92% of the glucose was catabolized via the glycolytic pathway and 8% by the hexose monophosphate pathway. The factor controlling the choice of pathways in this organism has been shown to be the availability of nicotinamide adenine dinucleotide phosphate (NADP) (25). The bacterium requires NADP for the operation of the hexose monophosphate pathway but lacks pyridine nucleotide transhydrogenase and reduced NADP ($NADPH_2$) oxidase, enzymes required for the reoxidation of $NADPH_2$. That this is not a phenomenon associated exclusively with marine bacteria is evident from the fact that the hexose monophosphate pathway in bacteria from other habitats as well as in mammalian tissues is rate-limited by the supply of NADP (25).

Ochynski and Postgate (59) compared the properties of freshwater and saltwater (though not necessarily marine) strains of *Desulfovibrio desulfuricans* and found that growth in a saline environment led to the production of a mucopolymannoside not chemically related to the cell wall. An increase in the content and change in the kind of "free amino acid material" within the cell was also noted. Adaptation of a freshwater strain to a saline environment led to the acquisition of these characters and a morphological change. For an adaptive change in the reverse direction, only the last character was studied and it was not lost.

OTHER FACTORS

Relation to Temperature

Most marine bacteria examined can be described as being facultatively psychrophilic because, according to Bedford (4) and others, the majority grow at 0 C, have a temperature optimum of 20 to 25 C, and do not grow above 30 C. ZoBell and Conn (87) reported that heating samples of seawater and marine mud to 30 C for 10 min killed about 25% of the bacteria, and only 20% survived 40 C for 10 min. Psychrophilic microorganisms are, of course, very widely distributed in nature, having been isolated in appreciable numbers from air, water, soil, plants, animals, and a great variety of foods (77).

Though psychrophily is not a characteristic unique for bacteria of marine origin, its physiological basis in marine bacteria is of considerable interest. Burton and Morita (15) found that 55 to 60% of the malic dehydrogenase activity of cell-free extracts of a marine facultative psychrophile (optimal temperature for growth, 24 C; maximal, 30 C) was lost by exposure of the extract to 30 C for 15 min. The rate of denaturation of the enzyme was much greater at 35 and 40 C. Heat stability of the enzyme was found to be greater in whole cells than in cell-free extracts (56). Heating the cells or treatment with a lysing agent apparently destroyed some regulatory factor for malic dehydrogenase activity. The data indicated that this regulatory factor was cell permeability. Additional support for the conclusion that at least two factors, heat lability of vital enzymes and membrane permeability, are involved in governing the maximal temperatures at which these organisms can grow arose from studies with an obligate psychrophile *Vibrio marinus* (optimal temperature for growth, 15 C; maximum, 20 C). Morita and Robison (57) found that temperatures from 20 to 30 C were sufficient to inactivate the metabolic systems involved in oxygen uptake, either endogenously or in the presence of glucose, in this organism. These temperatures also caused leakage of 260- to 280-mμ absorbing material. The amount of leakage was greater with increased exposure as well as increased temperature.

Evidence is accumulating that enzymes and enzyme-forming systems in other psychrophilic microorganisms are abnormally sensitive to heat (3, 30, 81).

Since more than 80% of the marine environment is perpetually colder than 5 C, factors permitting growth of marine bacteria at the lower end of their temperature range are also of concern. Of particular interest in this connection was the observation of Morita and Burton (56) that in whole cells of a marine facultative psychrophile there was a 50% decrease in malic dehydrogenase activity with each 10 C decrease in temperature down to 13.8 C. A further temperature drop to 5 C reduced the enzyme activity only 15%. Since malic dehydrogenase activity in cell-free extracts was reduced 64% over the same temperature range, the authors concluded that whole cells have some mechanism for permitting enzymes to function at low temperatures at rates which are higher than one would expect from their response to temperature in a cell-free system.

Relation to Pressure

Since the average depth of the world's oceans is more than 2 miles, and hydrostatic pressure increases roughly 1 atm for each 10 m of depth, much of the sea floor is subjected to pressures exceeding 300 atm. At the deepest points in the ocean, hydrostatic pressures approaching 1,100 atm prevail. Thousands to millions of bacteria are known to be present per gram of marine sediments (89). One might expect, therefore, that organisms able to survive and grow at the bottom of the sea would be more tolerant of pressure than terrestrial species. ZoBell and Johnson (90) compared the effects of pressure on representative species of terrestrial and marine bacteria. None of the terrestrial bacteria multiplied perceptibly at a pressure of 600 atm, and growth of most was slowed by 300 atm. Marine species from near the surface of the sea resembled the terrestrial bacteria in their sensitivity to pressure, whereas those isolated from depths, where the pressure approximated 500 atmospheres, grew readily at a pressure of 600 atm. Mixed microflora from muds of the same depth appeared to grow faster under pressure. Organisms whose growth was favored by pressure were referred to as barophiles. ZoBell and Morita (91) found bacterial populations ranging from 10^3 to 10^6 per gram of wet mud in samples taken from depths of 7,000 to 10,000 m. Counts made at a pressure of 1,000 atm (the approximate pressure prevailing at these depths) were in most cases appreciably higher than those conducted at 1 atm. The authors reported that a good many tests made on bacteria which grew at a pressure of 1,000 atm demonstrated their inability to grow in similar media incubated at 1 atm. Similarly, among the many cultures tested, none which grew at 1 atm did so when incubated at 1,000 atm.

Kriss and co-workers (38) isolated 146 strains of bacteria from deep-sea bottom deposits and from garden soil which had been subjected to high hydrostatic pressure. The organisms could be divided into two groups, those which remained viable but were unable to reproduce at 450 atm of pressure and those which were able to grow at this pressure. Only one strain was found which developed better at 450 atm of pressure than at atmospheric pressure. In general, strains growing well at 450 atm grew even better at 1 atm. These workers reported the isolation from the upper layers of the soil of cultures able to grow and reproduce at 1,500 atm of pressure.

The mechanism of action of pressure on biological systems has been extensively studied by Johnson and co-workers (34). The effects of pressure have been explained in terms of the molecular volume change accompanying a limiting reaction. The influence of pressure not only may be modified but even reversed in direction by a change in temperature. Below the normal optimal temperature, an increase in pressure may produce inhibition by opposing the molecular volume increase accompanying the limiting reaction. At temperatures above the optimum, the critical enzyme undergoes a reversible denaturation that proceeds with an even larger volume increase than the limiting reaction. At these temperatures, the net effect of pressure is to increase the rate of the reaction by reversing the denaturation of the enzyme to a greater extent than opposing the catalytic reaction. In keeping with this hypothesis, ZoBell and Johnson (90) observed that lower temperatures markedly accentuated the growth-retarding and disinfecting effects of pressure on bacterial cultures. At higher temperatures, pressure in some cases acted in the direction of opposing the unfavorable effects on growth and viability caused by high temperature. As a further example of the effect, Morita and Haight (55) observed malic dehydrogenase activity at 101 C under hydrostatic pressure.

There are enormous technical problems associated with obtaining quantitative information on the relation of the various types of marine bacteria in deep-sea bottom deposits to pressure. It takes 8 to 18 hr to bring samples to the surface from a depth of 10,000 m (91), and no way has yet been devised to maintain samples during collection and subsequent manipulation at the pressures and temperatures prevailing in the depths. If there are bacteria at the bottom of the sea which depend upon the particular temperature-pressure combination found there to maintain the conformation of vital polyelectrolytes, many of these bacteria could well be rendered nonviable by the decrease in pressure and increase in temperature associated with bringing samples to the surface.

Taxonomic Position

It is of interest to know whether bacteria in the sea differ in a sufficient number of characteristics from bacteria in other habitats to warrant their being placed in a separate taxonomic group. Miyamoto and co-workers (54), for instance, proposed that the gram-negative polarly flagellated rod forms found widely distributed in the ocean be grouped in a new genus *Oceanomonas*. The distinctive character of this genus would be the degree of halophilism usually exhibited by marine bacteria. Sakazaki et al. (71) could not accept this proposal, since the genus would include a group of enteropathogenic marine bacteria which they concluded were vibrios. Shewan and co-workers (74) have worked out a determinative scheme for gram-negative bacteria from the marine environment which groups these organisms into the genera *Pseudomonas*, *Xanthomonas*, *Aeromonas*, *Vibrio*, *Achromobacter*, *Alcaligenes*, *Flavobacterium*, *Cytophaga* and a peritrichously flagellated group referred to as "Paracolons." Colwell and Gochnauer (20) examined 60 bacterial cultures of marine origin for approximately 100 characteristics, including Na^+ and Mg^{++} requirement, amino acid growth response, and the standard bacteriological characters. The data were coded and analyzed by electronic computer by use of the Adansonian method. Also, these data were compared by computer with other data similarly obtained for 131 named strains of the Eubacteriales and Pseudomonadales. Results of the analyses showed four groupings within the marine strains, three *Pseudomonas* and one *Vibrio* cluster. No single characteristic was exclusive to any one of the groups which were based on overall similarity. This data suggested that separate genera should not be formed to describe marine species, and is the type of result one might expect to obtain if there is a close evolutionary relationship between marine and terrestrial species. Since life is believed to have originated in the sea, it is not unlikely that the common ancestor of both marine and terrestrial bacteria was a marine bacterium. Although much remains to be done to elucidate the differences between marine and terrestrial species at the molecular level, enough information is available to suggest that a quite limited number of successive mutations could convert a marine species to a form which would not be dependent on the sea for its survival.

Capacity to Survive in Seawater

It is evident that there are bacteria in the sea which depend upon the kinds and amounts of inorganic ions in seawater for their survival. Just what proportion of the bacteria in the sea have this dependence is not yet clear. That it is probably a high proportion, at least of those bacteria able to grow on laboratory media, was made evident very early in the study of marine microbiology by the large increases in counts obtained when marine materials were plated on seawater rather than freshwater media. It is also evident that there are bacteria having some of the special characteristics of marine bacteria in other environments. Such characteristics as Na^+ dependence and lytic susceptibility will not alone stamp bacteria as being uniquely marine. Nevertheless, bacteria dependent on the inorganic composition of seawater for their survival appear to predominate in the sea, even though the possession of these inorganic requirements confers no obvious competitive advantage on the organisms.

It has long been known that seawater possesses marked bactericidal activity for a variety of terrestrial organisms. This is not a simple matter of intolerance to the concentrations of salts that are present in the sea, because seawater can be rendered nontoxic for some organisms by autoclaving and for others by adding small amounts of appropriate organic materials. Because of its relation to sewage disposal, most of the investigations have dealt with the effects of seawater on *E. coli* (16). The loss of viability which occurs when cells of this organism are suspended in either natural or artificial seawater can be prevented by adding small amounts either of cysteine or of other amino acids having the capacity to form chelate complexes with metal ions (73). Jones (35) observed that the long lag phase which occurred when *E. coli* was grown in media prepared with seawater or 2.5% NaCl could be overcome by the addition of small amounts of compounds which had in common the capacity to act as chelating agents. It has been concluded by the various workers that the bactericidal action of seawater for *E. coli* is due to its content of toxic heavy metals in trace amounts. Saz et al. (72) report the presence in seawater of a nondialyzable, heat-labile compound having rapid bactericidal activity against both penicillin-sensitive and -resistant strains of *Staphylococcus aureus*. Under the conditions of the experiments, the substance exhibited no activity against *E. coli*.

The key, then, to the distinction between marine and terrestrial bacterial species may well be the mechanism or mechanisms which confer on bacteria the capacity to survive and grow in the sea. The fact that ability to survive in the sea is linked to the possession of particular inorganic requirements is probably not fortuitous, but a direct relationship between these characteristics, if it exists, remains to be elucidated. We are forced to conclude that there are bacteria which are uniquely marine because they are able to survive and grow in the sea, and we have yet to find out why.

Relation of Organisms Isolated to Indigenous Flora

It has been known for many years [see Waksman et al. (82)] that the numbers of bacteria in seawater or marine mud able to grow on laboratory media are as many as 1,000 times smaller than the numbers observed by direct microscopic examination. This has been emphasized recently by Kriss and co-workers (38), who reported that plating on standard laboratory media detected not more than 0.1 to 1% of the total numbers of microorganisms which can be observed microscopically in seawater or mud samples. That at least some of these forms must be viable was indicated by the fact that unusual morphological types never isolated from laboratory media were capable of forming microcolonies on glass slides submerged in seawater. Furthermore, deep-sea investigations in the Black Sea, Pacific, Atlantic, and Arctic oceans showed that some of the microbial forms revealed by direct microscopic examination were widely distributed. It would thus appear that present knowledge of the properties of marine bacteria has been gained from studies on representatives of the less than 1% of bacteria in the sea able to grow under ordinary laboratory conditions. To what extent their characteristics are common also to the types of organisms yet uncultured remains to be determined. This, of course, is a problem not confined to the marine environment. Only a small percentage of the bacteria observed by microscopy in soil and fresh water ever grow on laboratory media [see Gibson (27) for a review]. We are therefore faced with the very real possibility that, in the case of many natural environments, the organisms isolated and studied may not in fact be true representatives of the indigenous population.

SUMMARY

The marine bacteria which grow on media giving the highest plate counts on seawater and marine materials are largely gram-negative rod forms most of which are motile. The majority are facultatively psychrophilic and some, particularly those from deep-sea bottom deposits, can grow at high hydrostatic pressures. Many have a preference for amino acids as sources of carbon, nitrogen, and energy, and some require vitamins and other growth factors. Metabolic pathways in these organisms appear to be similar to those in other species.

Marine bacteria have special requirements for inorganic ions, partly to supply the needs of the organisms for growth and metabolism, partly to maintain the integrity of the cells. They have a highly specific need for Na^+ for growth, which has been shown in two species to reflect the presence of a Na^+-dependent mechanism for transporting substrates into the cells. Some of the bacteria fail to grow in the absence of halide ions, and this requirement can be satisfied either by chloride or bromide. Their need for Mg^{++} or for a combination of Mg^{++} and Ca^{++} exceeds that of most terrestrial species. For some marine bacteria, the effect of salts in maintaining the integrity of the cells has been shown to be due entirely to the capacity of the salts to interact directly with the cell envelopes. For other species of marine origin salts may also have an osmotic function.

Although the marine bacteria examined have a number of characteristics in common, the only one which clearly distinguishes them from bacteria in other habitats is a capacity to survive and grow in the sea. In this respect, then, marine bacteria are unique. Taxonomic studies show that the marine bacteria which have so far been studied fit well into genera which have already been defined. It should be remembered, however, that less than 1% of the bacteria observed in seawater and marine mud by microscopy grow under laboratory conditions. It is therefore quite possible that the organisms so far examined are not representative of the indigenous flora.

LITERATURE CITED

1. ABRAM, D., AND N. E. GIBBONS. 1961. The effect of chlorides of monovalent cations, urea, detergents and heat on morphology

and the turbidity of suspensions of red halophilic bacteria. Can. J. Microbiol. **7**:741–750.

2. BAARS, J. K. 1930. Over sulfaatreductie door bacteriën. Dissertation, Technische Hoogeschool, Delft, The Netherlands.
3. BAXTER, R. M., AND N. E. GIBBONS. 1962. Observation on the physiology of psychrophilism in a yeast. Can. J. Microbiol. **8**:511–517.
4. BEDFORD, R. H. 1933. Marine bacteria of the northern Pacific Ocean. The temperature range of growth. Contrib. Can. Biol. Fisheries **8**:433–438.
5. BROWN, H. J., AND N. E. GIBBONS. 1955. The effect of magnesium, potassium and iron on the growth and morphology of the red halophilic bacteria. Can. J. Microbiol. **1**:486–501.
6. BROWN, A. D. 1960. Inhibition by spermine of the action of a bacterial cell-wall lytic enzyme. Biochim. Biophys. Acta **44**:178–179.
7. BROWN, A. D. 1961. The peripheral structures of gram-negative bacteria. I. Cell wall protein and the action of a lytic enzyme system of a marine pseudomonad. Biochim. Biophys. Acta **48**:352–361.
8. BROWN, A. D. 1962. The peripheral structures of gram-negative bacteria. III. Effects of cations on proteolytic degradation of the cell envelope of a marine pseudomonad. Biochim. Biophys. Acta **62**:132–144.
9. BROWN, A. D. 1963. The peripheral structures of gram negative bacteria. IV. The cation-sensitive dissolution of the cell membrane of the halophilic bacterium, *Halobacterium halobium*. Biochim. Biophys. Acta **75**:425–435.
10. BRYANT, M. P., I. M. ROBINSON, AND H. CHU. 1959. Observations on the nutrition of *Bacteroides succinogenes*—a ruminal cellulolytic bacterium. J. Dairy Sci. **42**:1831–1847.
11. BUCKMIRE, F. L. A., AND R. A. MACLEOD. 1964. Mechanism of lysis of a marine bacterium. Bacteriol. Proc., p. 41.
12. BUKATSCH, F. 1936. Über den Einfluss von Salzen auf die Entwicklung von Bakterien. Sitzber. Akad. Wiss. Wien, Math. Naturw. Klasse Abt. I **145**:259–276.
13. BURKHOLDER, P. R., AND G. H. BORNSIDE. 1957. Decomposition of marsh grass by aerobic marine bacteria. Bull. Torrey Bot. Club **84**:366–383.
14. BURKHOLDER, P. R. 1963. Some nutritional relationships among microbes of sea sediments and waters. Symp. Marine Microbiol., p. 133–150. Charles C Thomas, Publisher, Springfield, Ill.
15. BURTON, S. D., AND R. Y. MORITA. 1963. Denaturation and renaturation of malic dehydrogenase in a cell-free extract from a marine psychrophile. J. Bacteriol. **86**:1019–1024.
16. CARLUCCI, A. F., AND D. PRAMER. 1959. Microbiological process report. Factors affecting the survival of bacteria in sea water. Appl. Microbiol. **7**:388–392.
17. CARSON, K. J., AND R. G. EAGON. 1964. EDTA-induced lysis of *Pseudomonas aeruginosa*: its relation to cell-wall structure and integrity. Bacteriol Proc. p. 32.
18. CHRISTIAN, J. H. B. 1956. The physiological basis of salt tolerance in halophilic bacteria. Ph.D. Thesis, University of Cambridge, Cambridge, England.
19. CHRISTIAN, J. H. B., AND M. INGRAM. 1959. Lysis of *Vibrio costicolus* by osmotic shock. J. Gen. Microbiol. **20**:32–42.
20. COLWELL, R. R., AND M. B. GOCHNAUER. 1963. The taxonomy of marine bacteria. Bacteriol. Proc., p. 40.
21. DIANOVA, E., AND A. VOROSHILOVA. 1935. Salt composition of medium and specificity of marine bacteria. Mikrobiologiya **4**:393–402.
22. DOUDOROFF, M. 1942. Studies on the luminous bacteria. I. Nutritional requirements of some species with special reference to methionine. J. Bacteriol **44**:451–459.
23. DRAPEAU, G. R., AND R. A. MACLEOD. 1963. Na^+ dependent active transport of α-aminoisobutyric acid into cells of a marine pseudomonad. Biochem. Biophys. Res. Commun. **12**:111–115.
24. EAGON, R. G., AND C. H. WANG. 1962. Dissimilation of glucose and gluconic acid by *Pseudomonas natriegens*. J. Bacteriol. **83**: 879–886.
25. EAGON, R. G. 1963. Rate limiting effects of pyridine nucleotides on carbohydrate catabolic pathways of microorganisms. Biochem. Biophys. Res. Commun. **12**:274–279.
26. FEENEY, R. E., AND J. A. GARIBALDI. 1948. Studies on the mineral nutrition of the subtilin-producing strain of *Bacillus subtilis*. Arch. Biochem. **17**:447–458.
27. GIBSON, J. 1957. Nutritional aspects of microbiol ecology. Symp. Soc. Gen. Microbiol. **7**:22–41.
28. GOLDMAN, M., R. H. DEIBEL, AND C. F. NIVEN, JR. 1963. Interrelationship between temperature and sodium chloride on growth of lactic acid bacteria isolated from meat-curing brines. J. Bacteriol. **85**:1017–1021.
29. GOUCHER, C. R., A. SARACHEK, AND W. KOCHOLATY. 1955. A time-course respiratory inactivation associated with Azotobacter cells deprived of Mg^{++}. J. Bacteriol. **70**: 120–124.
30. HAGEN, P. O., AND A. H. ROSE. 1962. Studies on the biochemical basis of low maximum temperature in a psychrophilic cryptococcus. J. Gen. Microbiol. **27**:89–99.
31. HARVEY, E. N. 1915. The effect of certain organic and inorganic substances upon light production by luminous bacteria. Biol. Bull. **29**:308–311.
32. HILL, S. E. 1929. The penetration of luminous bacteria by the ammonium salts of fatty

acids. I. General outline of the problem and the effects of strong acids and alkalis. J. Gen. Physiol. **12**:863–872.

33. JOHNSON, F. H., AND E. N. HARVEY. 1938. Bacterial luminescence, respiration and viability in relation to osmotic pressure and specific salts of sea water. J. Cell. Comp. Physiol. **11**:213–232.
34. JOHNSON, F. H., H. EYRING, AND M. J. POLISSAR. 1954. The kinetic basis of molecular biology. John Wiley & Sons, Inc., New York.
35. JONES, G. E. 1964. Effect of chelating agents on the growth of *Escherichia coli* in seawater. J. Bacteriol. **87**:483–499.
36. KORINEK, J. 1927. Ein Beitrag zur Mikrobiologie des Meeres. Zentr. Bakteriol. Parasitenk. Abt. II **66**:500–505.
37. KOTIN, J. 1963. On the effect of ionic strength on the melting temperature of DNA. J. Mol. Biol. **7**:309–311.
38. KRISS, A. E. 1963. Marine microbiology (deep sea). [Translation by J. M. Shewan and Z. Kabata. Oliver & Boyd, Edinburgh.]
39. LARSEN, H. 1962. Halophilism, p. 297–342. *In* I. C. Gunsalus and R. Y. Stanier [ed.], The bacteria: a treatise on structure and function, vol. 4. Academic Press, Inc., New York.
40. LITTLEWOOD, D., AND J. R. POSTGATE. 1957. Sodium chloride and the growth of *Desulphovibrio desulfuricans*. J. Gen. Microbiol. **17**:378–389.
41. MAGER, J. 1959. The stabilizing effect of spermine and related polyamines and bacterial protoplasts. Biochim. Biophys. Acta **36**:529–531.
42. MAGER, J. 1959. Spermine as a protective agent against osmotic lysis. Nature **183**: 1827–1828.
43. MACLEOD, R. A., E. ONOFREY, AND M. E. NORRIS. 1954. Nutrition and metabolism of marine bacteria. I. Survey of nutritional requirements. J. Bacteriol. **68**:680–686.
44. MACLEOD, R. A., AND E. ONOFREY. 1956. Nutrition and metabolism of marine bacteria. II. Observations on the relation of sea water to the growth of marine bacteria. J. Bactiol. **71**:661–667.
45. MACLEOD, R. A., AND E. ONOFREY. 1956. Nutrition and metabolism of marine bacteria. VI. Quantitative requirements for halides, magnesium, calcium and iron. Can. J. Microbiol. **3**:753–759.
46. MACLEOD, R. A., AND E. ONOFREY. 1957. Nutrition and metabolism of marine bacteria. III. The relation of sodium and potassium to growth. J. Cell. Comp. Physiol. **50**:389-401.
47. MACLEOD, R. A., H. HOGENKAMP, AND E. ONOFREY. 1958. Nutrition and metabolism of marine bacteria. VII. Growth response of a marine flavobacterium to surface active agents and nucleotides. J. Bacteriol. **75**: 460–466.
48. MACLEOD, R. A., C. A. CLARIDGE, A. HORI, AND J. F. MURRAY. 1958. Observations on the function of sodium in the metabolism of marine bacteria. J. Biol. Chem. **232**:829–834.
49. MACLEOD, R. A., AND A. HORI. 1960. Nutrition and metabolism of marine bacteria. VIII. Tricarboxylic acid cycle enzymes in a marine bacterium and their response to added salts. J. Bacteriol. **80**:464–471.
50. MACLEOD, R. A., A. HORI, AND S. M. FOX. 1960. Nutrition and metabolism of marine bacteria. X. The glyoxylate cycle in a marine bacterium. Can. J. Microbiol. **6**:639–644.
51. MACLEOD, R. A., AND T. I. MATULA. 1961. Solute requirements for preventing lysis of some marine bacteria. Nature **192**:1209–1210.
52. MACLEOD, R. A., AND T. I. MATULA. 1962. Nutrition and metabolism of marine bacteria. XI. Some characteristics of the lytic phenomenon. Can. J. Microbiol. **8**:883–896.
53. MACLEOD, R. A., AND E. ONOFREY. 1963. Studies on the stability of the Na^+ requirement of marine bacteria. Symp. Marine Microbiol., p. 481–489. Charles C Thomas, Publisher, Springfield, Ill.
54. MIYAMOTO, Y., K. NAKAMURA, AND K. TAKIZAWA. 1961. Pathogenic halophiles proposals of a new genus "Oceanomonas" and of the amended species names. Japan. J. Microbiol. **5**:477–486.
55. MORITA, R. Y., AND R. D. HAIGHT. 1962. Malic dehydrogenase activity at 101 C under hydrostatic pressure. J. Bacteriol. **83**:1341–1346.
56. MORITA, R. Y., AND S. D. BURTON. 1963. Influence of moderate temperature on growth and malic dehydrogenase activity of a marine psychrophile. J. Bacteriol. **86**: 1025–1029.
57. MORITA, R. Y., AND S. M. ROBISON. 1964. Moderate temperature effects on oxygen uptake of *Vibrio marinus*, an obligate marine psychrophile. Bacteriol. Proc., p. 38–39.
58. MUDRAK, A. 1933. Beitrage zur Physiologie der Leucht-bakterien. Zentr. Bakteriol. Parasitenk. Abt. II **88**:353–366.
59. OCHYNSKI, F. W., AND J. R. POSTGATE. 1963. Some biochemical differences between fresh water and salt water strains of sulphate-reducing bacteria. Symp. Marine Microbiol., p. 426–441. Charles C Thomas, Publisher, Springfield, Ill.
60. OSTROFF, R., AND B. S. HENRY. 1939. The utilization of various nitrogen compounds by marine bacteria. J. Cell. Comp. Physiol. **13**:353–371.
61. PAYNE, W. J. 1958. Studies on bacterial utilization of uronic acids. III. Induction of oxidative enzymes in a marine isolate. J. Bacteriol. **76**:301–307.
62. PAYNE, W. J. 1960. Effects of sodium and potassium ions on growth and substrate

penetration of a marine pseudomonad. J. Bacteriol. **80**:696–700.
63. PRATT, D. B., AND G. WADDELL. 1959. Adaptation of marine bacteria to media lacking sodium chloride. Nature **183**:1208–1209.
64. PRATT, D., AND F. C. HAPPOLD. 1960. Requirements for indole production by cells and extracts of a marine bacterium. J. Bacteriol. **80**:232–236.
65. PRATT, D., AND W. RILEY. 1955. Lysis of a marine bacterium in salt solutions. Bacteriol. Proc., p. 26.
66. PRATT, D., AND M. AUSTIN. 1963. Osmotic regulation of the growth rate of four species of marine bacteria. Symp. Marine Microbiol., p. 629–637. Charles C Thomas, Publisher, Springfield, Ill.
67. PRATT, D. 1963. Specificity of the solute requirement by marine bacteria on primary isolation from sea-water. Nature **199**:1308.
68. REPASKE, R. 1956. Lysis of gram negative bacteria by lysozyme. Biochim. Biophys. Acta **22**:189–191.
69. RHODES, M. E., AND W. J. PAYNE. 1962. Further observations on effects of cations on enzyme induction in marine bacteria. Antonie van Leeuwenhoek. J. Microbiol. Serol. **28**:302–314.
70. RICHTER, O. 1928. Natrium: Ein notwendiges Nährelement für eine marine mikroärophile Leuchtbakterie. Anz. Oesterr. Akad. Wiss. Math. Naturw. Kl. **101**:261–292.
71. SAKAZAKI, R., S. IWANAMI, AND H. FUKUMI. 1963. Studies on the enteropathogenic, facultatively halophilic bacteria, *Vibrio parahaemolyticous*. I. Morphological, cultural and biochemical properties and its taxonomical position. Japan. J. Med. Sci. Biol. **16**:161–188.
72. SAZ, A. K., S. WATSON, S. R. BROWN, AND D. L. LOWERY. 1963. Antimicrobial activity of marine waters. I. Macromolecular nature of antistaphylococcal factor. Limnol. Oceanogr. **8**:63–67.
73. SCARPINO, P. V., AND D. PRAMER. 1962. Evaluation of factors affecting the survival of *Escherichia coli* in sea water. VI. Cysteine. Appl. Microbiol. **10**:436–440.
74. SHEWAN, J. M. 1963. The differentiation of certain genera of gram negative bacteria frequently encountered in the marine environments. Symp. Marine Microbiol., p. 499–521. Charles C Thomas, Publisher, Springfield, Ill.
75. SISTROM, W. R. 1960. A requirement for sodium in the growth of *Rhodopseudomonas spheroides*. J. Gen. Microbiol. **22**:778–785.
76. STANIER, R. Y. 1941. Studies on marine agar-digesting bacteria. J. Bacteriol. **42**:527–559.
77. STOKES, J. L. 1963. General biology and nomenclature of psychrophilic micro-organisms, p. 186–192. *In* Recent progress in microbiology. Int. Congr. Microbiol., 8th, Montreal.
78. TAKACS, F. P., T. I. MATULA, AND R. A. MACLEOD. 1964. Nutrition and metabolism of marine bacteria. XIII. Intracellular concentrations of sodium and potassium ions in a marine pseudomonad. J. Bacteriol. **87**:510–518.
79. TOMLINSON, N., AND R. A. MACLEOD. 1957. Nutrition and metabolism of marine bacteria. IV. The participation of Na^+, K^+, and Mg^{++} salts in the oxidation of exogenous substrates by a marine bacterium. Can. J Microbiol. **3**:627–638.
80. TYLER, M. E., M. C. BIELLING, AND D. B. PRATT. 1960. Mineral requirements and other characters of selected marine bacteria. J. Gen. Microbiol. **23**:153–161.
81. UPADHYAY, J., AND J. L. STOKES. 1963. Temperature-sensitive hydrogenase and hydrogenase synthesis in a psychrophilic bacterium. J. Bacteriol. **86**:992–998.
82. WAKSMAN, S. A., H. W. REUSZER, C. L. CAREY, M. HOTCHKISS, AND C. E. RENN. 1933. Studies on the biology and chemistry of the Gulf of Maine. III. Bacteriological investigations of the sea water and marine bottoms. Biol. Bull. **64**:183–205.
83. WIEBE, W. J., AND J. LISTON. 1963. The effects of magnesium on growth of marine bacteria. Bacteriol. Proc., p. 2.
84. WILLIAMS, J., R. L. TODD, AND W. J. PAYNE. 1963. Isocitrate lyase in an alginolytic bacterium. Can. J. Microbiol. **9**:549–553.
85. YOUNG, E. G., R. W. BEGG, AND E. I. PENTZ. 1944. Inorganic nutrient requirements of *Escherichia coli*. Arch. Biochem. **5**:121–136.
86. ZOBELL, C. E., AND H. D. MICHENER. 1938. A paradox in the adaption of marine bacteria to hypotonic solutions. Science **87**:328–329.
87. ZOBELL, C. E., AND J. E. CONN. 1940. Studies on the thermal sensitivity of marine bacbacteria. J. Bacteriol. **40**:223–238.
88. ZOBELL, C. E., AND H. C. UPHAM. 1944. A list of marine bacteria including descriptions of sixty new species. Bull. Scripps Inst. Oceanogr. **5**:239–292.
89. ZOBELL, C. E. 1946. Marine microbiology. Chronica Botanica Co., Waltham, Mass.
90. ZOBELL, C. E., AND F. H. JOHNSON. 1949. The influence of hydrostatic pressure on the growth and viability of terrestrial and marine bacteria. J. Bacteriol. **57**:179–189.
91. ZOBELL, C. E., AND R. Y. MORITA. 1957. Barophilic bacteria in some deep sea sediments. J. Bacteriol. **73**:563–568.
92. ZOBELL, C. E., AND S. C. RITTENBERG. 1938. The occurrence and characteristics of chitinoclastic bacteria in the sea. J. Bacteriol. **35**:275–278.

Editor's Comments on Paper 22

22 BROWN
Effects of Salt Concentration During Growth on Properties of the Cell Envelope of a Marine Pseudomonad

IS THERE A STRUCTURAL RESPONSE BY BACTERIA TO THE MARINE ENVIRONMENT?

Much of the research on specific ion effects has focused on the cell wall and cell membrane because this seemed a logical primary site for ionic effects to be operative. Thus, it was only natural that many of the same investigators who examined ionic requirements would start to examine the structural aspects of marine bacteria.

Until about 1960 to 1962, gram-negative bacteria were considered to be nonsusceptible to lysozyme attack and generally incapable of forming true protoplasts that did not contain any contaminating peptidoglycan layer (29). Obviously, if one wanted to study the various external layers of a marine pseudomonad, one must separate the envelope quantitatively from the cytoplasmic membrane.

In 1961, Brown (2) demonstrated that there existed labile proteins in the outer layers of a gram-negative bacterium, that the acidic labile protein was turned over by an autolytic enzyme and was in the "outer layer," and finally that the diaminopimelic acid was released by this enzyme. He then proceeded (Paper 22) to study the effects of cations on structure. In this paper, Brown made the point that the requirement or function of salt in this pseudomonad is not for the maintenance of osmotic pressure or water activity, but, as he later pointed out, the membrane proteins required an ionic environment similar to the marine system (3).

MacLeod and co-workers in two papers published in 1970 (14, 15) finally succeeded in presenting a method for the separation of the cell wall layers of a marine pseudomonad, B16. By using chemical and physical techniques and electron microscopy, they demonstrated that the cell envelope could be separated into five distinct layers by the careful manipulation of the ionic strength of the suspending medium during differential centrifugation of the cells. Meanwhile, Buckmire and Murray (4, 5) were investigating the cell wall of a marine spirillum from which they isolated the hexagonal surface units, demonstrated their protein nature, and showed that, after separation into monomers, the units could reassemble themselves on a cell-wall-fragment template probably as trimers. Thus, we see a correlation between the detailed chemical studies and the elegant electron micrograph work of Murray and co-workers (4, 24) with the even earlier studies of Houwink (18).

The electron microscope has enabled us to delineate surface subunits and structures of these marine bacteria. The envelope structure of *Nitrosocystis oceanus* was also examined by electron microscopy and found to consist of approximately seven layers, with the surface covered by repeating hexagonal units of 50 Å (31). The marine photosynthetic bacteria were also subjected to freeze-etching and electron microscopy. Surface details of *Chromatium buderi* revealed a pentagonal arrangement; *Chromatium gracile* displayed the typical hexagonal subunits when observed via electron microscopy by Remsen et al. (28). Similar studies have been conducted on marine thiobacilli, in which the different stages in the growth cycle are reflected in marked changes in the ultrastructure, expecially as regards the rippling of the cell envelope and the presence of membrane vesicles (23).

Interpretation of the cleavage plane of freeze-etching required that one understand whether the cleavage pattern followed the membrane or some other layer within the highly complex cell wall structure. DeVoe et al. (8) examined the marine pseudomonad B16 and found that cleavage followed either the outer double track layer or the underlying soluble layer; Forge et al. (13) localized the cleavage plane as being dependent on the outer double track layer. At the same time they noted that the molecular nature of this layer permits its involvement in the activity of cell-wall-associated enzymes, which have been demonstrated by MacAlister et al. (21), Costerton (6), and Thompson et al. (30). Further studies by DeVoe and Oginsky (9, 10) have shown that divalent cations are necessary for envelope integrity, and O'Leary et al. (25) have noted that NaCl was required for the isolation of the lipopolysaccharide (LPS) fraction of the cell walls, which contained 0.1 percent 2-keto-3-deoxyoctulosonic acid (KDO) and 12.9 percent lipid A. The peptidoglycan layer of B16 was also isolated and found to consist mainly of glucosamine, alanine, muramic acid, glutamic acid, and diaminopimelic acid (15). Comparisons with other gram-negative bacteria indicate, then, that B16 (16), several vibrio species (7), and *Pseudomonas iridescens* (22) contain a typical gram-negative LPS fraction and a typical gram-negative peptidoglycan. Thus, the explanation for the marine nature of these organisms must lie elsewhere.

It has frequently been observed in bacterial strains growing on agar surfaces that dissociation into rough and smooth or large and small colonial types will occur spontaneously. The differences in the cell wall and membrane structure that might account for this are under examination now. So far, only differences in optimal Na^+ concentrations and the ability to form protoplasts have been documented (17), but, clearly, a biochemical change must occur that would account for such a high degree of colonial pleomorphism.

In many vibrio cultures, cellular pleomorphism is expressed as enlarged cells and round bodies. The ultrastructure of these was investigated and the hypothesis presented that their appearance in late log phase cells resulted from alterations in cell division (11). From later work with another isolate, though, the authors concluded that round-body formation is a normal process in the life cycle of this bacterium (19). Thus, the intracellular organization of marine bacteria has not been completely ignored during studies on cell walls and cell membranes. Extensive intracytoplasmic membrane structures have been observed in *Vibrio marinus* (12), although the appearance of these membranous structures is considered an expression of the growth stage and/or nutrient level in the medium according to the authors. However, Murray and Watson (24) demonstrated that in the nitrifying bacterium *Nitrosocystis oceanus* extensive intracytoplasmic membrane systems exist throughout the life cycle of the organism and are believed to be involved in the nitrification process. These cytoplasmic membranes were later isolated and their fine structure revealed (27).

The nonfruiting, gliding, gram-negative bacteria frequently isolated from the marine environment (20), the flexibacteria, have also been subjected to the study of environmental effects on their cell division. Poos et al. (26) found that elongation and cell size were dependent on temperature, and mesosomes were observed in the cells. This is the first report of true mesosomes in marine bacteria, as other authors (12) have commented on their failure to observe these specific internal membranous structures in marine vibrios.

Other structures such as flagella have been examined by electron microscopy (1), but so far no descriptions of their chemical compositions have been reported. Although not very common in the marine enviornment, a few spore-forming bacteria have been isolated. To date, neither chemical composition nor morphological details as revealed by electron microscopy of the whole cells or spores have been forthcoming. Unfortunately, even the more common gram-positive corynebacteria, brevibacteria, and arthrobacters have escaped comparative studies as to their structure, composition, and physiology. Certainly, there is a great deal of work remaining in these and other areas before one can make a general statement about the distinctive structural or chemical nature of marine bacteria.

REFERENCES

1. Allen, R. D., and Baumann, P. 1971. Structure and arrangement of flagella in species of the genus *Beneckea* and *Photobacterium fischeri*. *J. Bacteriol.* *107*:295–302.
2. Brown, A. D. 1961. The peripheral structures of gram-negative bacteria. I. Cell wall protein and the action of a lytic enzyme system of a marine pseudomonad. *Biochim. Biophys. Acta* *48*:352–361.
3. Brown, A. D., Drummond, D. G., and North, R. J. 1962. The peripheral structures of gram-negative bacteria. II. Membranes of bacilli and spheroplasts of a marine pseudomonad. *Biochim. Biophys. Acta* *58*: 514–531.
4. Buckmire, F. L. A., and Murray, R. G. E. 1970. Studies on the cell wall of *Spirillum serpens* . I. Isolation and partial purification of the outermost cell wall layers. *Can. J. Microbiol.* *16*:1011–1022.
5. Buckmire, F. L. A., and Murray, R. G. E. 1973. Studies on the cell wall of *Spirillum serpens*. II. Chemical characterization of the outer structured layer. *Can. J. Microbiol.* *19*:59–66.
6. Costerton, J. W. 1973. Relationship of a wall-associated enzyme with specific layers of the cell wall of a gram-negative bacterium. *J. Bacteriol.* *114*:1281–1293.
7. Costerton, J. W., Ingram, J. M., and Cheng, K. J. 1974. Structure and function of the cell envelope of gram-negative bacteria. *Bacteriol. Rev.* *38*:87–110.
8. DeVoe, I. W., Costerton, J. W., and MacLeod, R. A. 1971. Demonstration by freeze-etching of a single cleavage plane in the cell wall of a gram-negative bacterium. *J. Bacteriol.* *106*:659–671.
9. DeVoe, I. W., and Oginsky, E. L. 1969. Antagonistic effect of mono-

valent cations in maintenance of cellular integrity of a marine bacterium. *J. Bacteriol.* *98*:1355–1367.
10. DeVoe, I. W., and Oginsky, E. L. 1969. Cation interactions and biochemical composition of the cell envelope of a marine bacterium. *J. Bacteriol.* *98*:1368–1377.
11. Felter, R. A., Colwell, R. R., and Chapman, G. B. 1969. Morphology and round body formation in *Vibrio marinus*. *J. Bacteriol.* 99:326–335.
12. Felter, R. A., Kennedy, S. F., Colwell, R. R., and Chapman, G. B. 1970. Intracytoplasmic membrane structures in *Vibrio marinus*. *J. Bacteriol.* *102*:552–560.
13. Forge, A., Costerton, J. W., and Kerr, K. A. 1973. Freeze-etching and x-ray diffraction of the isolated double-track layer from the cell wall of a gram-negative marine pseudomonad. *J. Bacteriol.* *113*:445–451.
14. Forsberg, C. W., Costerton, J. W., and MacLeod, R. A. 1970. Separation and localization of cell wall layers of a gram-negative bacterium. *J. Bacteriol.* *104*:1338–1353.
15. Forsberg, C. W., Costerton, J. W., and MacLeod, R. A. 1970. Quantitation, chemical characteristics, and ultrastructure of the three outer cell wall layers of a gram-negative bacterium. *J. Bacteriol.* *104*:1354–1368.
16. Forsberg, C. W., Rayman, M. H., Costerton, J. W., and MacLeod, R. A. 1972. Isolation, characterization, and ultrastructure of the peptidoglycan layer of a marine pseudomonad. *J. Bacteriol.* *109*:895–905.
17. Gow, J. A., DeVoe, I. W., and MacLeod, R. A. 1973. Dissociation in a marine pseudomonad. *Can. J. Microbiol.* *19*:695–701.
18. Houwink, A. L. 1956. Flagella, gas vacuoles and cell wall structure in *Halobacterium halobium*; an electron microscopy study. *J. Gen. Microbiol.* *15*:146–150.
19. Kennedy, S. F., Colwell, R. R., and Chapman, G. B. 1970. Ultrastructure of a marine psychrophilic *Vibrio*. *Can. J. Microbiol.* *16*:1027–1031.
20. Lewin, R. A., and Lounsbery, D. M. 1969. Isolation, cultivation and characterization of *Flexibacteria*. *J. Gen. Microbiol.* *58*:145–170.
21. MacAlister, T. J., Costerton, J. W., Thompson, L., Thompson, J., and Ingram, J. M. 1972. Distribution of alkaline phosphatase within the periplasmic space of gram-negative bacteria. *J. Bacteriol.* *111*:827–832.
22. Mongillo, A., Deloge, K., Pereira, D., and O'Leary, G. P. 1974. Lipopolysaccharide from a gram-negative marine bacterium. *J. Bacteriol.* *117*:327–328.
23. Murphy, J. R., Girald, A. E., and Tilton, R. C. 1974. Ultrastructure of a marine *Thiobacillus*. *J. Gen. Microbiol.* *85*:130–138.
24. Murray, R. G. E., and Watson, S. W. 1965. Structure of *Nitrosocystis oceanus* and comparison with *Nitrosomonas* and *Nitrobacter*. *J. Bacteriol.* *89*:1594–1609.
25. O'Leary, G. P., Nelson, J. D., Jr., and MacLeod, R. A. 1972. Requirement for salts for the isolation of lipopolysaccharide from a marine pseudomonad. *Can. J. Microbiol.* *18*:601–606.
26. Poos, J. C., Turner, F. R., White, D., Simon, G. D., Bacon, K., and Russell, C. T. 1972. Growth, cell division, and fragmentation in a species of *Flexibacter*. *J. Bacteriol.* *112*:1387–1395.

27. Remsen, C. C., Valois, F. W., and Watson, S. W. 1967. Fine structure of the cytomembranes of *Nitrosocystis oceanus*. *J. Bacteriol.* *94*:422–433.
28. Remsen, C. C., Watson, S. W., and Trüper, H. G. 1970. Macromolecular subunits in the walls of marine photosynthetic bacteria. *J. Bacteriol.* *103*:255–258.
29. Stanier, R. Y., Doudoroff, M., and Adelberg, E. A. 1963. *The Microbial World*, 2nd ed., Prentice-Hall, Inc., Englewood Cliffs, N.J., 753 pp.
30. Thompson, J., Green, M. L., and Happold, F. C. 1969. Cation-activated nucleotidase in cell envelopes of a marine bacterium. *J. Bacteriol.* *99*: 834–841.
31. Watson, S. W., and Remsen, C. C. 1970. Cell envelope of *Nitrosocystis oceanus*. *J. Ultrastruct. Res.* *33*:148–160.

22

Reprinted from *Biochim. Biophys. Acta*, **49**, 585–588 (1961)

EFFECTS OF SALT CONCENTRATION DURING GROWTH ON PROPERTIES OF THE CELL ENVELOPE OF A MARINE PSEUDOMONAD

A. D. Brown
C.S.I.R.O., Marine Laboratory, Cronulla, Sydney (Australia)

Gram-negative marine and halophilic bacteria are notably pleomorphic. Marine pseudomonad No. 11 (ref. 1) is no exception and its pleomorphism together with a low content of amino sugar in the cell wall has led to the suggestion that the wall of this organism is mechanically weak[1]. The association of pleomorphism with an environment of relatively high salt concentration suggests that the latter factor has contributed to the cell-wall properties either at the evolutionary level or by directly influencing cell-wall synthesis, or both. The work to be described was undertaken to determine what effects variations of salt concentration of the growth medium have on quantitative cell-wall composition and, if possible, whether any such effects could be attributed to osmotic pressure, water activity or ionic strength.

The organism was grown at 30° with aeration for 24 h in "double strength" artificial sea water[1], or dilutions thereof, containing 1 % (w/v) peptone and 2 ml 0.2 *M* Na_2HPO_4/l. The artificial sea water is a mineral salts medium containing Na^+, K^+, Mg^{2+}, Ca^{2+}, Fe^{2+}, Cl^- and SO_4^{2-} and at "single strength" has a total molarity of 0.49. Some media were also supplemented as follows with sucrose, glycerol, Mg^{2+} or spermine: 1, artificial sea water (0.061 *M*) plus sucrose (0.64 *M*); 2, artificial sea water (0.49 *M*) plus sucrose (0.37 *M*); 3, artificial sea water (0.061 *M*) plus glycerol (0.64 *M*); 4, artificial sea water (0.49 *M*) plus glycerol (0.37 *M*); 5, artificial sea water (0.061 *M*) plus spermine (0.002 *M*); 6, artificial sea water (0.061 *M*) plus Mg^{2+} (0.031 *M*, to give a total salt concentration of 0.092 *M* and a total Mg^{2+} concentration of 0.038 *M*, which is equal to that obtained in 0.35 *M* artificial sea water). Cell walls were prepared in the cold as described previously[2]. This procedure gives a double structure[3] which has since been shown to consist of the units conventionally described as the cytoplasmic membrane and the cell wall. It has not so far been possible to isolate the outer membrane (the cell wall) of this organism but by making use of the cell-wall autolytic system[3], the inner one has been obtained. Its preparation and properties will be described elsewhere by BROWN, DRUMMOND AND NORTH. Henceforth in this communication the term "envelope" will designate the combined inner and outer membranes as isolated. Protein and total phosphorus were estimated on envelopes; "reducing substances" and amino sugars were estimated in centrifuged acid hydrolysates as previously described[3]. α,ε-Diaminopimelic acid was estimated in acid hydrolysates (6 *N* HCl for 16 h at 100°) by a method which will be described elsewhere. In essence it consisted of chromatographing the hydrolysate on Whatman No. 1 or 3MM paper in the solvent system of RHULAND *et al.*, dipping the chromatograms in a solution of ninhydrin, heating, eluting the coloured complex with aqueous acetone and measuring the absorbancy at 405 mμ.

Organisms grown within the range of concentrations, 0.98–0.25 *M* were "normal" rods as previously described[1]. With decreasing concentration below 0.18 *M* the organisms became shorter and rounder and the yields of cells decreased. At the lowest concentration used, 0.061 *M*, the organisms were either spherical (dia. 1–2 μ) or short elipses. The spherical cells retained their motility and were not characterized by withdrawal of the cytoplasm from sections of the cell wall as often noted in spheroplasts of this and other bacteria. Chemical properties of the cell envelopes associated with these changes are shown in Fig. 1. It will be seen that protein content was positively correlated while amino sugar, phosphorus and "reducing substances" were negatively correlated with salt concentration of the growth medium. The magnitude of the changes was about the same in each case and it is likely that all the results of Fig. 1 are a direct consequence of changes in the content of cell-envelope protein. Diaminopimelic acid fell within the range 0.30 % (at 0.061 *M*) to 0.43 % (at 0.98 *M*) but the scatter of values was such that no correlation with salt concentration was demonstrated. The cell-wall autolytic system of this organism is affected by ionic strength of phosphate buffers[3] and preparations were examined in the present instance for variations in the shape of the curves depicting lysis *versus* buffer concentration. Variations did occur but could not be related to salt concentration of the growth medium.

Sucrose is generally held to be unable to penetrate lipoprotein permeability barriers freely and can therefore directly influence the osmotic pressure to which a bacterial cell is subjected. Glycerol, on the other hand, apparently does penetrate such barriers freely and therefore does not directly influence osmotic pressure (see, for example, MITCHELL AND MOYLE[5]).

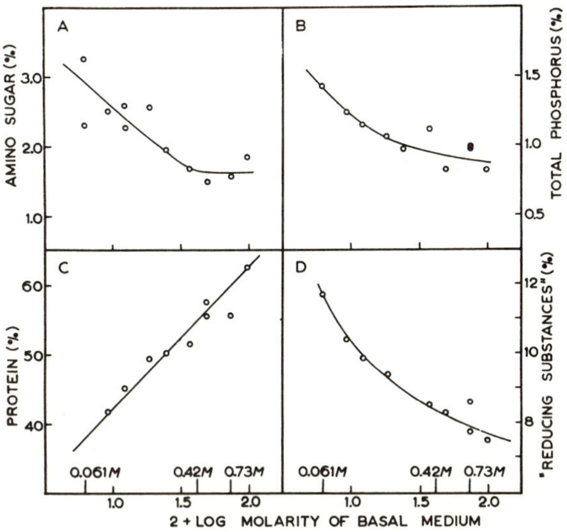

Fig. 1. Effects of salt concentration of the growth medium on chemical composition of the cell envelope. A, amino sugar (as glucosamine); B, total phosphorus; C, protein; D, "reducing substances" (as glucose).

In order to ascertain the effects of these solutes on the cell envelope, the organism was grown (or incubated) in Media 1–4. The organism failed to grow in Medium 1 but grew well in Medium 2 suggesting that growth in Medium 1 was prevented by osmotic effects. In Media 3 and 4 the organism grew with yields and morphology similar to those obtained in the same salt concentration without the addition of glycerol. Properties of the envelopes of such cells are listed in Table I which shows that, at the lower salt concentration the composition of the envelope resembled that of organisms grown at the same ionic strength rather than at the same water activity. "Reducing substances" were exceptional but it is possible that they were influenced by metabolism of some of the glycerol, perhaps to form more cell-envelope phospholipid. The contribution of glycerol to water activity is less in the medium of higher salt concentration and differences in chemical composition of the envelopes are not so marked in these cases.

Autolysis of cell-wall protein of this organism is inhibited by spermine[2], by divalent cations *e.g.* Mg^{2+} (BROWN, unpublished observation) and by salt concentrations of the order of 0.5 M (ref. 3). As it seemed likely that factors which prevent breakdown of this protein in the isolated cell envelope might also facilitate its formation during growth, the organism was grown in Media 5 and 6. Both spermine and Mg^{2+} increased the yield of cells; Mg^{2+}, but not spermine, altered morphology to give

TABLE I

COMPOSITION OF THE CELL ENVELOPES OF ORGANISMS GROWN WITH AND WITHOUT ADDED GLYCEROL

	Growth Medium					
*Substance**	*Medium 3 (0.061 M salts + 0.64 M glycerol) (%)*	*Artificial sea water of water activity equal to Medium 3 (0.42 M)** (%)*	*0.061 M salts (%)*	*Medium 4 (0.49 M salts + 0.37 M glycerol) (%)*	*Artificial sea water of water activity equal to Medium 4 (0.73 M)** (%)*	*0.49 M salts (%)*
Protein	39.7	55	38	59.5	60	56
Amino sugars	3.74	1.7	2.9	1.89	1.5	1.6
Reducing substances	8.55	8.3	11.7	6.0	7.7	8.2
Phosphorus	1.97	1.0	1.4	1.04	0.9	1.0

* Comparative values are derived from Fig. 1, which should be consulted for an indication of precision etc.
** Determined by freezing-point measurements.

fewer spherical forms than otherwise obtained in media of the same salt concentration (*cf.* TAKAHASHI AND GIBBONS[6]). Envelope composition of cells grown with added Mg^{2+}, however, resembled that obtained from media of the same total salt concentration rather than the same concentration of Mg^{2+} (protein, 40.9 %; amino sugar, 2.54 %; phosphorus, 1.82 %).

It is evident that in dilute media the cell envelopes resemble the cell walls of terrestrial Gram-negative bacteria more closely by having a higher content of amino sugar than they do in concentrated media. It is thought that the main factor determining this, however, is formation of cell-wall protein. Spermine and Mg^{2+} could increase the yield of cells by facilitating the formation of cell-wall protein without necessarily increasing appreciably the amount of protein in individual cell envelopes. The relation of Mg^{2+} and other factors to the stability and formation of the labile cell-wall protein will be the subject of another more detailed communication. Finally it is noted that the effects observed were caused principally by salt concentration (or ionic strength) *per se* and, insofar as it is possible to separate them, the contributions of the related properties of water activity and osmotic pressure appeared to be minor.

1 A. D. BROWN, *J. Gen. Microbiol.*, 23 (1960) 471.
2 A. D. BROWN, *Biochim. Biophys. Acta*, 44 (1960) 178.
3 A. D. BROWN, *Biochim. Biophys. Acta*, 48 (1961) 352.
4 L. E. RHULAND, E. WORK, R. F. DENMAN AND D. S. HOARE, *J. Am. Chem. Soc.*, 77 (1955) 4844.
5 P. MITCHELL AND J. MOYLE, in E. T. C. SPOONER AND B. A. D. STOCKER, *Bacterial Anatomy*, Cambridge University Press, 1956, p. 150.
6 I. TAKAHASHI AND N. E. GIBBONS, *Can. J. Microbiol.*, 5 (1959) 25.

Received January 31st, 1961

Editor's Comments on Papers 23 Through 25

23 MORITA
The Basic Nature of Marine Psychrophilic Bacteria

24 ZOBELL and MORITA
Deep-Sea Bacteria

25 MORITA
Effects of Hydrostatic Pressure on Marine Microorganisms

MARINE BACTERIAL RESPONSE TO IN SITU TEMPERATURES AND PRESSURES

Two physiologically distinguishing properties of many marine bacteria are their ability to grow at the low temperatures found in the oceans (average of 4 to 5°C) and their ability to grow at the greatly increased pressures found at oceanic depths. In fact, in all the cases studies so far, a direct correlation has been observed between survival at increased hydrostatic pressure, the temperature of incubation, and the salinity of the medium. Thus, any discussion of temperature and pressure must recognize the interrelatedness of these two parameters with salinity (2, 9, 18, 48, 58).

Those organisms which have an optimum growth temperature below 15°C, a maximum growth temperature below 20°C, and a minimal temperature for growth of 0°C are labeled *psychrophiles*; those which can grow at low temperatures but have an optimum growth temperature above 15°C are currently termed *psychrotrophic* (43). Because the majority of oceanic water is 4 to 5°C and some of the most productive waters, such as the Antarctic convergence, are even colder, there has been considerable interest in the psychrophilic marine bacteria. In 1892, while working with luminescent bacteria, Forester obtained isolates that grew best at around 0°C (22), thus confirming Fischer's earlier isolations (21). Others have isolated bacteria from snow caps (29), Issatschenko (28) obtained psychrophilic bacteria and yeasts from Arctic waters, and Frankland and Burgess (23) recorded their appearance in colder ocean waters in 1897.

In retrospect, it seems strange that, despite the work on temperature effects reported by Butkevich and Butkevich in 1936 (10) and that of other early workers (6, 22, 34, 35, 60), most investigators continued to ignore the importance of the incubation temperature either for initial isolation or for subsequent pure culture studies. During the last 15 years we have rediscovered the earlier truths of only 30 years ago. ZoBell and Conn (62) in 1940 again emphasized the importance of incubation temperature on the recovery of marine isolates on agar plates, but even ZoBell in subsequent papers (63, 64) was unable to complete his work at *in situ* temperatures and instead used 20 to 30°C. No doubt part of this was due to the difficulties of adapting incubators to temperatures that were frequently below ambient laboratory conditions. However, this is not so today as reasonably priced and reliable refrigerated incubators are generally available; low-temperature incubation and handling of all marine materials should by now be standard practice in all marine microbiological laboratories.

Inasmuch as a psychrophile grows at temperatures considered "holding temperatures" for other microbes, the biochemical or molecular properties that can explain the growth of these organisms at 0°C should have been actively studied. There are indeed several reviews of this subject concerning marine psychrophilic bacteria (41, 43, 46). The detailed biochemical studies up to 1968 of temperature effects on enzyme systems, amino acid pools, and protein synthesis in psychrophilic bacteria have been concisely summarized by Morita in Paper 23. In this review he examines the evidence showing leakage of protein $>$ RNA $>$ DNA $>$ amino acid pools from cells of *Vibrio marinus* exposed to 15, 22.3, and 29.7°C. Furthermore, certain enzymes from *V. marinus* appear to be heat sensitive and others less so; for example, dehydrogenases lose activity upon exposure to temperatures greater than 26 to 30°C, whereas phosphoisomerase, aldolase, and hexokinase are more resistant. In addition, greater stability is conferred on all the proteins by increased sodium chloride concentrations. It is interesting that the optimal growth temperature characteristic is distinctive for these psychrophiles (24).

One would predict that oxygen uptake rates would be drastically affected by the incubation temperature; but the observed effect does not in reality correspond to this prediction. With *V. marinus*, if the cells are grown at 4°C and the rates of oxygen uptake at 4 and 15°C are compared, it will be found that there is actually less oxygen consumed at the higher, though nonlethal, temperature (41). In addition, Zachariah and Liston have confirmed the impact of temperature on oxidation during their study of pseudomonads involved in fish spoilage (61). Here they noted an up to four times greater rate of oxidation of alanine by cells grown at 8°C over the rate obtained when the cells were grown at 22°C. Obviously, then, enzyme activation and inactivation as a consequence of changing temperature must be influential in determining the psychrophilic nature of these bacteria.

Another interesting study demonstrated that the K_m for L-serine deaminase is different in cells of *V. marinus* grown at 4°C from those grown at 15°C (3). Perhaps, then, in multimeric enzyme systems (dehydrogenases or amino acid oxidases or serine deaminases) the different isozymes have different temperature optima, and this enables the cells to tolerate wider fluctuations in temperature from 0 to 20 or 30°C.

Similar results on the effect of incubation temperature on the incorporation and uptake of labeled amino acids by *Flavobacterium* sp., *Pseudomonas* sp., and *Brevibacterium* have been reported by Ishida et al. (27).

That the psychrophilic character might confer competitive ecological advantage has not been ignored either. Harder and Veldkamp (25) showed that the obligate psychrophile has a faster growth rate at 4°C than the psychrotrophic organism. Meanwhile, Sieburth (57) demonstrated a delayed response in the autochthonous heterotrophic bacteria to changes in the ambient temperature. He further noted that temperature selection was not restricted to specific genera. Obviously, more comparative data of this nature are needed to understand seasonal cycling in marine coastal waters.

Meanwhile, other workers have investigated nonmarine psychrotrophic microbes and noted fundamental changes in cellular integrity (4) and in filament formation of *Bacillus* (20), ribosomal protein synthesis in *Clostridium* sp. and *Bacillus* sp. (8, 26), and increased fatty acid unsaturation in the yeast *Candida* sp. (31). Changes in the configuration of the aminoacyl-tRNA synthetase in *Micrococcus cryophilus* (36) and even in the type of amino acid accumulated in the intracellular pool in a *Brevibacterium* have been reported (44, 45). So far, comparative studies with marine psychrophilic and psychrotrophic bacteria grown at different temperatures have not been completed.

For each 10-m depth in the oceans, there is an increase of approximately 1 atm of pressure. Thus, for those organisms living in the abyssal depths, the importance of extreme hydrostatic pressure to their biochemistry and survival cannot be overlooked. In 1883, Certes (12) recognized the importance of pressure to these bacteria and proceeded to recover viable marine bacteria from samples exposed to 100 and 300 atm. However, all were killed by exposure to 600 atm for 24 hours (13, 14). He also noted that under these high hydrostatic pressures, the bacteria produced an acid medium rather than the normal alkaline conditions (14). Furthermore, after 21 days *Chlamydococcus pluvialis* was present only in the high-pressure tube (14). Devices for such studies had been developed earlier by Milne-Edwards (37), who assisted Certes in his first investigations on deep-sea sedimentary bacteria (see Paper 1).

However, Regnard (52) felt that infusoria and "ferments" exposed to high pressure (1000 atm) were placed in a resting state and were only revitalized when the pressure was released. This led to several articles by Certes, who attempted to explain the differences between his experimental results and those of Regnard. Generally, Certes felt that the inhibitory or bacteriostatic effects noted by Regnard were due to the culture medium, temperature, oxygen, or toxic metal products coming from his 5-week incubation periods (13, 14). As usual, no resolution to this mini-controversy was achieved, and Regnard's views were widely quoted to show a lack of life and lack of microbial activity at a depth of 4000 m (53, 54). A little later, Chlopin and Tammann (15) summarized their extensive studies on pressure effects on pure cultures of pathogenic bacteria, general bacteria (no description provided), and yeasts by concluding that 3000 atm was lethal, that pressure effects were related to the time at pressure and the means of applying the pressure, and that motility was among the first properties to be adversely affected.

By the turn of the century, many workers were studying the effects of pressure on the growth of bacteria, and *Pseudomonas aeruginosa* was frequently included in these studies (32). Bacterial luminescence was found to be partially dependent on both temperature and pressure by Johnson et al. (30). The physiological work up to 1936 was reviewed and evaluated by Cattell (11), who found that most of the emphasis had been placed on the determination of the pressure death point (time dependent) and the possibility of using high pressures to destroy pathogenic or food spoilage organisms (33). Except for the studies noted above on marine luminescent bacteria, investigations on the nature of high-pressure effects on marine bacteria seemed to diminish in interest until ZoBell and his co-workers revived the concept of barophilic and barophobic bacteria during the late 1940s. ZoBell and Johnson (64) drew attention to the interrelationship between temperature and survival at high pressures in 1949, and this was followed by a series of papers with Oppenheimer (47, 65) showing that morphological variants were common in bacteria exposed to high pressures, but that the microbes could survive higher pressures at higher temperatures. Restoration of the cultures to 1 atm also restored the original morphology; thus the observed elongation and pleomorphism were expressions of alterations in cell wall synthesis or inhibition of division by the cells (64).

At the same time, ZoBell and co-workers succeeded in isolating bacteria that grew well at 600 atm. In Paper 24, ZoBell and Morita studied bacteria from deep-sea sediments using their specially designed hydrostatic pressure devices. This study encompassed not only isolation techniques and the construction of hydrostatic pressure chambers, but it also advanced our concepts of bacteria in the sediments obtained from 10,000 m; their degradative abilities, heat sensitivity, and numbers (up to 2×10^6 per g wet weight). The predominant types were aerobic, nitrate reducers, or ammonifiers as determined by most-probable-number methods.

Drawing upon this series of publications and the delineation of the construction of hydrostatic pressure chambers by ZoBell and co-workers, many investigators began to study marine bacteria under pressure. Two major directions have been taken in these studies: attempts to understand the nature of the temperature sparing effect on pressure sensitivity, and attempts to explain the types of biochemical adaptations that are necessary for survival at these elevated pressures. A most succinct summary of the information available until 1967 was published by Morita (Paper 25), in which he described the development of pressure instrumentation and the effects of elevated pressures on fungi, higher organisms, proteins, and bacteriophage, as well as bacteria.

Since that time there has been quite literally an explosion in the number of papers on this subject as new techniques have been presented for the study of enzyme kinetics under high pressure (38) and for adding and removing substances to a reaction or incubation vessel under pressure (55). A recent review of pressure apparatus and effects was published in 1972 (42). Of late, additional studies on marine microorganisms have shown that the uptake of amino acids (49), sugar analogues (56), and uracil (5), and amino acid incorporation into proteins (59) and RNA-protein synthesis (1) are all inhibited by pressures ranging from 200 to 600 atm. At moderate pressures up to 400 atm, Ehrlich (19) found that manganous oxidation was not seriously impaired. Above this pressure, though, there was a marked reduction in enzyme activity. The inhibition by 600 atm of pressure of malic dehydrogenase in

Escherichia coli (40) was also found to be partially true with a marine vibrio (39). However, increased malate levels stabilized the enzyme, and the K_m was found to increase with increasing pressures (39). The rate of photosynthesis of several algae was also found to be adversely affected by increased pressure (51), but no discussion of possible temperature moderation of this inhibitory action was presented.

Several Soviet workers have also investigated pressure effects on metabolic pathways operative in barotolerant bacteria. A marine pseudomonad was found to produce less CO_2 and twice as much volatile organic acids when grown at 500 atm as when grown at 1 atm (16, 17). These studies have recently been extended to three additional strains of barotolerant marine bacteria, and it is postulated that formic dehydrogenase may be inhibited at elevated pressures (7).

Explanations of the specific effects of elevated pressure, as opposed to the descriptive effects listed above, are very few. Based on studies with ATPase from several animal sources, rabbit creatine kinase, bovine liver arginosuccinase, rat liver pyruvic carboxylase, chicken alkaline phosphatase, peroxidase from horseradish, *Neurospora* DPN glycohydrolase, rabbit skeletal myokinase, and snake venom 5′-nucleotidase, Penniston (50) has proposed that multimeric enzyme systems are more susceptible to high-pressure inhibition than are monomeric enzyme systems because of the dissociation of the noncovalently bonded subunits. Thus, for organisms to survive and function at deep-sea pressures, monomeric enzyme systems or stronger subunit bonding would be necessary. One potential explanation for the sparing effect elevated temperatures have on pressure sensitivity could reside in the fact that, while the pressure acts to decrease the specific volume of a protein (50), higher temperatures could compensate for this by increasing the specific volume or restoring the hydration of the molecules through greater "mobility" of the water molecules. The physical changes of water molecules in relation to macromolecules under increased pressure are poorly understood at this time. Obviously, there are many exciting and potentially rewarding studies still to be done in this field.

REFERENCES

1. Albright, L. J., and Hardon, M. J. 1974. Hydrostatic pressure effects upon protein synthesis in two barophobic bacteria, pp. 160–173. In:

R. R. Colwell and R. Y. Morita (eds.), *Effect of the Ocean Environment on Microbial Activities.* University Park Press, Baltimore, Md., 587 pp.

2. Albright, L. J., and Henigman, J. F. 1971. Seawater salts—hydrostatic pressure effects upon cell division of several bacteria. *Can. J. Microbiol. 17*:1246–1248.
3. Albright, L. J., and Morita, R. Y. 1972. Effects of environmental parameters of low temperatue and hydrostatic pressure on L-serine deamination by *Vibrio marinus. J. Ocean Soc. Jap. 28*:25–32.
4. Alsobrook, D., Larkin, J. M., and Sega, M. W., 1972. Effect of temperature on the cellular integrity of *Bacillus psychrophilus. Can. J. Microbiol. 18*:1671–1678.
5. Baross, J. A., Hanus, F. J., and Morita, R. Y. 1974. Effects of hydrostatic pressure on uracil uptake, ribonucleic acid synthesis, and growth of three obligately psychrophilic marine vibrios, *Vibrio alginolyticus* and *Escherichia coli*, pp. 180–203. In: R. R. Colwell and R. Y. Morita (eds.), *Effect of the Ocean Environment on Microbial activities.* University Park Press, Baltimore, Md., 587 pp.
6. Bertel, R. 1912. Sur la distribution quantitative des bactéries planctoniques des côtes de Monaco. *Bull. Inst. Oceanogr. Monaco No. 224*:1–12.
7. Blokhina, T. P. 1971. Peculiarities of the fermentation of glucose and galactose by barotolerant bacteria at increased hydrostatic pressures. *Microbiology USSR 40*:734–737.
8. Bobier, S. R., Ferroni, G. D., and Inniss, W. E. 1972. Protein synthesis by the psychrophiles *Bacillus psychrophilus* and *Bacillus insolitus. Can. J. Microbiol. 18*:1837–1843.
9. Brown, C. M., and Stanley, S. O. 1972. Environment-mediated changes in the cellular content of the "pool" constituents and their associated changes in cell physiology. *J. Appl. Chem. Biotechnol. 22*:363–389.
10. Butkevich, N. V., and Butkevich, V. S. 1936. Multiplication of sea bacteria depending on the composition of the medium and the temperature. *Microbiology USSR 4*:322–343.
11. Cattell, M. 1936. The physiological effects of pressure. *Biol. Rev. 11*: 441–476.
12. Certes, A. 1884. Note relative à l'action des hautes pressions sur la vitalité des micro-organismes d'eau douce et d'eau de mer. *C. R. Seances Soc. Biol. Fil. 36*:220–222.
13. Certes, A. 1884. De l'action des hautes pressions sur les phénomènes de la putréfaction et sur la vitalité des micro-organismes d'eau douce et d'eau de mer. *C. R. Hebd. Seances Acad. Sci. 99*:385–388.
14. Certes, A., and Cochin, D. 1884. Action des hautes perssions sur la vitalité de la levure et sur les phénomènes de la fermentation. *C. R. Seances Soc. Biol. Fil. 36*:639–640.
15. Chlopin, G. W., and Tammann, G. 1903. Ueber den Einfluss hoher Drucke auf Mikroorganismen. *Z. Hyg. Infektionskr. 45*:171–204.
16. Chumak, M. D., and Blokhina, T. P. 1964. Influence of high pressure on the accumulation of organic acids during glucose fermentation of barotolerant bacteria. *Microbiology USSR 33*:200–204.
17. Chumak, M. D., Tarasova, N. W., and Blokhina, T. P. 1964. Qualitative composition of organic acids formed in the fermentation of glucose by barotolerant bacteria. *Microbiology USSR 33*:510–513.

18. Cooper, M. F., and Morita, R. Y. 1972. Interaction of salinity and temperature on net protein synthesis and viability of *Vibrio marinus. Limnol. Oceanogr. 17*:556–565.
19. Ehrlich, H. L. 1974. Response of some activities of ferromanganese nodule bacteria to hydrostatic pressure, pp. 208–221. In: R. R. Colwell and R. Y. Morita (eds.), *Effect of the Ocean Environment on Microbial Activities*. University Park Press, Baltimore, Md., 587 pp.
20. Ferroni, G. D., and Inniss, W. E. 1973. Thermally caused filament formation in the psychrophile *Bacillus insolitus. Can. J. Microbiol. 19*: 581–584.
21. Fischer, B. 1888. Bakterienwachsthum bei 0°C sowie über das Photographiren von kulturen leuchtender Bakterien in ihrem eigenen Lichte. *Zentralbl. Bakteriol., Abt. II, 4*:89–92.
22. Forster, J. 1892. Ueber die Entwickelung von Bakterien bei niederen Temperaturen. *Zentralbl. Bakteriol. 12*:431–436.
23. Frankland, E., and Burgess, W. T. 1897. Sea-water microbes in high latitudes. *Chem. News 75*:1–2.
24. Hanus, F. J., and Morita, R. Y. 1968. Significance of the temperature characteristic of growth. *J. Bacteriol. 95*:736–737.
25. Harder, W., and Veldkamp, H. 1971. Competition of marine psychrophilic bacteria at low temperatures. *Antonie van Leeuwenhoek J. Microbiol. Serol. 37*:51–63.
26. Irwin, C. C., Akagi, J. M., and Himes, R. H. 1973. Ribosomes, polyribosomes, and deoxyribonucleic acid from thermophilic, mesophilic, and psychrophilic clostridia. *J. Bacteriol. 113*:252–262.
27. Ishida, Y., Nakayama, A., and Kadata, H. 1974. Temperature salinity effects upon the growth of marine bacteria, pp. 80–92. In: R. R. Colwell and R. Y. Morita (eds.), *Effect of the Ocean Environment on Microbial Activities*. University Park Press, Baltimore, Md., 587 pp.
28. Issatchenko, B. L. 1914. *Investigations on the Bacteria of the Glacial Arctic Ocean*. Petrograd, 300 pp.
29. Janowski, T., 1888. Ueber den Bakteriengehalt des Schnees. *Zentralbl. Bakteriol., Abt. II, 4*:547–552.
30. Johnson, F. H., Eyring, H., and Williams, R. W. 1942. The nature of enzyme inhibitions in bacterial luminescence: sulfanilamide, urethane, temperature and pressure. *J. Cell. Comp. Physiol. 20*:247–260.
31. Kates, M., and Paradis, M. 1973. Phospholipid desaturation in *Candida lipolytica* as a function of temperature and growth. *Can. J. Biochem. 51*:184–197.
32. Krause, P. 1902. Ueber durch Pressung gewonnenen Zellsaft des *Bacillus pyocyaneus* nebst einer kurzen Mitteilung über die Einwirkung des Druckes auf Bakterien. *Zentralbl. Bakteriol., Abt. I, 14*: 673–678.
33. Larson, W. P., Hartzell, T. B., and Diehl, H. S. 1918. The effect of high pressures on bacteria. *J. Infect. Dis. 22*:271–279.
34. Lloyd, B. 1929–1930. Bacteria of the Clyde Sea area: a quantitative investigation. *J. Mar. Biol. Assoc. U.K. 16*:879–907.
35. Lloyd, B. 1930–1931. Muds of the Clyde Sea area. II. Bacterial content. *J. Mar. Biol. Assoc. U.K. 17*:751–765.
36. Malcolm, N. L. 1968. A temperature induced lesion in amino acid-transfer ribonucleic acid attachment in a psychrophile. *Biochim. Biophys. Acta 157*:493–503.

37. Milne-Edwards, A. 1882. Explorations des grandes profondeurs de la mer faites à bord de l'aviso *Le Travailleur. Ann. Chim. Phys., Ser. 5, 27*:555–567.
38. Mohankumar, K. C., and Berger, L. R. 1972. A method to do rapid enzyme kinetic assays at increased hydrostatic pressure. *Anal. Biochem. 49*:336–342.
39. Mohankumar, K. C., and Berger, L. R. 1974. Kinetic regulation of enzymic activity at increased hydrostatic pressure: studies with a malic dehydrogenase from a marine bacterium, pp. 139–145. In: R. R. Colwell and R. Y. Morita (eds.), *Effect of the Ocean Environment on Microbial Activities*. University Park Press, Baltimore, Md., 587 pp.
40. Morita, R. Y., 1957. Effect of hydrostatic pressure on succinic, formic, and malic dehydrogenases in *Escherichia coli. J. Bacteriol. 74*:251–255.
41. Morita, R. Y. 1966. Marine psychrophilic bacteria. *Oceanogr. Mar. Biol. Ann. Rev. 4*:105–121.
42. Morita, R. Y. 1972. 8. Pressure. 8.1. Bacteria, fungi and blue-green algae, pp. 1361–1389. In: O. Kinne (ed.), *Marine Ecology*, Vol. 1, Environmental Factors, Pt. 3. Wiley-Interscience, New York, 529 pp.
43. Morita, R. Y. 1975. Psychrophilic bacteria. *Bacteriol. Rev. 39*:144–167.
44. Ogata, K., Kato, N., Osugi, M., and Tochikura, T. 1969. Studies on the low temperature fermentation. II. Amino acid formation by facultative psychrophilic bacterium. *Agric. Biol. Chem. 33*:711–717.
45. Ogata, K., Tochikura, T., Kato, N., and Osugi, M. 1969. Studies on the low temperature fermentation. I. Isolation and characterization of psychrophilic bacteria. *Agric. Biol. Chem. 33*:704–710.
46. Oppenheimer, C. H. 1970. 3. Temperature. 3.1. Bacteria, fungi and blue-green algae. In: O. Kinne (ed.), *Marine Ecology*, Vol. 1, Environmental Factors, Pt. 1, pp. 347–363. Wiley-Interscience, New York, 681 pp.
47. Oppenheimer, C. H., and ZoBell, C. E. 1952. The growth and viability of sixty-three species of marine bacteria as influenced by hydrostatic pressure. *J. Mar. Res. 11*:10–18.
48. Palmer, D. S., and Albright, L. J. 1970. Salinity effects on the maximum hydrostatic pressure for growth of the marine psychrophilic bacterium *Vibrio marinus. Limnol. Oceanogr. 15*:343–347.
49. Paul, K. L., and Morita, R. Y. 1971. Effects of hydrostatic pressure and temperature on the uptake and respiration of amino acids by a facultatively psychrophilic marine bacterium. *J. Bacteriol. 108*:835–843.
50. Penniston, J. T. 1971. High hydrostatic pressure and enzymic activity: inhibition of multimeric enzymes by dissociation. *Arch. Biochem. Biophys. 142*:322–332.
51. Pope, D. H., and Berger, L. R. 1974. Algal photosynthesis at constant pO_2 and increased hydrostatic pressure, pp. 203–208. In: R. R. Colwell and R. Y. Morita (eds.), *Effect of the Ocean Environment on Microbial Activities*. University Park Press, Baltimore, Md., 587 pp.
52. Regnard, P. 1884. Recherches expérimentales sur l'influence des très hautes pressions sur les organismes vivants. *C. R. Hebd. Seances Acad. Sci. 98*:745–747.
53. Regnard, P. 1891. *La Vie dans les eaux*. G. Masson. Libraire de L'Académie de Médicine, Paris, 500 pp.
54. Richard, J. 1907. *L'Océanographie*. Vuibert and Nony, Editeurs, Paris, 398 pp.

55. Schwarz, J. R., and Landau, J. V. 1972. Inhibition of cell-free protein synthesis by hydrostatic pressure. *J. Bacteriol.* *112*:1222–1227.
56. Shen, J. C., and Berger, L. R. 1974. Measurement of active transport by bacteria at increased hydrostatic pressure, pp. 173–180. In: R. R. Colwell and R. Y. Morita (eds.), *Effect of the Ocean Environment on Microbial Activities,* University Park Press, Baltimore, Md., 587 pp.
57. Sieburth, J. McN. 1967. Seasonal selection of estuarine bacteria by water temperature. *J. Exp. Mar. Biol. Ecol.* *1*:98–121.
58. Stanley, S. O., and Morita, R. Y. 1968. Salinity effect on the maximal growth temperature of some bacteria isolated from marine environments. *J. Bacteriol.* *95*:169–173.
59. Swartz, R. W., Schwarz, J. R., and Landau, J. V. 1974. Comparative effects of pressure on protein and RNA synthesis in bacteria isolated from marine sediments, pp. 145–160. In: R. R. Colwell and R. Y. Morita (eds.), *Effect of the Ocean Environment on Microbial Activities.* University Park Press, Baltimore, Md., 587 pp.
60. Trombetta, S. 1891. Die Fäulnissbakterien und die Organe und das Blut ganz gesund getodteter Thiere. *Zentralbl. Bakteriol.* *10*:664–669.
61. Zachariah, P., and Liston, J. 1973. Temperature adaptability of psychrotrophic *Pseudomonas. Appl. Microbiol.* *26*:437–438.
62. ZoBell, C. E., and Conn, J. E. 1940. Studies on the thermal sensitivity of marine bacteria. *J. Bacteriol.* *40*:223–238.
63. ZoBell, C. E., and Feltham, C. B. 1940. The influence of temperature on the activities of marine bacteria. *J. Bacteriol.* *36*:452.
64. ZoBell, C. E., and Johnson, F. H. 1949. The influence of hydrostatic pressure on the growth and viability of terrestrial and marine bacteria. *J. Bacteriol.* *57*:179–189.
65. ZoBell, C. E., and Oppenheimer, C. H. 1950. Some effects of hydrostatic pressure on the multiplication and morphology of marine bacteria. *J. Bacteriol.* *60*:771–781.

23

Reprinted from *Bull. Misaki Mar. Biol. Inst. Kyoto Univ.*, **12**, 163–177 (Feb. 1968)

THE BASIC NATURE OF MARINE PSYCHROPHILIC BACTERIA

Richard Y. Morita

Departments of Microbiology and Oceanography, Oregon State University, Corvallis, Oregon, U.S.A. 97331

Introduction

Bacteria that meet the textbook definition of psychrophiles were thought to be non-existent by many investigators working in microbiology until the recent isolation by Eimhjellen (see Hagen, Kushner, and Gibbons, 1964) and Morita and Haight (1964). Because there are so many incorrectly labeled psychrophilic bacteria in the literature Stokes (1962) defined psychrophiles as organisms which grow sufficiently at 0 C to be visible within a week and then subdivided the psychrophiles into two categories, facultative and obligate. The facultative psychrophiles have an optimum temperature for growth at 20 C or higher while obligate psychrophiles have an optimum temperature for growth below 20 C. This definition conveniently allows us to retain all the organisms designated as psychrophiles which are not truly psychrophilic in nature. Hence, many of the marine bacteria described in the literature fall into the category of facultative psychrophilic bacteria—mainly because they are capable of growth at 0 C but have an optimum growth temperature above 20 C.

Obligately psychrophilic bacteria should not be considered rare since they can be isolated from various sources such as water below the thermocline and at both polar regions. Obligately psychrophilic bacteria have been isolated by our group as well as by A. H. Rose, K. Eimhjellen and H. Veldkamp (personal communication). Stanley (personal communication) suggests that the reason why many investigators have not isolated obligately psychrophilic bacteria from the Antarctic is due to the use of the wrong source material. Source materials which are subjected to radiant energy undergo wide temperature variations, and as a result the obligately psychrophilic bacteria are rendered non-viable.

Because more than 90%, by volume, of the ocean is colder than 5 C, it can be expected that the obligately psychrophilic bacteria play an important role in the activities attributed to bacteria in the sea.

Although there are various reviews on psychrophilic bacteria (Witter, 1961; Campbell Soup Company's Symposium, 1962), it should be recognzed that these reviews deal with facultative psychrophiles according to the de-

---

1. Published as technical paper No. 2236, Oregon State Agricultural Experiment Station.

finition given by Stokes (1962). Morita's (1966) review does deal to some degree on the obligately psychrophilic marine bacteria.

Because the elucidation of obligately psychrophilic bacteria in the sea has been made only recently, not too much data have been reported on the basic nature of this thermal class of microorganisms. Our laboratory has been investigating the problem of obligately psychrophilic bacteria for the last few years. Therefore I would like to present at this Conference some of the latest data we have obtained. Much of this data still remains to be published in the various journals. Because of the time limitation for my presentation, I will present mostly data and not cover methodology or the significance of our research to any great extent.

Organism

For most of our studies we have employed the obligately psychrophilic marine bacterium, *Vibrio marinus* MP-1 (Colwell and Morita, 1964). This organism was isolated 165 miles off the coast of Oregon from a depth of 1200 m and a temperature of 3.24 C. The optimum temperature for growth of this organism is 15 C while the maximum temperature for growth is 20 C (Morita and Haight, 1964). This organism has been shown to be able to reproduce very rapidly at both 4 C and 15 C, giving a cell yield of 10^9 cells/ml and 10^{11} cells/ml respectively under optimal conditions (Morita and Albright, 1965). Our laboratory has numerous obligately psychrophilic marine bacteria, but we have chosen to work primarily on *V. marinus* MP-1 so that we can elucidate "why" and "how" temperature affects this organism. The basic principles that we learn from this organism may or may not be applicable to other obligately psychrophilic bacteria. However, we will leave this problem for investigations in the future.

V. marinus MP-1 cells were found to expire when exposed to temperatures above its maximum growth temperature of 20 C, depending upon the length of exposure as well as temperature employed (Morita and Haight, 1964). Respiration studies by Robinson and Morita (1966) demonstrated that exposure of the cells to temperatures above 20 C brought about a decrease in the rate of respiration, thereby indicating that some metabolic lesion(s) had taken place due to the increase in temperature. More specifically it pointed to an abnormal thermolabile enzyme(s).

Abnormal Heat Lability of Malic Dehydrogenase

The abnormal heat lability of malic dehydrogenase (MDH) was first demonstrated in a facultative psychrophile, *V. marinus* PS 207, by Burton and Morita (1963) and Morita and Burton (1963). The inactivation of this enzyme was initiated at approximately 30 C—in fact 20% of the original activity is retained when the cell-free extracts of the containing MDH was

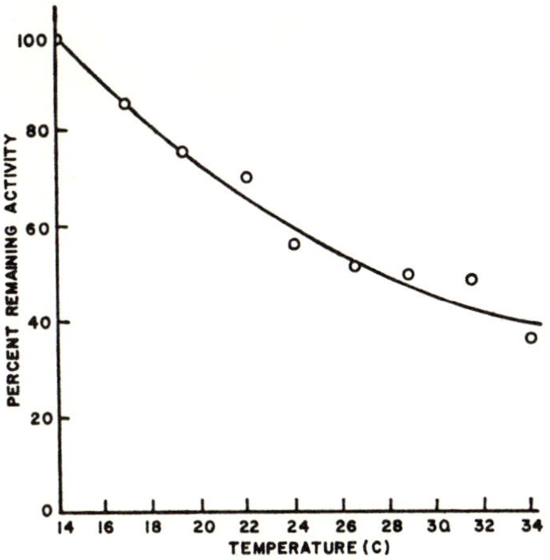

Fig. 1

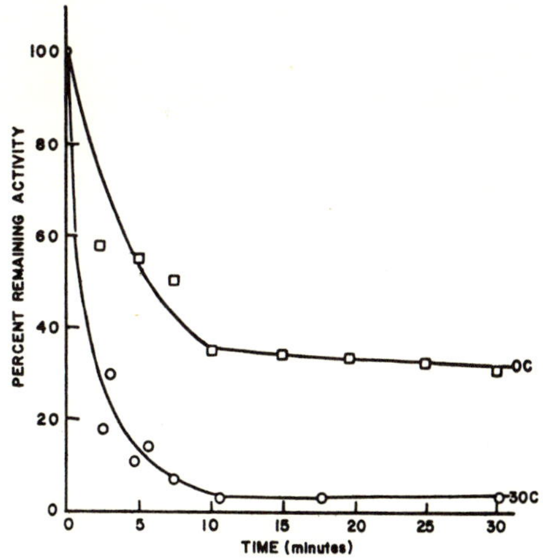

Fig. 2

Fig. 1. Malic dehydrogenase activity of washed cells of *Vibrio marinus* MP-1 after exposure to various temperatures for 60 minutes. Activity of cells treated at 14 C for minutes was arbitrarily set at 100%.
Reprinted by permission after Langridge and Morita, 1966.

Fig. 2. Effect of temperature on malic dehydrogenase activity of cell-free extracts. The cell-free extract contained a protein concentration of 0.89 mg/ml and was treated for various times and temperature, as shown on the graph, prior to assay at 20 C.
Reprinted by permission after Langridge and Morita, 1966.

heated to 30 for 40 minutes.

In the case of *V. marinus* MP-1, the obligate psychrophile, Langridge and Morita (1966) demonstrated that the MDH in whole cells becomes inactivated when exposed to temperatures above its optimal temperature for growth of 14 C (Fig. 1). Exposure of the cells to temperature between 0 C and 15 C did not inactivate the enzyme. The effect of moderate temperature on the MDH in cell-free extracts of the same organisms is more pronounced (Fig. 2). This abnormal thermolability is one of the reasons why *V. marinus* MP 1 expires when exposed to temperatures above its maximal growth temperature. The use of ammonium sulfate in the reaction mixture was found to aid the thermostability of the enzyme, hence it was employed in our studies. The effect of temperature on the partially purified MDH is shown in Fig. 3 & 4. This enzyme was purified 20 fold by ammonium sulfate fractionation and sephadex column chromatography. The results in Fig. 3 and 4 are biased toward a higher thermal stability because of the presence of ammonium sulfate. Nevertheless it was found that temperatures above or below the organism's optical temperature for growth inactivated the MDH. However, this enzyme inactivation due to a lower temperature (0 C) or elevated temperature of 33 C could be reactivated by placing the enzyme at 15 C in the presence of ammonium sulfate (Fig. 5). The partially purified MDH cannot be reactivated when exposed to a temperature of 40 C, which indicated that the inactivation process has resulted in a conformational change of the MDH

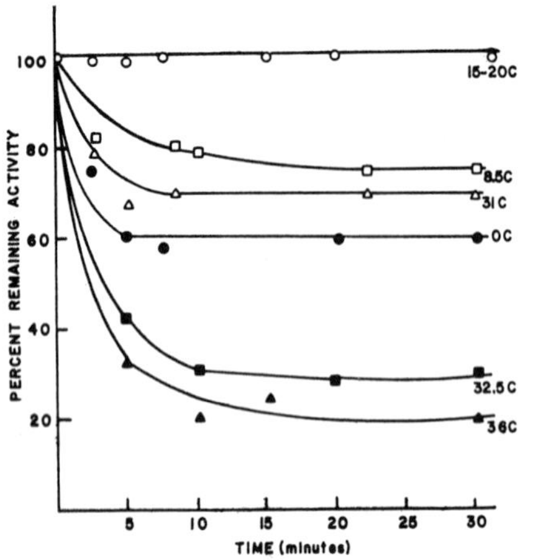

Fig. 3

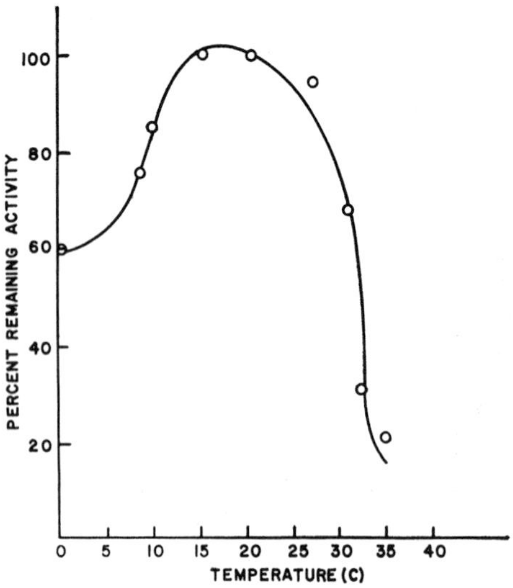

Fig. 4

Fig. 3. Effect of temperature on partially purified malic dehydrogenase activity. Inacitivation was performed in 0.2 M tris-sulfate (pH 7.4) and 0.42 mg of protein per ml.

Reprinted by permission after Langridge and Morita, 1966.

Fig. 4. Replot of Fig. 3 showing the effect of 30-minute exposure of partially purified malic dehydrogenase to increasing temperatures.

Reprinted by permission after Langridge and Morita, 1966.

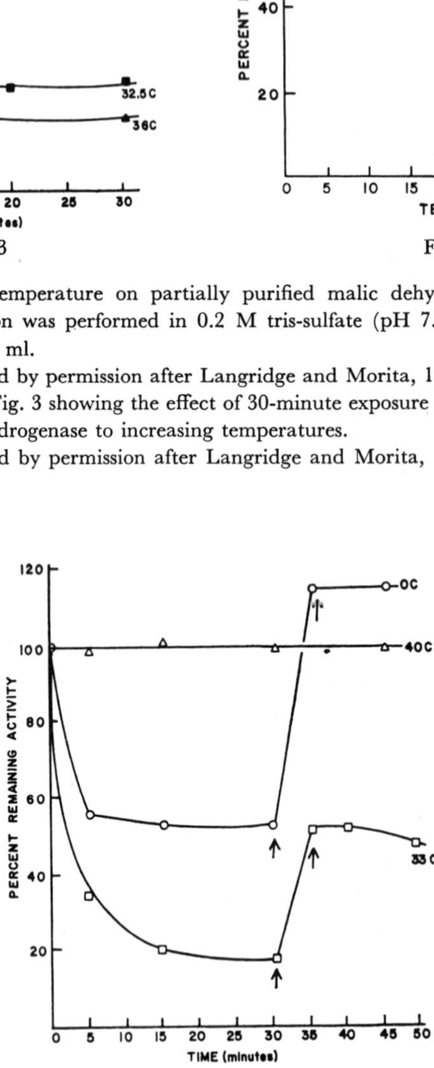

Fig. 5. Temperature reactivation of malic dehydrogenase. At the time indicated by arrows, the enzyme mixtures were placed at 15 C and assayed at 20 C. The 40 C curve represents enzyme heated in the presence of 1,250 u moles of ammonium sulfate per mg of protein.

Reprinted by permission after Langridge and Morita, 1966.

to a point where the original conformation cannot be assumed.

In this study we could not detect any isozymic forms of MDH by use of analogs of nicotinamide adenine dinucleotide. The lack of the isozymic forms of MDH may be the result of the purification procedure. Disc electrophoresis may indicate the presence of isozymic forms of the enzyme and each may have a different temperature characteristic.

Abnormally Thermolabile Enzymes in Cell-free Extracts of Vibrio Marinus MP-1

Succinic dehydrogenase activity was lost rapidly on exposure of the cell-free extract to temperatures above 0 C. Approximately one third of the activity was lost when the enzyme was exposed for one hour at 12 C when compared to the same time of exposure at 0 C. At 16 C for one hour, approximately two-thirds of the activity remained. However, a plateau of thermal stability occurred between 16 and 26 C, and above 26 C the enzyme continued to lose its catalytic capacity. No succinate oxidation could be detected when the enzyme was exposed to 33 C for one hour (Mathemeier, 1966).

Exposure of cell-free extracts between 0 C and 29 C for one hour did not result in any loss of hexokinase activity. A rapid drop of activity is noted on exposure of the cell-free extracts to temperatures higher than 29 C, resulting in a complete inactivation at 33 C (Mathemeier, 1966).

No loss of phosphoglucose isomerase activity could be detected when cell-free extracts were exposed to temperatures between 0 and 31 C, and again a rapid inactivation occurred at higher temperatures. Subjecting the cell-free preparation for one hour at 36 C or 38 C, depending upon constituents of the assay system, resulted in complete inactivation.

Aldolase in cell-free extracts on the other hand lost very little activity when subjected to temperatures from 20 to 26 C, but inactivation was very rapid between 26 and 32 C. Lactic dehydrogenase activity produced a rapid drop in activity when exposed to temperatures between 17 and 25 C, with a plateau of inactivation between 25 and 32 C. Complete inactivation resulted at 39.5 C.

It should be noted that the above mentioned enzymes in this section possessed higher enzymatic activity or were more thermostable in the presence of NaCl than in tris-HCl. The exact function of NaCl still remains to be investigated.

Thermally-Induced Leakage

The leakage of protein, DNA, RNA, and amino acid from the cells of *V. marinus* MP-1 when subjected to heat is shown in Fig. 6, 7, and 8 for 15.0

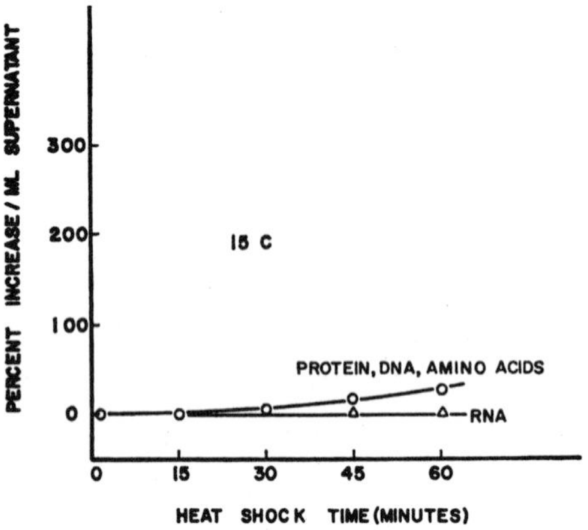

Fig. 6

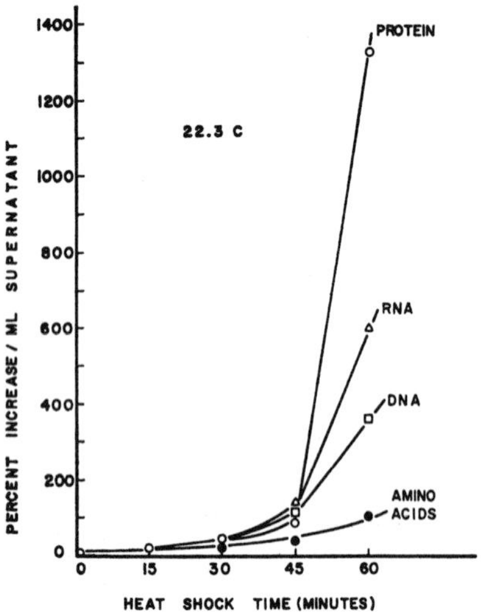

Fig. 7

Fig. 6. Heat-induced leakage from *Vibrio marinus* MP 1 at 15 C.
Reprinted by permission after Haight and Morita, 1966.

Fig. 7. Heat-induced leakage from *Vibrio marinus* MP 1 at 22.3 C.
Reprinted by permission after Haight and Morita, 1966.

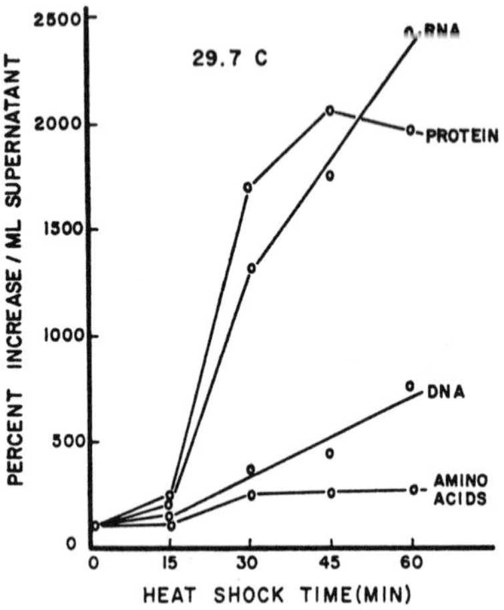

Fig. 8

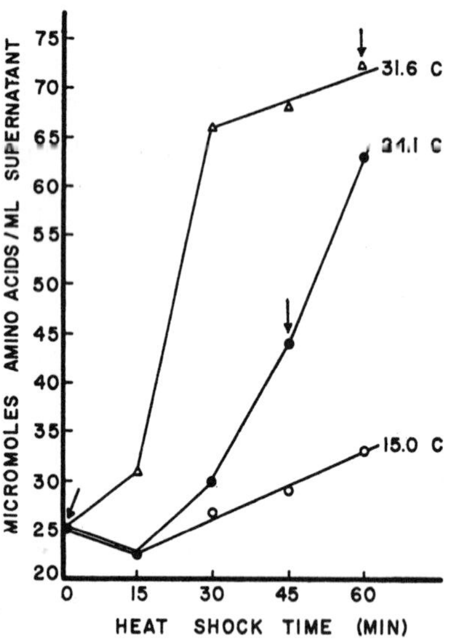

Fig. 9

Fig. 8. Heat-induced leakage from *Vibrio marinus* MP 1 at 29.7 C.
Reprinted by permission after Haight and Morita, 1966.

Fig. 9. Heat-induced leakage of amino acids from *Vibrio marinus* MP 1. Cell concentration equivalent to 22 mg of protein/ml.
Reprinted by permission after Haight and Morita, 1966.

Table 1. Amino acid analysis of deproteinized suprenatant fluids from heat-treated of *Vibrio marinus* MP-1.

Amino acid	0 Min	45 min at 24.1 C	60 min at 31.6 C
	μmoles	μmoles	μmoles
Lysine	0.392	1.714	2.300
Histidine	Trace	Trace	0.166
Arginine	0	0	0.193
Aspartic	0.533	2.501	4.324
Glutamic	12.100	20.679	39.740
Thronine	0.155	0.238	0.589
Serine	0.189	0.384	0.902
Glycine	0.096	0.403	3.542
Cystine/2	Trace	Trace	1.021
Proline	—[a]	—[a]	—[a]
Alanine	5.513	7.564	9.752
Valine	1.924	2.105	2.926
Methionine	0.581	0.659	0.828
Isoleucine	1.158	1.312	1.840
Leucine	0.740	0.897	1.426
Tyrosine	0.470	0.500	0.800
Phenylalanine	1.014	1.049	1.509

[a] Proline peak obscured by glutamic acid peak.
Reprinted by permission after Haight and Morita, 1966.

C, 22.3 C and 29.7 C respectively. At 15 C the amount of thermally induced leakage material after 60 minutes exposure is very slight. However, when the temperature is elevated to 22.3 C the protein, DNA, RNA, RNA and amino acids do not appear in the supernatant fluid unless the cells have been heated for 45 minutes. The rates of thermally-induced leakage differs so that protein<RNA<DNA<amino acids. The same pattern holds true when the cells are exposed to a temperature of 29.7 C but leakage appears after only 15 minutes of heating (Haight and Morita, 1966).

The total amount of amino acids released from thermally shocked cells was determined at 15 C, 24.1 C and 31.6 C (Fig. 9). Analysis of the amino acids is shown in Table 1. The basic and polar amino acids all tend to increase in level with increased in time and temperature of heat treatment. The non-polar amino acids appear to decrease with increased temperature and time of heat treatment. If the hydrophobic lipid of the cell was not released from the cells by heat shocking, the cell residues would have become more hydrophobic as the hydrophilic materials moved into the supernatant fluid. The change in hydrophilic-hydrophobic balance thereby shifts, resulting the distribution of the amino acids.

Since all the various intracellular compounds do not leak out at the same time, a simple rupture of the cell membrane is not the cause of loss of viability

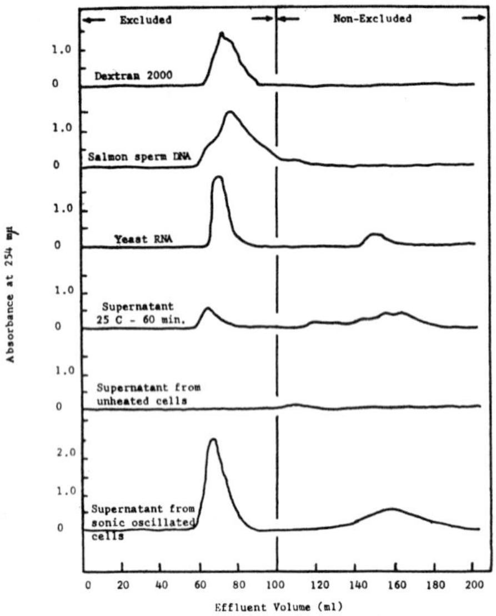

Fig. 10. Sephadex G-25 elution patterns of supernatant material prepared from heat-shocked and sonically treated cells of *Vibrio marinus* MP 1.
Reprinted by permission after Haight and Morita, 1966.

and leakage. This becomes very evident when it is realized that the smaller molecular weight compound (amino acids) appear last instead of first. The order of the appearance of the various compounds may be under cellular control since the results indicate subtle damage more than a simple rupture of the membrane.

Both polymeric and nonpolymeric RNA was present in the supernatant fluid due to heat shock (Fig. 10). At the present time it is not known whether the RNA was enzymatically hydrolyzed after leaving the cell or before.

It should also be mentioned that in the supernatant fluid malic dehydrogenase as well as glucose-6-phosphate dehydrogenase has been identified. If one accepts the concept that the various TCA enzymes are located in the membrane, then the occurrence of malic dehydrogenase due to thermal shock is a result of membrane damage (Langridge, Haight and Morita, unpublished data).

Studies on Gelatinase

The exoencyme, gelatinase, was obrained from another obligately psychrophilic marine vibrio (MP 41) (Weimer, 1966). Since the degradation of protein is an essential step in the mineralization process of proteins, this enzyme was studied for its relationship to temperature and hydrostatic pressure. The organism possessed an optimal temperature for growth at 18 C and a maximal growth temperature of 24 C.

When cells of MP 41 are washed twice and then placed in nutrient gelatin, it was found that the amount of gelatin hydrolyzed increased with increased temperature within a temperature range between 5 and 24 C. A short incubation period was employed so that a large amount of cell proliferation would not occur. The gelatinase activity would then be a result mainly of gelatinase synthesized by cells and its subsequent activity instead of the activity being due mainly to cell numbers. Gelatinase activity was found to take place at 5 C.

The ability of cells to produce gelatinase is influenced by both the temperature and pressure of incubation. At 15 C the effect of pressure between 1 and 600 atm are more pronounced showing that increased pressure decreased the gelatinase production and activity. Although less gelatinase is produced at 10 C than 15 C the effect of 1, 100 and 200 atm is not great, but a sudden decrease occurs between 200 and 300 atm.

A 28-fold purification of gelatinase was obtained by ammonium sulfate fractionation and column chromatography employing Sephadex G-200. This enzyme displays an optimum activity at 40 C and activity still remains at 55 C. The reaction rate at 40 C is rapid compared to the activity at 5 C.

If gelatinase was dialyzed to completely free it from salts, no activity could be demonstrated when assayed in the presence of gelatin substrate prepared with or without artificial sea water. If on the other hand employing the method of purification (which includes a small amount of salt) and assayed in the presence of gelatin made up with varying concentrations of sea salts, the higher concentrations gave less activity. The range of salt concentration employed was 0.4 to 15.4 percent sea salts.

In pressure-temperature sudies on purified gelatinase, it was noted that at 5 C and various pressures applied (1 to 600 atm), the rate of the enzyme reaction was so slow that contamination developed in the reaction tubes. However, at 40 C a slight progressive decrease is noted with increased pressures up to 600 atm, while at 25 C an initial drop is noted between 1 and 100 atm and pressure above 100 to 600 little influence.

If other proteases act in the same manner as gelatinase to temperature, then it is probable that proteolytic enzyme reactions can take place on marine food products without the organisms being present—in other words the enzyme is produced at low temperature and acts even though the temperature is elevated above the obligate psychrophile maximum growth temperature. This elevated temperature causes death of the obligate psychrophile.

Recognizing that not all the enzymes of obligately psychrophilic bacteria are abnormally thermolabile, this may be a reflection on the evolutionary development of this thermal class—especially if it evolved from mesophile. In other words, not all the enzymes of the organism has evolved to the extent that all of them are "psychrophilic" enzymes possessing abnormal thermolabile properties.

Some Physiological Differences in Cells When Grown at Environmental and Optical Temperatures

Various marine organisms are known to be capable of more growth at temperature 10 to 20 C above their natural habitats (ZoBell, 1946). Some of our studies have been directed to determining the physiological differences between the cells grown at optimum and the original environmental temperature. Differences between cells grown at the above two different temperatures are not noticeable unless non-nutrient menstruum is employed.

The themolability of 4 C and 15 C cells of *V. marinus* MP-1 was tested (Haight, J. J. and Morita, 1966). Table 2 shows that viability of 15 C grown cells was lost within 60 minutes at 26.6 C but not within 180 minutes at 24.6

Table 2. Viability of *V. marinus* at 15C following exposure to heat

Time	Exposure Temperature (C)								
(min)	5.0	15.0	20.5	22.8	24.6	26.6	28.4	30.6	32.3
0	+	+	+	+	+	+	+	+	+
20	+	+	+	+	+	+	+	+	0
60	+	+	+	+	+	0	0	0	0
80	+	+	+	+	+	0	0	0	0
100	+	+	+	+	+	0	0	0	0
120	+	+	+	+	+	0	0	0	0
140	+	+	+	+	+	0	0	0	0
160	+	+	+	+	+	0	0	0	0
180	+	+	+	+	+	0	0	0	0

\+ indicates visible growth within 96 hours.
0 indicates no visible growth within 96 hours.
Reprinted by permission after Haight, J. J. and Morita 1966.

Table 3. Viability of *V. marinus* at 4C following exposure to heat

Time	Exposure Temperature (C)								
(min)	5.6	15.1	20.7	22.8	24.7	26.6	28.4	30.6	34.2
0	+	+	+	+	+	+	+	+	+
20	+	+	+	+	+	+	+	0	0
60	+	+	+	+	+	0	0	0	0
80	+	+	+	+	0	0	0	0	0
100	+	+	+	+	0	0	0	0	0
120	+	+	+	+	0	0	0	0	0
140	+	+	+	0	0	0	0	0	0
160	+	+	+	0	0	0	0	0	0
180	+	+	+	0	0	0	0	0	0

\+ indicates visible growth within 96 hours.
0 indicates no visible growth within 96 hours.
Reprinted by permission after Haight, J. J. and Morita, 1966.

C. Table 3 shows similar data for 4 C grown cells but viability was lost within 80 minutes at 24.7 C and in 140 minutes at 22.8 C. A comparison of the two tables illustrates a difference in the thermal susceptibility of cells grown at 4 and 15 C. Differences in the amount of material (absorbing at 260 mμ) leaking into the suspending menstruum after heating for 3 hours at various temperatures between 4 C and 15 C grown cells are illustrated in Fig. 11. This 260 mμ absorbing material has been identified as RNA and indicates that the thermal damage to the cytoplasmic membrane is greater in the case of 4 C grown cells.

Warburg studies employing glucose as a substrate were also run to determine if the respiration rate of cells grown at two different temperatures were different. The respiratory rates of the 15 C grown cells assayed at 15 C (15@15) and the 15 C grown cells assayed at 4 C (15@4) are shown in Fig. 12. As would be expected 15@15 and 15@4 differ by a factor of 2 (Q_{10} rule). A similar Q effect is noted for 4@4 and 4@15. Striking differences are noted when the 15@15 as well as the 15@4 and 4@4 are compared. Although the

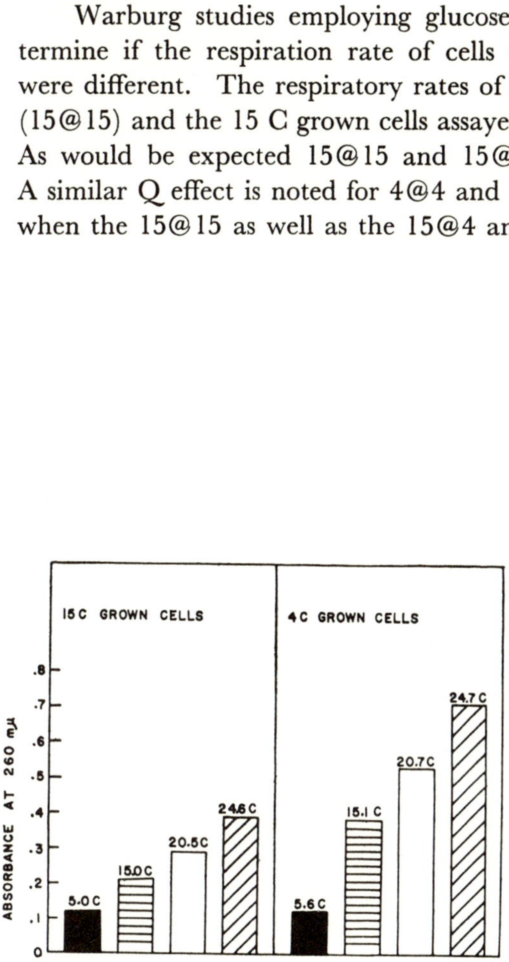

Fig. 11

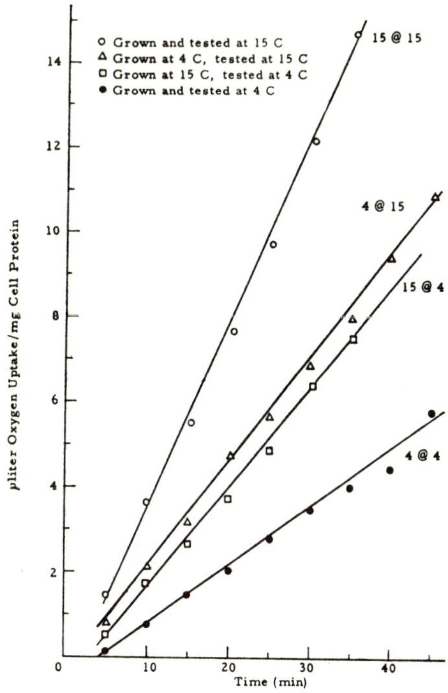

Fig. 12

Fig. 11. Absorbance of 260 mμ of supernatants from heat-shocked cells of *Vibrio marinus* MP-1 4 and 15 C. The cells were heated at the designated temperatures for 3 hours and were removed by centrifugation at 0 C.
Reprinted by permission after Haight, J. J. and Morita, 1966.

Fig. 12. Respiration in the presence of glucose by cells of *Vibrio marinus* MP 1 grown at 4 and 15 C. All values are corrected for controls.
Reprinted by permission after Haight, J. J. and Morita, 1966.

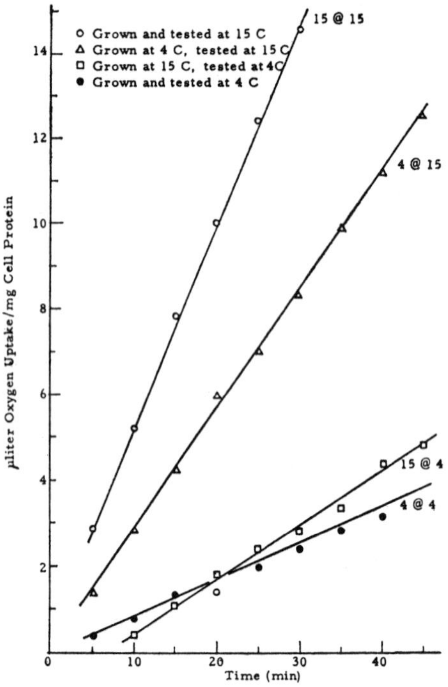

Fig. 13. Endogenous respiration by cell of *Vibrio marinus* MP 1 grown at 4 and 15 C. Reprinted by permission after Haight, J. J. and Morita, 1966.

reasons for the differences are not known, the following are possibilities:

1. If it is assumed that the cells are equally permeable to glucose at both temperatures, the differences in the slopes could be due to differences in the level of glucose-utilizing enzymes in the cells.
2. The differences could be in permeability to glucose.
3. Differences in intracellular organization.

The same pattern holds true in endogenous respiration of cells (Fig. 13). These studies also show that the original environmental temperature should be used in laboratory studies to insure proper understanding of the relationships between the marine organisms and their environment.

L-serine Deamination by Washed Cells of V. marinus MP-1

In the regeneration of nitrogenous compounds necessary for primary productivity, the proteolytic degradation of proteins and the deamination of amino acids are important. Former studies on gelatinase represent examples of proteolysis. It was found that *V. marinus* MP-1 had the ability to deaminate numerous amino acids. Although glutamine was found to be deaminated the greatest, it was not chosen for this study because of its instability at alkaline pH. The production of ammonia from L-serine by washed cells took place best at pH 8.4 (Albright, 1966).

Differences could be noted when cells were grown at 15 C or at 4 C. In terms of substrate saturation, the 15 C grown cells required approximately 50 μmoles ml of L-serine while 4 C grown cells required only 3 μmoles. This large difference may be a matter of permeability but nevertheless the amount of ammonia liberated by cells grown at the two different temperatures was approximately the same, provided the systems were saturated with substrate. The low substrate saturation of 4 C grown cells may be advantageous to the organism in the marine environment where low organic matter concentrations are found.

Maximal deamination occurred at 40 C with 15 C grown cells. Very little activity is noted at 50 C. This enzyme activity is not abnormally heat labile.

On the other hand, the 4 C grown cells displayed a different pattern of enzymic activity with temperature. Two peaks of activity were observed; one at 11 C and the other at 35 C. The first peak may be the result of cellular integrity since washed cells were employed. The second leak may be due to a destruction of permeability control or metabolic control mechanism. Since there are two known mechanisms (L-serine dehydrase and L amino oxidase) to yield ammonia from L-serine, the peaks in ammonia production may be the result of two different mechanisms. Isozyme production may be another answer in which each isozyme has its own temperature optimum.

In all cases (15 and 4 C grown cells) where hydrostatic pressure was added to the system, pressures from 130 to 300 atm caused stimulation of ammonia production from L-serine. This is close to the hydrostatic pressure from which the organism was originally isolated.

However, when 15 C grown cells were tested at 4 C, no pressure optimum was observed. A decrease in temperature as well as an increase in pressure both bring about a molecular volume decrease, hence a concomitant change in the conformational change of the enzyme or enzymes involved.

Significance of Research on Obligately Psychrophilic Marine Bacteria

Recognizing that over 90 per cent, by volume, of the oceans is colder than 5 C, the activities of these forms below the thermocline becomes important—especially when we attribute the minerilization processes of the various organic compounds catalyzed by bacteria. However, we have still yet to determine the exact numbers as well as their activities in the ocean. The Antarctic and Arctic waters are high in productivity. This organic rich water sinks and is eventually upwelled. Stanley (personal communication) has demonstrated obligately psychrophilic marine bacteria in the Antartic waters and Burton (personal communication) has demonstrated their existence in Arctic waters.

It is only obvious that in cold marine water where they have been demonstrated that they play an important role in the mineralization of the various types of organic matter present in the cold water.

Thermal studies on enzymes of obligately psychrophilic bacteria are important since the regeneration of nutrients in the sea depends upon the well-being of the entire organism and any metabolic lesion can cause a cessation in their activities. Our thermolability studies complement the reason why organisms have the ability to function at low temperature. An abnormally thermolabile enzyme only emphasizes that the enzyme is geared to low temperature for activity.

Acknowledgment

The author wishes to acknowledge his present and former graduate students (Dr. R. D. Haight, Dr. Paul F. Mathemeier, Mrs. Janet J. Haight, Mrs. Mary S. Weimer, Miss P. Langridge, Miss Sarah Robinson, Mr. L. J. Albright, and Mr. W. W. Miller) for the unpublished and published data used in this paper.

The research data presented in this paper was supported by NSF grant GB 2472, U.S. Public Health grant AM 06752 from the Institute of Arthritis and Metabolic Diseases, and a U.S. Public Health training grant 5 Tl GM 704 from the Institute of General Medical Sciences.

All Tables and Figures are reproduced by permission given by the Editors of the Journal of Bacteriology and Limnology and Oceanography.

References

ALBRIGHT, L. J. The effect of temperature and hydrostatic pressure on deamination of L-serine by *Vibrio marinus*, an obligate psychrophile. Oregon State Univ. M. S. Thesis (1966).

BURTON, S. D. and R. Y. MORITA. Denaturation and renaturation of malic dehydrogenase in a cellfree extract from a marine psychrophile. *J. Bacteriol.* **86:** 1019-1024 (1963).

CAMPBELL Soup Company. *In*, Proceedings low temperature microbiology symposium—1961. Campbell Soup Co., Camden, N. J. 322 pp. 1962.

COLWELL, R. R. and R. Y. MORITA. Reisolation and emendation of description of *Vibrio marinus* (Russell) Ford. *J. Bacteriol.* **88,** 831-837 (1964).

HAGEN, P. O., D. J. Kushner and N. E. GIBBONS. Temperature-induced death and lysis in a psychrophilic bacterium. *Can. J. Microbiol.* **10:** 813-822. 1964.

HAIGHT, J. J. and R. Y. MORITA. Some physiological differences of *Vibrio marinus* grown at environmental and optimal temperatures. *Limnol. Oceanog.* **11:** 470-474 (1966).

HAIGHT, R. D. and R. Y. MORITA. Thermally induced leakage from *Vibrio marinus* an obligate psychrophilic bacterium. *J. Bacteriol.* **92:** 1388-1393 (1966).

LANGRIDGE, P. and R. Y. MORITA. Thermolability of malic dehydrogenase from the obligate psychrophile, *Vibrio marinus. J. Bacteriol.* **92:** 418-423 (1966).

MATHEMEIER, P. F. Thermal inactivation studies on some enzymes from *Vibrio marinus.* Oregon State Univ. Ph.D. Thesis (1966).

MORITA, R. Y. Marine psychrophilic bacteria. *Oceanog. Mar. Biol.: an annual review.* **4:** 105-121 (1966).

MORITA, R. Y., and L. J. ALBRIGHT. Cell yields of *Vibrio marinus*, an obligate psychrophile, at low temperatures. *Can. J. Microbiol.* **11:** 221-227 (1965).

MORITA, R. Y. and S. D. BURTON. Influence of moderate temperature on the growth and malic dehydrogenase of a marine psychrophile. *J. Bacteriol.* **86:** 1025-1029 (1963).

MORITA, R. Y. and R. D. HAIGHT. Temperature effects on the growth of an obligate psychrophilic marine bacterium. *Limnol. Oceanog.* **9:** 103-106 (1964).

STOKES, J. L. pp. 187-192. *In* N. E. Gibbons (ed.). Recent progress in microbiology. Symposia, VIII International Congress for Microbiology, Univ. Toronto Press, Toronto Press, Toronto, Canada. 1962.

WEIMER, M. S. Purification and kinetics of gelatinase obtained obligately psychrophilic marine vibrio. Oregon State Univ. M. S. Thesis (1966).

WITTER, L. D. Psychrophilic bacteria—a review. *J. Dairy. Sci.* **44:** 983-1015 (1961).

ZoBELL, C. E. Marine Microbiology. Chronica Botanica. Walthan, Mass. 240 p. (1946).

24

Reprinted from *Galathea Rep. Copenhagen,* **1**, 139–154 (1959)

DEEP-SEA BACTERIA

By CLAUDE E. ZOBELL *and* RICHARD Y. MORITA

Professor of Microbiology, University of California, La Jolla, California, U.S.A.

Research Assistant, University of California. Presently Assistant Professor of Biology, University of Houston, Houston, Texas, U.S.A.

SUMMARY[1]

The presence of living bacteria in some of the deepest parts of the ocean was demonstrated on the Galathea Expedition. Sediment samples taken from the bottom of the Philippine Trench at depths exceeding 10 000 meters were found to contain from 10^4 to 10^6 living bacteria per ml. Large bacterial populations were also detected in sediment samples from the floor of the Kermadec-Tonga Trench (6790 to 9820 meters), the Sunda Deep (7020 meters) in the Java Trough, and the Weber Deep (7250 meters) in the Banda Sea. Several hundred microbial analyses were made on nine sediment samples taken from depths exceeding 10 000 meters.

The occurrence of bacteria in deep-sea sediment samples was demonstrated by direct microscopic examinations shortly after the samples were hauled aboard the Galathea. Growth or reproduction of the bacteria in nutrient sea water media proved that the bacteria were alive. Their indigenity was indicated by three lines of evidence: (1) Every possible precaution was exercized to prevent the contamination of the deep-sea samples, (2) More bacteria were found in the sediment samples than in the overlying water, and (3) The deep-sea bacteria were unique in their ability to grow preferentially or exclusively at *in situ* hydrostatic pressures (700 to 1 000 atmospheres) and low temperatures (3 to 5°C).

Apparatus is described which was designed to maintain sediment samples and bacterial cultures at any desired pressure in the ship's refrigerator. This apparatus made it possible to send by Air Express deep-sea sediment samples held at 1 000 atm to our laboratory at Scripps Institution for further observation.

The deep-sea bacteria appear to be even more heat-sensitive than pressure-sensitive. Many were killed by holding them for ten minutes at 30°C. Less than 20 per cent of the deep-sea bacteria survived for ten minutes at 40°C.

The unique pressure, temperature, and nutrient requirements of the deep-sea bacteria presents an enigma in bacterial taxonomy, because the bacteria cannot be characterized by conventional or standard method cultural procedures. Morphologically the deep-sea bacteria resemble common soil and water forms, but physiologically and culturally they show differences which suggest that most deep-sea bacteria are new and undescribed species or genera.

Though predominantly aerobic, some of the deep-sea bacteria develop anaerobically. Among the physiological types that were present and active at *in situ* pressures and temperatures were starch hydrolyzers, nitrate reducers, sulfate reducers, ammonifiers, and nitrifiers. The types of bacteria present as well as the organic and sulfate content of profile series of mud cores from the Kermadec-Konga Trench suggests that the bacteria have been active *in situ* in altering organic compounds and in reducing sulfate.

Besides affecting the non-conservative chemical composition of sea water and sediment, the deep-sea bacteria are believed to contribute to the nutrition of benthic fauna. The "standing crop" of organic carbon in the cells of living bacteria is estimated at from 0.2 to 2.0 mg per liter of sediment at the mud-water interface. There may be from 10 to 100 "crops" or bacterial generations per year. The bacteria are believed to obtain their carbon and energy requirements from organic detritus settling to the sea floor and to a much larger extent from dissolved or colloidal organic compounds carried by the movements of water masses.

1. Contribution from the Scripps Institution of Oceanography, University of California, La Jolla, California, U.S.A., New Series No. 926. These investigations were supported in part by grants from the Office of Naval Research, U.S.A. Department of Navy, and the Rockefeller Foundation.

INTRODUCTION

Observations on the Galathea Expedition in 1951 demonstrated for the first time the occurrence of living bacteria in some of the deepest parts of the ocean. Prior to this time, 5942 meters was the greatest depth at which bacteria had been found. Most microbiologists questioned whether bacteria could exist in oceanic abysses, although an abundant microflora has been shown to occur in near-shore sediments (ZoBell, 1946).

On the Talisman Expedition, Certes (1884a) found a few bacteria in bottom deposits taken from a depth of 5100 meters. The indigenous nature of these bacteria was indicated by their ability to grow in nutrient medium at hydrostatic pressures that were approximately isobaric with the depth from which they were taken (Certes, 1884b). While on the Humboldt Plankton Expedition to the West Indies, Fischer (1894) found a few bacterial colonies on nutrient agar plates inoculated with deep-sea sediment samples, one of which came from a depth of 5280 meters, but there was nothing to indicate whether these bacteria were indigenous or adventitious species.

In 1948, we obtained from a 5800-meter deep off Bermuda a mud sample in which numerous barophilic bacteria were found. The term "barophilic" was coined by ZoBell and Johnson (1949) to characterize microbes which grow preferentially or exclusively at high hydrostatic pressures. During the 1950 Mid-Pacific Expedition to the Marshall Islands, large bacterial populations were demonstrated in pelagic sediments, including 14 samples taken from depths exceeding 4000 meters; one of these samples containing bacteria came from a depth of 5942 meters (Morita and ZoBell, 1955).

Cultural methods, supplemented by direct microscopic observations, were employed to investigate the numbers and kinds of bacteria in marine materials collected by the Galathea. Immediately after being hauled aboard the ship, deep-sea sediment and water samples were transferred aseptically to steel vessels for compression to hydrostatic pressures approximately isobaric with the depth from which taken, and placed in a refrigerator at 3-5°C. This operation ordinarily required from 5 to 15 minutes. Portions of each sample were removed for detailed examination as time and facilities permitted.

Temperature Tolerance of Deep-Sea Bacteria.

Haste in handling the deep-sea samples proved to be of utmost importance owing to the temperature sensitivity of the bacteria. About 25 per cent of the marine bacteria collected off the coast of California by ZoBell and Conn (1940) were killed in 10 minutes when warmed to 30°C and only 20 per cent survived for 10 minutes at 40°C. Deep-sea bacteria have proved to be even more sensitive to heat than those from shoal waters.

The temperature of most of the deep-sea samples ranged from 5° to 10°C when hauled aboard the Galathea, having warmed up from about 3°C while being brought to the surface. The air temperature in the tropics, like that of the surface water, ranged from 28° to 30°C. Room temperature in the Galathea laboratory was usually a few degrees higher, and the temperature in the stateroom used as a microbiological laboratory often exceeded 40°C. Consequently we were working near or beyond the threshold of temperature tolerance of many marine microbes.

In spite of haste in handling materials and precautionary measures designed to keep them cool, many of the more sensitive organisms were probably killed by the heat. Lacking a cold room in which to make the necessary transfers involving minute amounts of inocula (often only 0.01 ml), part of the material got warmed up to near laboratory temperature during the measuring, transfer, dilution, and inoculation of the bacteria therein. Fortunately, a good many of the deep-sea bacteria did survive the extreme change in climate to which they were subjected during sampling and analytical operations, but we do not know how many may have perished. Data summarized in Table I illustrate the loss of viability of bacteria in deep-sea samples intentionally subjected to different temperatures for 10 minutes prior to plating on nutrient agar.

Only a small percentage (2.3 to 4.7) of the bacteria survived heating to 50°C for 10 minutes and less than 20 per cent survived for 10 minutes at 40°C. Most of those that survived proved to be spore formers. From the data in Table I, it might appear that most of the deep-sea bacteria tolerated a temperature of 30°C for 10 minutes, but this may not be true because, even at "room" temperature, the samples were subjected for several seconds to temperatures between 30° and 40°C or higher. Perhaps all of

Table I. Relative numbers of bacteria in sediment samples from Philippine Trench which developed on nutrient agar after small portion of sample was held for 10 minutes at stated temperature. The results are expressed as percentages of the number of colonies which developed from portions of the same samples subjected to "room" temperature during rapid routine manipulations.

Station Number	Water depth	Rapid room	Held 10 minutes at		
			30° C	40° C	50° C
	meters	per cent	per cent	per cent	per cent
413	10 060	100	91.6	18.0	4.7
418	10 190	100	82.5	7.4	2.3
419	10 210	100	93.7	11.8	3.0
420	10 160	100	78.4	6.5	2.8

the more heat-sensitive microbes were lost. This should stress the importance of providing properly air-conditioned rooms for microbiological manipulations on subsequent deep-sea expeditions. Maintaining the organisms at *in situ* temperatures and pressures during their transposition from the deep-sea floor to the surface may also be a useful innovation.

Effect of Hydrostatic Pressure on Marine Microbes.

Although bacteria appear to require for growth or reproduction a hydrostatic pressure which is near that of their native environment, it seems likely that few if any are killed by the decompression resulting from their transposition from the deep-sea floor to the surface. Such decompression may actually be advantageous to the organisms, because the resultant adiabatic cooling (about 1.6°C from a depth of 10 000 meters to the surface) helps counteract warming during transposition through warmer surface water. Of course, microbes may be killed mechanically by the sudden release of pressure in a system supersaturated with gas, but in spite of the pressure, there is no more gas dissolved in deep-sea water than in surface water.

Prolonged storage of deep-sea sediment samples at refrigeration temperatures has resulted in the gradual death of the bacteria at atmospheric pressure. Very few living bacteria could be detected in sediment samples from the Philippine Trench (depth 10 000 meters) following six weeks' storage at 3° to 5°C at atmospheric pressure, although large numbers of barophilic bacteria have persisted for 30 months in similar samples maintained in the refrigerator at 1000 atm. Whether this die-off is directly attributable to reduced pressure or is an indirect effect on the metabolism and nutrition of the bacteria under the experimental conditions can be determined only by further investigations.

Working with freshly collected deep-sea material (less than an hour after it was hauled aboard the Galathea), we found many more bacteria developing in nutrient media incubated at *in situ* pressures than at atmospheric pressure. The barophilic property of deep-sea bacteria is illustrated by representative results with samples taken from the Philippine Trench (Table II).

Table II. Most probable numbers (MPN) of bacteria per gram wet weight demonstrated in sediment samples from the Philippine Trench by the minimum dilution method in nutrient medium incubated in refrigerator at different pressures.

Station Number	Location of station		Water depth	Incubation pressure	
	Latitude	Longitude		1 atm.	1000 atm.
	North	East	meters	MPN	MPN
413	10° 20′	126° 36′	10 060	2 300	760 000
418	10° 13′	126° 43′	10 190	930	3 500 000
419	10° 19′	126° 39′	10 210	680	210 000
420	10° 24′	126° 40′	10 160	8 400	920 000
421	10° 29′	126° 05′	1 000	540 000	0
422	10° 49′	126° 01′	1 960	2 300 000	0
424	10° 28′	126° 39′	10 120	5 900	2 800 000

The most probable numbers of bacteria were estimated by "Standard Methods for the Examination of Water and Sewage" (American Public Health Assoc., 1936), as amplified by HALVORSEN and ZIEGLER (1933) and by PRESCOTT *et al.* (1947). Methods of inoculating the nutrient medium and of applying pressure are outlined below.

From the data in Table II, it will be observed that from 10 to 100 times as many bacteria from a depth of about 10 000 meters grew in nutrient medium incubated at 1000 atm as the number which grew at 1 atm. We now know that the effect of hydrostatic pressure on bacteria is partly a function of the composition of the medium, suggesting that with proper nutrients there may be as many bacteria from the deep sea that would grow at 1 atm as at 1000 atm, but we have not yet learned what these proper nutrients might be.

Although data like those summarized in Table II indicate the existence of strict barophiles (high pressure dependent bacteria), as this progress report is written we must admit that the pressure picture appears to be somewhat more complicated. Be this as it may, finding more bacteria in deep-sea sediments growing at *in situ* pressures (1000 atm) than at 1 atm is regarded as conclusive evidence that most, if not all, of the bacteria were indigenous to the deep and that they are physiologically active at high pressures. This conclusion is supported by finding in nearby shallow sediments from the sides of the Philippine Trench large numbers of bacteria which grew at 1 atm and none which grew at 1000 atm. This latter observation agrees with the findings of ZoBell and Johnson (1949) and ZoBell and Oppenheimer (1950), who reported that surface-dwelling microbes fail to grow at deep-sea pressures and many are slowly killed in nutrient media when held at pressures ranging from 400 to 600 atm. According to Oppenheimer and ZoBell (1952), pressures ranging from 200 to 600 atm inhibited the normal growth of nearly all of 63 different species of marine bacteria, mostly from shoal water habitats. Five of these were killed in four days by compression to 200 atm, 11 were killed at 400 atm, and 23 were killed in four days at 600 atm.

Nutrient Media and Enumeration Procedures.

The nutrient medium employed routinely for demonstrating the most probable number of bacteria by the minimum dilution method had the following composition:

Peptone (Difco)	5.0 gm
Ferric phosphate	0.1 gm
Potassium nitrate	1.0 gm
Soluble starch	2.0 gm
Yeast extract (Difco)	1.0 gm
Sea Water	1000.0 ml

Following autoclave sterilization its pH was 7.7. It was dispensed in 9.0-ml quantities in 15-ml screw-cap bottles.

Into the first bottle of sterile medium in each series was introduced aseptically 1.0 gm of the sample. After vigorously shaking to distribute the sediment (or water) sample evenly throughout the medium, a 1.0-ml aliquot was transferred, with a semi-automatic pipette, to the second bottle of sterile medium. This gave a 1:10 dilution or provided a 0.1-gm inoculum. After similarly shaking this second bottle, 1.0 ml was removed to inoculate the third bottle of sterile medium to give a 1:100 dilution or a 0.01-gm inoculum. The dilution procedure was repeated seriatim by powers of 10 until the original sample was diluted to 1:1 000 000.

Assuming a random distribution of the bacteria in the sediment sample and throughout the dilution series, it follows that bacterial growth in the 1:10 000 and all lower dilutions but no growth in the 1:100 000 dilution, for example, would indicate the presence of at least 10 000 viable bacteria per gram of sample but not 100 000 per gram. The most probable number (MPN) of bacteria in certain samples was determined with greater accuracy by preparing quintuplets of each series. For example, finding growth in all 5 tubes of medium inoculated with the 1:10 series, 3 positive in the 1:100 series, and none of the 5 positive in the 1:1000 series would indicate a bacterial population (MPN) of 79 per gram.

For incubation at atmospheric pressure, the diluted material in the nutrient medium was left in the screw-cap bottles. These bottles provided about 5 ml of air space for aerobes. To estimate the abundance of anaerobic bacteria in the sediment samples, the inoculated bottles were filled to capacity with O_2-free medium and the caps screwed tightly in place to exclude atmospheric oxygen. Following incubation at the desired temperature and pressure, the medium was examined for evidence of bacterial growth as manifested by increased turbidity of the medium or by finding large numbers of bacteria by direct microscopic examinations.

Ammonia liberation from peptone by ammonifiers:

$$R\text{-}NH_2 + HOH \rightarrow R\text{-}OH + NH_3$$

was detected by nesslerizing 0.2 ml of medium in a spot plate.

The presence of nitrate-reducing bacteria was indicated by the appearance of nitrite or by the disappearance of nitrate in the inoculated nutrient medium. Iodine was added to aliquots placed in a spot plate to test for the terminal presence of starch; its absence indicated the activity of starch-hydrolyzing bacteria. Sulfate-reducing bacteria were demonstrated in medium M-10-E:

Calcium lactate	3.5 gm
Magnesium sulfate, hydrated	0.2 gm
Potassium phosphate, dibasic	0.2 gm
Sodium sulfite	0.1 gm
Ferrous ammonium sulfate	0.1 gm
Ascorbic acid	0.1 gm
Peptone (Difco)	1.0 gm
Yeast extract (Difco)	1.0 gm
Bacto-agar (Difco)	3.0 gm
Sea water	1000.0 ml

It was adjusted to pH 7.5 following autoclave sterilization. The medium was dispensed in 15-ml screw-cap bottles, which were filled to capacity to displace all air, since the sulfate reducers are strict anaerobes. Following inoculation by serial dilutions and incubation, the presence of sulfate reducers was indicated by the blackening of the medium, resulting from the formation of ferrous sulfide:

$$SO_4^{--} + 10H \rightarrow H_2S + 4H_2O$$
$$H_2S + FeSO_4 \rightarrow FeS + H_2SO_4$$

Sea water enriched with 0.1 per cent ammonium phosphate (dibasic) and buffered with 0.2 per cent calcium carbonate was used to detect the presence of nitrifying bacteria. Following inoculation and incubation at the desired temperature and pressure, the enriched sea water was examined for the appearance of nitrite:

$$NH_4^+ + 2O_2 \rightarrow NO_2^- + 2H_2O$$

Nitrifiers which oxidize ammonium to nitrite were demonstrated in deep-sea sediment samples only when the inoculated medium was incubated at high pressures.

Inoculated media or samples to be held at high pressures were transferred to small glass vials, size 10 × 50 mm. After filling to capacity (5 ml) each vial was closed with a tapered No. 000 Neoprene stopper. When subjected to pressure in a steel vessel filled with hydraulic fluid, the stopper functioned as a piston, compressing the contents of the vial to approximately the same pressure as was built up in the steel vessel. Numerous tests have established that the compressed Neoprene piston stoppers provide an effective seal against contamination of any kind, the higher the pressure the better the seal.

High Pressure Apparatus.

The pressure cylinders (Figure 1) were constructed of 18-8 type 303 stainless steel. Such a vessel having an inside diameter of 1-3/8 inches and an inside length of 11 inches holds 25 10 × 50-mm stoppered culture tubes and weighs only 6 kg. The vessels were fabricated by drilling a radially central hole, with tapered bottom, in 12-inch lengths of 2-1/4-inch steel bar stock. The caps were prepared by machining 3-inch lengths of 3-inch steel bar stock as illustrated. Machining also provided for a male-threaded top extension for the permanent attachment of a needle valve and a perforated cone-shaped ceiling for exhausting air from the vessel. When fitted with an "O" ring and the vessel is filled level full with hydraulic fluid. the latter is displaced through the cap and attached needle valve when the cap is screwed on the cylinder.

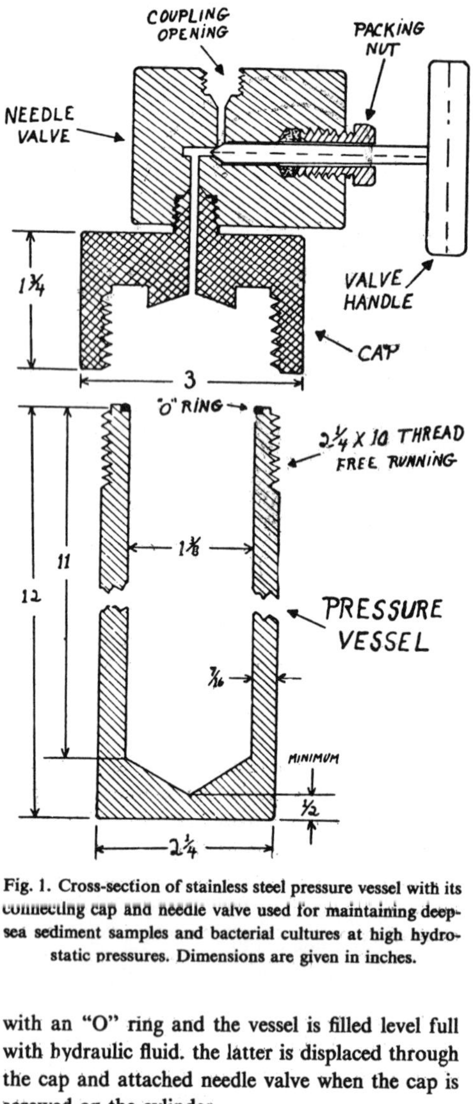

Fig. 1. Cross-section of stainless steel pressure vessel with its connecting cap and needle valve used for maintaining deep-sea sediment samples and bacterial cultures at high hydrostatic pressures. Dimensions are given in inches.

In practice the pressure vessel, equilibrated to the exact temperature at which it is to be incubated, is loaded with piston-stoppered culture tubes and filled level full with water at the same temperature. Hastily the cap is screwed on, an operation that can be performed by hand in a few seconds without the use of a wrench, and connected to the hydraulic pump (Figure 2). A few strokes with the handle of the pump produce a pressure up to 1000 atm in 20 to 30 seconds. After closing the needle valve, the pressurized vessel is disengaged from the pump for storage in the refrigerator or other thermostat. Con-

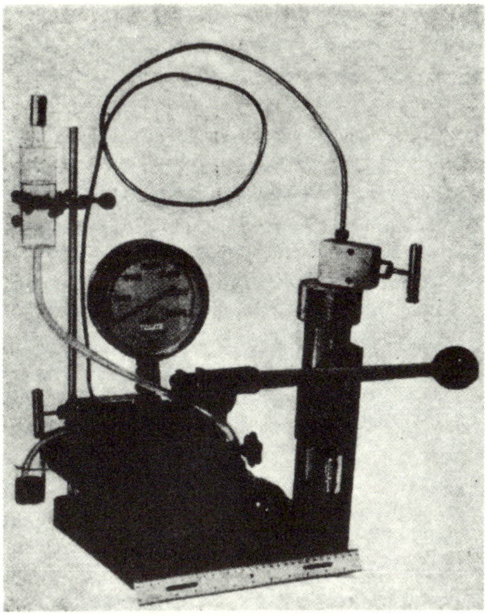

Fig. 2. High pressure vessel connected by semi-flexible steel tubing to hydraulic pump and Bourdon gauge. A rack of small culture tubes with piston stoppers is shown in foreground. ($^1/_{10}$ natural size).

stant pressure is maintained almost indefinitely provided the temperature is kept constant. The pressure changes between 6 and 7 atm per 1 °C change in temperature. Adiabatic heating amounts to about 2°C during compression to 1000 atm and it requires about 30 minutes for temperature equilibrium to be established by heat conduction in the thermostat, the half-life of the adiabatic temperature increase being between 5 and 10 minutes.

The portable pump pictured in Figure 2 was prepared from a hydraulic truck jack having a capacity of 30 tons or about 4000 atm. The pump is mounted on a block of wood along with the pressure gauge, high-pressure valves, hydraulic fluid reservoir, and semi-flexible steel tubing for connecting the pump to pressure vessel. The clamp on the corner of the assembly is for holding the pressure vessel in an upright position during pressurization. Immediately after applying pressure the needle valve on the pressure vessel is closed and the vessel is returned to the thermostat after disconnecting the top coupling.

The pumps were obtained from the Blackhawk Mfg. Co., Milwaukee, Wisconsin. Needles valves, steel tubing, couplings, and pressure gauges were supplied by the American Instrument Co., Silver Springs, Maryland. "O" rings were provided by The Parker Appliance Co., Cleveland, Ohio. The pressure vessels were prepared by the Special Developments Division, Scripps Institution of Oceanography, La Jolla, under the supervision of James M. Snodgrass. The hydraulic fluid employed in the pump was a 1:1 mixture of water and glycerol.

The pressure had to be released from the vessels before the latter could be opened for the examination of cultures or stored samples. For depressurization the vessel was reconnected to the pump and gauge assembly. Then pressure was applied by operating the pump until the pressure registered on the Bourdon gauge corresponded with the pressure initially applied to the vessel. If the pressure indicator did not move upon opening the needle valve, it showed that the pressure in the vessel was the same as that initially applied. Should the terminal pressure in the vessel be either more or less than that initially applied, the gauge would register an increase or decrease as the case might be. This procedure was always used to check the terminal pressures in the vessels. Interestingly, in thousands of cases the terminal pressure was always found to be the same as that initially applied, provided the temperature had been kept constant.

Air Express Shipment of Pressurized Samples.

Most of the microbiological analyses were made on the Galathea. However, because of their great significance, several sub-samples of sediment from the Philippine Trench were shipped by air express from Cebu to California, where they were examined in our laboratories at the Scripps Institution of Oceanography. This unique operation was made possible only through the cooperation of personnel from several organizations, including the officers and men on the Galathea, the East Asiatic Company, Royal Danish Consulate Service, Philippine Airlines, and the University of California.

Indispensable services were rendered by Mr. Kjeld Danoe, who had our high pressure (1000 atm) vessels, containing sediment samples recently taken from the bottom of the Philippine Trench, insulated and packed in dry ice at Cebu for sending by air express to Manila. There this unique package was received two hours later by Gustav Halberg and K. Raaschou Nielsen, who made arrangements for its immediate transfer to a refrigerator on a Philippine Airlines plane. Thirty hours later the P.A.L. hostess placed the refrigerated pressurized vessel in the hands of a colleague, who met the plane in San Francisco, thanks to Royal Danish Navy radiograms that had forwarded instructions from the Galathea. Neither the temperature nor the pressure of the vessels changed perceptibly from the time they left the Galathea until they were in the laboratory at La Jolla. Examination by direct microscopic and cultural procedures confirmed the presence of numerous living bacteria in sediment samples from depths exceeding 10 000 meters.

Direct Microscopic Observations.

As soon as time permitted on the Galathea, small sub-samples of deep-sea sediment were examined with a phase contrast microscope at a magnification of 970 diameters. By means of a sterile platinum loop designed to deliver approximately 0.01 ml, portions were spread evenly over an area of 1.0 cm^2 on a clean glass slide and protected with a cover glass. Each bacterium observed in the area (0.0002 cm^2) covered by one field of the microscope represents 500 bacteria per 0.01 ml or 50 000 per ml of the original sample. With samples containing fewer than 50 000 per ml, it was necessary to scan several fields in order to find any bacteria. It so happened, however, that a good many of the sediment samples had bacterial populations exceeding 10^5 per ml. Thus, in spite of the obscuring effects of sediment particles with which the bacteria were associated, several bacteria were observed in some fields.

The first significant microscopic observation was made on a sample of silty clay dredged from a depth of 8870 meters in the Philippine Trench on 13 July 1951 at Station No. 412. Twelve sub-samples were prepared for microscopic examination; 50 or more fields of each being scanned. In nearly every field were found definite discrete bacterial cells, including ten different morphological types. Most common were non-sporulating rods with rounded ends, size about 0.5 microns in width by from 2 to 3 microns in length. Fairly abundant were capsulated ovoid rods having an average width of 0.8 microns and a length of 2 microns. Slender rods, 0.6 microns in width and up to 6 microns in length were found attached to sediment particles. Also recognized were several small vibrio, two different types of sporulating bacilli, two kinds of spirilla, and a few diplococci. Motility was not observed in wet preparations. Ostensibly the deep-sea bacteria lose their powers of locomotion due to the change in climate (lower pressure and higher temperature), because it was established by subsequent observations that the vibrio and some of the rods are flagellated.

These microscopic observations on the sediment sample from Galathea Station No. 412 were considered as presumptive evidence for the occurrence of bacteria from a depth of 8870 meters. However, conclusive evidence that they were indigenous species in this, and, eventually, in sediment samples from even greater depths had to await cultural tests, the first of which were completed 24 July 1951. Also observed in this sediment sample were a few empty diatom tests and some fish scales with attached bacteria. There was no evidence of cellulose fibers or of any kind of particulate organic matter besides bacterial cells. The absence of plankton organisms helped to rule out the possibility of the bacteria's having been picked up inadvertently while the dredge was passing through the photosynthetic zone, where the bacterial population was of the order of 10^3/ml.

The microscope also revealed the presence of bacteria in the topmost strata of a clay core collected 15 July 1951 at Station No. 413 from a depth of 10 060 meters. The core was 75 cm long. Detecting no bacteria in the lower strata of the core either by cultural or by direct microscopic examination is highly significant; negative results proved that sediment samples could be collected and analyzed on the Galathea without contamination. The topmost 0.1 cm layer of sediment was estimated to contain at least 3×10^7 bacteria per ml. Cultural methods showed the presence of 7.6×10^5 bacteria which grew at 1000 atm and only 2.3×10^6 which grew at 1 atm.

Predominating in the clay sample from Station No. 413 were rod-shaped bacteria with only a few helicoidal-shaped (vibrio and spirilla) cells, and still fewer spherical-shaped (cocci) bacteria. A good many of the bacteria occured in pairs or short chains, suggesting that the bacteria were reproducing. No protozoans, or other form of microscopic life other than bacteria, were observed.

It was extremely difficult to distinguish bacteria from sediment particles in clay samples dredged from depths of 10 190 and 10 210 meters at Stations No. 418 and 419 respectively. The direct microscopic examination of material dissected from the interior of balls of clay revealed the presence of fewer than 10^5 bacteria per ml, but exterior material was primarily from the mud-water interface. None of the bacteria observed in wet preparations exhibited any motility. Most of the bacteria were small, 1 to 2 microns long by 0.5 to 0.8 microns in diameter.

That the majority of them came from the deep-sea floor was indicated by cultural tests; the MPN at 1000 atm being 10^5 to 10^6 per ml as compared with MPN of only 10^2 per ml at 1 atm (see Table II).

The microscopic examination of greyish-green clay from a depth of 10 120 meters at Station No. 424 was more rewarding. Besides showing the presence of numerous bacilli (order of 10^7 per ml) there were many large (1.6×8 microns) spirilla unlike any bacteria previously described. None of the latter appeared among the many bacteria that grew in nutrient media incubated at 1000 atm (MPN 2.8×10^6 per ml). The microscope also showed many bodies believed to be fungous spores. None of these germinated in nutrient medium incubated at 1000 atm, but in similar medium incubated at 1 atm bacteria were overgrown by fungi. This observation suggests the occurrence of viable fungous spores on the deep-sea floor. Failure to find fungous mycelial material in any of the deep-sea samples indicates that fungi are not growing on the deep-sea floor. Four different kinds of diatom tests were observed with no evidence of protoplasm.

A good many of the minimum dilution cultures prepared for MPN determinations were examined microscopically to confirm bacterial growth. Most of the deep-sea bacteria were morphologically similar to those from soil and shallow water environments with which we are more familiar, showing even less variety among the forms found. Short rods predominated. About 5 per cent were vibrio. Cocci were encountered only occasionally. Differential staining procedures proved that the majority of the deep-sea bacteria are Gram-negative. Capsulated forms were common. Endospores were observed in about 20 per cent of the bacilli.

Physiological Characteristics of Bacteria From Philippine Trench.

The deep-sea bacteria exhibited considerable physiological versatility in differential media at 1000 atm (Table III). Apparently there were almost as many anaerobes as aerobes and a good many reduced nitrate. Some seemed to be able grow in the absence of free oxygen by utilizing nitrate as the hydrogen acceptor. The sulfate reducers were the only obligate anaerobes demonstrated.

The deep-sea nitrifiers differ in their activity at high pressure from surface-dwelling forms as do the nitrate reducers. Nitrate reduction by several shoal water species, which have been tested by ZoBell and Budge (1954), is retarded by pressures of 400 to 600 atm and at 1000 atm their nitratase system is slowly inactivated.

Table III. Minimum numbers of different physiological types of bacteria detected per ml of sediment from the Philippine Trench.

Station number	418	419	420	424
Water depth in meters	10 190	10 210	10 160	10 120
Total aerobic bacteria	10^6	10^5	10^5	10^6
Total anaerobic bacteria	10^5	10^5	10^5	10^5
Starch hydrolyzers	10^3	10^2	10^2	10^3
Nitrate reducers	10^5	10^4	10^5	10^5
Sulfate reducers	10^2	10	10^2	0
Ammonifiers	10^5	10^4	10^5	10^5
Nitrifiers	10	0	0	10

Bacteria in Indian Ocean Deeps.

Two sediment samples, suitable for bacteriological analyses, were obtained from depths exceeding 7000 meters in the Indian Ocean. Large numbers of viable bacteria were found in both samples, but surprisingly not nearly as many as in sediment samples taken from the Philippine Trench at depths exceeding 10 000 meters. Moreover, almost as many bacteria grew in nutrient medium at 1 atm as at 700 atm, and only a few from the Sunda and Weber Deeps grew at 1000 atm (Table IV).

Finding in these deeps starch hydrolyzers, nitrate reducers, sulfate reducers, ammonifiers, and nitrifiers which were active in differential media at 700 atm indicates that they are indigenous species which are probably active *in situ*. The activity of the deep-sea bacteria *in situ* is probably limited by the nutrient content of sea water and sediment and not by the high hydrostatic pressure or low temperature.

Bacteria in Kermadec-Tonga Trench.

Barophilic bacteria were found in 4 different sediment samples from the Kermadec-Tonga Trench (Table V).

Unlike the microflora found predominating in the Philippine Trench, nearly as many and in one sediment sample from the Kermadec-Tonga Trench (Station No. 686) more bacteria developed in nutrient media incubated at 1 atm than at high *(in situ)* pressures when isolates were examined. The inability of these bacteria recovered from the deep-sea floor to reproduce at *in situ* pressures suggests that

Table IV. Most probable numbers (MPN) of bacteria per gram wet weight demonstrated in sediment samples from Indian Ocean by minimum dilution method in nutrient media incubated in refrigerator at different pressures.

Station number	463 (Sunda Deep)	492 (Weber Deep)
Station location	10°16′ S × 109°51′ E	5°31′ S × 131°01′ E
Water depth	7020 meters	7250 meters
MPN at 1 atm	690 000	810 000
MPN at 700 atm	1 050 000	2 300 000
MPN at 1000 atm	4 800	17 000
Starch hydrolyzers at 700 atm	1 000	1 000
Nitrate reducers at 700 atm	10 000	10 000
Sulfate reducers at 700 atm	100	10
Ammonifiers at 700 atm	100 000	10 000
Nitrifiers at 700 atm	0	10

Table V. Minimum numbers of bacteria per gram of sediment from the Kermadec-Tonga Trench as determined by minimum dilution method in differential media incubated in refrigerator at different pressures.

Station number	649	650	658	686
Latitude	35°15′ S	32°20′ S	35°51′ S	28°30′ S
Longitude	178°40′ W	176°54′ W	178°31′ W	176°53′ W
Water depth (meters)	8300	6620	6720	9820
Total aerobes at 1 atm	10^5	10^6	10^4	10^6
Total aerobes at near *in situ* pressure*	10^6	10^6	10^6	10^5
Total anaerobes at 1 atm	10^4	10^6	10^5	
Starch hydrolyzers at 1 atm	0	10		
Nitrate reducers at 1 atm	0	0		
Sulfate reducers at 1 atm	0	0		

* Since the pressure-depth gradient in the sea approximates 0.1 atm/M, nutrient media inoculated with these four sediment samples were incubated at about 850, 680, 790, and 1000 atm respectively.

Table VI. Minimum numbers of bacteria per gram wet weight of sediment from different core depth detected in nutrient media incubated in refrigerator at 1 atm.

Station number	677		686	
Location of station	28°30′ S × 175°53′ W		20°53′ S × 173°31′ W	
Water depth	9190 meters		9820 meters	
Core depth	Aerobes	Anaerobes	Aerobes	Anaerobes
0-5 cm	1 000 000	1 000	1 000 000	1 000 000
5-10 cm	1 000 000	100 000	1 000 000	1 000 000
10-15 cm	100 000	10 000	1 000 000	100 000
15-20 cm	1 000 000	100 000	1 000 000	100 000
20-25 cm	100 000	10 000	1 000 000	10 000
25-30 cm	10 000	1 000	1 000 000	1 000
45-50 cm			10 000	10 000
85-90 cm			10 000	1 000

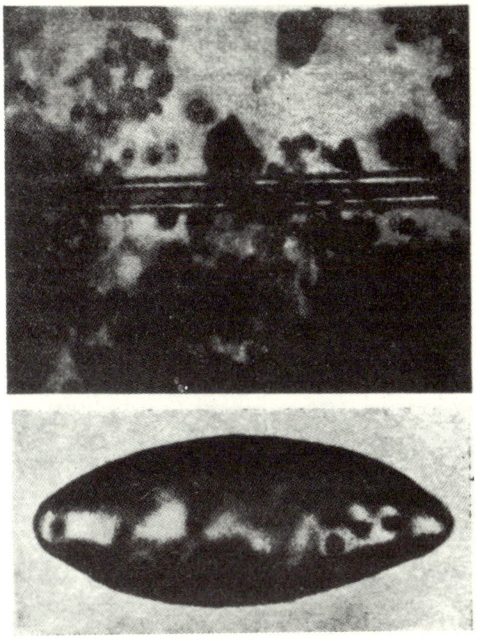

Fig. 3. Photomicrographs of two apparently intact diatoms found in deep-sea sediment following two years' storage in refrigerator at a pressure of 1000 atm.

they are not native to the deep-sea floor. The implication is that they have sunk to the sea floor where they may remain dormant for long periods of time. Such forms that tolerate high pressure without being able to reproduce are termed *baroduric*.

What appeared to be baroduric foraminifera and diatoms were detected in some of the Kermadec-Tonga Trench sediment samples examined by phase contrast microscope. This was confirmed by E. J. F. Wood (1956) working in our laboratory at La Jolla two years after the samples were taken. Since their collection the sediment samples had been stored in piston-stoppered tubes under conditions which were approximately isothermic and isobaric with the environment from which originally taken. Every one in the Scripps Institution laboratory who examined these preparations agreed that certain *Globigerina* contained protoplasm as did the *Coscinodiscus* and *Navicula* specimens. Extensive efforts to cultivate these foraminifera and diatoms have been unsuccessful. They are believed to be transient or adventitious forms which, like some of the baroduric bacteria, have sunk to the deep-sea floor. Regardless of their origin or significance, it is noteworthy that they have been preserved in refrigerated pressure vessels for many months. Two of the diatoms are shown in Figure 3.

The prolonged survival of baroduric bacteria buried at considerable depth was indicated by finding them in adundance to the bottom of cores collected from the Kermadec-Tonga Trench (Table VI). The vertical distribution of aerobes as well as anaerobes appeared to be sporadic, the general trend being decreasing abundance with core depth. Finding numerous aerobes buried at depths where there is probably no free oxygen constitutes further evidence of their dormancy. Some activity of bacteria is indicated by changes with core depth in the non-conservative chemical content of the sediment.

Chemical Content of Deep-Sea Cores.

Portions of two cores from the Kermadec-Tonga Trench were preserved in sealed bottles and brought back to the Scripps Institution laboratory for chemical analysis. Water content was determined by subtracting the constant dry weight (at 105°C) from the initial wet weight. The dried sediment was used for determining organic carbon, carbonate, Kjeldahl nitrogen, ammonia nitrogen, and sulfate.

Organic carbon was estimated by oxidizing at boiling point with a mixture of concentrated sulfuric acid, potassium iodate, and C. P. oxygen after first liberating carbonate carbon by treatment with phosphoric acid and flushing with a stream of pure nitrogen. The carbon dioxide resulting from both treatments was collected (separately) in standard barium hydroxide solution which was back-titrated with N/100 hydrochloric acid. The method of Bien (1952) for determining carbonate and organic carbon was used. This method employs a closed system which eliminates erroneous results from atmospheric CO_2.

Ammonia was driven off weighed sediment samples by sodium hydroxide treatment. Micro-Kjeldahl procedures were employed for determining organic nitrogen. Weighed samples were washed several times with distilled water accompanied by mechanical agitation to leach out sulfate, which was then determined gravimetrically as its barium salt. Results of the chemical analyses are summarized in Table VII.

The low carbonate content was expected, since carbonate in sediments usually decreases with depth in the ocean (Kuenen, 1950). This may be due to the increased solubility of calcium carbonate at the lower pH of sea water associated with the high hydrostatic pressure (Buch and Gripenberg, 1932).

Table VII. Certain non-conservative chemical constituents found in different strata of two mud cores from the Kermadec-Tonga Trench. Values are recored as milligrams per gram of dried (at 105°C) sediment.

Core from Galathea Station No. 677; water depth 9190 meters.

Core depth	Water content	Carbonate content	Organic carbon	Ammonia nitrogen	Kjeldahl nitrogen	Sulfate content
cm	per cent	mg/gm	mg/gm	mg/gm	mg/gm	mg/gm
0-5	46.2	0.7	0.5	0.8	0.5	1.9
5-10	43.8	0.0	0.2	0.2	0.3	1.5
10-15	40.1	0.2	0.1	0.1	0.3	1.2
15-20	40.0	0.6	0.2	0.1	0.3	1.6
20-25	41.8	0.2	0.2	0.1	0.4	1.2
25-28	38.4	0.4	0.1	0.1	0.3	0.9

Core from Galathea Station No. 686; water depth 9820 meters.

Core depth	Water content	Carbonate content	Organic carbon	Ammonia nitrogen	Kjeldahl nitrogen	Sulfate content
cm	per cent	mg/gm	mg/gm	mg/gm	mg/gm	mg/gm
0-10	53.7	0.2	0.3	0.1	0.2	2.0
15-20	51.7	0.5	0.3	–	0.2	1.6
25-30	50.7	0.5	0.2	0.2	0.2	1.3
35-40	52.5	0.3	0.3	0.1	0.3	1.1
45-50	53.0	–	0.1	0.0	0.1	–
55-60	50.8	0.2	0.2	0.1	0.3	0.2
65-70	53.4	0.4	0.2	0.0	0.4	0.4
75-80	52.0	0.2	0.2	–	0.2	0.2
85-90	47.7	–	0.2	0.1	0.2	–

The organic carbon as well as the organic nitrogen content of the deep-sea sediments was very low; in the deeper core strata the C/N ratio was 1:1. Total contents and the C/N ratios are considerably lower than in near-shore sediments, conditions which have been noted by WISEMAN and BENNETT (1940) and by ARRHENIUS (1952). Arrhenius postulates that the low C/N ratios in deep-sea clay deposits may be caused by some property, e.g., low calcium carbonate content, of the clay favoring conservation of nitrogen during the low rate of organic matter deposition, or it might be a reflection of differences in the way in which organic matter is decomposed in the deep sea. Most likely the relatively abundant ammonia-nitrogen content of the deep-sea deposits has resulted from the decomposition of proteinaceous materials. Bacteria (ammonifiers) which liberate ammonium from proteinaceous material were detected in virtually all deep-sea sediments tested for their presence.

There seems to be a general trend of decreasing organic content with core depth, which suggests its slow oxidation. In the absence of any obvious supply of free oxygen below the mud-water interface, the organic matter must be attacked by anaerobes, if by any kind of bacteria. The rate of attack appears to be very slow. Sulfate-reducing bacteria could account for the slow oxidation of organic matter in strictly anaerobic environments. Sulfate-reducing bacteria have been found in deep-sea sediments and, interestingly, there is a definite decrease in the sulfate content of the deep-sea cores.

Barophilic Sulfate-Reducing Bacteria.

Because of their importance as geochemical agents, special effort was devoted to the detection of sulfate-reducing bacteria in deep-sea sediment samples collected by the Galathea. Besides reducing sulfate to sulfide and oxidizing a great variety of organic compounds, certain sulfate reducers fix nitrogen (SISLER and ZOBELL, 1951), and utilize molecular hydrogen (ZOBELL, 1947; SISLER and ZOBELL, 1950). Sulfate-reducing bacteria, which commonly occur in near-shore sediments (ZOBELL and RITTENBERG, 1948), were also found by MORITA (1954) in several ocean basins. Populations exceeding 100 per gram of wet sediment were demonstrated in collections from a depth of 5032 meters at Mid-Pacific Station No. 36 (MORITA and ZOBELL, 1955).

As shown in Table III above, barophilic sulfate-reducing bacteria were demonstrated in Philippine Trench sediment samples from depths exceeding 10 000 meters. Such bacteria which grew at 700 atm were also demonstrated in bottom deposits from the Sunda Deep and the Weber Deep (see Table IV). Sulfate reducers from the Weber Deep have been maintained in pressure vessels at 3-5°C and 700 atm for more than 4 years and are still under observation at the Scripps Institution. Upon decompressing and opening the pressure vessel for the examination of some of the original sediment sample, a strong odor of hydrogen sulfide emanated, indicating activity of the organisms during storage. Small quantities of the mud were used to inoculate medium M-10-E in piston-stoppered tubes for enrichment in the refrigerator at 700 atm. After 60 days' incubation there was no evidence of sulfate reduction, but after 10 months sulfide had been produced in all tubes of medium M-10-E inoculated with this stored mud from the Weber Deep (7250 meters). The uninoculated controls were negative. There was no evidence of sulfate reduction in similar medium after 17 months' incubation at 1 atm in medium inoculated either with some of the original mud sample or with enrichment cultures that developed at 700 atm.

The obligate barophilic sulfate-reducing bacteria from the Weber Deep have been transplanted four successive times and kept growing at 3-5°C and 700 atm. With the cooperation of Dr. JAMES W. BARTHOLOMEW, electron micrographs were made. They show a small ovoid-shaped microbe about 0.3 microns in width by 0.5 to 0.8 microns in length, which differs in both shape and size from other known sulfate reducers commonly classified as *Desulfovibrio* species. Though highly pleomorphic, known *Desulfovibrio* species are usually comma-shaped organisms 0.5 to 1.0 by 3 to 5 microns in size. The barophilic sulfate reducer from the Weber Deep is believed to be a new genus, but further studies must precede its naming. Like most barophilic bacteria, it grows very slowly (ZOBELL and MORITA, 1957).

An Enigma in Bacterial Taxonomy.

In their unique ability to grow at high hydrostatic pressures, a good many of the bacteria isolated from the deep sea differ from known or described species whose habitat is soil, sewage, near-shore sediments, and other surface or shallow-water environments. However, pressure tolerance is probably not a sufficiently distinctive characteristic to delineate species, particularly since the deep-sea bacteria may have become acclimatized to high pressure while sinking from the surface, or their pressure tolerance may be associated with the composition of the medium.

Ordinarily, the taxonomic position of bacteria is based upon their physiological, cultural, and morphological characteristics as determined at 1 atm. Since some of these characteristics are altered and others are indeterminate at high pressure, we are confronted by an enigma in bacterial taxonomy. At least superficially a good many of the deep-sea bacteria appear to be new species or even new genera, but this can be established only by much more study.

The only deep-sea microbe to which we have applied a generic name is a small (0.4 micron) Gram-negative sphere, usually occurring singly but occasionally in pairs. It has exhibited no evidence of motility and does not form endospores. It liquefies gelatin and liberates ammonia from peptone in sea water medium in the refrigerator at 1000 atm. It oxidizes glucose with the formation of acid; sucrose, lactose, and starch are not attacked. In the refrigerator (3-5°C) it survives at 1 atm, but is killed in a few minutes when warmed to 25°C. It was isolated from two different mud samples from the Philippine Trench. Tentatively it has been named *Bathycoccus galathea*. The Greek suffix *bathy-* denotes great depth and *coccus* indicates that the cells are spherical.

Oceanographic Significance of Deep-Sea Bacteria.

What is the ecological and oceanographic significance of the bacterial populations found in oceanic deeps? Are they primarily passive inhabitants on the deep-sea floor or is their rate of metabolism and reproduction rapid enough for them to affect substantially chemical or biological conditions? Our observations prove only the presence of viable bacteria in some of the deepest parts of the ocean and show that these deep-sea bacteria are physiologically active in the laboratory under experimental conditions designed to duplicate deep-sea conditions, including high hydrostatic pressure and low temperature. Prerequisite to assessing the oceanographic significance of these deep-sea bacterial populations is much more information on the circulation of water masses, the amounts of organic matter reaching the deep-sea floor, the non-conservative chemical composition of the water and bottom deposits, abundance and growth rates of other organisms inhabiting the deeps, and more information on the characteristics of the bacteria.

An estimate of an average bacterial population of 10^6 per ml in the topmost cm of deep-sea sediment is believed to be conservatively low. Approximately this many were demonstrated in nutrient medium by the minimum dilution method, which does not detect all living microbes, since no one medium can provide for the nutrient requirements and essential environmental conditions for the growth of all bacteria. Moreover, groups of bacteria occurring in agglomerates or attached to solid particles register only as single individuals in the minimum dilution method. Extensive experience by bacteriologists who have been analyzing soil, sewage, milk, and similar materials for several decades shows that conventional cultural methods usually detect only about 10 per cent of the viable bacterial population. Actually bacterial populations of the order of 10^7 per ml of deep-sea sediment were indicated by the direct microscopic method, but the microscope fails to distinguish living from dead bacteria.

Taking 10^6 per ml as the order of magnitude of the bacterial population and considering these bacteria to contain an average of 2×10^{-13} gm of organic carbon per cell, it follows that 1.0 ml of such sediment would contain 2×10^{-7} gm of bacterial carbon, or 0.2 mg per liter. In a layer 1 cm thick there would be 2 mg of bacterial carbon per square meter.

Such a standing crop of bacteria may contribute substantially to the nutrition of benthic animals. Protozoans, worms, sponges, filter feeders, and mud eaters are among the animals known to ingest and digest bacteria (ZoBell and Feltham, 1938). Certain animals can survive and thrive on an exclusive diet of bacteria (ZoBell and Landon, 1937), and some can reduce the bacterial population of sea water to 10^2 to 10^3/ml. Bacteria consist largely of easily digestible proteins and lipids. Many are rich in vitamins and other accessory growth factors which may be beneficial to benthic fauna far removed from products of photosynthetic activity.

Kriss (1954) has stressed the importance of bacteria as food for marine animals and in the organic cycles in the sea. According to Kriss and Rukina (1952), the microbial biomass in marine sediments and overlying water column amounts to from 0.2 to 1.4 gm per square meter.

The extent to which bacteria contribute to the nutrition of deep-sea animals cannot be estimated from the standing crop or population of either, but must be based upon reproduction or growth rates. The population is only an expression of a dynamic balance between the rate of reproduction and the rate of death. Laboratory observations indicate that deep-sea bacteria reproduce (by transverse fission) once every 2 to 20 hours in nutrient medium incubated at 1000 atm and 3°C; somewhat slower in deep-sea sediment or sea water to which no nutrients have been added. They may be dormant *in situ,* as indeed many appear to be in the lower strata of cores from the Kermadec-Tonga Trench (Table VI), but at the mud-water interface they are probably reproducing. The low content of organic matter in deep-sea sediments and immediately overlying water must be attributable to bacterial activity, and there is pretty good evidence for the activity of sulfate-reducing bacteria. Should the rate of reproduction *in situ* be comparable to that found in the laboratory under conditions designed to simulate the *in situ* environment, we may think with confidence of an *in situ* reproduction rate which could provide a hundred or more bacterial "crops" per year, but other dynamic factors must be evaluated before concluding that this is a reasonable estimate.

The maximum amount of bacterial cell substance producible per unit of time is limited by the amount of available energy source. This is believed to consist largely of dissolved or colloidal organic matter reaching the bacteria. Chemosynthetic autotrophs may fix some carbon on the deep-sea floor, but such bacteria appear to play a very minor role. Of course, bacteria are highly efficient in recycling the organic waste products of animals, but there must be an influx of energy to keep the processes going. This energy, originating as fixed carbon in the photosynthetic zone, must reach the deep-sea floor primarily by settling or by the movements of water masses.

Ordinarily an average of one-third of the organic carbon assimilated by heterotrophic bacteria in dilute media is converted into bacterial cell substance, the other two-thirds being oxidized to CO_2. With similar efficiency a bacterial population of 10^6/ml representing a standing crop of 0.2 mg bacterial carbon per liter would require 0.6 mg organic carbon per liter and about the same amount of O_2. The complete oxidation of organic matter requires somewhat more than its own weight in O_2; 1.0 mg of various carbohydrates requiring 1.07 to 1.18 mg O_2, common fatty acids 1.06 to 2.97 mg O_2, triglycerides 2.40 to 2.92 mg O_2, amino acids and proteins 0.64 to 2.58 mg O_2/mg.

Replacing a bacterial population of 10^6/ml 100

times per year would require 60 mg of organic carbon per liter and somewhat more than this amount of O_2. This is much more than occurs in sea water at any one time, but throughout the year organic matter on the sea floor may be replenished by the movements of water masses and to a lesser extent by sedimentation; O_2 by diffusion and water movements. Apparently bacterial populations of the order of 10^6/ml occur only in the immediate vicinity of the mud-water interface, falling off rapidly with distance off the bottom. Only a few samples of sea water from depths exceeding 7000 meters and a meter or more off the bottom have been examined for the presence of bacteria, but all of these, like hundreds of water samples collected from depths exceeding 1000 meters, have bacterial populations of less than 10/ml. If this is representative of conditions throughout deep water masses, organic matter and O_2 from a water column of considerable thickness may be supplying the active biotic zone at the mud-water interface.

Besides oxidizing most kinds of organic matter with the consumption of O_2 and contributing to the nutrition of benthic fauna, bacteria may affect the non-conservative chemical composition of the deep-sea floor. The decomposition of organic matter probably results in the formation of complex marine humus and the liberation of phosphate, sulfate or sulfide, and nitrogenous compounds. Ammonifiers liberate ammonia from certain nitrogenous substances and nitrifiers may oxidize the ammonia to nitrite or nitrate. In the absence of free oxygen, sulfate reducers may produce hydrogen sulfide. If the latter diffuses into oxygenated waters, it may be oxidized with the formation of sulfate or sulfur. Though not detected on the Galathea Expedition, other bottom-dwelling bacteria may fix nitrogen or oxidize hydrogen or methane, which often result from the fermentation of organic compounds.

Bacteria in Water.

Although attention was concentrated on the microbial content of deep-sea sediments, several samples of sea water were examined for the presence of bacteria. Aseptic samples were collected in sterile evacuated bottles hermetically sealed with glass capillary tubes which could be sheared off at any desired depth to provide for the aspiration of water (ZoBell, 1941). Near-surface samples were collected in glass bottles tossed forward, with a hand line, off the bow of the Galathea in order to avoid contamination from the ship. Water samples from depths of 7000 to 8000 meters were collected in 300 ml collapsible rubber bulbs attached in properly spaced series to the hydrographic wire (Figure 4). At greater depths the elasticity of the rubber was adversely affected by the high pressure as has been described by ZoBell (1954).

From 12 July to 15 August 1951 a total of 44 surface water samples collected over the Philippine Trench were examined for bacteria by the minimum dilution method. In every one of these samples were detected from several hundred to a few thousand bacteria per ml, which grew in nutrient medium incubated at air temperature (ca. 30°C). There was no growth in similarly inoculated medium incubated at 1000 atm, and very little growth at 3-5°C, unlike the bacteria found in Philippine Trench sediments.

In organic-rich waters from Philippine Island bays and harbors visited by the Galathea were found

Table VIII. Minimum numbers of bacteria per ml of surface sea water collected on Galathea at different locations. The bacteria were demonstrated in nutrient medium incubated at air temperature and atmospheric pressure.

Total samples	General location of stations from which samples were collected	Date of collection	Bacteria per ml
2	Manila Bay	9-10 July	10^4 to 10^5
11	Over Philippine Trench	12-15 July	10^2 to 10^3
3	Tubajon Bay off Dinagat Island	17-19 July	10^3 to 10^4
12	Over Philippine Trench	21-24 July	10^2 to 10^3
2	Cebu Harbor	25 July	10^4 to 10^6
9	Over Philippine Trench	26-29 July	10^2 to 10^3
1	Kanlanuk Bay off Bucas Grande	30 July	10^4
2	Candos Bay off Lapinigan Island	31 July	10^4
8	Over Philippine Trench	2-8 August	10^2 to 10^3
6	Cebu Harbor	10-12 August	10^4 to 10^5
4	Over Philippine Trench	14-15 August	10^3

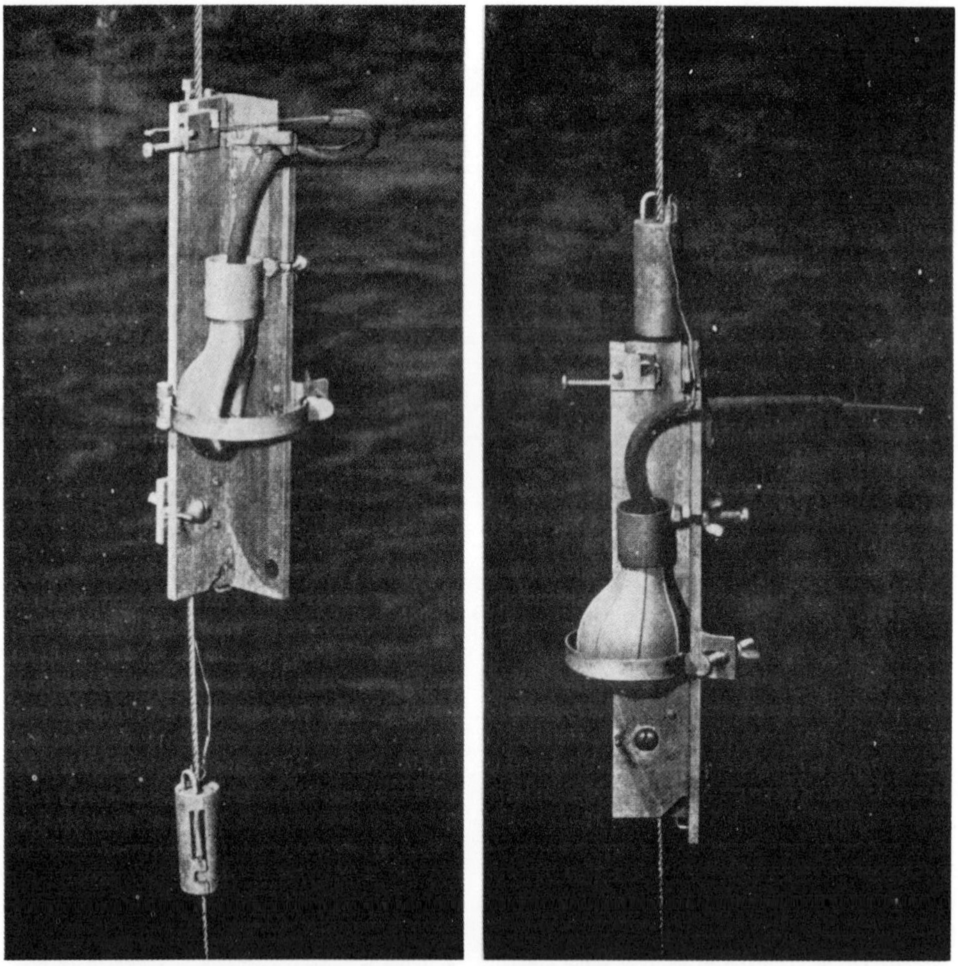

Fig. 4. Bacteriological sampler attached to hydrographic wire; at left the sterile rubber bulb is shown collapsed ready to be lowered into the water. Upon reaching the desired depth a messenger is dropped which shears off the capillary glass tubing thus permitting the thick-walled rubber tubing to flip out to the position shown at the right so water is aspirated from an area considerable distance from the apparatus in order to minimize self-contamination. Simultaneously the messenger suspended from the apparatus (left) is released, permitting it to drop for activating similar or other hydrographic apparatus at greater depth.

from 10 to 1000 times as many bacteria as in near-surface water collected over the Philippine Trench (Table VIII).

Throughout the photosynthetic zone over the Philippine Trench to a depth of 100 to 200 meters, the bacterial population ranged from 10^2 to 10^3 per ml. Not enough samples were analyzed to determine whether the vertical abundance of bacteria corresponded with vertical distribution of phytoplankton, as has been found to be the case in certain areas (ZoBELL, 1946). At depths exceeding 1000 meters fewer than 10 bacteria per ml were found. In northern Pacific Ocean deeps Kriss (1952) found few bacteria in water samples, but in bottom sediments he found from 10^6 to 10^8 bacteria per gram.

Acknowledgments.

The authors appreciate the financial support of the University of California Research Committee (Grant No. 1250), the U.S.A. Office of Naval Research (contract N6onr-27518), and The Rockefeller Foundation. It is also a pleasure to acknowledge the invaluable assistance of Dr. ANTON F. BBUUN together with all the officers, crew members, and scientists on the Galathea, who contributed in many ways to the success of the bacteriological investigations. Special thanks are due to Cmdr. TAGE FEDDERSEN, Lt. L. FERDINAND, Capt. SVEN GREVE, Cmdr. SIGURD BARFOED, TORBEN WOLFF, POUL JACOBSEN, ALF KIELERICH, and P. ANDREASEN for technical help. A special debt of gratitude is due the Royal Danish Navy and the people of Denmark who subsidized the Expedition.

REFERENCES

AMERICAN PUBLIC HEALTH ASSOCIATION, 1946: Standard Methods for the Examination of Water and Sewage. Amer. Pub. Health Assoc., New York City, 9th Ed., 286 pp.

ARRHENIUS, G., 1952: Sediment cores from the East Pacific. Part I. Properties of the sediment and their distribution. Rept. Swedish Deep-Sea Expedition, 5: 1-90.

BIEN, G. S., 1952: Marine Chemical Analytical Methods. Scripps Institution Oceanography, Progress Report 52-58: 1-9.

BUCH, K., and GRIPENBERG, S., 1932: Über den Einfluss des Wasserdruckes auf pH und das Kohlensäuregleichgewicht in grösseren Meerestiefen. Jour. du Conseil, 7: 233-245.

CERTES, A., 1884a.: Sur la culture, à l'abri des germes atmosphériques, des eaux et des sédiments rapportés par les expéditions du Travailleur et du Talisman; 1882-1883. Compt. rend. Acad. Sci., 98: 690-693.

– 1884b.: De l'action des hautes pressions sur les phénomènes de la putréfaction et sur la vitalité des micro-organismes d'eau douce et d'eau de mer. Compt. rend. Acad. Sci., 99: 385-388.

FISCHER, B., 1894: Die Bakterien des Meeres nach den Untersuchungen der Plankton-Expedition unter gleichzeitiger Berücksichtigung einiger älterer und neuerer Untersuchungen. Ergebnisse der Plankton-Expedition der Humboldt-Stiftung, 4: 1-83.

HALVORSEN, H. O., and ZIEGLER, N. R., 1933: Quantitative Bacteriology. Burgess Pub. Co., Minneapolis, Minn., 64 pp.

KRISS, A. E., 1952: Microbial life in the ocean depths. "Advances in Modern Biology." Moscow, pp. 194-218 (Russian article).

– 1954: The basic tasks of sea and ocean microbiology. Vestnik Akad. Nauk SSSR, 8: 22-34 (Russian article).

KRISS, A. E., and RUKINA, E. A., 1952: Microorganisms in the bottom deposits of ocean regions. Isvestiya Akad. Nauk SSSR, 6: 67-79 (Russian article).

KUENEN, P. H., 1950: Marine Geology. John Wiley & Sons, New York City, 568 pp.

MORITA, R. Y., 1954: Occurrence and significance of bacteria in marine sediments. Doctorate Dissertation, University of California, Los Angeles, 80 pp.

MORITA, R. Y., and ZOBELL, C. E., 1955: Occurrence of bacteria in pelagic sediments collected during the mid-Pacific Expedition. Deep-Sea Research, 3: 66-73.

OPPENHEIMER, C. H., and ZOBELL, C. E., 1952: The growth and viability of sixty-three species of marine bacteria as influenced by hydrostatic pressure. Jour. Mar. Res., 11: 10-18.

PRESCOTT, S. C., WINSLOW, C. A., and MCGRADY, M. H., 1946: Water Bacteriology. John Wiley & Sons, New York City, 368 pp.

SISLER, F. D., and ZOBELL, C. E., 1951: Nitrogen fixation by sulfate-reducing bacteria indicated by nitrogen/argon ratios. Science, 113: 511-512.

WISEMAN, J. D., and BENNETT, H., 1940: The distribution of organic carbon and nitrogen in sediment from the Arabian Sea. Sci. Rept. John Murray Expedition, 3: 160-200.

WOOD, E. J. F., 1956: Diatoms in the ocean deeps. Pacific Science, 10: 377-381.

ZOBELL, C. E., 1941: Apparatus for collecting water samples from different depths for bacteriological analysis. Jour. Mar. Res., 4: 173-188.

– 1946: Marine Microbiology. Chronica Botanica, Waltham, Mass., 240 pp.

– 1947: Microbial transformation of molecular hydrogen in marine sediments. Bull. Amer. Assoc. Petrol. Geol., 31: 1709-1751.

– 1952: Bacterial life at the bottom of the Philippine Trench. Science, 115: 507-508.

– 1952: Occurrence and importance of bacteria in Indonesian waters. Jour. Scient. Res. Indonesia, 4: 2-6.

– 1952: Dredging life from the bottom of the sea. Research Reviews, Office of Naval Research, October, pp. 14-20.

– 1953: Bakterier under tusind atmosfaerers tryk. Vor Viden, Copenhagen, No. 96: 563-568.

– 1954: Some effects of high hydrostatic pressure on apparatus observed on the Danish Galathea Deep-Sea Expedition. Deep-Sea Research, 2: 24-32.

ZOBELL, C. E., and BUDGE, K. M., 1954: Effect of high hydrostatic pressure on activity of marine nitrate-reducing bacteria. O.N.R. Progress Report, No. 7: 3-13.

ZOBELL, C. E., and CONN, J. E., 1940: Studies on the thermal sensitivity of marine bacteria. Jour. Bact., 40, 223-238.

ZOBELL, C. E., and FELTHAM, C. B., 1938: Bacteria as food for certain marine invertebrates. Jour. Mar. Res., 1: 312-327.

ZOBELL, C. E., and JOHNSON, F. H., 1949: The influence of hydrostatic pressure on the growth and viability of terrestrial and marine bacteria. Jour. Bact., 57: 179-189.

ZOBELL, C. E., and LANDON, W. A., 1937: Bacterial nutrition of the California mussel. Proc. Soc. Exper. Biol. & Med., 36: 607-609.

ZOBELL, C. E., and MORITA, R. Y., 1953: Dybhavets bakterieliv. Galatheas Jordomsejling 1950, under redaktion af A. F. Bruun, Sv. Greve, H. Mielche, R. Spärck, Copenhagen: 199-206. (Translated: Bacteria in the Deep Sea. The Galathea Deep Sea Expedition 1950-1952, described by Members of the Expedition, edited by Anton F. Bruun, Sv. Greve, Hakon Mielche and Ragnar Spärck, London 1956: 202-210).

– 1957: Barophilic bacteria in some deep sea sediments. Jour. Bact., 73: 563-568.

ZOBELL, C. E., and OPPENHEIMER, C. H., 1950: Some effects of hydrostatic pressure on the multiplication and morphology of marine bacteria. Jour. Bact., 60: 771-781.

ZOBELL, C. E., and RITTENBERG, S. C., 1948: Sulfate reducing bacteria in marine sediments. Jour. Mar. Res., 7: 602-617.

25

Reprinted from pp. 187–203 of *Oceanogr. Mar. Biol. Annu. Rev.*, **5** (1967), 653 pp.

EFFECTS OF HYDROSTATIC PRESSURE ON MARINE MICROORGANISMS*

RICHARD Y. MORITA

Departments of Microbiology and Oceanography,
Oregon State University, Corvallis, Oregon

Hydrostatic pressure, which is one of the main environmental parameters in the sea, is sorely neglected—mainly because of difficulties of the necessary instrumentation and the lack of qualified investigators interested in this field. It is, however, a most important parameter and must not be neglected if we are eventually to understand the relationship between hydrostatic pressure and life in the deep sea. Since this parameter is not an environmental factor for those working with terrestrial forms of life, most of the laboratory techniques, analyses, and so on, are not geared to its study. The study of hydrostatic pressure is an area of research which may be approached from a modern biological viewpoint and major contributions can be made concerning the effects of this parameter at the organism, and cellular, as well as molecular level. There are a multitude of questions which present themselves in relation to pressure when dealing with deep-sea organisms.

Conditions for the existence of life on the abyssal sea floor are discussed by Menzies (1965) but he does not consider in detail the intimate relationship of pressure and temperature to the physiological processes of life in the deep sea; the relationship between pressure and life in the deep sea still remains to be investigated. Menzies (1965) points out that publications concerning the ecology of the deep sea have generally minimized the role of pressure as a factor influencing the penetration of life into the deeps, while Bruun (1957) attributes the failure to recover living animals from the deep sea to temperature changes encountered by the animals and not to pressure changes or decompression; however, I hope in this paper to show that pressure should not be neglected and that it is an important factor in the life of organisms in the deep sea. Even pressures as great as those found in the hadal portions of the oceans do not exclude the existence of life as was first demonstrated by the Galathea Deep-Sea Expedition. Visual observations of life in the hadal portions have also been made by the bathyscaphe, TRIESTE.

All marine organisms, except those living at the surface, are subjected to varying degrees of hydrostatic pressure, which will depend upon the depth,

* Published as technical paper no. 2245, Oregon Agricultural Experiment Station.

TABLE I

Apparatus employed for the study of biological systems under hydrostatic pressure

Pressure apparatus	Reference	Remarks
Pressure centrifuge chamber	Brown, 1934	To measure relative consistency (gel strength) of gelated cortical protoplasm
Microscope pressure chamber	Marsland and Brown, 1936	To observe material under hydrostatic pressure
Pressure-temperature apparatus for luminescence studies	Brown, Johnson and Marsland, 1942	—
Pressure chamber for growing bacteria, incubating enzymes, and so on	Johnson and Lewin, 1946	—
Pressure centrifuge equipment	Marsland, 1950	Same as Brown's (1934) equipment except that temperature control was added
Pressure cylinder for growing bacteria, incubating enzymes, and so on	ZoBell and Oppenheimer, 1950	Improvement upon Johnson's model employing an O-ring seal, and so on
Pressure flask for studying plants under a few atm.	Ferling, 1957	Use of a mercury column to create hydrostatic pressures up to about 2 atm.
Optical pressure cylinder	Morita, 1957b	Employs optical white sapphire windows and a neutral piston to prevent the hydraulic fluid from contaminating the reaction mixture
Pulsating aeration under pressure	Hedén and Malmborg, 1961	Means by which small volumes of culture can be grown in continuous supply of air
pH apparatus	Distèche, 1959	Describes electrodes for measuring pH under pressure
Teflon and sapphire cell	Gill and Rummel, 1961	For optical absorption studies under high pressure
Pressure cell with electrodes for electrical discharges	Brandt, *et al.*, 1962	Discharges electrical current throughout aqueous system under pressure
Temperature controlled pressure cylinder	Suzuki and Suzuki, 1962	For protein denaturation studies and generally for pressures above 1000 atm.
Electrochemical measurements under hydrostatic pressure	Distèche, 1962	Buffers under pressure are discussed
Pressure cylinder to incubating systems under high temperature	Morita and Mathemeier, 1964	Employs a metal to metal seal
Pressure window design and packing of windows	Robertson, 1963	Description of various types of windows and housing construction, discussion of spectral measurements at high pressure
Optical rotation pressure cell	Rifkind and Applequist, 1964	For optical rotation studies up to 130 atm.
Pressure chamber for studying $CaCO_3$ solubility	Pytkowicz and Conners, 1964	Saturation of $CaCO_3$ is 2·7 times larger at 1000 atm. than at 1 atm., suggesting deep sea is undersaturated
Optical rotation cells	Gill and Glogovsky, 1964	Transmission of light for a given wavelength setting is measured
Pressure viscometer	Horne and Johnson, 1966	Method to measure the viscosity of fluids under pressure

temperature, and salinity. For biological purposes, a rough rule is to allow an increase of 1 atm for every 10 m water depth. The average depth of the ocean is estimated to be 3800 m which represents a hydrostatic pressure of approximately 380 atm, while at the bottom of the Challenger Deep (depth of 10,860 m) the hydrostatic pressure approximates to 1086 atm. The hydrostatic pressure at the bottom of the Challenger Deep is, in fact, greater than that indicated by the above rough rule because of the increased density of the sea water due to compression from the water above.

Different investigators employ various units to measure hydrostatic pressure. Unfortunately much of the work done on pressure by biologists in the U.S. has been reported in terms of pounds per square inch (psi). One atmosphere is equivalent to 14·696 psi, 1·033227 kg/cm^2, 1·01325 bars, 1·01325 dynes/cm^2 or 760 mm Hg at 0° C. It must be noted that hyperbaric studies should not be confused with hydrostatic pressure studies. Generally speaking research on hydrostatic pressure may be divided into two categories, namely, extremely high hydrostatic pressure and moderate hydrostatic pressure, the latter being concerned with pressures ranging from 1 atm to approximately 1000 atm. The pressures in the marine environment fall largely within the second category and this review will concern itself mainly with pressures in this range.

INSTRUMENTATION

Not too long ago Kalckar (1962) wrote that: "One of the fields of biology in which pressure studies are likely to make themselves felt is in the study of macromolecules. Pressure chambers with windows fitted to spectrophotometers, spectropolarimeters, and fluorometers will become available if they have not done so already." This statement only reflects the present instrumentation within the field. It is extremely difficult to design optical pressure cylinders for accurate biochemical or physiological measurements. Many of these difficulties are discussed by Robertson (1963), and so far no manufacturer has provided the biologist or biochemist with adequate hydrostatic pressure instruments. Because of this situation, individuals have designed and built various types of apparatus for pressure studies, and these are listed in Table I.

EARLY OBSERVATIONS ON HYDROSTATIC PRESSURE EFFECTS ON BIOLOGICAL SYSTEMS

Research on the effect of hydrostatic pressure on biological material began in the nineteenth century when organisms were dredged from the bottom (up to 6000 m depth) during the Talisman Expedition in 1882–1883 (Regnard, 1891). As a result of obtaining of organisms from the depths, research was conducted employing the Cailletet press to study the reaction of various organisms to pressure, the importance of the duration of pressurization, and the recovery time of pressure-treated organisms. Unfortunately, the importance of temperature in relation to pressure was not stressed and temperature was not measured or recorded.

When various types of material, such as eggs, milk, urine, blood, were subjected to pressure of 700 atm. they failed to undergo normal spoilage for periods from two to seven weeks (Regnard, 1884). Certes (1884a) showed that bacterial growth in various types of infusions was retarded when the suspension was subjected to 350 to 500 atm. Further work demonstrated that yeast fermentation was retarded by several hundred atmospheres of pressure (Regnard, 1884), and this has been subsequently verified by Morita (1965). The activities of various bacteria growing in nutrient media differed when the cultures were subjected to various pressures, according to Certes (1884a). Certes (1884b) also demonstrated that marine bacteria collected from depths of 5000 m were more resistant to pressure than terrestrial forms. No explanation for these observations was attempted.

EFFECTS OF HIGH PRESSURE

The effects of high hydrostatic pressure above the kilobar range (1000 atm) have been studied by various investigators. Although this is generally outside the range of pressure in the marine biosphere some results are of interest to this subject. When subjected to high hydrostatic pressure bacteria are generally killed (Roger, 1895; Chopin and Tammann, 1903; Hite, Giddings and Weakley, 1914; Larson, Hartzell and Diehl, 1918; Luyet, 1937a, b), viruses are generally inactivated (Basset, Lisbonne and Macheboeuf, 1933, 1935; Wollman, Macheboeuf, Bardach and Basset, 1936), and enzymes are completely or partially inactivated (Basset, Lisbonne, and Macheboeuf, 1933; Macheboeuf, Basset, and Levey, 1933; Macheboeuf and Basset, 1934; Matthews, Dow and Anderson 1940; Curl and Jansen, 1950a, b; Vignais, Barbu, Basset and Macheboeuf, 1951; Suzuki and Kitamura, 1963). The latest work on the denaturation of proteins above the kilobar range is being done by the Suzukis and their associates in Japan (Suzuki, K., 1960; Miyagawa and Suzuki, K., 1963a, b; Suzuki, C., Kitamura, Suzuki, K., and Osugi, 1963; Suzuki, C., and Suzuki, K., 1963; Suzuki, C., Suzuki, K., Kitamura and Osugi, 1963; Suzuki, K., and Kitamura, 1963; Suzuki, K., Miyosawa, Y. and Suzuki, C., 1963). The denaturation of proteins by various methods involves conformational changes of the proteins, but it should be remembered that the denaturation process involves different conformational changes depending upon the method employed to denature the protein. For instance, Suzuki and Suzuki (1962) have demonstrated that ovalbumin has an optical rotation of $-27{\cdot}6°$ in the native state, $-47{\cdot}9°$ when pressure denatured (9000 kg. cm^2, 2 hr, 30° C), $-100°$ when urea denatured, and $-60°$ when heat denatured (100° C for 30 min). Their interpretation of the results is that pressure denaturation may be due mainly to the α-helical structure rupture of ovalbumin, whereas urea denaturation is mainly due to rupture of the H bonds responsible for secondary structure. The high pressure denatured ovalbumin is fully precipitable in solubility tests and is fully hydrolyzable by proteolytic enzymes. The study of the action of high pressure may provide a clue to the action of moderate pressure on proteins of the cell.

Suzuki and Kitamura (1960) found that below 2000 atm, heat denaturation of haemoglobin is retarded by pressure, but above 2000 atm the denaturation process is accelerated. Retardation of heat denaturation by pressure is claimed to be due to a prevention of the unfolding of the protein molecule which involves a molecular volume change. Pressure is also thought to favour an increased ionization of the protein molecule. Enzymatically degraded protein has been reported by Bresler, Glikina, Selezneva and Finogeov (1952) to be resynthesized when treated with pressure in the presence of the proteolytic enzyme. The newly synthesized protein was found to have specific antigenicity analogous to the original protein and to possess some of its biological activity. Talwar and Macheboeuf (1954) were unable to reproduce these results.

When ribonucleic acid (RNA) was exposed to high pressure, Vignais and co-workers (1951) reported the synthesis of unstable polynucleotides. Degradation of RNA also occurred under pressure, both in the absence and presence of RNAase. Deoxyribonucleic acid (DNA) was found to be very stable under pressure. High pressure will retard its thermal denaturation (Hedén, Lindahl, and Toplin, 1964).

PRESSURE STUDIES ON ORGANISMS OTHER THAN BACTERIA

The effects of pressure on biological systems are manifold and pressure investigations dealing with organisms other than bacteria have been carried out by many investigators (Ebbecke, 1935, 1944; Cattell, 1936; Hardy and Bainbridge, 1951; Knight-Jones and Qasim, 1955; Digby, 1961; Rice, 1961; Entright, 1962, 1963; Vacquier, 1962; Schlieper, 1963). Many of these procedures are not useful in the study of the growth, metabolism, and function of microorganisms under pressure, but some of the investigations do have a bearing on any interpretation of the bacterial system.

Macdonald (1965), working with *Tetrahymena pyriformis*, confirmed the results of Fontaine (1929) on *Ulva lactuca*, that oxygen consumption was lowered when pressure was applied to the system. On the other hand Napora (1964) found that the prawn, *Systellaspis debilis*, increased its oxygen consumption as well as its phosphate excretion when subjected to 500 to 1500 psi. In field experiments these organisms increased their metabolic activity to compensate for the depressing effect of low temperature. The normal cleavage of *Arbacia* eggs, when exposed to 400 atm, is inhibited. Upon the release of pressure the eggs divide rapidly until they reach the same stage of cleavage as non-pressurized eggs, and they then continue to divide at the normal rate. It seems that nuclear division takes place without cell division when these cells are under pressure (Marsland, 1938).

It is recognized that cytoplasmic movement results from sol-gel reactions within cells (Marsland and Brown, 1936; Landau, Zimmerman and Marsland, 1954; Landau, 1959). When protoplasmic gels change from the sol to the gel state the process requires energy and is accompanied by molecular volume increase. Increased pressure favours the sol state and as a result amoeba under pressure do not form pseudopodia readily. It appears that the ATP system may be involved as the energy source (Marsland and Brown, 1942; Landau, Marsland and Zimmerman, 1955). Landau (1960), working with embryonic chick heart fibroblasts, has found that the maximal solation of the cytoplasmic gel occurs at a specific pressure level which, in turn, is dependent upon the temperature. This solation causes the loss of typical fibroblast shape and results in the formation of spherical cells. For each 5° C increment of temperature a proportional increase in hydrostatic pressure (750 psi) was needed for solation. From the data of Zimmerman (1963) it appears that in general DNA synthesis in *Arbacia* eggs is interrupted at some time between prophase and telophase when the eggs are subjected to increased pressure. Under pressure, fertilized eggs incorporate tritiated thymidine into DNA, and chromosomes progress through the DNA cycle although not at a normal rate. Further studies by Zimmerman and Marsland (1964) indicate that pressure brings about a drastic disorganization of the linear and radial structure of the spindle-aster complex and an irregular clumping of the chromosomes.

Rainford, Noguchi and Morales (1958) have studied the activity of myosin A and B upon addition of adenosine triphosphate and adenosine triphosphatase by measuring its "superprecipitation" (saturation of myosin by ATP) as indicated by turbidity. The rate of superprecipitation and the dephosphorylation of myosin B is halved under 500 atm. In the case of myosin A under the identical conditions, dephosphorylation is unaffected. These authors believe that the site of action is due to the pressure effect on the activated complex myosin B–Mg–ATP which, in contrast to the Mg–ATP complex, involves a rather large molecular volume. The activated complex, they believed, involved the release of electrostricted water.

PRESSURE EFFECT ON NON-BIOLOGICAL PARAMETERS

The non-biological parameters which affect organisms living in the depths of the ocean are also affected by hydrostatic pressure. Recognizing that there is an intimate relationship between the organism and the environment, the effect of pressure on these parameters should not be neglected if we are to understand the importance of hydrostatic pressure to life in the sea.

Since the early work of Buch and Gripenberg (1932), the effect of hydrostatic pressure on the pH of sea water has received considerable attention by Distèche (1959, 1962, 1964) and Ptykowicz and Conners (1964). Buch and Gripenberg (1932) found that increased pressure decreased the pH. The pH of sea water with a salinity of 35‰ at 0° C changed approximately −0·020 units at pH 8·5 for every 100 atm pressure. Distèche (1959), employing pH electrodes capable of withstanding high pressure, found that the pH of sea water decreases linearly with pressure at 0·3 pH units at 100 kg/cm^2. The intimate relationship between pH, alkalinity, and the solubility of calcium carbonate in sea water will be discussed by Pytkowicz in the next Volume of this Series. Employing pressure equipment, Pytkowicz and Conners (1964) have been able to demonstrate that the saturation of calcium carbonate is about 2·7 times larger after exposure to 1000 atm than it is when saturation is achieved at 1 atm. Their experiments suggest that the deep sea may be undersaturated with calcium carbonate.

Pressure also influences the ionization of water as well as other inorganic compounds. Generally, this ionization is increased as pressure increases (Owen and Brinkley, 1941). If pressure influences the ionization of various compounds, then it can be expected to influence the ionic volumes; these have recently been determined for the major constituents of sea by Duedall (1966). The fact that the ionization of water is greater at higher pressures, must be taken into account when dealing with macromolecules such as those present in the cell. The compressibility of water is another factor that must be taken into consideration. Much of the work on pressure effects on reaction rates and compression of the transition state dealing with inorganic compounds is above the kilobar range and marine chemists still have much to do with regard to the effects of pressure on chemical equilibria and reaction rates below the kilobar range.

OCCURRENCE OF BAROPHILIC BACTERIA

In 1949 ZoBell and Johnson proposed the term "barophilic" to describe bacteria that grow at pressures higher than 500 atm while the term "barotolerant" or "baroduric" was reserved for those organisms which were not injured by prolonged incubation under, or subjection to, high pressures. The occurrence of barophilic and baroduric bacteria in the hadal portions of the biosphere was observed by ZoBell (1952), ZoBell and Morita (1957, 1959), and Kriss (1963). More recently ZoBell and Morita (unpublished data) in 1964, have cultured bacteria from the Challenger Deep. Many bacteria are capable of growth only when subjected to the isobaric and isothermic conditions from which they are obtained (ZoBell and Morita, 1959). In addition, there are quite a few bacteria capable of growth at 1 atm. This latter type can be termed baroduric. Viable sulphate-reducing bacteria were found in sediment samples from the Sunda Deep and Weber Deep which had been kept at a pressure of 700 atm and a temperature of 3°–5° C for 4 years. This material was subcultured in the medium used for *Desulphovibrio* and incubated at 5° C and 700 atm for many months. No sulphate reduction was noted after 2 months incubation; however, after a prolonged incubation of 10 months sulphate reduction was detected. All inoculated controls incubated at 1 atm and 5° C were negative—even after several years of incubation; barophilic bacteria apparently grow very slowly. At the present time barophilic bacteria remain an academic curiosity and research on them awaits the development of instrumentation able to supply a constant level of air to the culture medium and rid the system of carbon dioxide or other waste gas if produced. Until such instrumentation is operational, large quantities of cells cannot be cultured for metabolic studies.

GROWTH OF BACTERIA AT HIGH PRESSURES

In all studies of the growth of bacteria at increased pressures, a pressure-temperature relationship can be demonstrated. This aspect will be discussed in more detail in the section (p. 197) dealing with molecular volume changes.

Johnson and Lewin (1946) working with *Escherichia coli* found that in the logarithmic growth phase it is slightly retarded by 68 atm at temperatures below the optimum for growth. Retardation of growth is much more pronounced at 300 atm. Incubating the culture at elevated temperatures and increased pressures does not bring about an increase in the amount of growth. ZoBell and Johnson (1949) demonstrated that pressures of 200–600 atm generally inhibited the growth of most freshwater and terrestrial bacteria, while Oppenheimer and ZoBell (1952) found that most marine bacteria, originally isolated from shallow marine sources, also failed to grow at pressures from 200–600 atm. In the latter study, a few of the bacteria were actually killed at elevated pressures, and filamentous growth often occurred. ZoBell and Johnson (1949) demonstrated that sudden compression and decompression did not injure the microorganisms. Since hydrostatic pressure and not gaseous pressure is involved, the compressibility of the fluid is not too great (about 3% at 1000 atm). Employing gas, Fraser (1951) subjected cultures to about 60 atm and released the pressure suddenly; this resulted in the rupture of up to 90% of the cells.

ZoBell and Oppenheimer (1950) have shown that *Serratia marinorubra* (now *Serratia marcescens*) grows forming long filaments when incubated under pressure, but no reproduction of the cells occurred. When the pressure was released, the long filaments divided into the normal length cells one would see at 1 atm. Pressure from 100–500 atm was found to retard growth and reproduction in *Escherichia coli*, although some growth was indicated by the formation of long filaments which showed little cell division (ZoBell and Cobet, 1962). As a result of compression, the lag phase of growth was retarded. The lethal effect of pressure was greater at the higher incubation temperature (40° C greater than 30° C, 30° C greater than 20° C). ZoBell and Cobet (1964) demonstrated that the biomass resulting from growth at elevated pressure had the same content of nucleic acid and protein as the biomass growth at 1 atm. At elevated pressures, more RNA and less DNA appeared per unit mass of cellular material.

ZoBell and Budge (1965) examined thirty marine bacteria for their ability to grow and reduce nitrate at increased pressure. Although these bacteria varied in their ability to grow at various pressures and reduce nitrate, all were found to be capable of growth and nitrate reduction at 300 atm. Inactivation of the nitrate-reducing system occurred on prolonged incubation at elevated pressures at 10° C with *Pseudomonas perfectomarinus*; at 10° C the amount of nitrate reduction with this organism decreased with increased pressure. Furthermore, at 1000 atm, the amount of nitrate reduction was greatly decreased. Working with an obligately psychrophilic marine bacterium, *Vibrio marinus* MP 1, Morita and Albright (1965) obtained better growth at 100, 200, and 300 atm than at 1 atm in 24 hours at 15° C, while at 800 atm, the organism died. At 3° C, better growth occurred at 100 and 200 atm in 24 hours. Bacterial spores are much less susceptible to the action of pressure than vegetative cells, and thermal inactivation of spores is also retarded by pressure (Johnson and ZoBell, 1949a, b).

The foregoing only illustrates that each species has its own pressure-temperature tolerance and, therefore, gross differences are to be expected in the response of various bacteria to hydrostatic pressure. Morita (1965) contends that studies on growth, reproduction, and death in bacteria under various pressures are perhaps the best type of experiments performed to date which reflect the sum of all factors affected by pressure.

PRESSURE EFFECTS ON BACTERIOPHAGE AND MUTATION

Retardation of thermal inactivation of several bacteriophages have been reported (Foster, Johnson and Miller, 1949). In bacteria infected with the *Escherichia coli* T series of phages, pressure treatment was found to diminish the burst size as well as to prolong the latent period (Foster and Johnson, 1951). In this situation, it was suggested that the pressure exerts its influence by preventing some template molecule(s) (probably DNA) to unfold: Foster and Johnson's (1951) results were confirmed by Rutberg (1964a). Inactivation of phage at constant pressure was found to be exponential with time (Rutberg, 1964b). Rutberg (1964c) has also shown that pressure has an inducing effect on *Escherichia coli* K, lysogenic for bacteriophage lambda, and at constant pressure, the number of induced bacteria is directly proportional to the time of exposure. This induction can be achieved with either actively growing or with starved cultures.

Palmer (1961) has reported a three-fold increase in the mutation rate to streptomycin resistance in *Serratia marinorubra* when exposed to 300 atm. Negative results were obtained at several different pressures. The frequency of the occurrence of morphological mutants of *Neurospora* is further retarded when nitrogen mustard is employed in conjunction with pressure than when nitrogen mustard is used alone (McElroy and Haba, 1949).

The effects of pressures greater than 1000 atm on DNA are reviewed by Hedén (1964).

EFFECT OF PREVIOUS PRESSURIZATION ON CELLS

When lyophilized cells of *Escherichia coli* were pressurized and then analysed for succinic dehydrogenase activity, Morita and ZoBell (1956) found that the amount of inactivation increased with the time of compression: half the enzyme system was inactivated after 4 hours at 600 atm and 30° C, and virtually all were irreversibly inactivated after 4 hours at 1000 atm at the same temperature. The inactivation effects of pressure were greater at temperatures above or below the temperature at which the organisms were grown. Morita (1957a) has subjected washed cells in phosphate buffer to various pressures and then determined the amount of ammonia produced from various substrates at 1 atm. Cells previously pressurized at 600 atm and 30° C for 3 hours, produced more ammonia from aspartic acid, alanine, and glutamic acid than cells held at 1 atm and 30° C for 3 hours. When serine, cysteine, and histidine were used as substrates, cells held at 1 atm produced more ammonia than cells held at 600 atm. It is not known what effect, directly or indirectly, pressure had on other enzymes, cell permeability, or transport systems of the cells. Although no explanation is given for the above results, it clearly demonstrates that pressure has an effect on the cells.

Berger (1959) reported that pressure affects the permeability of the cell, and Britten and McClure (1962) have stated that when cells of *Escherichia coli* are exposed to 1300 atm, a large part of the amino-acid pool leaks out of the cell.

PRESSURE EFFECTS ON METABOLIC PATHWAYS

Unfortunately, the effect of pressure on the various enzymes of the main metabolic pathways (glycolytic scheme, TCA cycle, and so on) have not been investigated to any great extent. So far, most of the attention has been centred round some of the various dehydrogenases of the TCA cycle largely because the cyclic operation of the TCA cycle coupled with the hydrogen and electron transport system is the main source of energy in aerobic organism. Pressure has been shown to reduce the rate of enzymatic action of a few of the dehydrogenases of the TCA system.

Morita (1957b) has studied the effect of pressure on the succinic and malic dehydrogenases in cells of *Escherichia coli*. This work involved the use of an optical pressure cell made to fit into a Beckman DU Spectrophotometer, and the reaction could be observed during the pressurization. It was found that pressures of 120, and 600 atm did not greatly affect the reaction rate of malic dehydrogenase but the level of activity between these pressures was not very great. When the pressure was increased to 1000 atm, however, a marked reduction in activity occurred. In the case of succinic dehydrogenase, increasing pressures brought about decreasing activities until at 1000 atm there was very little dehydrogenase activity.

Hill and Morita (1964) have observed that *Allomyces macrogynus* died under prolonged pressurization (600 atm for 72 hours) at 27° C. In order to by-pass any permeability factors of the cell, isolated mitochrondria were prepared and used as the source of the enzymes in these experiments. Succinic dehydrogenase activity did not appear to be affected when the mitochondria were subjected to 1200, and 400 atm; however, at 600 and 1000 atm very little activity could be noted. When α-ketoglutarate dehydrogenase was assayed, a slight progressive decrease in activity was noted with increasing pressures, but when the material was subjected to 1000 atm, no activity could be detected. Progressively decreasing activity with increasing pressures was noted with oxalosuccinic dehydrogenase, but activity was still detected at 1000 atm. Isocitric dehydrogenase gave a similar pattern as oxalosuccinic dehydrogenase, but in this case, very little activity was observed at 1000 atm. Recognizing that death of a cell is a complicated process, Hill and Morita (1964) postulated that one of the mechanisms causing death was the cessation of the cyclic activity of the TCA cycle, so depriving the organism of one of its key mechanisms for obtaining energy. Although the primary biochemical lesion may be at the site of the dehydrogenases (especially isocitric and succinic dehydrogenases) it is still not known whether pressure affects the hydrogen and electron transport mechanisms. This is an important question since energy is derived from electron transport. The above data cannot be explained in any simple way. Mitochondria are complex structures, and pressure may also influence the sol-gel relations of this system. In all likelihood, there are pressure-sensitive components of the mitochondria. Some of the findings may be explained as a result of the adverse effect of pressure on enzyme-substrate formation, since this complex involves a molecular volume increase. Hill (1962) could detect no difference in the electron micrographs of ultra-thin sections of mitochondria held at 1 atm or at elevated pressures.

PRESSURE EFFECTS ON PROTEINS

Molecular volume change, which can be considered as the increase or decrease in the volume of a molecule, is affected by both temperature and pressure. In general, molecular volume increases when the temperature is raised and decreases when the hydrostatic pressure is increased. Pressure can play a role in many physical and chemical properties as a result of its effect on the molecular volume change. The amount of pressure may influence the degree of hydration, ionization, electroconductivity, dipolar interaction, chemical reaction rates and water structure. Since pressures in the marine biosphere lie approximately between 1 and 1000 atm, most of the moderate pressure effects are on macromolecules such as proteins, which play a major role in maintaining and regulating life processes.

Kauzmann (1959) defines protein denaturation as a process or sequence of processes in which the spatial arrangement of the polypeptide chains within a molecule is changed from that typical of the native protein to a more disordered arrangement. This disorderly arrangement results in a volume change in the protein. Temperature increases, alkaline pH, and the presence of certain compounds (alcohol, urethane, sulphonamide, and so on) favour protein denaturation or enzyme inactivation (Johnson, Brown and Marsland, 1942; Johnson and Chase, 1942; Johnson, Eyring and Williams, 1942; Campbell and Johnson, 1946; Eyring, Johnson and Gensler, 1946; Johnson and Campbell, 1946; Johnson, Baylor and Fraser, 1948; Johnson, Kauzmann and Gensler, 1948). In many instances, moderate pressure often inhibits denaturation of proteins or enzyme inactivation. The amount of ionization, of zwitterion formation, solvent structure changes, weakening of bonds (hydrogen), and degree of ionization of the buffer system are all listed by Johnson, Eyring and Polissar (1954) as means by which pressure may influence the stability of proteins and the molecular volume change. Ionization of a protein would appear to favour protein stability since it would theoretically increase the attraction between areas of the protein. Ionization of the buffer system and solvent structure would naturally affect the ionization of the protein. Weakening of hydrogen bonds results in unfolding (molecular volume increase) of the molecule. A compilation of volume change values for amino acids, simple peptides, and commercially available enzymes in purified form are given by Johnson, Eyring and Polissar (1954).

The specificity of macromolecules (proteins, nucleic acids, enzymes, and so on) depend on how they are folded (Tanford, 1961); Eyring, Johnson and Gensler (1946) point out that a folded globular molecule cannot serve as a template in the asymmetric synthesis of one particular optical isomer over the other. In order to serve as a template the molecule must unfold. Since a volume change would probably occur in the unfolding process, pressure will affect this process. It takes only a little imagination to see how this could affect the growth, reproduction, and metabolic process of the cell. Although individual enzyme systems have been studied in relation to temperature-pressure relationships as well as to their volume change, the reason why various cells respond differently to pressure is still not in sight.

In enzyme systems a molecular volume increase results from the formation of enzyme-substrate complexes (Laidler, 1951). In general, the process by which the enzyme unfolds to accept the substrate and so form an enzyme-substrate complex involves a molecular volume increase which is opposed by pressure. The formation of the enzyme-substrate complex is usually the rate-limiting step in enzyme action under pressure.

The effect of temperature and pressure on the bioluminescence of marine bacteria has received considerable attention—mainly because it has proved to be an enzyme reaction that can readily be studied kinetically and the results analysed using the theory of absolute reaction rates (Johnson, Eyring and Polissar, 1954; Johnson, 1957). The effect of moderate pressure on the degree of bioluminescence depends on the bacterial species, the temperature, and the chemical composition of the medium on which the organism is grown (Johnson and Eyring, 1948). Brown, Johnson and Marsland (1942) have demonstrated that at temperatures above the optimum for bioluminescence, increased pressures give increased bioluminescence while below the optimum bioluminescence temperature, increased pressures decrease the amount of bioluminescence. Further studies (Johnson *et al.*, 1951) have shown that members of the carbamate series have an effect on the pressure relations of bioluminescence. At constant temperature, the lower members of the series (methyl and ethyl) reduced the inhibition of bioluminescence caused by pressure, while the intermediate members of the series had no effect on bioluminescence; the higher members of the carbamate series (*n*-hexyl and *n*-octyl) further increased the inhibition of bioluminescence resulting from the application of pressure.

Thermal inactivation of the bioluminescent system in *Photobacterium phosphoreum* at 34° C can be retarded by pressure up to 330 atm (Johnson and Eyring, 1948). The optimum for the system is 21° C. The action of certain disinfectants and drugs on bioluminescence is reversed by hydrostatic pressure, providing the temperature is near to the optimum. The theory of absolute reaction rates (Johnson, Eyring and Polissar, 1954; Johnson, 1957) as applied to biological phenomena has not been as successful when applied to biological systems other than bioluminescence (Lamanna and Mallette, 1959), and this condition may reflect deficiencies in the theory or experimental difficulties in making the required precise measurements. Berger (1958) has found that at increased pressure, the rate of hydrolysis of a phenylglycoside by a glycosidase is inhibited over the range of 20°–50° C at 1 to 1000 atm. Pressure was also found to decrease the rate of thermal denaturation; when, however, substrate was absent, the thermal denaturation was accelerated by the applied pressure.

The phosphatase activity of various marine bacteria has been studied by Morita and Howe (1957), who found wide variations from one species to another. Some marine bacteria increase their phosphatase activity with increasing pressure while others decrease their activity under these conditions.

No malic dehydrogenase activity was found to occur at 101°C at pressures from 1 atm to 700 atm (Morita and Haight, 1962), but activity was demonstrated when the pressure was raised above 700 atm. Using a temperature of 105° C, Morita and Mathemeier (1964) showed an inorganic pyrophosphatase activity under pressure. In the former case, malic dehydrogenase activity is lost at 1 atm at 78° C, and in the latter case, no activity of inorganic pyrophosphatase could be demonstrated at 100° C at 1 atm. These are extreme examples by which pressure counteracts the volume change due to temperature.

Washed cell suspensions of *Escherichia coli* were examined by Haight and Morita (1962) for the deamination of aspartic acid under various temperature-pressure conditions. At 37° and 45° C, increasing pressure produced progressive decreases in the amount of aspartase activity but, at 56° C, the least amount of aspartase activity occurred at 1 atm, whereas increasing pressure resulted in increasing activity. Pressure was also found to protect the enzyme system from thermal denaturation (56° C for 1 hour), but the most interesting feature was that substrate had to be present to prevent such thermal denaturation; this would suggest that the binding effect of enzyme-substrate complex was necessary for thermal protection. Studies by Morita and Mathemeier (1964) indicate that the inorganic pyrophosphatase-Mg complex is necessary for thermal protection rather than the enzyme and substrate, indicating that the enzyme-cofactor complex is the more stable form. Furthermore, Mathemeier and Morita (1964) have demonstrated that the enzyme-cofactor ratio is temperature dependent.

Since all of the above mentioned studies involve proteins obtained from organisms normally living at 1 atm, the question as to whether the proteins, protoplasmic gels, and cellular structures of deep-sea organisms differ from those organisms living near the surface of the sea is of considerable importance.

PRESSURE AS AN ECOLOGICAL PARAMETER

Menzies and Wilson (1961) state that pressure should receive more consideration as a factor in the recovery of organisms from the deep sea. This statement was based upon their field experiments in which they exposed shallow-water crabs and mussels to graduated depths (increased hydrostatic pressures) in the ocean. They found mussels would survive twice the pressure lethal to crabs (219 atm against 109 atm). (Controls for temperature were also employed so that pressure was the main factor in their survival.) The fact that various bacteria isolated from shallow-water environments respond differently to pressure, and that barophilic bacteria do not grow at 1 atm is a further reason for taking into consideration pressure as a factor in the vertical distribution of organisms in the sea.

Although a few hundred atmospheres of pressure may not bring about drastic changes in the molecular volume of cellular components and result in conformational changes, any change which affects each of the various macromolecules within the cell may be such as to unbalance the synchronous metabolic processes, so that either death of the cell results or at least a 'cessation of its growth and metabolism'. It must be stressed that biomass studies on the distribution of microorganisms in the ocean are not sufficient since they do not indicate whether or not the microorganisms are active in any given environment.

It is clear from the foregoing that pressure is to be regarded as an important parameter of the marine environment. So far relatively little has been done on marine bacteria themselves and much of the work lacks a cohesive approach. Far more integrated studies are required.

ACKNOWLEDGEMENTS

This paper was supported by U.S. Public Health Service grant AM 06752 of the National Institute of Arthritis and Metabolic Diseases, NSF grant GB 2472, and training grant 5 T1 GM 704–04 from the National Institute of General Medical Sciences.

REFERENCES

Basset, J., Lisbonne, J. and Macheboeuf, M. A., 1933. *C.r. hebd. Séanc. Acad. Sci., Paris*, **196**, 1540–1542.
Basset, J., Wollman, E., Macheboeuf, M. A. and Bardach, M., 1933. *C.r. hebd. Séanc. Acad. Sci., Paris*, **196**, 1138–1139.
Basset, J., Wollman, E., Machebouef, M. A. and Bardach, M., 1935. *C.r. hebd. Séanc. Acad. Sci., Paris*, **200**, 1247–1248.
Berger, L. R., 1958. *Biochim. biophys. Acta*, **30**, 522–528.
Berger, L. R., 1959. *Bact. Proc.*, 129 only.
Brandt, B., Edebo, L., Hedén C. G., Hjortzberg-Nordlund, B., Selin, I. and Tigerschiold, M., 1962. *TVF*, **33**, 222–229.
Bressler, C. E., Glikina, M. V., Selezneva, N. A. and Finogeov, P. A., 1952. *Biokhimiya*, **17**, 44–55.
Britten, R. J. and McClure, F. T., 1962. *Bact. Rev.*, **26**, 292–335.
Brown, D. E., 1934. *J. cell. comp. Physiol.*, **5**, 335–346.
Brown, D. E., Johnson, F. H. and Marsland, D. A., 1942. *J. cell. comp. Physiol.*, **20**, 151–168.
Bruun, A. F., 1957. *Mem. geol. Soc. Am.*, **67**, 641–672.
Buch, K. and Gripenberg, S., 1932. *J. Cons. perm. int. Explor. Mer*, **7**, 233–245.
Campbell, D. H. and Johnson, F. H., 1946. *J. Am. chem. Soc.*, **68**, 725 only.
Cattell, McK., 1936. *Biol. Rev.*, **11**, 441–476.
Certes, A., 1884a. *C.r. hebd. Séanc. Acad. Sci., Paris*, **99**, 385–388.
Certes, A., 1884b. *C.r. Soc. biol., Paris*, **36**, 220–222.
Chopin, G. W. and Tammann, G., 1903. *Z. Hyg.*, **45**, 171–204.
Curl, A. L. and Jansen, E. F., 1950a. *J. biol. Chem.*, **184**, 45–54.
Curl, A. L. and Jansen, E. F., 1950b. *J. biol. Chem.*, **185**, 713–724.
Digby, P. S. B., 1961. *Nature, Lond.*, **191**, 366–368.
Distèche, A., 1959. *Rev. scient. Instrum.*, **30**, 474–478.
Distèche, A., 1962. *J. electrochem. Soc.*, **109**, 1084–1092.
Distèche, A., 1964. *Bull. Inst. océanogr. Monaco*, **64**, No. 1320, 10 pp.
Duedall, I. W., 1966. *The partial equivalent volumes of salts in sea water*, M. S. Thesis, Oregon State University, Corvallis, 47 pp.
Ebbecke, U., 1935. *Pflügers Arch. ges. Physiol.*, **236**, 648–657.
Ebbecke, U., 1944. *Ergebn. Physiol.*, **45**, 34–183.
Eyring, H., Johnson, F. H. and Gensler, R. L., 1946. *J. phys. Chem.*, **50**, 453–464.
Entright, J. T., 1962. *Comp. Biochem. Physiol.*, **7**, 131–145.
Entright, J. T., 1963. *Limnol. Oceanogr.*, **8**, 382–387.
Ferling, E., 1957. *Planta*, **49**, 235–270.
Fontaine, M., 1929. *C.r. hebd. Séanc. Acad. Sci., Paris*, **189**, 647–649.
Foster, R. A. C., Johnson, F. H. and Miller, V. K., 1949. *J. gen. Physiol.*, **33**, 1–16.
Foster, R. A. C. and Johnson, F. H., 1951. *J. gen. Physiol.*, **34**, 529–550.
Fraser, D., 1951. *Nature, Lond.*, **167**, 33–34.

Gill, S. J. and Glogovsky, R. L., 1964. *Rev. scient. Instrum.*, **35**, 1281–1283.
Gill, S. J. and Rummel, S. D., 1961. *Rev. scient. Instrum.*, **32**, 752 only.
Haight, R. D. and Morita, R. Y., 1962. *J. Bact.*, **83**, 112–120.
Hardy, A. C. and Bainbridge, R., 1951. *Nature, Lond.*, **167**, 354–355.
Hedén, C.-G., 1964. *Bact. Rev.*, **28**, 14–29.
Hedén, C.-G. and Malmborg, A. S., 1961. *Sci. Rep. Inst. Super. Sanita*, **1**, 213–221.
Hedén, C.-G., Lindahl, T. and Toplin, I., 1964. *Acta Chem. scand.*, **16**, 1150–1156.
Hill, E. P., 1962. *Some effects of hydrostatic pressure on growth and mitochondria of* Allomyces macrogynus, Ph.D. Thesis, University of Nebraska, 50 pp.
Hill, E. P. and Morita, R. Y., 1964. *Limnol. Oceanogr.*, **9**, 243–248.
Hite, H., Giddings, N. J. and Weakley, C., 1914. *Bull. W. Va Univ. agric. Exp. Stn*, **146**, 1–67.
Horne, R. A. and Johnson, D. S., 1966. *J. Phys. Chem.*, **70**, 2182–2190.
Johnson, F. H., 1957. In, *Microbial Ecology*, edited by R. E. O. Williams and C. C. Spicer, Cambridge Univ. Press, London and New York. 134–167.
Johnson, F. H., Baylor, M. B. and Fraser, D., 1948. *Archs Biochem.*, **19**, 237–245.
Johnson, F. H., Brown, D. E. and Marsland, D. A., 1942. *Science, N.Y.*, **95**, 200–203.
Johnson, F. H. and Campbell, D. H., 1946. *J. biol. Chem.*, **163**, 689–698.
Johnson, F. H. and Chase, A. M., 1942. *J. cell. comp. Physiol.*, **19**, 151–161.
Johnson, F. H., Eyring, H. and Polissar, M. J., 1954. *The Kinetic Basis of Molecular Biology*, John Wiley and Sons, Inc., New York, 874 pp.
Johnson, F. H. and Eyring, H., 1948. *Ann. N.Y. Acad. Sci.*, **49**, 376–396.
Johnson, F. H., Eyring, H. and Williams, R. W., 1942. *J. cell. comp. Physiol.*, **20**, 274–268.
Johnson, F. H., Flagler, E. A., Simpson, R. and McGeer, K., 1951. *J. cell. comp. Physiol.*, **37**, 1–14.
Johnson, F. H., Kauzmann, W. J. and Gensler, R. L., 1948. *Archs Biochem.*, **19**, 229–236.
Johnson, F. H. and Lewin, I., 1946. *J. cell. comp. Physiol.*, **28**, 23–45.
Johnson, F. H. and ZoBell, C. E., 1949a. *J. Bact.*, **57**, 353–358.
Johnson, F. H. and ZoBell, C. E., 1949b. *J. Bact.*, **57**, 359–362.
Kalckar, H. M., 1962. In, *Horizons in Biochemistry*, edited by M. Kasha and B. Pullman, Academic Press, Inc., New York, 513–522.
Kauzmann, W., 1959. *Adv. Protein Chem.*, **14**, 1–63.
Knight-Jones, E. W. and Qasim, S. Z., 1955. *Nature, Lond.*, **175**, 941 only.
Kriss, A. E., 1963. *Marine Microbiology (Deep Sea)*, translated by J. M. Shewan and Z. Kabata, Oliver and Boyd, London, 536 pp.
Lamanna, C. and Mallette, M. F., 1959. *Basic Bacteriology*, Williams and Wilkins, Baltimore, Maryland, 2nd ed., 536 pp.
Laidler, K. J., 1951. *Archs Biochem.*, **30**, 226–236.
Larson, W. P., Hartzell, T. B. and Diehl, H. S., 1918. *J. infect. Dis.*, **22**, 271–279.
Landau, J. V., 1959. *Ann. N.Y. Acad. Sci.*, **78**, 487–500.
Landau, J. V., 1960. *Expl Cell Res.*, **21**, 78–87.
Landau, J. V., Marsland, D. A. and Zimmerman, A. M., 1955. *J. cell comp. Physiol.*, **45**, 309–329.
Landau, J. V., Zimmerman, A. M. and Marsland, D. A., 1954. *J. cell. comp. Physiol.*, **44**, 211–232.
Luyet, B., 1937a. *C.r. hebd. Séanc. Acad. Sci., Paris*, **204**, 1214–1215.
Luyet, B., 1937b. *C.r. hebd. Séanc. Acad. Sci., Paris*, **204**, 1506–1508.
Macdonald, A. G., 1965. *Expl Cell Res.*, **40**, 78–84.
Macheboeuf, M., Basset, J. and Levey, G., 1933. *Annls Physiol. Physicochim. biol.*, **9**, 713–722.
Macheboeuf, M. and Basset, J., 1934. *Ergebn. Enzymforsch.*, **3**, 303–308.
Marsland, D. A., 1938. *J. cell. comp. Physiol.*, **12**, 57–70.

Marsland, D. A., 1950. *J. cell. comp. Physiol.*, **36,** 205–227.
Marsland, D. A. and Brown, D. E. S., 1936. *J. cell. comp. Physiol.*, **8,** 167–178.
Marsland, D. A. and Brown, D. E. S., 1942. *J. cell. comp. Physiol.*, **20,** 295–305.
Mathemeier, P. F. and Morita, R. Y., 1964. *J. Bact.*, **88,** 1661–1666.
Matthews, J. E., Jr., Dow, R. B. and Anderson, A. K., 1940. *J. biol. Chem.*, **135,** 697–705.
McElroy, W. D. and de la Haba, G., 1949. *Science, N.Y.*, **110,** 640–642.
Menzies, R. J., 1965. *Oceanogr. Mar. Biol. Ann. Rev.*, **3,** 195–210.
Menzies, R. J. and Wilson, J. B., 1961. *Okios*, **12,** 302–309.
Miyagawa, K. and Suzuki, K., 1963a. *Rev. phys. Chem. Japan*, **32,** 43–50.
Miyagawa, K. and Suzuki, K., 1963b. *Rev. phys. Chem. Japan*, **32,** 51–56.
Morita, R. Y., 1957a. *J. Bact.*, **74,** 231–233.
Morita, R. Y., 1957b. *J. Bact.*, **74,** 251–255.
Morita, R. Y., 1965. In, *The Fungi*, edited by S. Ainsworth and A. Sussman, Academic Press, Inc. New York, **1,** 551–557.
Morita, R. Y. and Albright, L. J., 1965. *Can. J. Microbiol.*, **11,** 221–227.
Morita, R. Y. and Haight, R. D., 1962. *J. Bact.*, **83,** 1341–1346.
Morita, R. Y. and Howe, R. A., 1957. *Deep Sea Res.*, **4,** 254–258.
Morita, R. Y. and Mathemeier, P. F., 1964. *J. Bact.*, **88,** 1667–1671.
Morita, R. Y. and ZoBell, C. E., 1956. *J. Bact.*, **71,** 668–672.
Napora, T. A., 1964. *Proc. of Symposium on Experimental Mar. Biol.*, Occasional Publication 2, Graduate School Oceanogr. Univ. Rhode Island, 92–94.
Oppenheimer, C. H. and ZoBell, C. E., 1952. *Mar. Res.*, **11,** 10–18.
Owen, B. B. and Brinkley, S. R., 1941. *Chem. Rev.*, **29,** 461–473.
Palmer, F. E., 1961. *The effect of moderate hydrostatic pressure on the mutation rate of* Serratia marinorubra *to streptomycin resistance*, M.S. Thesis, Univ. Calif., La Jolla. 57 pp.
Pytkowicz, R. M. and Conners, D. N., 1964. *Science, N.Y.*, **144,** 840–841.
Rainford, P., Noguchi, H. and Morales, M., 1958. *Biochemistry*, **4,** 1958–1965.
Regnard, P., 1884. *C.r. Soc. Biol.*, **36,** 187–188.
Regnard, P., 1891. *Recherches experimentales sur les conditions physiques de la vie dans les eaux*, Libraire de l'Academie de Medecine, Paris, 500 pp.
Rice, A. L., 1961. *J. exp. Biol.*, **38,** 391–401.
Rifkind, J. and Applequist, J., 1964. *J. Am. chem. Soc.*, **86,** 4207–4208.
Robertson, W. W., 1963. In, *Techniques of Organic Chemistry*, edited by H. B. Jonassen and A. Weisberger, Interscience Publ., New York, **1,** 157–172.
Roger, H., 1895. *Arch. Physiol. Norm. Pathol.*, 5th ser., **7,** 12–17.
Rutberg, L., 1964a. *Acta path. microbiol. scand.*, **61,** 81–90.
Rutberg, L., 1964b. *Acta path. microbiol. scand.*, **61,** 91–97.
Rutberg, L., 1964b. *Acta path. microbiol. scand.*, **61,** 98–105.
Rutberg, L. and Hedén, C.-G., 1960. *Biochem. biophys. Res. Commun.*, **2,** 114–116.
Schlieper, C., 1963. *Veröff. Inst. Meeresforch. Bremerh.*, Sonderband, 31–48.
Suzuki, C., Kitamura, K., Suzuki, K. and Osugi, J., 1963. *Rev. phys. Chem. Japan*, **32,** 30–36.
Suzuki, C. and Suzuki, K., 1962. *J. Biochem., Tokyo*, **52,** 67–71.
Suzuki, C. and Suzuki, K., 1963. *Archs Biochem. Biophys.*, **102,** 367–372.
Suzuki, C., Suzuki, K., Kitamura, K. and Osugi, J., 1963. *Rev. phys. Chem. Japan*, **32,** 37–42.
Suzuki, K., 1960. *Rev. phys. Chem. Japan*, **29,** 91–98.
Suzuki, K. and Kitamura, K., 1960. *Rev. phys. Chem. Japan*, **29,** 81–85.
Suzuki, K. and Kitamura, K., 1963. *J. Biochem., Tokyo*, **54,** 214–219.
Suzuki, K., Miyosawa, Y. and Suzuki, C., 1963. *Archs Biochem. Biophys.*, **101,** 225–228.
Talwar, G. P. and Macheboeuf, M., 1954. *Annls Inst. Pasteur, Paris*, **86,** 169–179.

Tanford, C., 1961. *The Physical Chemistry of Macromolecules*, John Wiley and Sons, Inc., New York, 710 pp.
Vacquier, V. Jr., 1962. *Science, N.Y.*, **135**, 724–725.
Vignais, P., Barbu, E., Basset, J. and Macheboeuf, M., 1951. *C.r. hebd. Séanc. Acad. Sci., Paris*, **232**, 2364–2366.
Wollman, E., Macheboeuf, M., Bardach, M. and Basset, J., 1936. *C.r. Séanc. Soc. Biol.*, **123**, 588–592.
Zimmerman, A. M., 1963. *Expl Cell. Res.*, **31**, 39–51.
Zimmerman, A. M. and Marsland, D., 1964. *Expl Cell Res.*, **35**, 293–302.
ZoBell, C. E., 1952. *Science, N.Y.*, **115**, 507–508.
ZoBell, C. E. and Budge, K. M., 1965. *Limnol. Oceanogr.*, **10**, 207–214.
ZoBell, C. E. and Cobet, A. B., 1962. *J. Bact.*, **84**, 1228–1236.
ZoBell, C. E. and Cobet, A. B., 1964. *J. Bact.*, **87**, 710–719.
ZoBell, C. E. and Johnson, F. H., 1949. *J. Bact.*, **57**, 179–189.
ZoBell, C. E. and Morita, R. Y., 1957. *J. Bact.*, **73**, 563–568.
ZoBell, C. E. and Morita, R. Y., 1959. *Galathea Rep*, **1**, 139–194.
ZoBell, C. E. and Oppenheimer, C. H., 1950. *J. Bact.*, **60**, 771–781.

Part IV

MICROBIAL CYCLING OF ORGANIC MATTER

Editor's Comments on Papers 26 Through 29

Since the early 1930s, when Waksman and his co-workers noted the increase in numbers of bacteria in stored seawater (34) and the decomposition of plant and animal residues (35–37), microbiologists have been attempting to obtain some quantitative estimates of the rates and amount of bacterial activity in the oceans. The papers selected for inclusion in Part IV are all concerned with the rate of microbial mineralization or synthesis in the sea and sediments.

Emphasis on the mineralization of organic matter to carbon dioxide, inorganic nitrogen, and phosphate stem from the earlier recognition that in soils microorganisms are the primary agents for the recycling of these compounds. Hence, it was logical to look to marine bacteria to play the same roles in the oceans. Bacteria that could decompose agar were quickly isolated by Waksman, who also attempted to determine the rate of carbon dioxide evolution and ammonia production during the decomposition of zooplankton (35). During the course of these investigations, he noted the correlation between rate of decomposition and the available nitrogen levels (35). It should be noted here, too, that in all their experiments Waksman and his co-workers added milligram-per-liter quantities of organic matter and inorganic nitrogen. These lowered nutrient levels were found to affect favorably the increases in bacterial oxidation rates and the decomposition of glucose, and still cause an increase in bacterial numbers. Again, as discussed previously, the necessity for low nutrient levels as shown in the first third of this century was ignored until its rediscovery near the end of the 1960s.

A method, though, for quantitatively measuring microbial decomposition was still needed. Then, in 1962, Parsons and Strickland presented the procedures for heterotrophic uptake–mineralization (Paper 26). The authors presented the idea that the assimilation of very low concentrations of ^{14}C-labeled material by microbial populations could provide an index of the heterotrophic potential of a water mass in a manner analogous to that currently in use for photosynthesis and primary productivity measurements (32). In fact, if one makes several major assumptions, a double reciprocal plot can be drawn from the amount of isotope retained within the cells versus the concentration of the organic substrate added. From this it is frequently possible to predict the turnover time and natural concentration of the compound in question (41, 43). The ^{14}C-labeled heterotrophic uptake method was immediately seized upon by marine microbiologists who had been searching for just such a means of quantitating microbial activities in the seas. However, all too frequently the basic assumptions implicit in the method and especially in the kinetic approach to the data were ignored (33, 42).

Several of these assumptions are critical to an understanding of the procedure: the substrate added is at a saturating concentration but at a level that does not significantly affect the natural concentration; there is no significant change in substrate concentration during the time course of the experiment; there is no uptake and then release of dissolved radioactively labeled compounds (i.e., labeled lactic acid being excreted following the uptake of labeled glucose); there is no significant change in the level of enzyme(s) or numbers of bacteria during the experiment; there are no activators or inhibitors such as metal ions, planktonic grazers, or end products that might limit or stimulate the uptake; and representative samples of the test system have been used.

By 1969, it became apparent that a major correction needed to be applied to the heterotrophic uptake studies; we had to account for the respired $^{14}CO_2$, which was not a constant amount but depended on the substrate in use. Consequently, Hobbie and Crawford published Paper 27, in which they noted the differences in values obtained when one included this correction factor. The method has now been applied widely to the seas (1, 11, 24, 26, 27, 29, 38–40) and to a few studies on marine sediments (23, 40). Simply because of the problems of filtering sediments and the absorption of organic molecules on sediment particles, leading to poor recovery of unmetabolized labeled material, less work has been done on sediments with this system. Of necessity, application has been confined to the determination of the "mineralization rate," that is, the amount of CO_2 released.

Recently, though, Griffiths et al. (10) have reemphasized the errors introduced by both the filtration of the water samples and by the method used to stop the reaction. The potential release of radioactively labeled cellular material suggested by Wright (41) was found indeed to be a property of natural marine bacterial populations, especially as regards the glutamate pool (10), and hence a major error may be included when considering absolute values for the heterotrophic potential. Despite the problems, mostly due to an overextrapolation of data and attempts at the formation of absolute values, the determination of relative heterotrophic potential for water bodies and relative mineralization rates for sediments does provide a comparative basis for estimating changes in various ecosystems. It is a useful and convenient tool, but caution must be exercised in both the execution of the experiments and the interpretation of the results.

Since growth rates are an indication of bacterial activity, procedures have been sought that would permit us to estimate them in natural systems. In a series of papers starting in 1953, Jannasch (12–14) developed the concept of the chemostat as a laboratory model for natural aquatic systems, because the flow-through characteristics of the chemostat seemed to simulate certain aspects of the natural aquatic environment. In Paper 28, Jannasch presented the theoretical and mathematical basis for these studies. He shows how bacterial competition can be simulated in the chemostat, and accounts for the survival of "slow growers" by their greater efficiency of function at low nutrient concentrations. In a later paper, Jannasch showed that generation times of as long as 200 hours may be expected in natural populations (14). He has also cautioned against the direct comparison of chemostat studies with the natural situation (15).

Further refinements on the flask chemostat have been proposed, most notably by Caldwell and Hirsch (5). They have developed a two-dimensional steady-state diffusion gradient that provides for nutrient gradients inoculated with natural mixed populations of bacteria. Competition, differential growth rates, and environmental stress can thus be tested with this apparatus (5).

An *in situ* type of chemostat has also been devised to study algal population changes (8). Submerged culture chambers are continually exposed to fresh nutrients in the surrounding water, and population changes can be induced by adding substances to the attached reservoirs. Despite having been developed for algal studies, there is no reason why this method should not prove useful for assaying in bacterial systems, especially in estuarine or littoral zone studies. Another interesting approach to ecosystem, growth rate, decomposition studies is the membrane filter method of Kunicka-Goldfinger and Kunicka-Goldfinger (19). In this technique, a semiconstant source of fresh nutrients is available without using the complex chemostat apparatus; this allows one to measure decomposition rates of selected organic materials or bacterial growth rates in a simple fashion. Again, the use of this technique with natural, mixed aquatic populations has not been reported, but it would seem useful in studies of competition and selective degradation.

The problems of rationalizing bacterial degradative abilities and slow growth rates in nature have been attempted by Kuznetsov (20), who combined direct counting procedures, ^{14}C-labeled organic production by phytoplankton, utilization of the labeled compounds by bacteria, and the effects of zooplankton grazing. Although working with various lakes, he was able to estimate ratios of production to biomass for bacteria ranging from 0.15 for oligotrophic Lake Baikal to 1.35 for eutrophic lakes (20); other such integrated studies would contribute greatly to a better understanding of marine microbiology.

Some of the theoretical considerations of this problem were clearly elucidated by Brock (3). He discussed the advantages and disadvantages of autoradiography, colonial development on immersed slides, and the capillary tubes of Perfil'ev. In addition, he addressed himself to the question of the microbial microenvironment and the fact that microbial distributions in nature are not homogeneous but are concentrated in microenvironments, hence the theoretical problems of sampling, frequency, size, and applicability of steady-state conditions (chemostat) as imitators of the "real" world.

Despite years of study, two schools seem to be developing now regarding the contribution of bacteria to the oceans. One school regards bacterial involvement in the recycling of organic matter and as symbionts for algal growth as essential, extensive, and active. In contrast, the second school seems skeptical of microbial activities in the sea, feeling that at best they are inadequate for the job. The former school is represented by the relatively early work of Starr, Jones, and Martinez in Paper 29. They demonstrated the extracellular release by 10 marine bacteria of vitamin B_{12}-types of compounds into their culture medium. This medium could then support the growth of *Euglena gracilis*, Z strain. Prior to this, many workers had noted the requirement for this vitamin in pure algal culture studies (21, 22) and had determined that the vitamin and other biologically important substances existed in seawater (7, 9, 18, 25). A likely source for this series of vitamin B_{12}-type of chemicals, though, was shown by Starr and co-workers (30, 31) to be either excretion by bacteria or release from their cellular residues. Attempts to correlate algal growth rates, size, and growth patterns under laboratory conditions with the concentration of vitamin B_{12} as found in natural algal blooms were sufficiently successful that Carlucci and Silbernagel (6) could propose vitamin B_{12} as a major contributory factor to the growth of natural populations of *Cyclotella nana, Monochrysis lutheri*, and *Amphidinium carteri*. In addition, the existence of the B vitamins in the marine environment was shown by Burkholder and Burkholder (4) to be related to the bacterial population in sediments; they noted that the productivity of the Bahia Fosforescente, Puerto Rico, could be dependent on marine sedimentary bacteria, which seemed to be releasing large quantities of vitamin B_{12}, biotin, and niacin for the growth of the dinoflagellates.

Besides these biosynthetic activities, other workers have noted the degradative activities of marine bacteria. One aspect of this function has been described above in the heterotrophic potential of marine waters and sediments as measured by the uptake of labeled compounds. An objection to these studies is that they do not repeat the deep-sea pressures and temperatures in which the organisms are normally found. Work recently reported by Baross et al. (2), however, has shown that, in both pure culture and in mixed natural populations (personal communication), the rates of uptake and mineralization are drastically affected by testing at lower *in situ* temperatures and pressures. With more complex and less refined substrates, Seiburth and Dietz (28) showed that normal foods could be decomposed by organisms in Naraganssett Bay, but that it was a cooperative effort between the macroorganisms and the microorganisms.

The studies described above are in sharp contradiction to the other school of thought, which is developing based upon the report of Jannasch et al. (16), who found little decomposition of sandwiches, soup, and apples when the sunken deep submersible ship, the *Alvin*, was recovered from the bottom of the sea. The results reported by these workers have forced a rethinking and a reexamination of our approaches to microbial activities. Whether it will, indeed, be typical of what is happening in the ocean or whether special environmental circumstances could account for the relatively pure state of the foods upon their recovery, only much further investigation can resolve. In expanded studies, Jannasch and Wirsen (17, 40) have shown mineralization and heterotrophic uptake not to be completely inhibited, as may have been inferred from their previous paper, but they did note a substantial decrease in the rates of decomposition compared with values reported for coastal areas.

Thus, an exciting area involving major contributions to our understanding of bacterial processes is developing. The question to be answered in the near future is not how many or which kinds of microorganisms exist in the deep sea, but what are they doing there and how fast? If we are to understand the biology of the higher organisms that survive in the marine environment, we must also understand the synthetic and degradative role microorganisms may play in this cycle.

REFERENCES

1. Andrews, P., and Williams, P. J. LeB. 1971. Heterotrophic utilization of dissolved organic compounds in the sea. III. Measurement of the oxidation rates and concentration of glucose and amino acids in sea water. *J. Mar. Biol. Assoc. U.K. 51*:111–125.
2. Baross, J. A., Hanus, F. J., and Morita, R. Y. 1974. Effects of hydrostatic pressure on uracil uptake, ribonucleic acid synthesis, and growth of three obligately psychrophilic marine vibrios, *Vibrio alginolyticus* and *Escherichia coli*, pp. 180–203. In: R. R. Colwell and R. Y. Morita (eds.), *Effect of the Ocean Environment on Microbial Activities*. University Park Press, Baltimore, Md., 587 pp.
3. Brock, T. D. 1971. Microbial growth rates in nature. *Bacteriol. Rev. 35*:39–58.
4. Burkholder, P. R., and Burkholder, L. M. 1958. Studies on B vitamins in relation to productivity of the Bahia Fosforescente, Puerto Rico. *Bull. Mar. Sci. Gulf Caribbean 8*:202–223.
5. Caldwell, D. E., and Hirsch, P. 1973. Growth of microorganisms in two-dimension steady-state diffusion gradients. *Can. J. Microbiol. 19*:53–58.
6. Carlucci, A. F., and Silbernagel, S. B. 1969. Effect of vitamin concentrations on growth and development of vitamin-requiring algae. *J. Phycol. 5*:64–67.
7. Daisley, K. W., and Fisher, L. R. 1958. Vertical distribution of vitamin B_{12} in sea water. *J. Mar. Biol. Assoc. U.K. 37*:683–686.
8. De Noyelles, F., Jr., and O'Brien, W. J. 1974. The *in situ* chemostat—a self-contained continuous culturing and water sampling system. *Limnol. Oceanogr. 19*:326–331.

9. Droop, M. R. 1955. A suggested method for the assay of vitamin B_{12} in sea water. *J. Mar. Biol. Assoc. U.K. 34*:435–440.
10. Griffiths, R. P., Hanus, F. J., and Morita, R. Y. 1974. The effects of various water-sample treatments on the apparent uptake of glutamic acid by natural marine microbial populations. *Can. J. Microbiol. 20*:1261–1266.
11. Herbland, A. M., and Bois, J. F. 1974. Assimilation et mineralisation de la matière organique dissoute dans la mer: méthode par comptage en scintillation liquide. *Mar. Biol. 24*:203–212.
12. Jannasch, H. W. 1953. Zur Methodik der quantitativen Untersuchung von Bakterienkulturen in flüssigen Medien. *Arch. Mikrobiol. 31*:114–124.
13. Jannasch, H. W. 1965. Continuous culture in microbial ecology. *Lab. Pract. 14*:1162–1167.
14. Jannasch, H. W. 1969. Estimations of bacterial growth rates in natural waters. *J. Bacteriol. 99*:156–160.
15. Jannasch, H. W. 1974. Steady state and the chemostat in ecology. *Limnol. Oceanogr. 19*:716–720.
16. Jannasch, H. W., Eimhjellen, K., Wirsen, C. O., and Farmanfarmaian, A. 1971. Microbial degradation of organic matter in the deep sea. *Science 171*:672–675.
17. Jannasch, H. W., and Wirsen, C. O. 1973. Deep-sea microorganisms: *in situ* response to nutrient enrichment. *Science 180*:641–643.
18. Johnston, R. 1955. Biologically active compounds in the sea. *J. Mar. Biol. Assoc. U.K. 34*:185–195.
19. Kunicka-Goldfinger, W., and Kunicka-Goldfinger, W. J. H. 1972. Semi-continuous culture of bacteria on membrane filters. I. Use for the bioassay of inorganic and organic nutrients in aquatic environments. *Acta Microbiol. Pol. Ser. B., 4*:49–60.
20. Kuznetsov, S. I. 1958. A study of the size of bacterial populations and of organic matter formation due to photo- and chemosynthesis in water bodies of different types. *Verh. Int. Ver. Limnol. 13*:156–169.
21. Lewin, R. A. 1954. A marine *Stichococcus* sp. which requires vitamin B_{12} (cobalamin). *J. Gen. Microbiol. 10*:93–96.
22. Lucas, C. E. 1955. External metabolites in the sea. *Deep-Sea Res. Oceanogr. Abst. Suppl. 3*:139–148.
23. Martin, E. L., Litchfield, C. D., Vreeland, R., and Nakas, J. P. 1976. Microbial mineralization of urea in continental shelf sediments. In preparation.
24. Okutani, K., Okaichi, T., and Kitada, H. 1972. Mineralization activity and bacterial population in the Hiuchi-nada. *Bull. Jap. Soc. Sci. Fish. 38*:1041–1049.
25. Provasoli, L., and Pinter, I. J. 1953. Assay of vitamin B_{12} in sea water. *Proc. Soc. Protozool. 4*:10.
26. Seki, H. 1968. Relation between production and mineralization of organic matter in Aburatsubo Inlet, Japan. *J. Fish. Res. Board Can. 25*:625–637.
27. Seki, H., Wada, E., Koike, I., and Hattori, A. 1974. Evidence of high organotrophic potentiality of bacteria in the deep ocean. *Mar. Biol. 26*:1–4.
28. Sieburth, J. McN., and Dietz, A. S. 1974. Biodeterioration in the sea and its inhibition, pp. 318–327. In: R. R. Colwell and R. Y. Morita

(eds.), *Effect of the Ocean Environment on Microbial Activities.* University Park Press, Baltimore, Md., 587 pp.

29. Sorokin, Yu. I. 1970. Determination of the activity of heterotrophic microflora in the ocean using C^{14}-containing organic matter. *Microbiology USSR 39*:133–138.
30. Starr, T. J. 1956. Relative amounts of vitamin B_{12} in detritus from oceanic and estuarine environments near Sapelo Island, Georgia. *Ecology 37*:658–664.
31. Starr, T. J., Jones, M. E., and Martinez, D. 1957. The production of vitamin B_{12}-active substances by marine bacteria. *Limnol. Oceanogr. 2*:114–119.
32. Strickland, J. D. H., and Parsons, T. R. 1968. *A Practical Handbook of Seawater Analysis.* Fish. Res. Board Can. Bull. No. 167, 311 pp.
33. Thompson, B., and Hamilton, R. D. 1974. Some problems with heterotrophic uptake methodology, pp. 566–575. In: R. R. Colwell and R. Y. Morita (eds.), *Effect of the Ocean Environment on Microbial Activities.* University Park Press, Baltimore, Md., 587 pp.
34. Waksman, S. A., and Carey, C. L. 1935. Decomposition of organic matter in sea water by bacteria. I. Bacterial multiplication in stored sea water. *J. Bacteriol. 29*:531–543.
35. Waksman, S. A., and Carey, C. L. 1935. Decomposition of organic matter in sea water by bacteria. II. Influence of addition of organic substances upon bacterial activities. *J. Bacteriol. 29*:545–561.
36. Waksman, S. A., Carey, C. L., and Reuszer, H. W. 1933. Marine bacteria and their role in the cycle of life in the sea. I. Decomposition of marine plant and animal residues by bacteria. *Biol. Bull. 65*:57–79.
37. Waksman, S. A., and Hotchkiss, M. 1937–1938. On the oxidation of organic matter in marine sediments by bacteria. *J. Mar. Res. 1*:101–118.
38. Williams, P. J. LeB. 1970. Heterotrophic utilization of dissolved organic compounds in the sea. I. Size distribution of population and relationship between respiration and incorporation of growth substrates. *J. Mar. Biol. Assoc. U.K. 50*:859–870.
39. Williams, P. J. LeB., and Askew, C. 1968. A method of measuring the mineralization by micro-organisms of organic compounds in seawater. *Deep-Sea Res. Oceanogr. Abst. 15*:365–375.
40. Wirsen, C. O., and Jannasch, H. W. 1974. Microbial transformations of some ^{14}C-labeled substrates in coastal water and sediment. *Microb. Ecol. 1*:25–37.
41. Wright, R. T. 1973. Some difficulties in using ^{14}C-organic solutes to measure heterotrophic bacterial activity, pp. 193–217. In: H. L. Stevenson and R. R. Colwell (eds.), *Estuarine Microbial Ecology,* Belle W. Baruch Library in Marine Science, Vol. 1. University of South Carolina Press, Columbia, S.C., 536 pp.
42. Wright, R. T. 1974. Mineralization of organic solutes by heterotrophic bacteria, pp. 546–565. In: R. R. Colwell and R. Y. Morita (eds.), *Effect of the Ocean Environment on Microbial Activities.* University Park Press, Baltimore, Md., 587 pp.
43. Wright, R. T., and Hobbie, J. E. 1966. Use of glucose and acetate by bacteria and algae in aquatic eco-systems. *Ecology 47*:447–464.

26

Reprinted from *Deep-Sea Res. Oceanogr. Abst.*, **8**, 211–222 (1962)

On the production of particulate organic carbon by heterotrophic processes in sea water

T. R. PARSONS and J. D. H. STRICKLAND

(*Received* 10 *April* 1961)

Abstract—It is suggested that the amount of heterotrophic carbon assimilation by micro-organisms in the sea may be a significant fraction of the photosynthetic production, when considering the whole water column in a deep ocean.

A method is described, allied to the carbon-14 method for measuring photosynthesis, whereby the heterotrophic uptake of any chosen substrate in sea water can be measured using the substrate labelled by carbon-14. The method is rapid and convenient and can give values of 'relative-heterotrophic potential.' As with the radiochemical method for measuring marine photosynthesis, the exact interpretation of results presents many problems but the technique should prove a useful exploratory tool. Tests have been made using fully labelled acetate and glucose.

By a method of progressive radio-isotopic dilution it is possible to measure either the concentration or organic substrate in sea water or obtain information of the enzyme kinetics of the micro-organisms assimilating a given substrate. A description is given of this method. A few preliminary trials to test its applicability have been encouraging.

INTRODUCTION

MARINE primary productivity is usually thought of as photosynthetic production and, in terms of the planet as a whole, this definition is correct as the contribution of chemolithotrophic processes is negligible. However, if *non-phagotrophy* be used as a criterion to define a plant then the heterotrophic (chemo-organotrophic) growth of bacteria, eumycetes, protozoa and algae could well be classed as a form of ' primary productivity,' namely the first production of particulate food-stuff.

The amount of ' soluble ' organic matter in the sea, which would form the substrate for such heterotrophic growth, nearly always exceeds the amount of ' particulate ' organic material by a factor of ten or more; fifty or more if we consider only living cells. The total amount of soluble organic carbon is as great as 20,000 mg/m^3 at the surface (FOX, ISAACS and CORCORAN, 1952) although 2000–3000 mg C/m^3 is a more common figure (KAY, 1954; PLUNKETT and RAKESTRAW, 1955; SKOPINTSEV, 1959). DUURSMA (1960) has reviewed the literature and recently contributed what are probably the most reliable figures. He has shown that even in the deep ocean the amount of organic carbon is rarely less than 500 mg/m^3. A round figure of 1000 mg C/m^3 in the bulk of the ocean seems a reasonable approximation for rough calculations, with as much as five to ten times this amount occurring in the surface waters and near to the land.

A wide variety of dissolved organic compounds occurs in sea water, especially at coastal locations. There are the simple acids, formic, acetic, glycollic and lactic (CRÉAC'H, 1955b; KOYAMA and THOMPSON, 1959) and the Krebs-cycle-acids, malic and citric (CRÉAC'H, 1955a, 1955b). In deposits of the Weddell Sea the latter acid apparently reached a sufficient concentration to precipitate the calcium salt (BANNISTER and HEY, 1936) and doubtless most of the other tricarboxylic-acid-cycle compounds will eventually be detected. Carbohydrates and amino acids have been reported, mainly in polymer form (LEWIS and RAKESTRAW, 1955; JEFFREY and HOOD, 1958; WANGERSKY, 1959) and there are long chain acids (WILLIAMS, 1961). It is to be inferred from the literature that an appreciable fraction of the total organic carbon in the ocean is tied up as a fairly resistant humus-like material (SKOPINTSEV (1959) mentions a pectin-protein complex) so that in both fresh and salt water only 10 per cent or less of the organically combined carbon may be available as a substrate for micro-organisms (KEYS, *et al* 1935; PLUNKETT and RAKESTRAW 1955; KUSNEZOW, 1959).

In addition to marine bacteria the sea has been found to contain eumycetes (mold fungi, yeasts, etc.) and the recently described Krassilnikoviae (BARGHOORN and LINDER, 1944; ZOBELL, 1946; HÖHNK, 1956; VISHNIAC, 1956; WOOD, 1958; KRISS and MITZKEVICH, 1958; RITCHIE, 1959; HULBURT *et al.*, 1960; FELL, *et al.*, 1960, and references cited therein). KRISS and co-workers have plotted the distribution of heterotrophic micro-organisms at all depths in the ocean (KRISS and MARKIANOVICH, 1959; KRISS, 1960; KRISS, *et al.*, 1960) giving a remarkable picture of the abundance and wide distribution of micro-heterotrophs. In fresh water the biomass of such material is sometimes comparable with that of the phytoplankton (KUSNEZOW, 1954, 1959).

It must be remembered that many of the phytoplankters themselves are capable of heterotrophism. The ability to grow heterotrophically has been reported for species from all the main divisions of the algae, although only studied extensively in the Chlorophyta. There can be little doubt that in nature certain diatoms, coccolithophores and micro-flagellates are capable of prolonged growth in complete darkness and contribute significantly to the pelagic as well as to the benthic hetrotrophic populations (RODHE, 1955; BERNARD, 1948, 1958; WOOD, 1956a, 1959).

Recent work by JANNASCH and JONES (1959) has indicated that the number of bacterial cells in the sea is considerably higher than once supposed. From their work and the data given by VISHNIAC (1956), KRISS (1960) and KRISS *et al.* (1960), together with reasonable assumption as to the size and carbon content of marine micro-organisms, we estimate the mean standing crop of particulate carbon, in the form of micro-heterotrophs to be *about* 0·1 mg/m³ in the oceans of the world between 50 °N and 50 °S. Of course, large variations occur both horizontally and vertically. In the arctic regions amounts are less and there the organisms appear to be concentrated mainly in the top 200 m. Thus the dissolved organic matter of the seas, which has probably little value as a direct nutrient (KROGH, 1931; BOND, 1933) can be utilized for the sustenance of higher marine animals, even in the pelagic environment, by a food chain initiated by the heterotrophic production of micro-organisms. In the open ocean photosynthetic productivity is relatively low (STRICKLAND, (1960) for typical values). If one assumes a value of *ca.* 0·1 mg C/m³ for the standing crop of heterotrophs and a growth rate roughly comparable with that of the phytoplankers, then the heterotrophic production of organic carbon per m³ could be as much as 0·5–1 per cent of the photosynthetic production. When it is remembered that the depth of the total water column in the ocean will be fifty times or more that of the euphotic zone, it will be seen that the heterotrophic production beneath a unit area of the *open ocean* could well be of the same order as photosynthetic production. (By contrast the production of coastal areas, with large standing crops of phytoplankton in shallow water, will always be dominated by photosynthesis except perhaps in mid-winter in arctic and subarctic regions).

The production of much of the organic particulate matter in the ocean can therefore result from a cycle of heterotrophic processes, with the photosynthetic production of the upper layer acting as a replenishing mechanism for energy losses (OHLE, 1956; WOOD, 1956b). These heterotrophic cycles have the effect of considerably increasing the final utilization of any photosynthetic primary production.

The question of immediate importance to marine ecologists is the relative contribution of phototrophic and heterotrophic production *in the open ocean.* In the work described below we have outlined a new technique which may be of use for the rapid measurement of heterotrophic growth and substrate concentrations, at least on a relative scale. The technique was evaluated in coastal areas in winter, but should be applicable to most marine environments. It will shortly be applied by us to offshore areas in the northeast Pacific Ocean.

In common with terrestrial micro-organisms, marine bacteria can utilize a large variety of substrates. Although the versatility of any one species may be limited, there are probably organisms present in the sea which are capable of using any substrate which may exist in the sea. The more widely used substrates for bacteria and fungi are simple compounds such as ethanol, glycerol, acetate, lactate, glucose, succinate, citrate and the amino acids (BARGHOORN and LINDER, 1944; ZOBELL, 1946; MACLEOD *et al.*, 1954).

A similar variety of substrates can be used by algae, when growing heterotrophically, although there is generally considerable selectivity. For example, the ' acetate flagellates,' described in HUTNER and LWOFF (1951) grow well on acetate and some other acids but have little ability to use glucose. Conversely, *Chlorella pyrenoidosa* is only sustained in the dark by glucose and, less efficiently, by galactose and acetate, other carbon sources (hexoses, pentoses, acids and even sugar phosphates) having no effect (SAMEJIMA and MYERS, 1958). Glucose and, to a lesser extent, acetate can be used by most Chlorophyceae and cryptomonads, often called ' sugar flagellates ' (BRISTOL-ROACH, 1928; MYERS *et al.*, 1947; PRINGSHEIM, 1954; and SCHLEGEL, 1959). Glucose was the favoured substrate for the fresh water and marine diatoms studied by LEWIN (1953) and LEWIN and LEWIN (1960), with lactate and acetate being less widely employed. Many diatoms could be sustained by the simpler acids, but showed no growth. Useful data have been collected by SAUNDERS (1957).

The study of heterotrophic growth by the use of substrates labelled with carbon-14 would seem a logical development, using a technique in many ways analogous to the carbon-14 method for marine photosynthesis developed by STEEMANN NIELSEN (1952). Two difficulties arise, however, which are not encountered with the Steemann Nielsen method; neither the nature nor the amount of substrate is known in a sample of sea water. The latter difficulty can be overcome by adding so much substrate to a sample that the total amount present is virtually equal to the added amount, (i.e. the original amount can be neglected) The number of possible substrates in sea water is so great, however, that the addition of all of them, fully labelled with carbon-14, is impracticable. In the present preliminary investigations we have chosen uniformly labelled glucose and acetate, and labelled carbon dioxide.

MATERIALS AND METHODS

Radioactive solutions were obtained by adding suitable weights of inactive glucose or acetate to fully labelled stock sources of the radioactive chemicals, adding 5 per cent sodium chloride solution and packaging 2·0 ml aliquots into glass ampoules which were immediately sealed and sterilized as in Steemann Nielsen's carbon-14 method (Strickland, 1960).

The quantities in each ampoule were such that when all the 2·0 ml were added to a B.O.D.* bottle of seawater sample, assumed to be exactly 300 ml, the bottle contained about 5 micro-curies of activity with an added substrate concentration equivalent to exactly 250 mg C/m^3. This amount was sufficiently great, see later, for any naturally occuring substrate to be neglected. The radioactive carbon dioxide additions were made using 2·0 ml of a solution containing 25 micro-curies of activity with no added carbonate. The total carbonate was calculated from a knowledge of pH and alkalinity, as in photosynthesis work. A high carbonate activity was added in order to bring counting rates to a suitable level, because the total carbonate-carbon concentration was about one hundred times greater than that present in the acetate and glucose experiments. Amounts of 50 or 100 micro-curies of carbon-14 carbonate would have been better but these were not available.

Seawater samples were collected in scrubbed-clean 5-litre Van Dorn plastic bottles and filtered within minutes through a clean 150-micron nylon net (to remove any larger organisms) into a large glass flask for subsequent aliquotting into blackened B.O.D. bottles. The glass flask and bottles were cleaned before every experiment with chrome-sulphuric acid and subsequently rinsed copiously with freshly distilled water. (The use of polyethylene-ware and of alcohol for sterilization is to be discouraged because of possible contamination by organic matter). B.O.D. bottles were filled to the top with the well-shaken seawater sample and exactly 3 ml poured out of each bottle to make way for the radioactive solution and preservative. The bottles were allowed to come to a temperature similar to that of the sea from whence the samples were taken, by allowing them to stand for 15 to 30 minutes in a cooled water bath. The samples were then ' innoculated ' with radioactive substrate solution, incubated for a period not exceeding 4 hours in a water bath, 'killed' by adding 1 ml of neutral 40 per cent formalin and filtered through a HA Millipore filter.

The filters were washed well with 3 per cent sodium chloride solution or filtered sea water, dried and the radioactivity ' counted.' All details of the experimental technique were identical with those used for carbon-14 photosynthesis experiments (Strickland, 1960). Under these conditions, with a 3–4 hour incubation period, the resulting counts from filters were always greater than 200 c.p.m.** above ' blank ' values.

A suitable ' blank ' procedure for this method is difficult to devise as there is no equivalent to the ' dark blank ' used in carbon-14 photosynthesis experiments. When the stock radioactive solutions were added to a sample of sea water, followed *immediately* by the addition of formalin and filtration, an appreciable count above background was observed. This was roughly 150 c.p.m. in the case of glucose, 40 c.p.m. with acetate and 50 c.p.m. with carbon dioxide and appeared to be a true ' blank ' in each case. Evidently it was the result of some form of rapid adsorption by the membrane or organisms. We have not done sufficient work to determine just how constant such blank counts may be. If sample counts exceed 1000 c.p.m. any blank corrections would be negligible, considering the general precision of the method. However, for

*Biological oxygen demand.
**Counts per minute

smaller uptake values, some form of blank correction must be attempted, at least on a few selected samples.

The heterotrophic uptake of carbon in the form of the added substrate is given by the formula :

$$\text{mg C/m}^3\text{/hr} = \frac{c.f. (S + A)}{C \mu t} \tag{1}$$

where μ is the micro-curies of carbon-14 added in the form of labelled substrate, A the added concentration of substrate carbon, S the concentration of any of the same substrate carbon already present in the sea water sample (units of mg C/m^3), t the time in hours of the incubation, c the radioactivity of the filtered organisms (c.p.m.) and C the c.p.m. from 1 micro-curie of carbon-14 as a *weightless source* in the same geiger counter assembly period, f is a factor to compensate for the effect of any isotopic discrimination that may occur against the C^{14} isotope as compared with the normal C^{12} atom.

The value of f may vary and is not known. It can either be neglected or assumed to be the same as the factor used for the phototrophic assimilation of carbon dioxide, which is thought to be about 1·05 (STRICKLAND, 1960 for discussion).

RESULTS

Some idea of the magnitude of the uptake of glucose and acetate, as calculated from equation 1 with $(S + A)$ put equal to 250 mg C/m^3, is given in TABLE 1. The source activities stated by the manufacturer were assumed to be correct and the same calibration factor was used as employed in photosynthetic work. Errors thus incurred are considered to be insignificant for this preliminary work. The data are for winter water in a sea area just off the coast near Nanaimo, British Columbia, Canada (49° 12′ N, 123° 58′ W).

Table 1. *Some apparent heterotrophic carbon uptake rates, using acetate and glucose in samples of coastal waters*

Substrate carbon concentration 250 mg/m^3				
Date	*Temperature* (°C)	*Depth* (m)	*Substrate*	*Rate* (mg C/m^3 per hr)
28.10.60	10	5*	Acetate	0·02
	10	5*	Glucose	0·03
	9	35**	Acetate	0·010
	9	35**	Glucose	0·019
3.11.60	10	20	Acetate	0·018
	10	20	Glucose	0·027
9.11.60	9·5	35	Acetate	0·004
	9·5	35	Glucose	0·008
	8·5	175	Acetate	0·0031
	8·5	175	Glucose	0·007
2.12.60	7·5	5	Acetate	0·012
9.12.60	7	5	Acetate	0·010
21.12.60	7	7·5	Acetate	0·008
	7	10	Glucose	0·01

*Photosynthetic rate on same sample at approx. optimum light. 2 mg C/m^3/hr.
**Photosynthetic rate on same sample at approx. optimum light. 1·4 mg C/m^3/hr.

In general the rate of uptake of glucose-carbon was about 50–100 per cent greater than that of acetate-carbon by the same population (the glucose uptake rate was 30–50 per cent *less* than acetate on a substrate molecule basis). The uptake of carbon dioxide in the dark, as calculated from photosynthesis ' blank ' experiments at about the same time of year, was around 0·01–0·02 mg C/m^3/hr. and, where a direct comparison was made with glucose and acetate, the carbon dioxide-carbon uptake rate was found to be almost the same as that for glucose-carbon (about six times *greater* on a substrate molecule basis).

Counting rates in all cases were several hundred c.p.m. above background and often greater than one thousand c.p.m. Agreement was generally good between duplicates. There seems no doubt that a definite and reproducible uptake of substrate was being measured. The *apparent* rate of uptake increased regularly with the duration of experiments, which varied from 1 to 10 hr, in a manner suggestive of a population growing exponentially. At 9 °C the doubling time was about 15 hr.

In no experiment were the sides of a B.O.D. bottle wiped with a ' policeman.' Any population proliferating on the glass surfaces did not contribute materially to the observed rates which were thus better representative of the rates to be expected in the sea. It was thought wise to restrict incubation periods as much as possible and a time of 2–4 hr seemed to be suitable.

Temperature effects were very marked and necessitated temperature control to at least $\pm$ 1 °C during incubation. In one experiment, with acetate in early December, a good Arrhenius plot was obtained from 7·5 °C to 26 °C giving an activation energy of 27,500 calories, corresponding to a Q_{10} at 10 °C of 5·2. Later a Q_{10} of 4·3 was found and in mid-December the value was about 1·7. In each case different water masses and doubtless different populations were being observed and there seems to be no reason to suppose that any constant Q_{10} values can be assumed. The Q_{10} will likely be sufficiently high, however, to make temperature control essential. The temperature should be kept as close to that of the initial sample as is practicable (cf. the results on marine bacteria by JOHNSON, (1936).

KINETIC ASPECTS

If varying volumes of a solution containing active and inactive substrate are added to samples of sea water, each of the same volume, then :

$$\mu\, C = \alpha\, A$$

where α is a proportionality constant.

Let v be the velocity of substrate utilization and make t sufficiently small that v changes but little during the course of an incubation period (or changes in a reproducible manner). From simple isotopic dilution it follows that :

$$c = \frac{\beta\, v\, \mu\, C}{(S + A)} = \frac{\alpha\, \beta\, A\, v}{(S + A)} \tag{2}$$

where β is a proportionality constant depending upon the conditions of the experiment and radioactive counting equipment used.

The dependence of v on substrate concentration $(S + A)$ will be of the form :

$$v = \frac{k\,(S + A)}{K + (S + A)} \tag{3}$$

where k is a velocity constant, depending on the microbiological system and its concentration in the water sample, and K is a constant (units mg C/m^3) of the form of a Langmuir isotherm constant or a Michaelis-Menton constant (HINSHELWOOD, 1946; JOHNSON *et al*, 1954; and other texts on microbiology and biochemical kinetics for further treatment).

Combining equations 2 and 3

$$c = \frac{Z A}{K + (S + A)} \tag{4}$$

where $Z = \alpha \beta k$.

If $1/c = y$ and $1/A = x$ equation 4 reduces to :

$$y = \frac{(K + S) x}{Z} + \frac{1}{Z} \tag{5}$$

a linear function from which $(K + S)$ can be evaluated from the value of x when y is zero, i.e.:

$$(K + S) = -\frac{1}{x} \tag{6}$$

If $K \ll S$, the intercept on the abscissa enables one to calculate the amount of substrate carbon initially present in the sample of sea water. If S is zero or much less than K, the intercept is a (temperature and pH dependent) constant characteristic of some enzyme system in the micro-organisms present in the sample of sea water.

A few preliminary trials were made to determine whether equation 5 could be used to evaluate $(K + S)$. The contents of ampoules were diluted with various volumes of sterile 5 per cent sodium chloride solution and various aliquots added to B.O.D. bottles, filled from the same source of sea water. FIG. 1. shows the result of two experiments with acetate at different temperatures, using samples taken 3 days apart, and also the plot of an experiment using glucose. Substrate and activity were added so that A ranged from 500 mg C/m^3 to about 3 mg C/m^3. The values of $(K + S)$ for several experiments are collected in TABLE 2.

Table 2. Kinetic parameters

Day	*Depth* (m)	*Temperature of Incubation* (°C)	*Duration of Incubation* (hr)	$(K + S)$ mg C/m^3	*Remarks*
6.12.60	5	26	4	15·6	Acetate Curve 1 Fig. 1
9.12.60	5	7	8·5	14·7	Acetate Curve 2 Fig. 1
21.12.60	10	16	4	0·75	Glucose Curve 3 Fig. 1
22.12.60	7·5	17	4·5	3·0	Acetate
22.12.60	7·5	7	9	3·0	Acetate

The rate of uptake of acetate was unaffected by the addition of up to ten times as much glucose. It would appear that both the acetate and glucose fixation in a sample of water are independent. Whether or not they occur independently in the same organism was not proven.

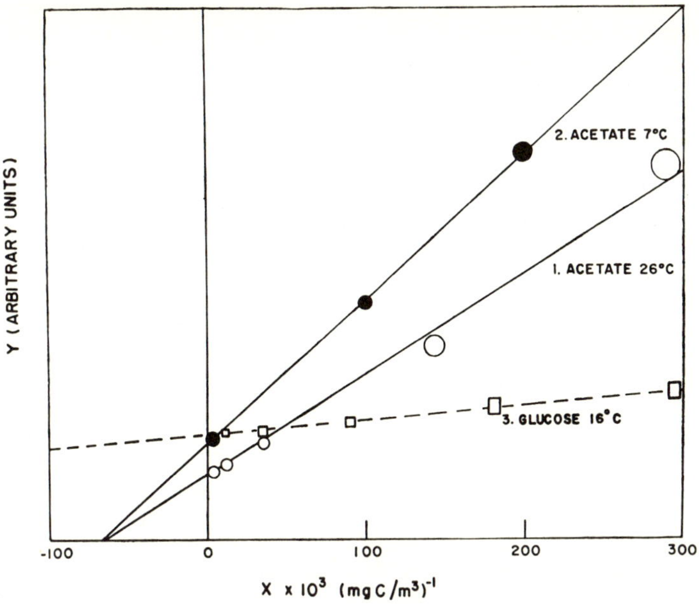

Fig. 1. Plot of Equation 5 with data for acetate and glucose.

DISCUSSION

Although the carbon-14 method of measuring marine photosynthesis constitutes a 'break-through' in the measurement of open ocean productivity it has been the cause of much controversy because of difficulty in interpreting results (STRICKLAND, 1960, for a review). The present complementary technique on heterotrophic growth measurement is even more difficult to interpret and it may be worthwhile to consider the possible ways in which glucose, acetate and carbon dioxide can be taken up by micro-organisms.

In all cases it is necessary to distinguish between maintenance and growth. An external substrate, and hence carbon-14, may enter the cell of a micro-organism and be used solely as fuel for the energy requirements of the cell. In this case the substrate is rapidly oxidized and lost. The radioactive carbon atoms may or may not be counted, depending on their residence time in the organism, but even if they are recorded, there will have been no net increase in cell carbon because a corresponding number of unlabelled atoms will have been lost, probably as carbon dioxide. Alternatively, a molecule of substrate may enter the cell and be incorporated permanently into the organism by synthesis into a major metabolic end-product. A corresponding amount of radio-activity is then fixed into the organism. The energy for this synthesis can be obtained by the (oxidative) breakdown of substrate originating either inside or outside the cell. The internal mechanism is unlikely, except when glutted cells are suddenly placed in a nutrient deficient environment. There would then be a gain of radioactivity but no net gain of carbon; a state of affairs which might loosely be termed ' exchange.' If the energy were obtained from an external substrate two possibilities can be envisaged. If the substrate used as an energy source were the same as the labelled substrate there would be a net increase of carbon in the cell but ' over-labelling ' might take place due to some of the labelled atoms in the energy-producing cycles being retained in the organism. If the substrate used were different from the labelled one (present in the sea water along with the latter) then a net increase of cellular carbon would result which could be at least as great as that measured by Equation 1, and probably greater.

It is clear from the present experiments that at least two substrates can be taken up simultaneously by the same marine population. Doubtless there are also other organic compounds in the water being attacked and hence the growth rate measured by any one labelled compound is probably a *minimum* estimate of the true net increase of cellular carbon. It is conceivable that only the ' basal metabolism ' of the micro-organisms is being satisfied and a form of ' exchange ' is occurring, but this is most unlikely to be a general rule.

The pathway of glucose assimilation by marine heterotrophs is straight-forward, in principle. Glucose enters early in the metabolic patterns and net assimilation could be relatively large.

Acetate is in a comparable oxidation state to glucose but its uptake mechanism is less certain. Acetate could combine with oxaloacetic acid, *via* the acetyl-CoA complex, to give citric acid which could then be used for protein synthesis or to give a supply of energy and reductant *via* the Krebs cycle. Alternatively, two molecules of acetate might combine by the Thunberg-Knoop condensation to give succinate, which could also be used either for cell synthesis or continue in the Krebs cycle to furnish energy. Recently MacLeod *et al.*, (1960) have shown a marine bacterium to possess a Krebs cycle shunt whereby acetate and isocitrate can produce succinate and malate directly. This provides a convenient mechanism for cell growth without oxidation losses in the tricarboxylic acid cycle.

Other possible mechanisms exist but growth on either glucose or acetate has in common the need for a supply of energy (ATP)* and reductant (DPNH or TPNH).** These can be obtained from oxidative or fermentative reactions of the two substrates themselves, but in the case of the fixation of carbon dioxide in the dark by marine heterotrophs the supply of energy and reductant must be obtained solely at the expense of organic matter inside or outside the cell. The numerous mechanisms that can be used by micro-organisms to fix carbon dioxide have been outlined in many texts (Werkman and Wilson, 1951; Hill and Whittingham, 1957). Possibly the ' malic enzyme ' reaction and the Ochoa reactions are the more common and the classical Wood-Werkman reaction itself may have little importance. In any case it seems to us that to refer to the Wood-Werkman and other mechanisms for the uptake of carbon dioxide as an ' exchange ', involving no true gain of cell carbon (Wood, 1956b; Steemann Nielsen, 1960) is over-extending the normal concept of isotopic exchange. The fractional assimilation of carbonate-carbon, even to an end product

*Adenosine triphosphate
**Reduced Di- or Tri-pyridine nucleotide

such as protein, can be quite high *in the presence of external organic substrates* (ABELSON *et al.*, 1952) and the heterotrophic uptake of carbon dioxide is, in this respect, no different in principle from the uptake of an organic molecule.

The present method enables one to obtain what we will term the ' relative heterotrophic potential ' of a water sample. This ' relative heterotrophic potential ' measured by one labelled substrate, is likely to be a *minimum* value for the true heterotrophic uptake of carbon provided that rates in the sea are not limited by the low concentrations of naturally occurring substrates.

It will be seen from TABLE 2 that values for $(K + S)$ and hence S alone are small compared with the added substrate concentration (A) of 250 mg C/m^3, and hence calculations from Equation 1, assuming $(S + A)$ equal to 250, are valid. However, the *observed* rates with 250 mg C/m^3 will only equal the actual *in situ* rates in the sea if the velocity in both cases is independent of substrate concentration, i.e. if K is much smaller than S and hence smaller than $(S + A)$ (ref. Equation 3). If this is the case, the values found in TABLE 2 are also the true substrate concentrations, and the technique forms a rapid and elegant method for the determination of certain marine trace organics.

If K is large compared with S the method has less value, and the ' relative heretotrophic potential ' data have only a qualitative significance. Nevertheless, they are still useful. An indication of the relative amounts of *viable* organisms and their biochemical characteristics can be obtained by a technique which is much less tedious than the conventional plating and counting methods. Furthermore, by the use of labelled amino acids, carbohydrates, etc. it should be possible to map the distribution of marine heterotrophs classified on the basis of their trophic habits.

It has been tacitly assumed that S in Equation 1 or Equation 5 is chemically identical with A. Although this seems most probable, it should be remembered that one cannot, by the present method, distinguish between A and any other compound which might compete directly with A in the rate-determining step that governs the uptake of this substrate (ref. constant k in Equation 3). As the rate-determining step is almost certainly the initial one in the metabolic pathway taken by A, only compounds very closely related biochemically (glucose and a glucose phosphate, etc.) would be expected to compete with each other. Glucose and acetate certainly did not.

One would suppose that with the substrate concentrations found in nature of 10^{-7} molar or less (TABLE 2) the natural uptake rate of a heterotroph in the marine environment would be affected and be very much lower than the rate measured in the presence of an excess of added substrate. This is expressed mathematically by Equation 3 when K is large when compared with S.

There is some evidence, however, that K may be small even when compared with S values as low as 10^{-7} M. Without speculating at this stage as to the exact significance of K, the constant may be assumed with some certainty to be temperature dependent and thus the value of $(K + S)$ would not be independent of temperature, as suggested by the results in TABLE 2, unless K were, in fact, appreciably smaller than S.

Values for a Michaelis–Menton type constant of less than 10^{-7} M have been reported but are uncommon. Yet it should be reasoned that marine micro-organisms must have evolved so as to utilize substrates at the concentrations occurring in nature, with possibly the mediation of an adsorptive concentration onto particulate matter. More work with a wider variety of samples and substrates is required to decide the value of Equation 5 for the calculation of natural substrate levels but the method looks promising.

Fisheries Research Board of Canada,
Pacific Oceanographic Group,
Nanaimo, British Columbia.

REFERENCES

ABELSON, P. H., BOLTON, E. T. and ALDONS, E. (1952) Utilization of carbon dioxide in the synthesis of proteins by *Escherichia coli*. *J. Biol. Chem.*, **198**, 165–172.

BANNISTER, F. A. and HEY, M. H. (1936) Report on some crystalline components of the Weddel Sea deposits. *Discovery Repts.*, **13**, 60–69.

BARGHOORN, E. S. and LINDER, D. H. (1944) Marine fungi, their taxonomy and biology. *Farlowia*, **1**, 395–467.

BERNARD, M. F. (1948) Récherches sur le cycle de *Coccolithus fragilis* Lohm, flagellé dominant des mers chaudes. *J. Cons., Cons. Perm. Int. Explor. Mer.*, **15**, 177–188.

BERNARD, M. F. (1958) Comparison de la fertillité élémentaire entre l'Atlantique tropical Africain, l'Océan Indien et la Méditerranée. *C. R., Acad. Sci., Paris*, **247**, 2045–2048.

BOND, R. M. (1933) A contribution to the study of the natural food cycle in aquatic environments. *Bull. Bingham Oceanogr. Coll.*, **4**, 1–89.

BRISTOL-ROACH, B. M. (1928) On the influence of light and of glucose on the growth of a soil alga. *Ann. Bot.*, **42**, 317–345.

CRÉAC'H, P. V. (1955a) Sur la présence des acides citrique et melique dans les eaux marines littorales. *C. R., Acad. Sci., Paris*, **240**, 2551–2553.

CRÉAC'H, P. V. (1955b) Quelques composants de la Matière organique de l'eau de mer littorale. Hélio-oxydation dans le milieu marin. *C. R., Acad. Sci., Paris*, **241**, 437–439.

DUURSMA, E. K. (1960) Dissolved organic carbon, nitrogen and phosphorus in the sea. Doctorate thesis, University of Amsterdam. *Netherlands J. Mar. Res.*, **1**, 1–148.

FOX, D. L., ISAACS, J. D. and CORCORAN, E. F. (1952) Marine leptopel, its recovery, measurement and distribution. *J. Mar. Res.*, **11**, 29–46.

FELL, J. W., AHEARN, D. G., MEYERS, S. P. and ROTH, F. J., Jr. (1960). Isolation of yeasts from Biscayne Bay, Florida, and adjacent benthic areas. *Limnol. and Oceanogr.*, **5**, 366–371.

HILL, R. and WHITTINGHAM, C. P. (1957) *Photosynthesis*. Methuen, London. 175 pp.

HINSHELWOOD, C. N. (1946) *The chemical kinetics of the bacterial cell*. Oxford University Press, London. 284 pp.

HÖHNK, W. (1956) Studien zur Brack- und Seewasser-mykologie. VI. Über die pilzliche Besiedlung verschieden salziger submerser Standorte. *Veröff. Inst. f. Meeresforschung, Bremerhaven*, **4**, 195–213.

HULBURT, E. M., RYTHER, J. H. and GUILLIARD, R. R. L. (1960) The phytoplankton of the Sargasso Sea off Bermuda. *J. Cons., Cons. Perm. Int. Explor. Mer.*, **151**, 115–127.

HUTNER, S. H. and PROVASOLI, L. (1951) The Phytoflagellates., In: *Biochemistry and Physiology of Protozoa*, A. LWOFF, Academic Press, New York. 27–128.

JANNASCH, H. W. and JONES, G. E. (1959) Bacterial populations in sea water as determined by different methods of enumeration. *Limnol. and Oceanogr.*, **4**, 128–139.

JEFFREY, L. M. and HOOD, D. W. (1958) Organic matter in sea water; an evaluation of various methods for isolation. *J. Mar. Res.* **17**, 247–271.

JOHNSON, F. H. (1936) The oxygen uptake of marine bacteria. *J. Bacteriol.* **31**, 547–556.

JOHNSON, F. H., EYRING, H. and POLISSAR, M. J. (1954) *The kinetic basis of molecular biology*. Wiley, New York. 874 pp.

KAY, H. (1954) Untersuchungen zur Menge und Verteilung der organischen substanz in Meerswasser. *Kiel. Meeresforch.*, **10**, 202–213.

KEYS, A., CHRISTENSEN, E. and KROGH, A. (1935) Organic metabolism of sea water with special reference to the ultimate food cycle in the sea. *J. Mar. Biol. Ass., U.K.*, **20**, 181–186.

KOYAMA, T. and THOMPSON, T. G. (1959) Organic acids of sea water. *Amer. Assoc. Adv. Sci., Int. Oceanogr. Congress, New York, Preprints of Abstracts*, 925–926.

KRISS, A. E. (1960) Micro-organisms as indicators of hydrological phenomena in seas and oceans. I. Methods. *Deep-Sea Res.*, **6**, 88–94.

KRISS, A. E., LEBEDEVA, M. N. and MITZKEVICH, I. N. (1960) Micro-organisms as indicators of hydrological phenomena in seas and oceans II. Investigation of the deep circulation of the Indian Ocean using microbiological methods. *Deep-Sea Res.*, **6**, 173–183.

KRISS, A. E. and MARKIANOVICH, E. M. (1959) On the utilization of water humus in the sea by micro-organisms. *Mikrobiologiia*, **28**, 399–406.

KRISS, A. E. and MITZKEVICH, I. N. (1958) " Krassilnikoviae." A new class of micro-organism found in the sea and ocean depths. *J. Gen. Microbiol.*, **19**, 1–12.

KROGH, A. (1931) Dissolved substances as food for aquatic organisms. *Biol. Rev.*, **6**, 412–442.

KUZNEZOW, S. L. (1954) Basic approaches to the study of correlation between the primary production of organic matter and the biogenous mass in a water reservoir. *Trudy, Problem. i. Temat. Sveshschanii, Akad. Nauk. SSSR*, **2**, 202–212 (C.A. **50**: 3679). 1956,

KUZNEZOW, S. L. (1959) *Die Rolle der Mikroorganismen im Stoffkreislauf der Seen.* VEB. Deutscher Verlag der Wissenschften, Berlin. 301 pp.

LEWIN, J. C. (1953) Heterotrophy in Diatoms. *J. Gen. Microbiol.* **9**, 305–313.

LEWIN, J. C. and LEWIN, R. A. (1960) Auxotrophy and heterotrophy in marine littoral diatoms. *Canad. J. Microbiol.*, **6**, 127–133.

LEWIS, G. L. JR. and RAKESTRAW, N. W. (1955) Carbohydrates in sea water. *J. Mar. Res.*, **14**, 253–258.

MACLEOD, R. A., HORI, A. and FOX, S. M. (1960) Nutrition and metabolism of marine bacteria. X. The glyoxylate cycle in a marine bacterium. *Canad. J. Microbiol.*, **6**, 639–644.

MACLEOD, R. A., ONOFREY, E. and NORRIS, M. E. (1954) Nutrition and metabolism of marine bacteria. I. Survey of nutritional requirements. *J. Bacteriol.*, **68**, 680–686.

MYERS, J., CRAMER, M. and JOHNSTON, J. (1947) Oxidative assimilation in relation to photosynthesis in *Chlorella. J. Gen. Physiol.*, **30**, 217–227.

OHLE, W. (1956) Bioactivity, production and energy utilization of lakes. *Limnol. and Oceanogr.*, **1**, 139–149.

PLUNKETT, M. A. and RAKESTRAW, N. W. (1955) Dissolved organic matter in the sea. Pap. Mar. Bid. Oceanogr., *Deep-Sea Res.*, Suppl. to Vol. **3**, 12–19.

PRINGSHEIM, E. G. (1954) Sugar flagellates. *Naturwiss.*, **41**, 380–381.

RITCHIE, D. (1959) Cultural characteristics as indicators of forces affecting geographical distribution of marine Imperfecti fungi. *Amer. Assoc. Adv. Sci., Int. Oceanogr. Congress, New York, Preprints of Abstracts*, 201–202.

RODHE, W. (1955) Can plankton productions proceed during winter darkness in subarctic lakes? *Ver. Int. Ver. Limnol.*, **12**, 117–119.

SAMEJIMA, H. and MYERS, J. (1958) On the heterotrophic growth of *Chlorella pyrenoidosa. J. Gen. Microbiol.*, **18**, 107–117.

SAUNDERS, G. W. (1957) Interrelation of dissolved organic matter and phytoplankton. *Botan. Rev.*, **23**, 389–410.

SCHLEGEL, H. G. (1959) Utilization of acetic acid by *Chorella* in light. *Z. Naturforsch.*, **14**b, 246–253 (C.A. 54 : 14370). 1960.

SKOPINTSEV, B. A. (1959) Organic matter of sea water. Amer. Assoc. Adv., Sci., *Int. Oceanogr. Congress, New York, Preprints, Abstracts*, 953.

STEEMANN, NIELSEN E. (1952) The use of radiactive carbon for measuring organic production in the sea. *J. Cons., Cons. Perm. Int. Explor. Mer.*, **18**, 117–140.

STEEMANN, NIELSEN E. (1960) Dark fixation of CO_2 and measurements of organic productivity, with remarks on chemo-synthesis. *Physiol. Plant.*, **13**, 348–357.

STRICKLAND, J. D. H. (1960) Measuring the production of marine phyto-plankton. *Bull. Fish. Res. Bd. Canad.*, **122**, 1–172.

VISHNIAC, H. (1956) On the ecology of the lower marine fungi. *Biol. Bull.*, **111**, 410–414.

WANGERSKY, P. J. (1959) Dissolved carbohydrates in Long Island Sound, 1956 1958. *Bull. Bingham Oceanogr. Coll.*, **17**, 87–94.

WERKMAN, C. H. and WILSON, P. W. (1951) *Bacterial Physiology*. Academic Press, New York. 707 pp.

WILLIAMS, P. M. (1961) Organic acids found in Pacific Ocean waters. *Nature Lond.* **189**, 219–220.

WOOD, E. J. F. (1956a) Diatoms in the ocean deeps. *Pacific Sci.*, **10**, 377–381.

WOOD, E. J. F. (1956b) Considerations on productivity. *J. Cons., Cons. Perm. Int. Explor. Mer.*, **21**, 280–283.

WOOD, E. J. F. (1958) The significance of marine microbiology. *Bacteriol. Rev.*, **22**, 1–19.

WOOD., E. J. F. (1959) Some aspects of marine microbiology. *J. Mar. Biol. Assoc., India*, **1**, 26–32.

ZOBELL, C. E. (1946) *Marine Microbiology.* Chronica Botanica, Waltham, Mass., U.S.A. 240 pp.

27

Reprinted from *Limnol. Oceanogr.*, **14**(4), 528–532 (1969)

RESPIRATION CORRECTIONS FOR BACTERIAL UPTAKE OF DISSOLVED ORGANIC COMPOUNDS IN NATURAL WATERS[1]

John E. Hobbie and Claude C. Crawford
Department of Zoology, North Carolina State University, Raleigh 27607

ABSTRACT

The uptake of ^{14}C-labeled organic compounds has been used by many workers to study heterotrophic microorganisms in natural waters. However, if flux rates of organic compounds are to be measured, the loss of $^{14}CO_2$ during incubation becomes an important source of error. A method is proposed in which the experiment is run in a closed system and the $^{14}CO_2$ collected after killing and acidification. Phenethylamine on chromatographic paper is the absorbing agent and the paper is counted by liquid scintillation. Studies of 19 compounds from pond water showed that 60% (aspartic acid) to 8% (arginine) of the labeled material entering the microorganisms was respired.

INTRODUCTION

Dissolved organic compounds exist in natural waters at such low concentrations (a few μg/liter) that only radioactive tracer techniques are sensitive enough to follow their flux. Aquatic bacteria appear to be chiefly responsible for the removal of these compounds from the water, and various modifications of the basic Parsons and Strickland (1963) technique have been developed to quantify the kinetics of uptake (Wright and Hobbie 1966; Vaccaro and Jannasch 1967). These methods all underestimate the total uptake because of loss of $^{14}CO_2$. Hamilton and Austin (1967) estimated that more than half of the labeled ^{14}C (as glucose) taken up by a culture of bacteria was immediately respired. Respiration loss in natural plankton was measured with an ion chamber and electrometer by Hobbie, Crawford, and Webb (1968), but the method allowed only six or seven experiments a day. The method presented here can be used with more than a hundred samples per day, yet is as accurate as ion chamber counting.

METHODS

The basic technique consists of adding different amounts of a labeled compound, for example, glucose-, acetate-, or amino acid-^{14}C, to each of a series of subsamples of natural water. After dark incubation on a shaking table at environmental temperature, the plankton is filtered onto membrane filters and the incorporated activity measured. In the technique proposed here, the incubation is made in a closed system and the $^{14}CO_2$ given off is absorbed by phenethylamine on the chromatographic paper which is kept for later liquid scintillation counting. This procedure was suggested by Smith (1967).

A 5-ml sample of natural water is placed in a 25-ml erlenmeyer flask and the substrate added with a micropipette. The flask is immediately sealed with a rubber serum stopper that has a plastic cup suspended from it, containing a 25- × 51-mm piece of accordion-folded chromatographic paper (Whatman No. 1). After incubation for an hour or more, 0.2 ml of a 2 N H_2SO_4 solution is injected through the septum. Next, still working through the septum, 0.2 ml of phenethylamine is slowly added to the folded paper and the flask then shaken for another hour at room temperature. The paper is immediately placed in a scintillation vial with 15 ml of toluene, 2,5-diphenyloxazole (PPO), and 1,4-2-(5-phenyloxazolyl)-benzene (POPOP) mixture. The plankton is filtered onto Millipore filters (0.45-μ pore size), rinsed with 10 ml of tap water, air-dried, and counted with the same scintillation mixture (Wolfe

---

[1] This work was aided by the Office of Water Resources Research, U.S. Department of the Interior, and by National Science Foundation Grant GB-5678. Contribution No. 5 from the Pamlico Marine Laboratory.

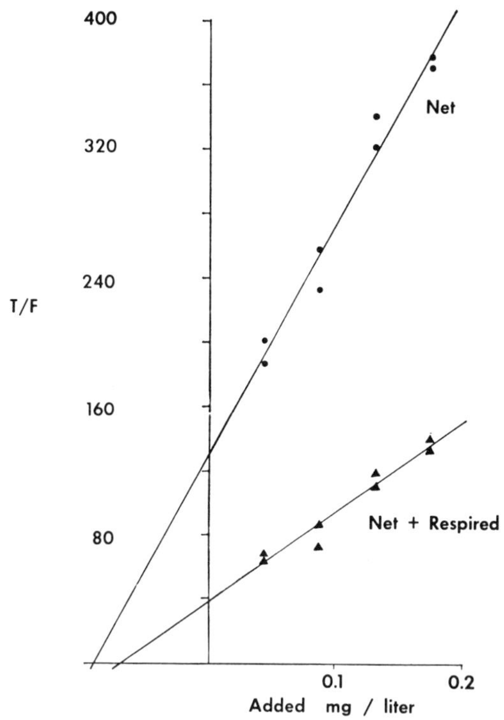

FIG. 1. Plot of the uptake of aspartic acid by bacteria according to a modified Michaelis-Menten equation (*see* text). In this plot, the units of the ordinate are hours. The upper line is calculated from isotope remaining in the bacteria (net) while the lower line is the net plus the amount of isotope respired. Sample collected from the Dairy Pond, Raleigh, N.C., 22 March 1968; incubation was at 10C.

and Schelske 1967). After counting both the plankton filter and the absorbing paper, an internal standard (10 μliter of toluene-^{14}C) is added, and all the vials recounted. The activity of incubated controls, consisting of natural water sterilized before addition of the isotope, is subtracted before calculations are made.

To determine the recovery efficiency of the $^{14}CO_2$, standard $Na_2{}^{14}CO_3$ solution (U.S. National Bureau of Standards) was injected into a *p*H 10 NaOH solution and treated as above. The absorbing paper was then wet-oxidized in a Van Slyke apparatus and the gas passed into an ion chamber for counting (Wetzel 1964).

Kinetic parameters are calculated from a modified Michaelis-Menten equation (*see* Wright and Hobbie 1966):

$$t/f = [(K+S)/V] + A/V.$$

The incubation time (t), divided by the fraction of the added activity taken up (f), is plotted against the amount of substrate added in the experiment (A). The maximum velocity of uptake (V) and the sum of the concentration of naturally occurring substrate (S) plus an uptake constant (K) can be calculated from such a plot. The intercept of the resulting straight line is $(K+S)/V$ or the time (T) required for the bacteria to remove an amount of substrate equal to S from solution (assuming equal rates of supply and removal).

RESULTS AND DISCUSSION

The plankton of a small eutrophic pond took up aspartic acid by transport systems whose kinetics can be described by the above equation. When the amount of substrate retained by the plankton (net uptake) was plotted, a V of 0.028 μg liter^{-1} hr^{-1}, a T of 131 hr, and a $(K+S)$ of 92 μg/liter were calculated (Fig. 1). When the activity represented by the respired $^{14}CO_2$ was added, V increased to 0.072 μg liter^{-1} hr^{-1}, T was reduced to 39 hr, and $(K+S)$ was reduced to 70 μg/liter. Theoretically, there should have been no change in $(K+S)$. Since such large changes of V and T were introduced, the respiration correction is certainly necessary if true uptake values are to be obtained.

Having established the importance of these corrections, we then tested the efficiency of the evolution, absorption, and counting of the $^{14}CO_2$. The gas phase counting system (Wetzel 1964) was calibrated by adding standard $Na_2{}^{14}CO_3$ solution, acidifying, and passing the evolved gas into the counting chambers. Some of the same solution was also added to the closed system and the $^{14}CO_2$ absorbed by the phenethylamine-moistened paper. The paper was then wet-oxidized in the gas phase system and 99% of the ^{14}C was recovered (2 runs). However, when parallel papers were prepared and counted with

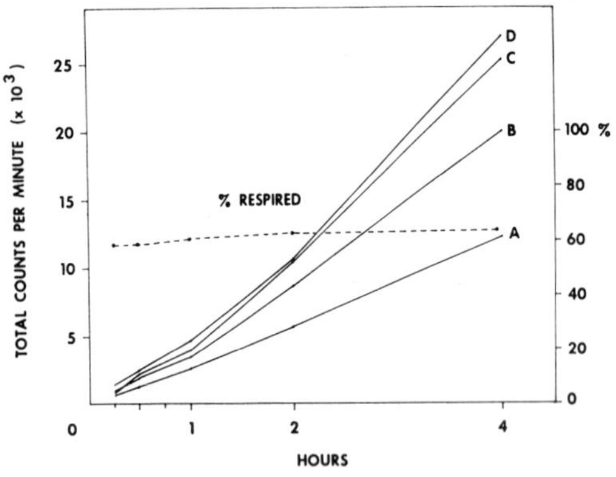

FIG. 2. Uptake of aspartic acid-U-^{14}C from pond water, Raleigh, N.C., 26 August 1968. Incubation temperature was 24C. The counts/min are the total of the activity remaining in the organisms after filtration plus the activity absorbed into phenethylamine on paper. The substrate concentrations in mg/liter were: A. 0.044; B. 0.087; C. 0.131; D. 0.175.

TABLE 1. *Kinetic analysis of the total uptake of aspartic acid-U over time. There were four substrate concentrations for each incubation period (0.044, 0.087, 0.131, 0.175 mg/liter). Sample from the Dairy Pond, Raleigh, N.C., 26 August 1968; incubation was at 24C*

Duration of incubation (hr)	V (mg liter^{-1} hr^{-1} $\times 10^{-3}$)	T (hr)	$K+S$ (mg/liter)	Resp. as % of total uptake	SD of resp. %
0.25	4.55	7.7	0.037	58.4	1.8
0.50	4.55	11.4	0.051	58.7	2.4
1.00	4.55	11.4	0.051	61.1	4.6
2.00	7.13	12.8	0.094	62.8	1.4
4.00	9.51	12.2	0.120	63.1	2.2

liquid scintillation (6 runs), only 82% of the ^{14}C was measured after correction with an internal standard. Because absorption approached 100%, it appears that there was masking or self-absorption of the activity by the paper itself. For this reason, all data reported here have been corrected by a factor of 1.23. This factor should be redetermined for differing procedures, counting machines, or absorption papers.

In any uptake studies of plankton, it is important to keep incubation time short and also to demonstrate that there is little change over time in the kinetic parameters of the plankton. Thus, it is important to study the timecourse of uptake at different substrate concentrations for each environment and substrate tested. Such a study of aspartic acid uptake in the same pond (Fig. 2) revealed fairly good linearity during the first hour but increases in uptake thereafter. The respiration rate of the substrate taken up changed only slightly during incubation. When the same data are plotted and the kinetic parameters calculated (Table 1), it appears that there is little change during the first hour in V and $(K+S)$, but a doubling of both of these occurs by 4 hr. There is little change in T because both V and $(K+S)$ have increased by the same factor. From these data, percentage respired appeared to be constant over the 4 hr, and a 1-hr incubation time would be best. Although it is difficult to prove, it seems that there is no immediate bottle effect on the planktonic bacteria. This may not be true in other situations, and a great sensitivity to bottles may occur in oligotrophic environments (H. Jannasch, personal communication).

Several different substrates were also tested to see if any patterns were detectable and if temperature had a large effect (Table 2). A temperature change of 15C had only minor effect on the percentage respired of glucose-6 and aspartic acid-U-^{14}C. There was, however, a large difference between the percentage of glucose-6 respired in March and in September and between the glucose-6 and the glucose-U-^{14}C. The latter can be explained by the greater tendency of the 6-carbon to remain in the cell in the Entner-Doudoroff pathway (Hamilton and Austin 1967). Acetate in solution is somewhat volatile and there is a possibility that some was picked up on the absorbing paper along with the $^{14}CO_2$.

The amino acids can be serially arranged with glutamic acid-U (61%) and aspartic acid-U (60%) having the highest respiration percentages. These two amino acids can enter the citric acid cycle through a single step involving a transaminase, so that data here agree with theory. Incidentally, Hobbie et al. (1968) also reported

TABLE 2. *Percentage respired of the total uptake of various substrates. All samples from the Dairy Pond, Raleigh, N.C.; counted by liquid scintillation*

Date	Compound	Samples	Concn (mg/liter)	Avg % respired of total uptake	SD	Temp (°C)
22 Mar						
	Glucose-6	8	variable	23.9	4.35	10
	Glucose-6	8	variable	24.2	3.59	25
	Aspartic acid-U	8	variable	58.3	4.94	10
	Aspartic acid-U	8	variable	56.7	8.43	25
12 Sep						
	Glucose-6	3	0.033	18.1	0.616	24
	Glucose-U	3	0.007	31.6	8.28	24
	Acetate-U	3	0.018	38.1	2.50	24
	Glutamic acid-U	3	0.196	61.3	0.79	24
	Aspartic acid-U	3	0.218	60.0	1.16	24
	Methionine-U	3	0.042	39.9	3.15	24
	Serine-U	3	0.143	31.7	1.82	24
	Tyrosine-U	3	0.329	29.7	1.55	24
	Glycine-U	3	0.006	28.0	0.70	24
	Proline-U	3	0.140	27.6	2.28	24
	Isoleucine-U	3	0.150	26.7	5.89	24
	Phenylalanine-U	3	0.237	26.7	4.33	24
	Alanine-U	3	0.041	26.3	1.09	24
	Threonine-U	3	0.222	26.2	6.94	24
	Leucine-U	3	0.197	25.6	0.14	24
	Valine-U	3	0.169	25.1	3.42	24
	Lysine-U	1	0.240	11.6	—	24
	Arginine-U	3	0.406	8.4	6.14	24
3 Oct						
	Aspartic acid-U	8	variable	46.8	7.80	24

that 60% of the aspartic acid total uptake was lost as CO_2 for a sample of estuarine water examined with the gas phase counting method.

The next most active amino acid is methionine-U (40%), and then 10 amino acids were found with rates between 32 and 25%. Finally, lysine-U (12%) and arginine-U (8%) had the lowest percentages; these two can be metabolized by passing through many intermediate steps. It is possible that they are being used mainly for protein synthesis.

The respiration of sodium-glycollate-1-^{14}C averaged 49% over a series of substrate concentrations. However, the controls for this one experiment gave unexpectedly high values for the absorbed $^{14}CO_2$. Glycollate added to different types of water gave the following counts/min for the filter paper and absorbing paper, respectively: added to distilled water, 62 and 83; added to sterile pond water, 82 and 444; and added to natural pond water, 654 and 1,015. All incubations were for 2 hr at 24C and sterilization was with 0.1 ml of a 2 N H_2SO_4 and 5% $HgCl_2$ solution. This loss of ^{14}C may be caused by enzymes released by the killed bacteria or present in the pond water.

In conclusion, it is important to measure the respirational loss of $^{14}CO_2$ to arrive at true estimates of the uptake of the dissolved organic compounds. The percentage of the total uptake that is respired ^{14}C varies with type of substrate, location of the label, and time of year. However, duration of incubation, temperature of incubation, and concentration of substrate does not matter much over the range of conditions presented here. The phenethylamine absorption method gives excellent recovery of the $^{14}CO_2$, but correction must be made for self-absorption effects when using liquid scintillation counting. Percentages of the total uptake that were respired could be interpreted partially on the basis of known metabolic pathways.

REFERENCES

HAMILTON, R. D., AND K. E. AUSTIN. 1967. Assay of relative heterotrophic potential in the sea: the use of specifically labelled glucose. Can. J. Microbiol., **13**: 1165–1173.

HOBBIE, J. E., C. C. CRAWFORD. AND K. L. WEBB. 1968. Amino acid flux in an estuary. Science, **159**: 1463–1464.

PARSONS, T. R., AND J. D. H. STRICKLAND. 1963. On the production of particulate organic carbon by heterotrophic processes in sea water. Deep-Sea Res., **8**: 211–222.

SMITH, D. E. 1967. Location of the estrogen effect on uterine glucose metabolism. Proc. Soc. Exptl. Biol. Med., **124**: 747–749.

VACCARO, R. F., AND H. W. JANNASCH. 1967. Variations in uptake kinetics for glucose by natural populations in seawater. Limnol. Oceanog., **12**: 540–542.

WETZEL, R. G. 1964. A comparative study of the primary productivity of higher aquatic plants, periphyton, and phytoplankton of a large, shallow lake. Int. Rev. Gesamten Hydrobiol., **49**: 1–61.

WOLFE, D. A., AND C. L. SCHELSKE. 1967. Liquid scintillation and Geiger counting efficiencies for carbon-14 incorporated by marine phytoplankton in productivity measurements. J. Conseil, Conseil Perm. Intern. Exploration Mer, **31**: 31–37.

WRIGHT, R. T., AND J. E. HOBBIE. 1966. Use of glucose and acetate by bacteria and algae in aquatic ecosystems. Ecology, **47**: 447–464.

28

Reprinted from *Limnol. Oceanogr.*, **12**(2), 264–271 (1967)

GROWTH OF MARINE BACTERIA AT LIMITING CONCENTRATIONS OF ORGANIC CARBON IN SEAWATER[1]

Holger W. Jannasch
Woods Hole Oceanographic Institution, Woods Hole, Massachusetts 02543

ABSTRACT

Growth responses of several species of heterotrophic marine bacteria to limiting concentrations of lactate, glycerol, and glucose in seawater were determined in a chemostat. In all cases, threshold concentrations of the limiting substrates were found below which the organisms were unable to grow. This phenomenon is explained on the basis of a positive feedback mechanism, which is abolished below a certain minimum population density. The resulting inhibitory effect on growth leaves a corresponding concentration of the limiting substrate unattacked.

According to their growth parameters, two types of species could be distinguished, one adapted to the marine environment by its ability to grow at low substrate concentrations and one inactive in natural seawater but surviving.

The implications of these results on the rates of microbial transformations and on the occurrence and concentrations of dissolved organic material in the sea are discussed.

INTRODUCTION

Rates of microbial activities in the sea are of essential interest in biological oceanography. Although detailed information on physiological and biochemical properties of various marine microorganisms under *laboratory* conditions is available, there are no quantitative data describing growth in nutrient concentrations characteristic of their *natural* environment. While recent studies on microbial growth in the sea have been focused largely on stimulating or inhibiting substances, this study is concerned with population behavior based on nutritional growth limitation.

Natural seawater is characterized by extremely low nutrient concentrations, compared with media commonly used for studies of bacterial growth. Most results obtained by the classical techniques for assessing microbial activities in such impoverished environments are not amenable to a meaningful interpretation. This study represents an attempt to consider two major problems:

1) The turnover rates of organic matter in seawater are usually too low to permit *direct* measurements of specific microbial activities. When substrates are added in measurable concentrations, the resulting turnover rates cannot be extrapolated to values presumably valid at lower substrate concentrations. This has been shown by studies on regulatory mechanisms in microbial cells. Kjeldgaard (1963) has shown that biosynthetic abilities diminish as the growth rate decreases. In turn, minimum requirements of the bacterial cell are known to vary with the synthetic ability (Herbert 1961*a*). Holme (1958) and Herbert (1961*a*) reported that under pronounced growth limitation the ratio between respiratory and biosynthetic metabolism is strongly affected by the nature of the limiting substrate, such as its function as a source of energy or as an essential nutrient. Thus, microbial activities measured after the addition of a nutrient can be interpreted only in terms of potentialities.

2) Incubating water samples of limited volume changes drastically the environmental conditions for growth of microorganisms. The natural habitat represents an open system where the environmental conditions do not alter with time. Immediately after sampling, the population becomes affected by the inherent properties of the closed growth system (batch culture). Depending upon the volume of the sample, the natural population will react to the decrease of the limiting substrate, to shifts from one limiting factor to another,

[1] Contribution No. 1882 from the Woods Hole Oceanographic Institution. This work was supported by National Science Foundation Grants GB 5199 and GB 861.

and to the increase of metabolic products. Characteristic of this situation is the immediate change of colony numbers as determined on a complex agar medium and of the species composition as determined on specific media (ZoBell and Anderson 1936; Potter 1960). These effects are less pronounced in natural waters with high turnover rates and high population densities.

In natural waters, exhaustion of the limiting nutrient and accumulation of metabolic products are assumed to be eliminated largely by metabolic interactions between species of different metabolic types within the natural population. These interactions, which contribute to the more or less steady state of the population, are destroyed when the sample is transferred to a closed system of relatively small volume. The technique of employing plastic bags submerged in the natural habitat (McAllister et al. 1961) represents an interesting compromise. It considers the delicate balance of natural populations while still taking advantage of conducting quantitative measurements in a closed system.

In short, most techniques for the determination of a microbial reaction in samples of natural water suffer from disregarding the mixed microbial population as a dynamic system. Under experimental steady-state conditions of a chemostat, these difficulties are eliminated to a certain degree by the use of time-independent relationships between culture conditions and growth properties of the organisms. Comparative studies of microbial growth in both closed and open culture systems (Herbert, Elsworth, and Telling 1956; Pfennig and Jannasch 1962; Jannasch 1965*b*) demonstrate their applicability in microbial ecology. In this paper the properties of a chemostat culture are used to assess growth responses of some marine bacteria to extremely low concentrations of the limiting substrate in seawater.

THEORETICAL

The chemostat is a refined continuous culture system designed to meet the technical requirements for the establishment of a steady state. The steady state is generally defined as a time-independent state where growth is counterbalanced by the removal of cells; that is, the exponential growth rate μ (in hr^{-1}) is equal to the exponential dilution rate D. Consequently, the population density $\bar{x}$ (in mg dry wt/liter) and the concentration of the limiting substrate in the culture $\bar{s}$ (in mg/liter) remain constant. The mathematical relationship between culture conditions, growth response, and growth parameters (Herbert 1961*b*) apply during steady state only.

Mixed populations in the chemostat undergo selective processes. The species achieving the fastest growth rate under the particular conditions will approach a steady-state population, while the competing species are displaced (Powell 1958; Jannasch 1965*a*). Thus, a chemostat population cannot be considered a model of a natural, mixed population. Contois and Yango (1964) and Bungay (1966) discussed experiments where steady states in mixed populations of two species could be obtained.

For the current study, it is of importance that during steady state the substrate concentration in the chemostat $\bar{s}$ is dependent on the growth rate μ but independent of the concentration of the limiting substrate in the reservoir s_0. The population density $\bar{x}$ is in proportion to s_0:

$$s_0 = \frac{\bar{x}}{y} + \left(K_s \frac{D}{\mu_m - D}\right), \qquad (1)$$

and

$$\bar{s} = K_s D / (\mu_m - D), \qquad (2)$$

where μ_m is the maximum growth rate at excess concentration of the limiting substrate; K_s is the concentration of the limiting substrate that gives rise to one-half of the maximum growth rate (saturation constant); and y is the weight of limiting substrate consumed/weight of cell substance produced (yield coefficient).

The theoretical relationship between population density and substrate concentration during steady state is shown as a

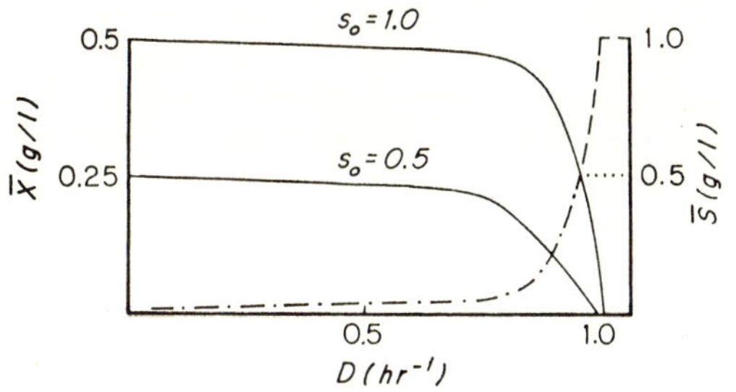

FIG. 1. Theoretical steady-state relationship between population density (solid line) and concentration of limiting substrate (broken line) in relationship to the dilution rate and at two concentrations of the limiting substrate in the reservoir, according to equation (1). Growth parameters are: $\mu_m = 1.0$ hr^{-1}, $K_s = 0.04$ g/liter, $y = 0.5$ (after Herbert et al. 1956).

function of the dilution rate in Fig. 1 for two substrate concentrations in the reservoir. It illustrates that, theoretically, $\bar{s}$ is independent of s_0, while the $\bar{x}$ is proportional to s_0. This relationship was checked in the following experiments.

EXPERIMENTAL

The technical requirement for growing microorganisms at steady state is maintaining the constancy of: medium flow, volume of culture, medium composition, temperature, and the degree of mixing and aeration. For this study, low and accurate flow rates for long periods of time were especially important. This was achieved by a gravity flow system (Fig. 2), which proved to be superior to all pump systems tested. This rather simple arrangement permitted feeding a number of chemostats from one medium reservoir and facilitated series of simultaneous experiments. The pressure heads were counterbalanced by a capillary resistance of Teflon tubing (diameter 0.5–0.8 mm, length 2–8 m) placed in a water bath of constant temperature. This resistance damped slight pressure changes in the chemostat due to the continuous removal of culture medium by the aerating gas mixture. Positive pressure in the culture vessel provided a rapid removal of the medium through a capillary and prevented back-contaminations. The inlet tube was heated to 80–90C with a 0.1 ohm/cm ni-

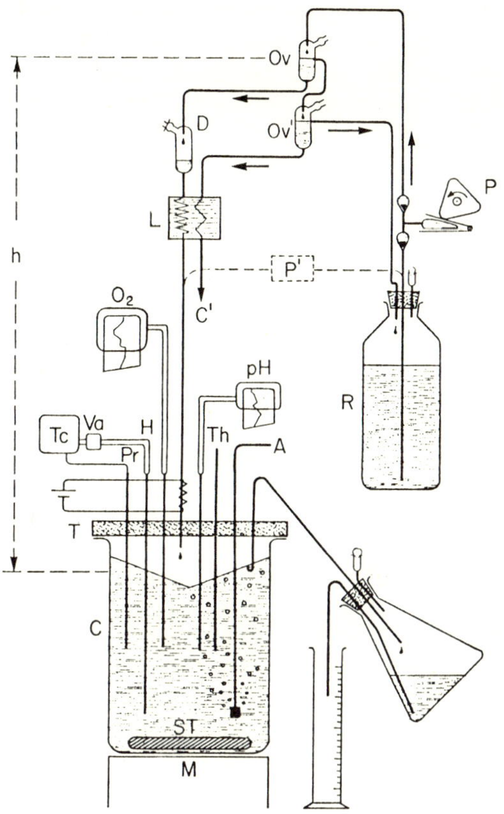

FIG. 2. Chemostat as used in this study (up to four of these units were used simultaneously). Symbols (sequence following the medium flow): R, reservoir (10 to 20 liters); P, pump for keeping overflow vessels, Ov, (pressure heads) filled; P', peristaltic pump, used for short time experiments and high flow rates only; Ov', second overflow vessel for second chemostat unit; D, drop counting device; L, capillary resistance in constant temperature water bath; O_2 and pH, measuring and recording device for dissolved oxygen and *p*H; Tc, thermoregulator; Va, variac; H, submersible heater (quartz); Pr, temperature probe; Th, thermometer; A, aeration device; T, Telfon-covered lid; C, culture vessel (150 to 1,200 ml); C', tube connection to second culture vessel; M, magnetic stirrer; St, stirring bar; h, niveau difference.

chrome resistance wire 15 cm long to prevent contamination from the culture.

The culture vessel was closed with a Teflon-covered stainless steel lid 2.5 cm thick. Tapered openings received silicone stoppers with bores for tubing and electrodes. In order to avoid zones of stationary growth, the air space above the surface of the culture (splash zone) was kept as small as possible and self-rinsing. According to the analytical procedures to follow, Dry Ice or a disinfectant was used in the receiving vessel.

The dilution rates of the chemostats could be independently adjusted by a) lifting or lowering the overflow vessel, b) choice of culture volume, and c) changing the length of the capillary resistance. The dilution rate was measured periodically in the receiving vessel. Its constancy was checked occasionally in the drop counting device. Rubber membrane pumps and ball flowmeters were used for aeration. The air was sterilized in a series of dry filters (cotton and glass wool) and then water saturated.

As the medium, prefiltered offshore seawater was filter-sterilized and supplemented with either Na-lactate, glycerol, or D-glucose in amounts of 0.5 to 100 mg/liter. Phosphate buffer (7.8 *p*H) and ammonium chloride were added to provide an initial C : N : P ratio of 10 : 4 : 1 by weight in order to secure growth limitation by the carbon and energy source. All experiments were done at 20C.

When $\bar{s}$ could not be measured directly, the growth parameters were determined as described by Jannasch (1963*a*) for the indirect calculation of $\bar{s}$. Enzymatic analyses were applied whenever known procedures (Bergmeyer 1963) could be adapted to seawater. Oxygen saturation was controlled by using a stationary gold amalgam–zinc electrode system (Tödt 1958) which was periodically exchanged when instability occurred due to calcium carbonate precipitations.

The following strains were isolated either from chemostat enrichments (Jannasch 1965*a*) or from batch culture enrichments (Vaccaro and Jannasch 1966): *Achromobacter aquamarinus* (208), *Achromobacter* sp. (317), *Pseudomonas* sp. (201), *Spirillum lunatum* (102), *Spirillum* sp. (101), and *Vibrio* sp. (204). All strains grew in synthetic media and required no vitamins. The purified cultures were kept in unsupplemented seawater for 2 to 6 months during which period plate counts decreased by 10 to 60%. For inoculation, the populations were filtered, washed, and resuspended in the culture vessel. As soon as plate counts indicated a distinct increase of cell numbers, flow was started at the chosen rate.

Dilution rates ranged from 0.4 to 0.01/hr, corresponding to retention times of 2.5 to 100 hr. Longer retention times did not result in steady-state populations due to wall growth or aggregation of cells.

Twice during a retention time, the population density x and, if possible, the concentration of the limiting substrate in the chemostat s were determined. When a steady state was established ($x \rightarrow \bar{x}$, $s \rightarrow \bar{s}$), a new reservoir with a lower concentration of the limiting substrate s_0 was connected to the chemostat. The s_0 was decreased stepwise (100, 50, 20, 10, 5, 2, 1, and 0.5 mg/liter) until washout of the population occurred. This procedure was carried out at the following relative dilution rates: $D/\mu_m = 0.5, 0.3, 0.2, 0.1$, and in some cases, 0.05. The specific maximum growth rates for each strain and medium were determined in stationary culture.

Cells were counted as units of equal size in counting chambers or on membrane filters (Jannasch 1953) and were calibrated on a dry weight basis at high population densities (Jannasch 1963*a*). These values were used for extrapolations at low population densities. This method fails with coccoid organisms and in the case of an abundant production of "involution"-forms. Best results were obtained with Spirilla, where cell diameter and amplitude did not change under starving conditions in the presence of a delayed division rate.

Death rate determinations must be done on the basis of viable counts. Here, exponential growth is no longer described by the *e*-function but by the base 2, because each cell, regardless of size, is assumed to give rise to one colony (doubling time of cell units regardless of size). The growth rate μ then becomes the division rate $\nu = \mu/\ln 2$ (Pfennig and Jannasch 1962).

RESULTS

The broken line in Fig. 3 represents the theoretical function given in equation (1)

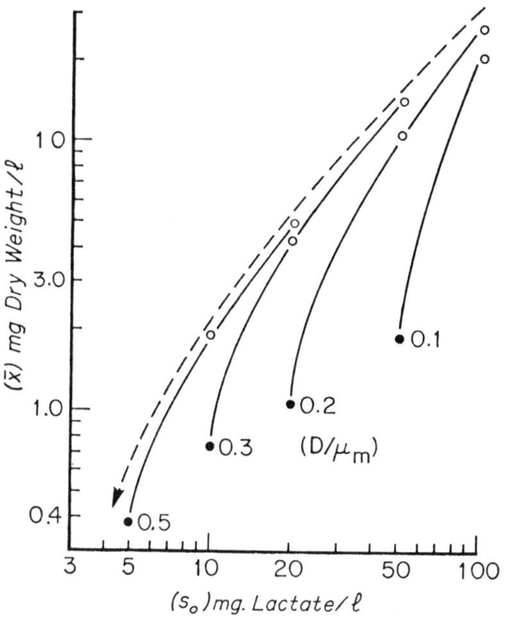

FIG. 3. Steady-state population density plotted vs. concentration of the limiting substrate in the reservoir at four different growth rates (strain 101).

for $D/\mu_m = 0.5$. When strain 101 was grown at this particular growth rate at various lactate concentrations in the reservoir, the decrease of the steady-state population density deviated from the theoretical curve. This deviation became larger with decreasing growth rates (Fig. 3).

Furthermore, below a certain minimum lactate concentration in the reservoir, no steady-state population could be maintained and outwash occurred. The resulting minimum population densities appear as dots in Fig. 3. The minimum value of s_0 will be almost identical with the threshold value of s_0 at which outwash occurred ($\bar{x} \rightarrow 0$) and at which $\bar{s}$ becomes equal to s_0.

Threshold concentrations for lactate, glycerol, and glucose and the maximum growth rates are given in Table 1 for six strains of marine bacteria.

According to equation (2) $\bar{s}$ is equal to K_s at $D/\mu_m = 0.5$. Therefore, if outwash occurs at this particular growth rate on account of a decreasing s_0, the resulting threshold concentration cannot be smaller than K_s. If it is considerably larger, it must

TABLE 1. *Threshold concentrations of three growth-limiting substrates (in mg/liter) in seawater at several relative growth rates of six strains of marine bacteria (see p. 267) and the corresponding maximum growth rates (in hr^{-1})*

Strain	D/μ_m	Lactate	Glycerol	Glucose
208	0.5	0.5	1.0	0.5
	0.1	0.5	1.0	0.5
	0.05	1.0	5.0	1.0
	μ_m	0.15	0.20	0.34
102	0.3	0.5	no growth	0.5
	0.1	1.0		5.0
	0.05	1.0		10.0
	μ_m	0.45		0.25
101	0.5	5	5	
	0.3	10	10	no growth
	0.2	20	100	
	0.1	50		
	μ_m	0.45	0.60	
204	0.3	1	5	1
	0.1	5	5	5
	0.05	10	10	20
	μ_m	0.15	0.35	0.40
317	0.5	20	50	
	0.3	50	50	no data
	0.1	100	>100	
	0.05	>100		
	μ_m	0.85	0.70	
201	0.5	20	50	20
	0.3	50	50	50
	0.2	100	>100	>100
	0.1	>100		
	μ_m	0.80	0.65	0.80

be assumed that growth is affected. As shown in Table 2 for two strains, the threshold concentrations of lactate exceed the corresponding K_s-values that were obtained in continuous culture at high population densities. Strain 317 (Table 1) represents a similarly "inefficient" species as far as growth at low substrate concen-

TABLE 2. *Growth constants (determined at high population densities) of two strains of marine bacteria (see p. 267) compared with threshold concentrations of lactate in seawater at two relative growth rates*

Strain	K_s (mg/liter)	μ_m (hr^{-1})	Threshold concn. (mg/liter) at $D/\mu_m = 0.5$	Threshold concn. (mg/liter) at $D/\mu_m = 0.1$
101	3.0	0.45	5	50
201	8.0	0.80	20	100

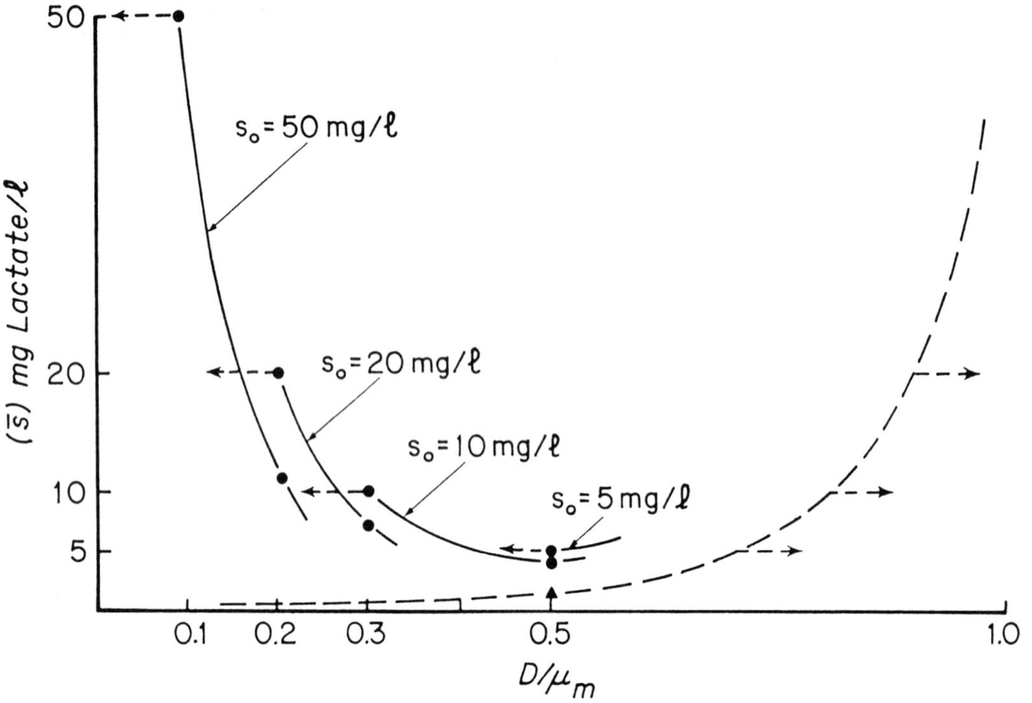

FIG. 4. Steady-state concentrations of limiting substrate plotted vs. relative growth rate at four different substrate concentrations in the reservoir for strain 101. Horizontal parts of the curves indicate washout of the culture ($\bar{s} = s_0$). The broken line indicates the theoretical relationship (▲ = K_s).

trations is concerned. In all cases, growth efficiency decreased with decreasing growth rate. This phenomenon can be expressed by a change of the yield coefficient in equation (2) (Jannasch 1963*b*).

The effect of starving on bacterial cultures often has been interpreted as an increase of a proposed death rate (Postgate and Hunter 1963). My experiments show that the ratio between viable count and direct count does not change appreciably with decreasing s_0. In other words, if a death rate existed, it did not change and it could not account for the phenomenon of premature washout.

DISCUSSION

The washout of the population from the chemostat at a certain minimum value of s_0 clearly indicates an unexpected change of growth limitation. The only factor changing during the experimental decrease of s_0 is the population density. Therefore, since no direct inhibitory action of the limiting substrate upon dilution is conceivable, the inhibitory effect, indicated by the increase of $\bar{s}$ towards the ordinate (Fig. 4), must be related to a population effect. In other words, whatever factor is limiting, the inhibitory effect will occur as a corollary of decreasing population density.

A comparison between Figs. 4 and 1 shows that, in the present case, the theoretical relationship between $\bar{s}$ and s_0 does not hold at relatively low growth rates and substrate concentrations. If the population density were plotted in Fig. 4, it would show a mirror image of the function of $\bar{s}$, as in Fig. 1.

An explanation of the phenomenon is suggested by the role of starter populations in batch cultures. In initially suboptimal media, metabolic products may act as conditioning agents for growth (Lodge and Hinshelwood 1943; Meyrath and McIntosh 1963). This positive feedback mechanism depends upon population density and on metabolic activity (growth rate). The current study shows that the phenomenon observed varies with both of these factors (Fig. 3). When s_0 and, consequently, $\bar{x}$ reach their minimum value, the assumed feedback action is abolished. At this point, the production of a proposed metabolite, for instance, does not meet the demand for maintenance of the present growth rate. The resulting drop of the growth rate at a constant dilution rate leads to complete washout. When the experiment is done at a lower dilution rate, the washout occurs in the presence of a higher population density. Evidence for such a phenomenon and its mathematical description has been given earlier by cultivating *Spirillum serpens* in a mineral medium with lactate as the limiting carbon and energy source (Jannasch 1963*a*). It was shown that the inhibitory effect at low population densities could be partly eliminated by external adjustment of the media. At higher population densities, this adjustment took place internally.

The threshold values reported in Table 1 are relative, of course, and not directly applicable to seawater. They depend upon 1) the species, 2) the experimental growth rate, and 3) the given growth conditions.

1) All species studied showed marked threshold concentrations of the limiting substrates. The enrichment processes were selective for species of high growth efficiency at low substrate concentrations, so most of them exhibited relatively low K_s-values. The two exceptions, strains 201 (*see* Table 2) and 317, were isolated from 2% lactate agar. Their high K_s-values, measured directly for all carbon sources tested, indicated low growth efficiency at low substrate concentrations.

It has been observed repeatedly (Jannasch 1965*a*) that in a given species high K_s-values always seem to be combined with high μ_m-values or low K_s-values with low μ_m-values (Table 2). If this is a fact, it follows that species efficiently growing in seawater will escape detection and isolation by the usual techniques of batch culture enrichments, plating, and so forth, and that species isolated from seawater on nutrient agar of the usual strength are not representative for the active marine microbial flora. Thus, Winogradsky's early concept of "autochthonous" and "zymogenous" types of microorganisms reappears in terms of growth parameters.

2) As shown in Fig. 3, the threshold concentrations reported depend on the growth rate of the organism. This latter parameter is not directly assessable in the natural environment. It could be indefinitely low. Theoretically, $\bar{s}$ approaches zero with decreasing growth rate (Fig. 1). However, the plot of the actual data shows that $\bar{s}$ reaches a minimum around 0.3 to 0.5 D/μ_m (Fig. 4). In other words, threshold concentrations can only be larger at lower growth rates.

3) The change of growth limitation below a certain population density must be attributed to the general growth conditions. In this sense, seawater must be considered a suboptimal medium.

The threshold concentrations reported are at least 1 to 2 orders of magnitude too high to account for endogenous respiration or maintenance metabolisms (Lamanna 1963). Furthermore, the inhibitory effect of the abolished feedback interaction is clearly indicated by the increase of $\bar{s}$ at low growth rates (Fig. 4). Thus, the phenomenon observed can be described as truly environmental.

The smallest threshold concentrations found in this study are still larger than the concentrations of similar compounds reported for natural seawater (Duursma 1965). This discrepancy must be explained by the fact that the steady-state experiments had to be done with pure cultures. The models made it possible to study and to exemplify a principle rather than to reproduce data of a more complex system.

The relatively high concentrations of dissolved organic carbon in the sea and their constancy (Menzel, in prep.; Walsh and Douglass 1966) are primarily explained by an unavailability of the material to microbial breakdown. The current study suggests that this unavailability must not be identical with chemical resistance against biological decomposition. The standing (steady-state or equilibrium) concentration of organic substrates not utilized by bacterial oxidation may be characteristic for seawater as a suboptimal environment for microbial life.

REFERENCES

BERGMEYER, H. U. 1963. Methods of enzymatic analysis. [Transl. from German.] Academic, New York, N.Y. 1064 p.

BUNGAY, H. R. 1966. Population oscillations in symbiotic continuous cultures. Am. Chem. Soc., 152nd Meeting, Q36.

CONTOIS, D. E., AND L. D. YANGO. 1964. Studies of mixed steady state microbiol populations. Am. Chem. Soc., 148th Meeting, Q17.

DUURSMA, E. K. 1965. The dissolved organic constituents of sea water, p. 433–475. *In* J. P. Riley and G. Skirrow [eds.], Chemical oceanography, v. 1. Academic, New York, N.Y.

HERBERT, D. 1961*a*. The chemical composition of micro-organisms as a function of their environment. Symp. Soc. Gen. Microbiol., **11**: 391–416.

———. 1961*b*. A theoretical analysis of continuous culture systems. Soc. Chem. Ind. (London), Monograph 12, p. 21–53.

———, R. ELSWORTH, AND R. C. TELLING. 1956. The continuous culture of bacteria: a theoretical and experimental study. J. Gen. Microbiol., **14**: 601–622.

HOLME, T. 1958. Glycogen formation in continuous culture of *Escherichia coli* B, p. 67–74. *In* Symp. Continuous Cult. Microorg. Prag.

JANNASCH, H. W. 1953. Zur Methodik der quantitativen Untersuchung von Bakterienkulturen in flüssigen Medien. Arch. Mikrobiol., **31**: 114–124.

———. 1963*a*. Bakterielles Wachstum bei geringen Substratkonzentrationen. Arch. Mikrobiol., **45**: 323–342.

———. 1963*b*. Bacterial growth at low population densities (II). Nature, **197**: 1322.

———. 1965*a*. Eine Notiz über die Anreicherung von Mikroorganismen in Chemostaten, p. 498–502. *In* Anreicherungskultur und Mutantenauslese. Zentr. Bakteriol., v. 1. (Suppl.)

———. 1965*b*. Continuous culture in microbiol ecology. Lab. Pract., **14**: 1162–1167.

KJELDGAARD, N. O. 1963. Dynamics of bacterial growth. Nyt Nord. Arnold Busck, Copenhagen.

LAMANNA, C. 1963. Endogenous metabolism with special reference to bacteria. Ann. N.Y. Acad. Sci., **102**: 515–793.

LODGE, R. M., AND C. N. HINSHELWOOD. 1943. Physicochemical aspects of bacterial growth. Part IX. The lag phase of *Bact. lactis aerogenes*. J. Chem. Soc., **1943**: 213–219.

MCALLISTER, C. D., T. R. PARSONS, K. STEPHENS, AND J. D. H. STRICKLAND. 1961. Measurements of primary production in coastal sea water using a large-volume plastic sphere. Limnol. Oceanog., **6**: 237–258.

MEYRATH, J., AND A. F. MCINTOSH. 1963. Size of inoculum and carbon metabolism in some Aspergillus species. J. Gen. Microbiol., **33**: 47–56.

PFENNIG, N., AND H. W. JANNASCH. 1962. Biologisch Grundfragen bei der homokontinuierlichen Kultur von Mikroorganismen. Ergeb. Biol., **25**: 93–135.

POSTGATE, J. R., AND J. R. HUNTER. 1963. The survival of starved bacteria. J. Appl. Bacteriol., **26**: 295–306.

POTTER, L. F. 1960. The effect of *p*H on the development of bacteria in water stored in glass containers. Can. J. Microbiol., **6**: 257–263.

POWELL, E. O. 1958. Criteria for growth of contaminants and mutants in continuous culture. J. Gen. Microbiol., **18**: 259–268.

TÖDT, F. 1958. Elektrochemische Sauerstoffmessungen. de Gruyter, Berlin. 212 p.

VACCARO, R. F., AND H. W. JANNASCH. 1966. Studies on heterotrophic activity in seawater based on glucose assimilation. Limnol. Oceanog., **11**: 596–607.

WALSH, G. E., AND J. DOUGLASS. 1966. Vertical distribution of dissolved carbohydrate in the Sargasso Sea off Bermuda. Limnol. Oceanog., **11**: 406–407.

ZOBELL, C. E., AND D. Q. ANDERSON. 1936. Observations on the multiplication of bacteria in different volumes of stored sea water and the influence of oxygen tension and solid surfaces. Biol. Bull., **71**: 324–342.

29

Reprinted from *Limnol. Oceanogr.*, **2**(2), 114–119 (1957)

The Production of Vitamin B_{12}-Active Substances by Marine Bacteria

THEODORE J. STARR, MARY E. JONES, AND DOMINGO MARTINEZ

Gulf Fishery Investigations, U. S. Fish and Wildlife Service, Galveston, Texas

ABSTRACT

The production of members of the vitamin-B_{12} family of compounds by 34 marine bacteria that were grown in a B_{12}-deficient medium was assayed. Using aliquots of the same bacterial culture, 70 per cent and 30 per cent had activity for the assay organisms *Escherichia coli* 113-3 and *Euglena gracilis*, *z* strain, respectively. Using the *Euglena* assay, no apparent relation was evident between the activities of the cell residues or corresponding supernatants and the relative amount of growth of each culture. However, of the 24 bacterial cultures tested, the supernatants of 42 per cent had activity, whereas the corresponding cell residues of 63 per cent had activity.

INTRODUCTION

Extensive reviews are available on members of the vitamin-B_{12} family of compounds dealing with their chemistry, functions in animal and microbial metabolism, distribution, production, and methods of measurement (McNutt 1952, Jukes *et al.* 1954, Hoff-Jørgensen 1954, and Ford and Hutner 1955). Burton and Lochhead (1951) and Lochhead and Burton (1955) have demonstrated that vitamin B_{12} is required by certain soil bacteria. Vitamin B_{12}-active compounds have been shown to be essential or stimulatory for the growth of various marine microorganisms (Hutner and Provasoli 1953, Droop 1954, Lewin 1954, Sweeney 1954, Droop 1955, and Wilson and Collier 1955). Such observations suggest that vitamin B_{12}-active compounds may have a vital role in the economy of the oceans, in production of algal blooms, and in population successions.

In previous studies (Starr 1956), the vitamin-B_{12} content of the suspended matter of estuarine and oceanic waters was assayed. The present investigation was undertaken to learn more about the vitamin-B_{12} cycle in marine environments. Information gained from this and related studies encompassing the vitamin cycles and the roles of organic compounds of the oceans may add materially to our understanding of marine ecology and its application to marine fisheries.

In this investigation microbiological assay procedures were used to measure the production of vitamin B_{12}-active compounds by 34 different bacteria isolated from marine environments. Relative levels of production were compared using two different microbiological assay organisms having different spectra of activity for members of the vitamin-B_{12} family of compounds.

EXPERIMENTAL

The 34 bacteria screened for production of vitamin B_{12}-active substances represent different morphological and physiological groups that are maintained in our permanent stock culture collection. They have not as yet been identified. They were isolated from a variety of marine samples by techniques described in a previous publication (Starr and Jones 1957).

The bacteria were grown in a vitamin B_{12}-deficient medium of the following composition: NaCl, 0.3 g; $MgSO_4 \cdot 7H_2O$, 0.02 g; KNO_3, 0.02 g; NH_4Cl, 0.01 g; K_2HPO_4, 0.04 g; Na acetate, 0.01 g; sucrose, 0.1 g; glycine, 0.01 g; yeast extract 0.02 g; and distilled water to 100 ml. The pH was adjusted to 7.5–7.7 with 1N NaOH. After 10 days' incubation at 26°C ± 2°C, aliquots of each culture were diluted 1:10 with a buffer at pH 4.5. The buffer contained 8.2 g/L of KH_2PO_4 and 0.5 g/L of $Na_2S_2O_5$. The diluted cultures were hydrolyzed in a boiling water bath for 20 min. After cooling, a second 1:10 dilution was made from the first with the same buffer. These dilutions, 1:10 and 1:100, of each culture were diluted further with distilled water to give the following final assay dilutions: 1:10, 1:20, 1:50, 1:100, 1:200, 1:500, 1:1000.

The following procedures were used in order to determine if vitamin B_{12}-active substances were confined to the bacterial cells or released into the medium. Twenty-four of the 34 bacteria were grown in 400 ml of medium in 1000 ml Erlenmeyer flasks. The cultures were incubated at 26° ± 2°C on a rotary shaker for 6 to 8 days. Twenty-ml aliquots of each culture were centrifuged at 10,000 rpm for 10 min at 4°C. The supernatant was decanted, and final dilutions were prepared from it. The cell residues were suspended in 10 ml of the phosphate-metabisulfite buffer, and final dilutions were made. Other aliquots of each culture were used for turbidimetric growth measurements using a Klett-Summerson colorimeter and for pH measurements.

Samples were assayed for vitamin-B_{12} activity using *Euglena gracilis*, *z* strain, (Pringsheim's) which was obtained through the courtesy of Dr. L. Provasoli. It is now available from the American Type Culture Collection. The methods and procedures were essentially those of Hutner *et al.* (1956) with minor modifications. Stock cultures of *Euglena* were maintained in screw-cap tubes containing: K_2HPO_4, 2 mg, Na_2·citrate·$2H_2O$, 2 mg; $FeCl_3 \cdot 6H_2O$, 0.2 mg; peptone, 60 mg; trypticase, 16 mg; yeast extract, 5 mg; and made to volume with 100 ml of distilled water. The pH was adjusted to 6.5. Assays were performed in screw-cap tubes (size 125 x 20 mm) using the basal medium, "dry mix", described by Hutner *et al.* (1956). Two and one-half ml of the double-strength basal medium were added to 2.5 ml of each dilution of the sample. Duplicate standard curves were made using commercial vitamin B_{12} to give the following final concentrations (mμg/100 ml): 0.0, 0.05, 0.12, 0.2, 0.3, 0.4, 0.5, 0.6, 0.8, and 1.0. All media were autoclaved at 15 lb for 10 min prior to inoculation. A 6 to 8 day-old *Euglena* culture grown in the stock culture maintenance medium was used for inocula. A bacteriological loop was used to transfer approximately 0.05 ml to 12 ml of the basal medium. One drop of the resulting suspension (approximately 0.08 ml) was used as inoculum per assay tube. Although it has been shown (Østergaard Kristensen 1954) that the supernatant fluid of well-grown *Euglena* cultures contains material which inhibits the growth of *Euglena* in concentrations as low as 1 in 2000 (Hutner *et al.* 1956), our dilution represented a safe margin with a concentration of approximately 1:15,000. Cultures were incubated under a battery of fluorescent lights at 27° ± 1°C for 6 days. Growth was measured using a Klett-Summerson colorimeter.

The other assay organism was the mutant strain *Escherichia coli* 113-3 obtained through the courtesy of Dr. B. D. Davis. The methods of assay were essentially those described by Burkholder (1951) and modified as described previously (Starr 1956). Stock cultures of *E. coli* were maintained on nutrient agar slants and were transferred weekly. Assays were performed as described for *Euglena*. Standard curves were made in duplicate to give the following final concentrations of commercial vitamin B_{12} (mμg/100 ml): 0.0, 1.0, 2.0, 5.0, 10.0, 15.0, 20.0, 30.0, and 40.0. All media were autoclaved at 10 lb for 3 min prior to inoculation. Inocula were prepared from a 24-hr slant of *E. coli* by carefully transferring a small amount of growth with a fine platinum needle to 12 ml of single strength basal medium. One drop (approximately 0.08 ml) of this cell suspension was used per tube. Experiments were incubated for 18 to 20 hr at 30°C on a tissue-culture apparatus rotating 20 to 24 rpm. Growth was measured turbidimetrically with a Klett-Summerson colorimeter.

RESULTS

Typical assay response curves for *Euglena* and *E. coli* to increasing concentrations of commercial vitamin B_{12} are shown in Figure 1.

The results of the preliminary screen to determine the numbers of bacteria producing vitamin B_{12}-active substances are given in Table 1. Using aliquots of the same culture, 70 per cent and 30 per cent of the isolates

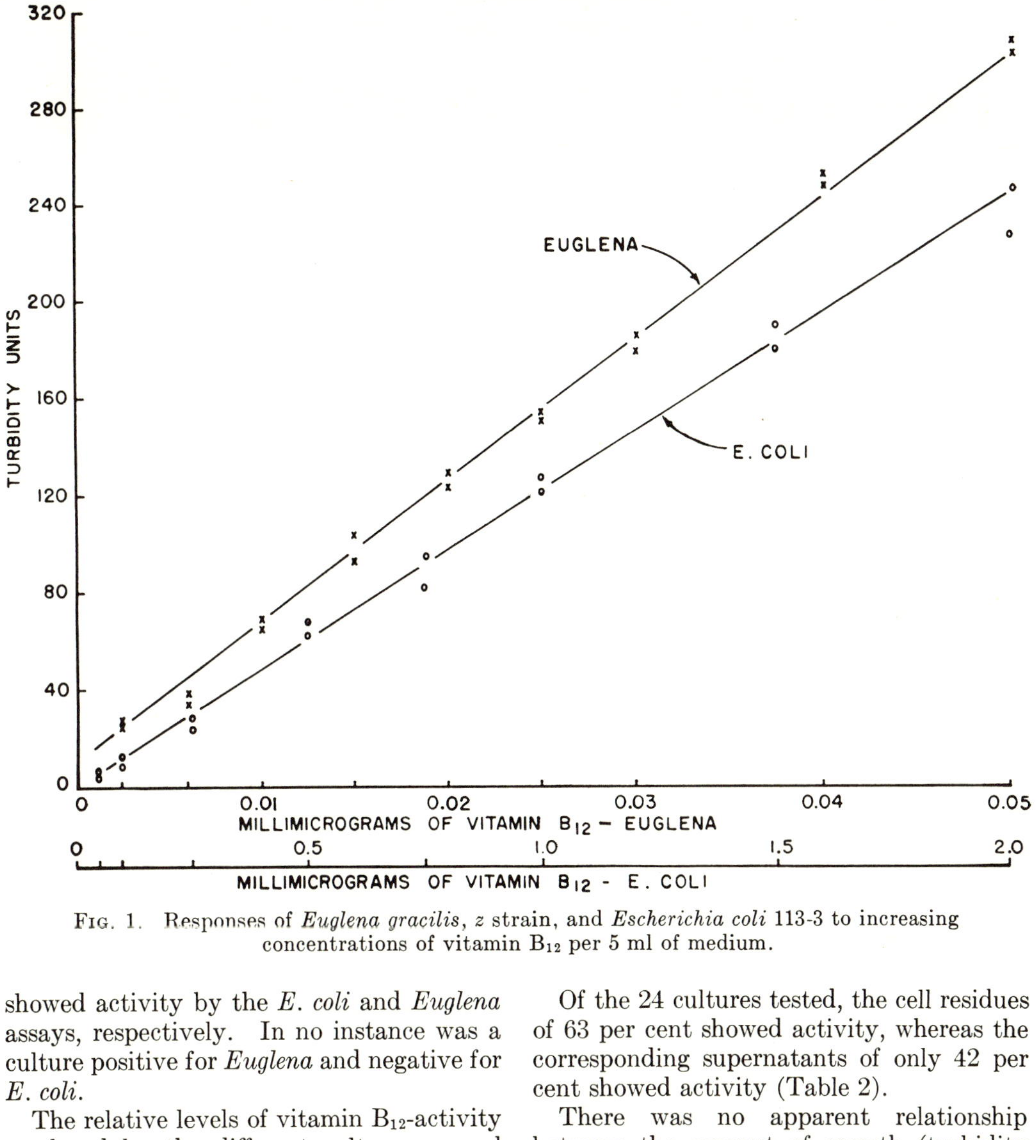

FIG. 1. Responses of *Euglena gracilis*, *z* strain, and *Escherichia coli* 113-3 to increasing concentrations of vitamin B_{12} per 5 ml of medium.

showed activity by the *E. coli* and *Euglena* assays, respectively. In no instance was a culture positive for *Euglena* and negative for *E. coli*.

The relative levels of vitamin B_{12}-activity produced by the different cultures ranged from 0 to 5.0 mμg/ml as measured by the *Euglena* assay and from 0 to 18.4 mμg/ml as measured by the *E. coli* assay.

Other experiments were designed to determine if vitamin-B_{12} activity was confined to the bacterial cells or released into the medium. The cell residues and supernatants of mass cultures were obtained and processed as described. Aliquots of them were assayed with *Euglena*. For comparison, the same cultures used in the previous experiments (Table 1) were screened.

Of the 24 cultures tested, the cell residues of 63 per cent showed activity, whereas the corresponding supernatants of only 42 per cent showed activity (Table 2).

There was no apparent relationship between the amount of growth (turbidity readings) or final pH of each culture and the levels of vitamin-B_{12} activity of either the cell residues or supernatants. For example, the vitamin-B_{12} activity of culture 266 which grew meagerly (turbidity, 15) was confined to the cell residue. Other cultures (126 and 222B) which grew prolifically produced relatively less activity. In no instance did a supernatant have activity and its corresponding cell residue have none, although the reverse situation occurred a number of times.

TABLE 1. *Comparative responses of* Escherichia coli *113-3 and* Euglena gracilis, z *strain, to aliquots of the same bacterial cultures*

Units of activity in terms of mμg vitamin B_{12} per ml.

Culture No.	*E. coli*	*Euglena*
Control	0	0
112	1.2	0.50
116	3.4	1.36
117	3.8	1.38
121	13.0	4.20
122	0	0
126	1.2	0
127A	0	0
127B	0	0
143	1.1	0
144	0.70	0
148	0	0
149	0.98	0
202	1.4	0
203	0	0
205	2.8	1.36
209	0.75	0.24
215	0.70	0
222A	0	0
222B	1.3	0
233	5.0	1.90
238	1.0	0
245	16.7	4.30
249	1.3	0
250	0	0
255	0	0
259	1.1	0
260	0.40	0
266	1.4	0
276	1.4	0.60
270	0.05	0
286	0	0
287	0	0
288	3.4	0.36
306	18.4	5.00

TABLE 2. *Response of* Euglena gracilis, z *strain, to preparations of the supernatants and cell residues of bacterial cultures, and accompanying data on pH and Klett units of growth*

Units of assay activity in terms of mμg vitamin B_{12} per ml.

Culture No.	Klett units	Final pH	Cell residues	Supernatants
Control	0	7.4	0	0
112	107	7.0	7.00	0.048
116	64	8.4	4.50	0.200
117	66	8.2	2.68	0.035
121	35	7.4	16.2	0.550
122	48	8.4	0	0
126	295	6.3	0.16	0
127A	45	8.2	0.02	0
127B	45	8.2	0	0
143	289	6.4	0	0
144	54	6.9	0	0
148	40	5.9	0	0
203	20	8.2	0	0
205	168	6.6	13.3	0.059
209	110	7.3	1.50	0.004
222A	29	8.3	0	0
222B	222	5.8	0.05	0
233	65	6.8	1.88	0.038
238	107	8.0	0	0
249	259	6.7	0.10	0
250	29	8.5	0	0
266	15	7.7	11.9	0
276	215	7.7	14.0	0.055
279	182	5.0	0.21	0.022
306	30	7.5	3.84	0.870

DISCUSSION

The responses of *E. coli* and *Euglena* to members of the naturally occurring B_{12}-vitamins are discussed in detail by Coates and Ford (1955), Ford and Hutner (1955), and Hutner *et al.* (1956). *E. coli* responds to non-specific substances such as methionine and to Factor B, whereas *Euglena* does not respond to either of them. Factor B (Kon 1955) is the non-nucleotide moiety common to several of the B_{12}-vitamins. If this factor or any one of a number of nucleotides are present, some organisms can complete the synthesis of a form of "vitamin B_{12}" containing that particular nucleotide (Coates and Ford 1955). From the standpoint of the requirements of man, the clinically active vitamin B_{12} or cyanocobalamin is of particular importance. From the standpoint of the growth requirements of marine microorganisms, the pseudovitamin B_{12} and related factors may prove to be as important as cyanocobalamin. Various investigators have demonstrated that cyanocobalamin is a growth factor required by some diatoms and flagellates (Provasoli and Pintner 1953). In addition, Droop (1955) showed that the cyanocobalamin required by the pelagic diatom *Skeletonema costatum* was spared by pseudovitamin B_{12}, Factor A, and Factor B. His noteworthy observations on the lack of specificity of the form of vitamin B_{12} preferred by this diatom indicates the probable importance of "cobalamins" (members of the vitamin-B_{12} family of compounds in addition to cyanocobalamin) in the economy of the sea.

In this investigation vitamin B_{12}-activity produced by marine bacteria was measured with *Euglena* and *E. coli* both of which have different spectra of activity. Of the 34 cultures screened, 30 per cent and 70 per cent had activity for *Euglena* and *E. coli*, respectively. These data are in agreement with those of Ericson and Lewis (1953) who screened 34 cultures using the *E. coli* 113-3 cup-plate assay and found that 70 per cent were producers. Burton and Lochhead (1951) found that 65 per cent of their bacterial cultures were producers. They examined terrestrial isolates using the *Lactobacillus lactis* assay. A table describing the responses of the different assay organisms to vitamin B_{12}-active compounds is given in Ford and Hutner (1955).

In all cases involving aliquots of the same culture which were tested with both assay organisms, the *E. coli* assay gave decidedly higher values (Table 1). This observation indicates the broader spectrum of response of *E. coli* to vitamin B_{12}-active substances compared to *Euglena* and the wide variation of specific vitamin-B_{12} activity produced by different cultures under similar conditions. The maximum levels of vitamin-B_{12} activity produced by culture no. 306 were 5.00 and 18.4 mμg/ml for *Euglena* and *E. coli*, respectively. Using the *L. lactis* assay Burton and Lochhead (1951) found terrestrial bacteria which produced over 500 mμg activity per ml of culture medium. Burkholder and Burkholder (1956) found marine bacteria which produced up to 150 mμg/ml. They used the *E. coli* agar plate method. Using the *E. coli* 113-3 cup-plate assay, Ericson and Lewis (1953) found marine bacteria which produced 0.005–140 mμg/ml. They stated that a level of approximately 10 mμg/ml is more comparable to those obtained for bacteria isolated from seawater and seaweed. This figure is in agreement with the levels (0.40–18.4 mμg/ml by *E. coli*) found during this investigation.

In experiments designed to determine if vitamin B_{12}-activity was confined to the cell residues or released into the medium, results varied with each culture as shown in Table 2. In general intracellular levels of vitamin B_{12} were higher than extracellular. No apparent relationship was evident between the levels of activity of either the cell residues or supernatants and the final bacterial growth or pH of the medium.

As shown in this and related studies by other investigators, the amount and specificity of B_{12}-activity produced by a particular bacterial culture varied with the conditions of the experiment and methods of measurement.

REFERENCES

Burkholder, P. R. 1951. Determination of vitamin B_{12} with a mutant strain of *Escherichia coli*. Science, **114:** 459–460.

Burkholder, P. R., and L. M. Burkholder. 1956. Vitamin B_{12} in suspended solids and marsh muds collected along the coast of Georgia. Limnol. & Oceanogr., **1:** 202–208.

Burton, M. O., and A. G. Lochhead. 1951. Studies on the production of vitamin B_{12}-active susbstance by microorganisms. Can. J. Bot., **29:** 352–359.

Coates, M. E., and J. E. Ford. 1955. Methods of measurement of vitamin B_{12}. In: R. T. Williams, "The Biochemistry of Vitamin B_{12}." Biochem. Soc. Symposia, **13:** 36–51.

Droop, M. R. 1954. Cobalamin requirement in Chrysophyceae. Nature, Lond., **174:** 520.

Droop, M. R. 1955. A pelagic marine diatom requiring cobalamin. J. Mar. Biol. Ass. U. K., **34:** 229–231.

Ericson, L. E., and L. Lewis. 1953. On the occurrence of vitamin B_{12}-factors in marine algae. Ark. Kemi, **6:** 427–442.

Ford, J. E., and S. H. Hutner. 1955. Role of vitamin B_{12} in the metabolism of microorganisms. Vitam. & Horm., **13:** 101–136.

Hoff-Jørgensen, E. 1954. Microbiological assay of vitamin B_{12}. In: B. Glick, "Methods of Biochemical Analysis." vol. 1. Interscience, New York. pp. 81–113.

Hutner, S. H., M. K. Bach, and G. I. M. Ross. 1956. A sugar-containing basal medium for B_{12}-assay with *Euglena*; application to body fluids. J. Protozool., **3:** 101–112.

Hutner, S. H., and L. Provasoli. 1953. A pigmented marine diatom requiring vitamin B_{12} and uracil. News Bull. Phycol. Soc. Amer., **6:** 7–8.

Jukes, T. H., and W. L. Williams; also D. E. Wolf and K. E. Folkers. 1954. In: "The Vitamins." Academic Press, New York. pp. 395–523.

Kon, S. K. 1955. Other factors related to vitamin B_{12}. In: R. T. Williams, "The Biochemistry of Vitamin B_{12}." Biochem. Soc. Symposia, **13:** 17–35.

Lewin, R. A. 1954. A marine *Stichococcus* sp. which requires vitamin B_{12} (cobalamin). J. Gen. Microbiol., **10:** 93–96.

Lochhead, A. G., and M. O. Burton. 1955. Qualitative studies of soil microorganisms. XII. Characteristics of vitamin-B_{12}-requiring bacteria. Can. J. Microbiol., **1:** 319–330.

McNutt, W. S. 1952. Nucleosides and nucleotides as growth substances for microorganisms. Progr. Chem. Org. Nat. Prod., **9:** 401–442.

Østergaard Kristensen, H. P. 1954. Investigations into the *Euglena gracilis* method for quantitative assay of vitamin B_{12}. Acta Physiol. Scand., **33:** 232–237.

Provasoli, L., and I. J. Pintner. 1953. Ecological implications of *in vitro* nutritional requirements of algal flagellates. Ann. N. Y. Acad. Sci., **56:** 839–851.

Starr, T. J. 1956. Relative amounts of vitamin B_{12} in detritus from oceanic and estuarine environments near Sapelo Island, Georgia. Ecology, **37:** 658–664.

Starr, T. J., and M. E. Jones. 1957. The effect of copper on the growth of bacteria isolated from marine environments. Limnol. & Oceanogr., **2:** 33–36.

Sweeney, B. M. 1954. *Gymnodinium splendens*, a marine dinoflagellate requiring vitamin B_{12}. Amer. J. Bot., **41:** 821–824.

Wilson, W. B., and A. Collier. 1955. Preliminary notes on the culturing of *Gymnodinium brevis* Davis. Science, **121:** 394–395.

Part V

THE MICROBIAL ROLE IN THE NITROGEN CYCLE OF THE SEA

Editor's Comments on Papers 30 Through 33

30 DUGDALE, GOERING, and RYTHER
High Nitrogen Fixation Rates in the Sargasso Sea and the Arabian Sea

31 WAKSMAN, HOTCHKISS, and CAREY
Marine Bacteria and their Role in the Cycle of Life in the Sea

32 WATSON
Characteristics of a Marine Nitrifying Bacterium, Nitrosocystis oceanus *sp.n.*

33 PAYNE, RILEY, and COX
Separate Nitrite, Nitric Oxide, and Nitrous Oxide Reducing Fractions from Pseudomonas perfectomarinus

The role of microorganisms in the nitrogen cycle in the sea has been intensively studied over the last 90 years. Dissolved nitrogen exists in the seas primarily in the forms of nitrate, nitrite, ammonia, and organically complexed forms. The papers selected for inclusion in Part V are concerned with the historical aspects, the isolation of the microbes involved, and more recent technical developments for the analysis of microbial activities.

Based upon the studies of Winogradsky and others on soil microorgansims, Brandt (7, 8) postulated that the same microorganisms, or their relatives, were responsible for the nitrogen cycle in the seas. Indeed, there was some basis for this postulate because of the work in 1905 by Keutner (30) on nitrogen fixation of marine bacteria. He found species of *Clostridium* and *Azotobacter* to be primarily responsible for nitrogen fixation and further that these organisms were generally attached to algae or plankton (30). This confirmed his work two years earlier with Benecke (4), which had revealed that both the anaerobic clostridia and the aerobic *Azotobacter* were responsible for nonsymbiotic nitrogen fixation. Keutner then tested for nitrogen fixation in the Baltic, the North Sea, the Indian Ocean, along the coast of Africa, and the

Malaysian Archipelago and found between 0.9 and 16 mg of nitrogen fixed per 200-ml sample, depending upon the amount of glucose and calcium carbonate he added. Benecke (3) returned to the Gulf of Naples, so well studied by Russell (50) and de Giaxa (18), and demonstrated here that nitrogen fixation was occurring in the water, mud, and the algae. Issatchenko (28, 29) showed that the nitrogen-fixing strains of *Azotobacter* and *Clostridium* could be isolated from cultures inoculated with the alga *Fucus*, and he concluded that therefore this was a likely occurrence in the ocean.

In all the attempts to isolate the nitrogen-fixing bacteria, however, two types of compounds had to be present: an energy source and calcium carbonate. Waksman et al. (58) found glucose and mannitol to be excellent carbon sources in their studies. Similar studies have been conducted most recently by Pshenin (45, 46), who investigated the distribution of nitrogen-fixing bacteria in the Black Sea and isolated both *Azotobacter* and *Clostridium* species from water, sediments, and algal thalli.

Since 1962, the demonstration of nitrogen fixation in the marine and coastal environments has progressed beyond the older methods of chemical analysis for the appearance of nitrate, nitrite, and ammonia (1, 7, 8, 12, 28, 30, 39, 42). Two methods are generally in use today. The acetylene reduction method applies only to nitrogen fixation and was developed by Stewart et al. in 1968 (52). It was later modified for field use by the inclusion of Pankhurst tubes (11). A critical evaluation of this technique and a comparison with Kjeldahl and ^{15}N procedures in soils led Rice and Paul (47) to caution that the assay must be calibrated for each ecosystem studied, and allowances must be made for limitations in nitrogen diffusion. Otherwise, gross discrepancies arise in the conversion of moles of acetylene reduced to ethylene to the moles of nitrogen fixed (47). Further restrictions on the interpretation of data were noted by Maruyama et al. (41), who found that organic compounds (especially glucose), pH, NaCl concentration, and temperature greatly influenced the rates of nitrogen fixation.

The other recently developed procedure involves the use of ^{15}N and the determination of the ratio $^{14}N : {}^{15}N$ via mass spectroscopy (21). The usefulness of this method for field determinations of the fate of the various forms of nitrogen has been amply demonstrated. By this technique, Dugdale, Goering, and Ryther found high nitrogen fixation rates in both the Sargasso and Arabian Seas (Paper 30). The method has been applied also to the tropical Atlantic Ocean and even denitrification processes when ^{15}N-labeled nitrate is added (17).

More recently, attention has been directed to estuarine areas (9) and salt marshes as the primary sources of nitrogen fixation because of the extensive presence of both heterotrophic bacteria and blue-green bacteria (25, 52) in this environment. Because of contamination of these areas by fertilizer runoff and sewage, Van Raalte et al. (54) have demonstrated that nitrogen fixation as measured by acetylene reduction can be severely inhibited by the presence of organic nitrogen. Finally, Wynn-Williams and Rhodes (62) have demonstrated that nitrogen fixation can be accomplished by nonsulfur photosynthetic bacteria in shallow bays and Iceland fjord waters. However, they also caution that the amount of bacterial fixation in the open sea is probably insignificant, so that the primary means by which nitrogen becomes bound in the sea is still largely unknown despite the work of the previous 90 years.

That much of the nitrogen fixation occurs in bays, estuaries, and salt marshes around the world becomes an even more likely possibility. These areas can serve then as an explanation for the major point of input into the nitrogen cycle in the oceans.

Once the nitrogen has been fixed into ammonia, it is generally converted via bacterial processes into nitrite and nitrate. The bacterial genera involved in these steps are primarily in the *Nitrosomonas* and *Nitrosocystis* groups (5).

An excellent summary by Waksman, Hotchkiss, and Carey of the numerous studies completed primarily by German and Russian marine scientists until 1930 is presented in Paper 31. In their discussion on nitrification in the sea, the authors present the standard technique for demonstrating both nitrogen fixation and nitrification in which seawater and/or marine sediments are used and enrichment culture conditions are applied. At various time intervals, chemical analysis for nitrite formation is completed on aliquots of the culture medium (58). As has been repeatedly verified, most of the nitrification appears to occur at subsurface depths, and it is especially rapid in the sediments.

The findings by Waksman et al., that there are apparently more ammonia oxidizers than nitrite oxidizers at subsurface depths, have been confirmed by Sugahara et al. (53). Carlucci and Strickland (13) were able to isolate ammonia oxidizers from surface waters in the North Pacific Ocean and from off the Scripps pier. Vargues and Brisou (55) noted an average 10-day lag in ammonia oxidation in samples from the Algerian coastal waters and sediments, but found that the curves for nitrate production matched those for nitrite production. One possible explanation for the discrepancies often encountered in these studies may be found in the work of Hooper and Terry (27), who noted that ammonia oxidation by *Nitrosomonas* could be inhibited by light. They also reported that anaerobic conditions, very high ammonia levels, or methanol, hydroxylamine, or various types of urea could spare this photoinhibitory effect on ammonia oxidation (27). A previous report by Spencer (51) on the requirement for iron during the laboratory cultivation of nitrifiers should not be ignored as a possible cause for the varying degrees of success by different investigators (6, 12, 26, 34–36, 42). In many of the failures to detect nitrification, it was later noted that the addition of diatom extracts improved the chances for obtaining nitrite formation (12, 63). Meanwhile, Spencer (51) had successfully substituted iron for diatom extracts to achieve nitrification. Certainly if this promising lead had been more widely applied, there might be fewer conflicting reports in the literature on the distributions of nitrifiers.

The importance of isolation conditions was also demonstrated by Watson when he was able to isolate and culture *Nitrosocystis oceanus*, an ammonia-oxidizing bacterium, for 600 m in the open Atlantic Ocean. Based upon his earlier work on the importance of pH and ammonia concentration to nitrification processes (59), he successfully obtained pure cultures of this marine ammonia oxidizer (60). His cultural and structural studies on this organism are presented in Paper 32. In this study he describes the internal cytomembrane structures that may be responsible for the ammonia-oxidizing ability of the organism, as well as its physiological requirements for oxidation and growth. Watson and co-workers (61) continued their studies on *N. oceanus* and demonstrated ammonia oxidation to nitrite in a cell-free system, which had been prepared in seawater and supplemented with ATP, magnesium, Tris buffer, and phosphate. Until this publication, few marine bacteria had been so thoroughly studied in both the physiological and morphological aspects.

The reverse stage of the nitrogen cycle, denitrification, has also been a subject of intense investigation. Unfortunately, two processes have been used, almost interchangeably, in describing denitrification. Actually, there is assimilatory nitrate reduction to nitrite, and then ammonia formation which is carried out by a wide variety of bacteria, algae, and fungi. True denitrification involves the respiratory decomposition of nitrate successively to nitrite, nitric and nitrous oxides, and, finally, nitrogen gas (44).

The production of nitrite maxima in the oceans has been ascribed to bacteria because of the widespread distribution of nitrate reducers. Beijerinck (2) and Fischer (24) found that the luminescent bacteria were also nitrate reducers; Russell (49) and Vernon (56) found nitrate reducers in the waters and sediments of the Gulf of Naples; and Brandt (7, 8) and Baur (1) completed extensive studies on "denitrifying" bacteria in the seas. One of the more interesting theories presented in the early part of the twentieth century was proposed by Drew (19, 20), who isolated, in almost pure culture from the seas around the Bahamas, a nitrate-reducing bacterium, *Bacterium calcis*. Because of its presence in extremely high numbers, Drew believed that this organism accounted for the decreased levels of nitrate in tropical waters. As a result, tropical waters were lacking in phytoplankton. In addition, because the ionic balance was upset in the water as a result of these bacterial metabolic activities, an increase in calcium carbonate precipitation resulted, which was mediated by *B. calcis*. Lipman (35), however, showed that calcium carbonate precipitation could occur without nitrate reduction. Furthermore, the fact that *B. calcis* could only grow at elevated temperatures and not at temperatures associated with northern waters led Drew (20) to conclude that this one bacterium was responsible for the differences in nitrate and phytoplankton between the productive northern waters and the nonproductive waters of the tropics. Besides isolation from the water column, it has been amply demonstrated that nitrate-reducing bacteria occur in all marine sediments: Atlantic (12, 26, 33, 43), tropics (30, 33, 48), Irish Sea (37), Clyde Sea (39), Mediterranean Sea (15, 16), Pacific (22, 33), and along the Japanese coast (31, 32).

Despite the ubiquitous nature and ease of isolation of nitrate-reducing bacteria, the ecological importance of this aspect of bacterial activities is still a matter of some debate. Partly, no doubt, this results from the recognition that phytoplankton can utilize forms of nitrogen other than nitrate (22, 38, 40). In Pacific waters, Wada and Hattori (57) noted that nitrite formation was a result of both nitrate reduction and ammonia oxidation. When production by these processes exceeded consumption, nitrite maxima occurred in the water column. Using both light and dark bottles, they demonstrated that consumption and production included algae as well as bacteria (57). In the oxygen-poor waters of Peru, Carlucci and Schubert (14) found that the accumulation of nitrite could be accounted for by facultatively anaerobic bacterial reduction of nitrate. They also isolated the bacteria that were actively reducing nitrate in these waters (14). In pure culture studies with 12 marine pseudomonads, Brown et al. (10) observed that nitrate was reduced to ammonia, which could then be assimilated by the cells for protein synthesis and growth. This assimilation proceeded by the glutamine pathway, involving the enzymes glutamine synthetase and glutamine transferase. These enzymes were not inhibited by the presence of nitrate, nitrite, or ammonia; but the presence of both ammonia and amino acids resulted in complete cessation of nitrate reduction (10). Therefore, increases in the organic nitrogen level in seawater could have the side effect of increasing the nitrate available to phytoplankton for growth, if indeed bacterial nitrate reduction is of ecological importance in the water column.

The isolation of true denitrifying bacteria has been less frequent, although Waksman believed that this group of microorganisms was of great ecological importance, because the end products of their metabolism resulted in a net loss of nitrogen to the oceans (58). One difficulty in proving the existence of true denitrifiers has been a method to demonstrate the gaseous end products. The traditional method has been to use an inverted vial in tubed media and estimate denitrifiers via most-probable-number technique. Making use of advances in gas chromatography, Payne, Riley, and Cox developed a procedure to overcome this problem. In Paper 33 they demonstrate the subterminal and final events in the denitrification process. The utility of this method for field analysis was reported in 1974 by Patriquin and Knowles (43), who found that the number of true denitrifiers in sediments was approximately 10^2 when using the classical, inverted tube method; became 10^4 when using a plate isolation technique; and became over 10^6 per g of wet sediment when they used gas chromatography to test for N_2 and N_2O release. They also noted that the process of denitrification was much more rapid in the presence of calcium carbonate and organic matter, especially with sand from the tropics (Barbados). More extensive studies of this nature will need to be completed before the ecological significance of bacterial denitrification can be established. If these levels of high numbers of denitrifiers in tropical marine sediments should be generally true, one could agree with both Waksman and Drew that nitrogen-depleting bacterial action could account for the lowered productivity of the tropics.

REFERENCES

1. Baur, E. 1901. Ueber zwei denitrificirende Bakterien aus der Ostsee. *Wiss. Meeresunters. Kiel* 6:9–23.

2. Beijerinck, M. W. 1891. Sur l'aliment photogène et l'aliment plastique des bactéries lumineuses. *Arch. Neerl. Sci. Exactes Nat. Haarlem 24*: 369–442.
3. Benecke, W. 1907. Uber stickstoffbindende Bakterien aus den Golf von Neapel. *Ber. Dtsch. Bot. Ges. 25*:1.
4. Benecke, W., and Keutner, J. 1903. Uber stickstoffbindende Bakterien aus der Ostsee. *Ber. Dtsch. Bot. Ges. 21*:333–346.
5. *Bergey's Manual of Determinative Bacteriology,* 8th ed. 1974. The Williams & Wilkins Co., Baltimore, Md., 1246 pp.
6. Berkeley, C. 1919. A study of marine bacteria, Straits of Georgia, B.C. *Trans. Roy. Soc. Can., Sect. 5*:15–40.
7. Brandt, K. 1899. Uber den Stoffwechsel im Meere. *Wiss. Meeresunters. Kiel (N.F.) 4*:213–230.
8. Brandt, K. 1912. Uber den Stoffwechsel im Meere. *3 Abh. Wiss. Meeresunters. Abt. Kiel (N.F.) 18*:186–253.
9. Brooks, R. H., Jr., Brezonik, P. L., Putnam, H. D., and Keirn, M. A. 1971. Nitrogen fixation in an estuarine environment: the Waccasassa on the Florida Gulf Coast. *Limnol. Oceanogr. 16*:701–710.
10. Brown, C. M., MacDonald-Brown, D. S., and Stanley, S. O. 1972. Inorganic nitrogen metabolism in marine bacteria: nitrogen assimilation in some marine pseudomonads. *J. Mar. Biol. Assoc. U.K. 52*:793–804.
11. Campbell, N. E. R., and Evans, H. T. 1969. Use of Pankhurst tubes to assay acetylene reduction by facultative and anaerobic nitrogen-fixing bacteria. *Can. J. Microbiol. 15*:1342–1343.
12. Carey, C. L. 1937–1938. The occurrence and distribution of nitrifying bacteria in the sea. *J. Mar. Res. 1*:291–304.
13. Carlucci, A. F., and Strickland, J. D. H. 1968. The isolation, purification and some kinetic studies of marine nitrifying bacteria. *J. Exp. Mar. Biol. Ecol. 2*:156–166.
14. Carlucci, A. F., and Schubert, H. R. 1969. Nitrate reduction in seawater of the deep nitrite maximum off Peru. *Limnol. Oceanogr. 14*:187–205.
15. Chamroux, S. 1972. Etude de bactéries marines réduisant le nitrate. I. Isolement. *Ann. Inst. Pasteur 122*:475–481.
16. Chamroux, S. 1972. Étude de bactéries marines réduisant le nitrate. II. Caractère halophile. *Ann. Inst. Pasteur 122*:483–488.
17. Chan, Y. K., and Campbell, N. E. R. 1974. A rapid gas-extraction technique for the quantitative study of denitrification in aquatic systems by N-isotope ratio analysis. *Can. J. Microbiol. 20*:275–281.
18. de Giaxa, Prof. 1889. Ueber das Verhalten einiger pathogenes Mikroorganismen im Meerwasser. *Z. Hyg. Infectionskr. 6*:162–224.
19. Drew, G. H. 1913. On the precipitation of calcium carbonate in the sea by marine bacteria, and on the action of denitrifying bacteria in tropical and temperate seas. Papers from the Marine Biological Laboratory at Tortugas. *Carnegie Inst. Wash. Publ. 182*:2–45.
20. Drew, G. H. 1913. Report of preliminary investigations on the marine denitrifying bacteria, made at Port Royal, Jamaica, and at Tortugas during May and June 1911. *Yearbook, Carnegie Inst. Wash. 10*:136–141.
21. Dugdale, R. C., Goering, J. J., and Ryther, J. H. 1964. High nitrogen fixation rates in the Sargasso Sea and the Arabian Sea. *Limnol. Oceanogr. 9*:507–510.

22. Eppley, R. W., Carlucci, A. F., Holm-Hansen, O., Kiefer, D., McCarthy, J. J., Venrick, E., and Williams, P. M. 1971. Phytoplankton growth and composition in shipboard cultures supplied with nitrate, ammonium, or urea as the nitrogen source. *Limnol. Oceanogr. 16*:741–751.
23. Feitel, R. 1902. Beiträge zur Kenntniss denitrifizirender Meeresbakterien. *Wiss. Meeresunters. Kiel 7*:91–105.
24. Fischer, B. 1894. Die Bakterien des Meeres nach den Untersuchungen den Planktonexpedition unter gleichzeitiger Berücksichtigung einiger älterer und neuerer Untersuchungen. *Zentralbl. Bakteriol. 15*:657–666.
25. Goering, J. J., Dugdale, R. C., and Menzel, D. W. 1966. Estimates of *in situ* rates of nitrogen uptake by *Trichodesmium* sp. in the tropical Atlantic Ocean. *Limnol. Oceanogr. 11*:614–620.
26. Gran, H. H. 1902. Havets Bakterier og deres Stofskifte. *Nat. Bergens 27*:72–84.
27. Hooper, A. B., and Terry, K. R. 1974. Photoinactivation of ammonia oxidation in *Nitrosomonas*. *J. Bacteriol. 119*:899–906.
28. Issatchenko, B. L. 1908. Zur Frage von der Nitrifikation in den Meeren. *Zentralbl. Bakteriol., Abt. II, 21*:430.
29. Issatchenko, B. L. 1926. Sur la Nitrification dans les mers. *C. R. Hebd. Seances Acad. Sci. 182*:185–186.
30. Keutner, J. 1905. Über das Vorkommen und die Verbreitung stickstoffbindender Bakterien im Meere. *Wiss. Meeresunters. Kiel 8*:29–45.
31. Kimata, M., Yoshida, Y., and Taniguchi, M. 1968. Studies on the marine microorganisms utilizing inorganic nitrogen compounds. I. On the marine denitrifying bacteria. *Bull. Jap. Soc. Sci. Fish. 34*:1114–1117.
32. Kimata, M., Yoshida, Y., and Taniguchi, M. 1969. Studies on the marine microorganisms utilizing inorganic nitrogen compounds. II. On the marine heterotrophic bacteria assimilating inorganic nitrogen compounds. *Bull. Jap. Sco. Sci. Fish. 35*:211–214.
33. Kriss, A. E. 1963. *Marine Microbiology (Deep Sea)*. J. M. Shewan and Z. Kabata (trans.). John Wiley & Sons, Inc., New York, 536 pp.
34. Lipman, C. B. 1922. Does nitrification occur in sea water? *Science 56*: 501–503.
35. Lipman, C. B. 1924. A critical and experimental study of Drew's bacterial hypothesis on $CaCO_3$ precipitation in the sea. *Carnegie Inst. Wash. Publ. 19*:179–191.
36. Lipman, C. B. 1929. Further studies on marine bacteria with special reference to the Drew hypothesis on $CaCO_3$ precipitation in the sea. *Carnegie Inst. Wash. Publ. 26*:231–257.
37. Litchfield, C. D. 1972. Interactions of amino acids and marine bacteria, pp. 145–169. In: L. H. Stevenson and R. R. Colwell (eds.), *Estuarine Microbial Ecology*, Belle W. Baruch Library in Marine Science, Vol. 1. University of South Carolina Press, Columbia, S.C., 536 pp.
38. Liu, M. S., and Hellebust, J. A. 1974. Utilization of amino acids as nitrogen sources and their effects on nitrate reductase in the marine diatom *Cyclotella cryptica*. *Can. J. Microbiol. 20*:1119–1125.
39. Lloyd, B. 1931. A marine denitrifying organism. *J. Bacteriol. 21*:89–96.
40. MacIsaac, J. J., and Dugdale, R. C. 1969. The kinetics of nitrate and ammonia uptake by natural populations of marine phytoplankton. *Deep-Sea Res. Oceanogr. Abst. 16*:45–57.

41. Maruyama, Y., Suzuki, T., and Otobe, K. 1972. Nitrogen fixation in the marine environment: the effect of organic substrates on acetylene reduction, pp. 341–353. In: L. H. Stevenson and R. R. Colwell (eds.), *Estuarine Microbial Ecology*, Belle W. Baruch Library in Marine Science, Vol. 1. University of South Carolina Press, Columbia, S.C., 536 pp.
42. Nathansohn, A. 1906. Ueber die Bedentung vertikales Wasserbewegungen fur die Produktion des Planktons in Meere. *Abh. Math.-Phys. Kl., K. Sachs. Ges. Wiss.* *29*:355–441.
43. Patriquin, D. G., and Knowles, R. 1974. Denitrifying bacteria in some shallow-water marine sediments: enumeration and gas production. *Can. J. Microbiol.* *20*:1037–1041.
44. Payne, W. J., Riley, P. S., and Cox, C. D., Jr. 1971. Separate nitrite, nitric oxide, and nitrous oxide reducing fractions from *Pseudomonas perfectomarinus*. *J. Bacteriol.* *106*:356–361.
45. Pshenin, L. N. 1959. The quantitative distribution of nitrogen-fixing bacteria and their ecology in the region of the Zernov Phyllophora field of the Black Sea. *Microbiology USSR* *28*:866–902.
46. Pshenin, L. N. 1963. Distribution and ecology of *Azotobacter* in the Black Sea, pp. 383–391. In: C. H. Oppenheimer (ed.), *Marine Microbiology*. Charles C Thomas, Publisher, Springfield, Ill., 769 pp.
47. Rice, W. A., and Paul, E. A. 1971. The acetylene reduction assay for measuring nitrogen fixation in waterlogged soil. *Can. J. Microbiol.* *17*:1049–1056.
48. Richards, F. A., and Broenkow, W. W. 1971. Chemical changes, including nitrate reduction in Darwin Bay, Galapagos Archipelago, over a 2-month period, 1969. *Limnol. Oceanogr.* *16*:758–765.
49. Russell, H. L. 1892. Untersuchungen über in Golf von Neapel lebende Bakterien. *Z. Hyg. Infectionskr.* *11*:165–204.
50. Russell, H. L. 1892. Bacterial investigation of the sea and its floor. *Bot. Gaz.* *17*:312–321.
51. Spencer, C. P. 1956. The bacterial oxidation of ammonia in the sea. *J. Mar. Biol. Assoc. U.K.* *35*:621–630.
52. Stewart, W. D. P., Fitzgerald, G. P., and Burris, R. H. 1968. Acetylene reduction by nitrogen fixing blue-green algae. *Arch. Mikrobiol.* *62*: 336–348.
53. Sugahara, I., Sugiyama, M., and Kawai, A. 1974. Distribution and activity of nitrogen cycle bacteria in water-sediment systems with different concentrations of oxygen, pp. 327–341. In: R. R. Colwell and R. Y. Morita (eds.), *Effect of the Ocean Environment on Microbial Activities*. University Park Press, Baltimore, Md., 587 pp.
54. Van Raalte, C. D., Valiela, I., Carpenter, E. J., and Teal, J. M. 1974. Inhibition of nitrogen fixation in salt marshes measured by acetylene reduction. *Estuarine Coastal Mar. Sci.* *2*:301–305.
55. Vargues, H., and Brisou, J. 1963. Researches on nitrifying bacteria in ocean depths on the coast of Algeria, pp. 415–425. In: C. H. Oppenheimer (ed.), *Marine Microbiology*. Charles C Thomas, Publisher, Springfield, Ill., 769 pp.
56. Vernon, H. M. 1898. The relations between marine, animal and vegetable life. *Mitt. Zool. Sta. Neapel* *13*:341–425.
57. Wada, E., and Hattori, A. 1971. Nitrite metabolism in the euphotic

layer of the Central North Pacific Ocean. *Limnol. Oceanogr. 16*:766–772.

58. Waksman, S. A., Hotchkiss, M., and Carey, C. L. 1933. Marine bacteria and their role in the cycle of life in the sea. II. Bacteria concerned in the cycle of nitrogen in the sea. *Biol. Bull. 65*:137–167.
59. Watson, S. W. 1963. Autotrophic nitrification in the ocean, pp. 73–84. In: C. H. Oppenheimer (ed.), *Marine Microbiology.* Charles C Thomas, Publisher, Springfield, Ill., 769 pp.
60. Watson, S. W. 1965. Characteristics of a marine nitrifying bacterium, *Nitrosocystis oceanus* sp. n. *Limnol. Oceanogr. 10*:R274–R289.
61. Watson, S. W., Asbell, M. A., and Valois, F. W. 1970. Ammonia oxidation by cell-free extracts of *Nitrosocystis oceanus. Biochem. Biophys. Res. Commun. 38*:1113–1119.
62. Wynn-Williams, D.D., and Rhodes, M. E. 1974. Nitrogen fixation by marine photosynthetic bacteria. *J. Appl. Bact. 37*:217–224.
63. ZoBell, C. E. 1935. The assimilation of ammonium nitrogen by *Nitzschia closterium* and other marine phytoplankton. *Proc. Natl. Acad. Sci. USA 21*:517–522.

30

Reprinted from *Limnol. Oceanogr.*, 9(4), 507–510 (1964)

HIGH NITROGEN FIXATION RATES IN THE SARGASSO SEA AND THE ARABIAN SEA[1]

Richard C. Dugdale and John J. Goering
Institute of Marine Science, University of Alaska, College

and

John H. Ryther
Woods Hole Oceanographic Institution, Woods Hole, Massachusetts

ABSTRACT

Nitrogen fixation rates have been measured in the Sargasso Sea and the Arabian Sea using the N^{15} method. Heavy enrichments in N^{15} were found in a number of experiments in which the blue-green alga, *Trichodesmium*, was used for the experimental material.

INTRODUCTION

In a previous communication (Dugdale, Menzel, and Ryther 1961), we reported the results of an experiment in which low nitrogen fixation rates associated with the blue-green alga, *Trichodesmium*, were measured using the N^{15} method. We have recently obtained confirmation of large-scale nitrogen fixation in the Sargasso Sea, where the original experiment was carried out, and in the Arabian Sea.

METHODS

The N^{15} method for detecting nitrogen fixation has been adapted by Neess et al. (1962) to the measurement of fixation rates in aquatic communities. The following procedure is carried out after placing water containing the desired organisms in a 1-liter flask having a standard taper joint at the neck and into which a specially designed gas flushing unit is fitted: 1) Atmospheric N_2 (as well as other gases) is flushed from the water with a mixture of 80% He and 20% O_2 at a pressure of 0.8 atm; 2) 0.2 atm N_2 enriched to 95% N^{15} is added to the flask and equilibrated with the aqueous phase by shaking; 3) the sample is incubated in the flask under the conditions selected for the experiment; 4) the particulate fraction is captured on a glass filter (Hurlburt 984H) and converted to molecular nitrogen by a Dumas combustion (Dugdale and Barsdate 1964); 5) the N^{15}/N^{14} ratio of the resulting N_2 is determined with a mass spectrometer, the ratio converted to atom per cent N^{15}, and the enrichment over the normal atom per cent N^{15} of the organic material calculated.

Isotope ratios were measured with either a Consolidated-Nier 21-201 magnetic mass spectrometer or with a Bendix 17-210 time-of-flight mass spectrometer. The atom per cent N^{15} values in Table 2 that have two significant figures to the right of the decimal point were obtained with the time-of-flight spectrometer. All other measurements were made with the magnetic spectrometer. The atom per cent excess N^{15} associated with each flask has been calculated by subtracting the control (normal atom per cent N^{15} of *Trichodesmium*) or by subtracting the average of the five Bermuda controls (0.349) given in Table 1.

RESULTS

Results for the Sargasso Sea nitrogen fixation experiments conducted at Bermuda are given in Table 1. Material for these experiments was collected about 4 km from Bermuda by towing a No. 8 or No. 20 plankton net at the surface. Upon return to the laboratory, individual colonies of

---

[1] Contribution No. 3 from the Institute of Marine Science and No. 1510 from the Woods Hole Oceanographic Institution. This research was supported by National Science Foundation Grant GB-24 and by the United States Biological Program of the International Indian Ocean Expedition. We wish to thank Dr. F. A. Richards, Department of Oceanography, University of Washington, for the Consolidated-Nier mass spectrometer used in these studies.

TABLE 1. *Sargasso Sea fixation results*

Date	Sample	Light	Length of incubation (hr)	Atom % N^{15}	Atom % N^{15} excess
4 Sept 1962	Control	—	—	0.348	—
	Flask 1	Artificial light	13	0.350	0.002
11 Sept 1962	Control	—	—	0.348	—
	Flask 1	Artificial light	18	0.349	0.001
	Flask 2	Artificial light	24	0.356	0.008
20 Sept 1962	Control	—	—	0.350	—
	Flask 1	Direct sunlight	8	0.518	0.169
	Flask 2	Direct sunlight	8	0.566	0.217
1 Oct 1962	Flask 1	Dark	24	0.353	0.004
	Flask 2	Artificial light	24	0.358	0.009
12 Oct 1962	Control	—	—	0.350	—
	Flask 1	50% of incident light	4.5	0.399	0.049
	Flask 2	80% of incident light	4.5	0.377	0.027
	Flask 3	100% of incident light	4.5	0.384	0.034
19 Oct 1962	Control	—	—	0.351	—
	Flask 1	5 hr direct sunlight, 15 hr artificial light	20	0.559	0.208
	Flask 2	5 hr direct sunlight, 15 hr artificial light	20	0.447	0.096

Trichodesmium were separated from the other plankton and placed in fixation flasks containing Millipore®[2] (0.45 μ)-filtered surface seawater. After treatment with N_2^{15}, the flasks were incubated for various lengths of time either in artificial light (16,000 lux) at 20C or in a seawater-cooled box exposed to direct sunlight. The box was fitted with neutral density filters for the 12 October 1962 experiment.

In some experiments (20 September 1962, 19 October 1962), the rate of fixation was about 20 times the highest rate reported in the earlier communication (Dugdale et al. 1961). In the remainder, fixation was low or undetectable. The highest rates observed here are comparable to fixation rates observed in lakes during periods of nitrogen-fixing blooms (Dugdale and Dugdale 1962). The 19 October 1962 experiment, in which 1 μg-at. NH_4^+-N/liter was added to flask 2, suggests that NH_4^+-N inhibits fixation. That a certain degree of variability occurs within a given experiment may be seen in the 20 September 1962 results, in which like treated flasks showed a difference in rate of fixation. Some of the experimental conditions that may affect fixation will be discussed later.

The results of nitrogen fixation measurements made during Cruise 4A of the U.S. Biological Ship for the International Indian Ocean Expedition, the *RV Anton Bruun*, are summarized in Table 2. The majority of the stations is located in the northern Arabian Sea; Stations 174 and 176 lie farther south along the coast of Saudi Arabia. The enrichments in N^{15} observed are similar to those shown in Table 1, the value for the 4-hr experiment on Station 188, 4.65 atom per cent excess, being the highest measured in the sea to date.

Colonies of *Trichodesmium* were used for experimental material in all the experiments except at Station 183, where a large quantity of *Rhizosolenia* was collected from the surface, and at Station 194, where a heavy bloom of *Noctiluca* occurred. Two distinct forms of *Trichodesmium* were observed in the No. 20 mesh nets normally towed for 10 min or less at 1, 10, 20, 30, and 40 m, one occurring as green bundles of filaments and the other as larger brown spherical clumps of filaments corresponding to the form we have worked with in the

[2] Registered trademark, Millipore Filter Corporation, Bedford, Massachusetts.

TABLE 2. *Indian Ocean fixation results*

Date	Station	Latitude	Longitude	Length of incubation (hr)	Atom % N^{15}	Atom % N^{15} excess
17 Oct 1963	174	16°27′ N	54°40′ E	32	0.719	0.370
18 Oct 1963	176	16°28′ N	57°09′ E	5.5	0.387	0.038
28 Oct 1963	183	23°42′ N	66°21′ E	9	0.366	0.017
29 Oct 1963	184	22°34′ N	65°50′ E	7	1.024	0.675
31 Oct 1963	188	23°21′ N	64°52′ E	4	5.00	4.65
1 Nov 1963	190	24°47′ N	61°39′ E			
	Flask *a*	10 colonies		5	0.81	0.46
	Flask *b*	Numerous colonies		5	0.47	0.12
1 Nov 1963	191	23°57′ N	60°58′ E	10	0.352	0.003
1 Nov 1963	192	23°08′ N	60°36′ E	30	0.902	0.553
3 Nov 1963	194	22°22′ N	60°06′ E	28	0.354	0.005
3 Nov 1963	195	21°32′ N	60°40′ E	38	0.354	0.005

Sargasso Sea. At six stations, the former type was observed at virtually all the depths mentioned above; the latter appeared and became abundant in the collections at Stations 191, 192, 193, and 195, all in northwestern Arabian Sea near the Gulf of Oman.

When the negative results with *Rhizosolenia* (Station 183) and *Noctiluca* (Station 194) are removed from consideration, it becomes clear that high enrichments occurred in a large proportion of the *Trichodesmium* experiments, that is, in five out of eight possibilities. Cloudiness, suggesting the presence of bacteria, developed quickly in the incubation flasks from Stations 191 and 195, a possible explanation for the failure to observe nitrogen fixation at these stations, which showed heavy concentrations of the brown form of *Trichodesmium*. Strong fixation occurred in experiments using the brown form exclusively (Station 192) and in those using the "green bundles" form (Station 190).

DISCUSSION

The variability observed in these measurements (*see* below) precludes certain levels of speculation regarding the significance of these data. However, we consider it a virtual certainty that the large-scale blooms of *Trichodesmium* reported from tropical oceanic regions are indeed nitrogen-fixing blooms analogous to those observed in lakes associated with several species of *Anabaena* by Dugdale and Dugdale (1962) and by Goering (1962).

Certain reservations must be made; for example, our experiments do not prove that *Trichodesmium* is itself able to fix nitrogen. However, this is unimportant from the point of view of the ecologist, since the ability to fix nitrogen has been clearly shown to lie with the *Trichodesmium* colonies (that is, the alga and any associated bacteria or fungi).

Up to this point we have not been concerned with the species composition of the experimental material beyond noting obvious macroscopic differences. Three species, *T. erythraeum* Ehrenb., *T. hildebrantii* Gom., and *T. thiebautii* Gom., are reported for the Indian Ocean by Desikachary (1959), and McLeod, Curby, and Bobblis (1962) suggested that two or more species may have been present in their collections at Bermuda.

The data obtained so far have been characterized by a disturbing lack of consistency, that is, experiments at Bermuda separated by only a few days give highly divergent results, and the same is true for the cruise data. We suspect that a large portion of this discrepancy may lie in the experimental method. Now that we have obtained N^{15} enrichments of a high order, replication and suitable experimental design should yield insight into the problem. McLeod et al. (1962) report a twentyfold variation in the rate of photosynthesis from cells of a given collection of *Trichodesmium*. In some collections, cells were present that would not photosynthesize, and other cells would do so only after a period of adaptation to light. Therefore, some of the variability from day to day and within any single experiment may be the result of differences in the physiological condition of the colonies. Menzel (1962) has also shown an autoinhibition of photosynthesis by *Trichodesmium* at Bermuda; at Station 190 flask *a* contained only 10 colonies and fixed nitrogen at double the rate of flask *b*, that contained numerous colonies. The organisms may also be sensitive to other features of the technique such as the length of sparging and composition of the sparging gas.

REFERENCES

DESIKACHARY, T. V. 1959. Cyanophyta. Academic, New York. 686 p.

DUGDALE, R. C., AND R. J. BARSDATE. 1964. Rapid conversion of organic nitrogen to N_2 for mass spectrometry by an automated Dumas procedure. (Unpublished manuscript.)

———, D. W. MENZEL, AND J. H. RYTHER. 1961. Nitrogen fixation in the Sargasso Sea. Deep-Sea Res., **7**: 298–300.

DUGDALE, V. A., AND R. C. DUGDALE. 1962. Nitrogen metabolism in lakes II. Role of nitrogen fixation in Sanctuary Lake, Pennsylvania. Limnol. Oceanog., **7**: 170–177.

GOERING, J. J. 1962. Studies of nitrogen fixation in natural fresh waters. Ph.D. Thesis. Univ. of Wisconsin. 133 p.

MCLEOD, G. C., W. A. CURBY, AND F. BOBBLIS. 1962. The study of the physiological characteristics of *Trichodesmium thiebautii*. A.E.C. Rept. Contrib. AT(30-1) 2646. Bermuda Biological Station. 13 p.

MENZEL, D. W. 1962. Inhibition of photosynthesis by *Trichodesmium* in the Sargasso Sea. A.E.C. Rept. Contrib. AT(30-1) 2646. Bermuda Biological Station. 6 p.

NEESS, J. C., R. C. DUGDALE, V. A. DUGDALE, AND J. J. GOERING. 1962. Nitrogen metabolism in lakes. I. Measurement of nitrogen fixation with N^{15}. Limnol. Oceanog., **7**: 163–169.

31

Reprinted from pp. 138–146 of *Biol. Bull.*, **65**, 137–167 (1933)

MARINE BACTERIA AND THEIR ROLE IN THE CYCLE OF LIFE IN THE SEA

S. A. Waksman, M. Hotchkiss, and C. L. Carey

[*Editor's Note:* In the original, material precedes this excerpt.]

Occurrence of Nitrifying Bacteria in the Sea

Historical

The formation of nitrate in the sea, as a result of oxidation of ammonia, is usually considered as the final step in the transformation of nitrogen, before it is made again available for assimilation by green plants; this does not exclude, of course, the probability that ammonia itself, the final nitrogenous product of decomposition of organic matter by bacteria, and nitrite, the first product of oxidation of ammonia, can also be utilized by marine algæ and by the chlorophyl-containing members of the plankton, as sources of nitrogen. Among the various marine problems, for which bacteria are believed to be responsible, none has aroused more discussion and greater interest than the process of nitrate formation in the sea, with the possible exception of nitrate-reduction by bacteria. This interest is due to the importance of nitrate formation in the metabolism of the sea and to the difficulty of studying the bacterial agents responsible for this process. An exact analogy is found in the study of the agents of nitrification on land: even forty years after the respective organisms have been isolated and cultivated by Winogradsky, papers still continue to appear which not only question the rôle of these organisms in the process but frequently doubt their very existence.

Boussingault looked upon the ocean, in 1860, as an immense reservoir of nitrogen in a combined form. Schlösing (43) demonstrated, in 1875, that while land waters are richer in nitrate, sea waters are richer in ammonia. It was recognized, however, that both ammonia and nitrate result from the decomposition of nitrogenous organic matter in the sea. Natterer (36) measured the NH_3, NO_2, and NO_3 content of sea water and found that the first was most abundant, while the last two were present in mere traces or were entirely absent; he believed that nitric acid produced by electric discharges sooner or later reaches the sea, enters there into organic combination, and is finally transformed to ammonia; this diffuses then into the atmosphere, and contributes again to the growth of plants on land; these results were not confirmed, however, by subsequent investigators. Brandt (9, 10) stated emphatically that the cycle of nitrogen in the sea is essentially not very different from that on land.

The first suggestion concerning the existence in the sea (Gulf of Naples) of bacteria responsible for the process of nitrification was made in 1898 by Vernon (45). Baur and Brandt (2) are believed to have demonstrated in 1900 the presence of nitrifying bacteria in the sea: two out of three mud samples inoculated into a solution containing ammonium salts gave active nitrification; however, sea water itself seemed to be free from the organisms concerned in this process. These results could not be confirmed by Gran (19) and Nathanson (37), who were unable to demonstrate these organisms either in the Norwegian fjords or in the Gulf of Naples. Gran found nitrifying bacteria only close to the shore. This led Nathanson to conclude that when bacteria are found in the sea not far from land, it is due to their introduction from the land soils by streams and land drainage. Gran and Nathanson adhered, therefore, to the earlier hypothesis of Schlösing that nitrates are not formed in the open sea, but are brought there from the outside, either from the atmosphere or from land.

Brandt argued that, if the nitrate comes into the sea either from the atmosphere or from land by means of streams and rivers, one would expect to find nitrate more abundant in the surface layers of the sea rather than in the lower depths, which is contrary to actual facts. The abundance of nitrate in the deeper layers of water led Brandt to conclude that nitrification takes place chiefly in the sea bottom or close to it. It was recognized, however, that close to the mouth of the rivers the relative concentration of nitrate was greater than in the open sea, but this was also found to hold true for other forms of nitrogen, namely ammonia and protein.

Thomsen (44) demonstrated the presence of nitrite-forming bacteria in considerable abundance in the sea bottom, although they were absent in the sea water and on algal material; the nitrate-forming organisms were also present in the bottom material, but only close to shore. The organisms responsible for the two processes, namely *Nitrosomonas* and *Nitrobacter,* were isolated from the sea and were found to be morphologically the same as the corresponding forms isolated by Winogradsky from land soil. The marine nitrite bacteria were considered as adaptation forms, their optimum temperature being similar to that of the bacteria from land. Thomsen found nitrifying bacteria not only in the mud from the Gulf of Naples, but also from the Kiel Bay and the North Sea. He believed that the negative results of

Nathanson were due to the nature of the medium which he had used for demonstrating the presence of these organisms, and to the fact that the cultures were not incubated sufficiently long and that the temperature was unfavorable. The occurrence of these bacteria on the bottom of the sea and not in the free water itself was explained by the fact that the water contains only traces of ammonia, while the continuous decomposition of plant and animal residues in or on the bottom supplies the necessary substrate for their action.

Issatchenko (24, 25) reported in 1908 that he found nitrifying bacteria in the water of the northern Arctic Sea; these organisms were present in the bottom material of the Catherine Coast (Murmansk) and of the North Ice Sea, as well as of the high seas; they were absent, however, in the surface water. Only nitrite-forming bacteria were found in the sea bottom, but not the nitrate-formers. In a later contribution (26), Issatchenko has shown that the nature of the sea bottom material is of importance in determining the abundance of nitrifying bacteria; these organisms were more abundant and could be more readily demonstrated in sandy bottoms and in shell-rich bottoms than in clay bottoms.

Liebert (30) attempted, in 1915, to isolate nitrifying bacteria from the water of high seas and from the ocean bottom; his results were entirely negative. He concluded that sea waters at a distance from land contain no bacteria capable of oxidizing ammonia and nitrate, due to the low content of these nitrogenous compounds in the sea. Marine mud from the North Sea also gave negative results, except in close proximity to shore. However, both nitrite and nitrate formation took place in the Zuyder Sea. Liebert believed that the FeS present so abundantly in the bottom of the ocean may plav an important rôle in the oxidation of the ammonia in marine bottoms. Berkeley (14), in 1919, tested sea water for the presence of nitrifying bacteria with negative results, even after three months incubation of the cultures. As a result of a series of investigations, Lipman (31) concluded that although nitrifying bacteria are absent in the water of the open sea, they are present in the sea bottom, such as calcareous sand. Similar results were reported by Harvey (21–23).

On the basis of these results, Brandt (11) concluded in 1926, with much justification, that the results so far obtained are sufficient to establish definitely the fact that bacteria capable of oxidizing ammonium salts are completely lacking in surface waters, but are present in marine bottoms.

The possibility of photo-chemical oxidation in the sea of ammonia to nitrite and even to nitrate has recently been suggested (51). It has been known (39) that solutions of ammonia and ammonium salts exposed to sunlight, in the presence of small quantities of a photo-sensitizer, will give rise to nitrite, especially in alkaline solutions. TiO_2, ZnO, CdO, and others act as photo-sensitizers. The photochemical formation of nitrite and nitrate can, at best, however, explain only partly the origin of nitrates in the sea, since it must take place in the surface layers, with the formation of ammonia preceding this process. The fact that the nitrate is largely found in the lower layers of water would tend to emphasize the probable limitation of this process in the formation of nitrate in the sea.

Experimental

In an attempt to study the occurrence and activities of nitrifying bacteria in the sea, it was deemed essential to establish at first the conditions favorable for the growth of these organisms. One deals here with bacteria highly selective in their metabolism, very sensitive to environmental conditions, and specific in their food requirements, as was amply shown for the corresponding organisms universally active and abundant in land soils. Whether or not the bacteria responsible for nitrite and nitrate formation in the sea are as highly sensitive to conditions as the land bacteria, the first prerequisite in such investigations was to select a medium favorable for their development under artificial laboratory conditions. Several preliminary experiments were, therefore, carried out in order to test sea water as a medium and the most optimum conditions for the growth of these organisms.

At first, fresh sea water, to which 0.05 per cent K_2HPO_4, some $CaCO_3$, and varying amounts of ammonium salts had been added, was used. These experiments have shown that neither nitrate nor nitrite was formed in such a medium. However, when fresh sand or mud was introduced into the flasks, active nitrite formation took place. The results of a typical experiment are reported in Table I. The medium used in this experiment consisted of,

Fresh sea water	1000 cc.
K_2HPO_4	1 gram
$(NH_4)_2SO_4$	1 gram
$CaCO_3$	5–10 grams

This medium was placed in a series of flasks; these were plugged with cotton and the medium left unsterilized. Some of the flasks received 30-gram portions of fresh bottom mud taken off the shore of Gay Head, while others did not receive any mud. At frequent intervals nitrite tests were made by removing 1-cc. portions of the culture, by the use of sterile pipettes. The test was made with a mixture of sulphanilic acid and diethyl β-naphthylamine solutions. One cc. of culture was diluted with 5 cc. of distilled water and 1-cc. portions of each of the two reagents added. The color was read after 5 or 10 minutes.

The results brought out in Table I clearly demonstrate the fact that sea water either does not contain any bacteria capable of oxidizing ammonium salts to nitrite or is not a favorable medium for the development of these bacteria. In the culture to which marine mud has been added active nitrite formation took place, even after 20 days incubation; the amount of nitrite formed increased rapidly on further incubation. This can be due either to the presence of nitrifying bacteria in the mud or to the fact that the mud made the sea water medium more favorable for the development of these organisms.

TABLE I

Nitrite formation in sea water medium and in sea water medium to which fresh marine mud has been added.

Tr = trace of nitrite; 0 indicates negative test; + = positive nitrite test; +++ = extensive nitrite formation; ++++ = maximum nitrite reaction; – indicates culture discarded.

Treatment of culture	Days of incubation of cultures						
	6	15	20	24	28	33	38
Water alone	0	0	0	0	tr	tr	tr
Water alone	0	0	0	0	tr	–	–
Water and mud	0	0	+	+++	++++	++++	++++
Water and mud	0	0	+	+++	++++	–	–

Another important point to be noted from this experiment is that only the tests for nitrite were positive, while the tests for nitrate were all negative, even in the mud cultures. Attention has been called, in the review of the literature, to the comparative rarity in the sea of the nitrate-forming bacteria, as compared with the nitrite-formers. This would seem to be an anomaly, since there is very little nitrite present in the sea as compared with nitrate. If one remembers, however, the fact that the nitrate bacteria are highly sensitive to traces of free ammonia, which would be produced from the ammonium salt in an alkaline medium, one would expect to find these organisms, if they are present at all, only after all the ammonia has been oxidized to nitrite. The tests were continued for a longer period, but no trace of nitrate could be detected within 57 days of incubation; however, after 60 to 62 days, the mud containing cultures gave a definite test for nitrate, and after 77 to 84 days, the test became strongly positive.

An experiment was then started to determine the influence of the initial concentration of ammonium salt added to the original medium upon the development of nitrite and nitrate-forming bacteria, since some of the previous investigators (31, 33) considered this to be an important factor. This experiment was carried out as follows: Twelve 250-cc. Erlenmeyer flasks received 40-gram portions of washed sea sand and 2 grams $CaCO_3$; the flasks were plugged with cotton and sterilized for 1 hour under pressure. Sixty-cc. portions of sea water containing 0.1 per cent K_2HPO_4, previously heated for 1 hour at 70° C., were then added to the flasks, as well as varying amounts of an ammonium sulfate solution, also heated previously at 70° C. Five of the flasks were inoculated with fresh surface water taken from Great Harbor,

TABLE II

Effect of varying concentrations of ammonium sulfate upon nitrite formation in sea water-sand medium.

0 indicates negative test; + = positive nitrite test; ++ and +++ = extensive nitrite formation; ++++ = maximum nitrite formation.

Concentration of $(NH_4)_2SO_4$ in 60 cc. of medium	Sea water inoculum					Mud inoculum					Enriched culture inoculum			
	Days of incubation					Days of incubation					Days of incubation			
	3	7	12	17	20	3	7	12	17	20	3	7	12	17
mgm.														
0	0	0	0	0	0	0	0	+	++++	++++				
10	0	0	0	0	0	0	0	++	++++	++++				
25	0	0	0	+	+++	0	0	0	+	++	0	0	+	++++
50	0	0	0	+	++	0	0	+	+++	++++	0	0	+	++++
100	0	0	0	0	0	0	0	0	++	++++				

near the Oceanographic Institution wharf, five flasks were inoculated with sea bottom mud from off the shore of Gay Head and 2 flasks with 4 drops of an enriched culture of nitrite-forming bacteria grown in a sand-sea water medium.

The concentration of the ammonium salt and the nature of the inoculum were found to have an important effect upon the rapidity of nitrite formation, as shown in Table II. The use of sea water as an inoculum gave no nitrite formation with either the highest or the lowest concentrations of the ammonium salt; however, a positive reaction was obtained in the cultures containing 25 and 50 mgm. of the salt, after 17 days incubation. The cultures inoculated with mud gave a positive nitrite test in some cultures in 12 days and in all cultures in 17 days. The mud cultures behaved in a manner similar to the enriched culture of the nitrite-forming organism.

Some of the cultures in the above experiment were incubated for a longer period than that reported in Table II. The culture containing 10 milligrams of ammonium sulfate and inoculated with mud gave very abundant nitrite formation up to 34 days; after 42 days, however, the nitrite disappeared. It had been completely and rapidly converted to nitrate.

In the previous experiments, both the water and the marine bottom material were obtained in the proximity of land. The following ex-

TABLE III

Presence of nitrite-forming bacteria in the sea water of the Gulf of Maine (Stations 1329, 1330, 1331, and 1332) and Georges Bank (Stations 1333 and 1334) at different depths.

Tr = trace; 0 indicates negative test; + indicates positive reaction; ++ = extensive nitrite formation; +? indicates doubtful reaction. For map showing location of stations, see Rakestraw, 1933, *Biol. Bull.*, **64**: 150.

Station No.	Depth of water	Days of incubation			
		10	15	20	23
	meters				
1329	Surface water	0	+?	0	0
1329	Above bottom	0	0	0	0
1330	Above bottom	0	0	0	0
1331	Surface water	0	0	0	0
1331	30	tr	tr	tr	+
1331	50	tr	+	+	+
1331	100	tr	0	0	+
1331	215	tr	+	+	+
1331	Above bottom	0	0	0	0
1332	Surface water	0	+	+	++
1332	50	0	+	+	+
1332	100	0	+	+	+
1333	Surface water	0	tr	tr	tr
1334	Surface water	0	tr	0	tr
1334	Deep water	0	tr	tr	tr

periments deal with the occurrence of nitrifying bacteria in the Gulf of Maine and on George's Bank, at a considerable distance from shore. Material for this experiment was obtained on a cruise of the "Atlantis," which took place during August 1–5, 1932, and which has been described in detail elsewhere (50). For this purpose a series of flasks were prepared containing the following materials:

Sand	15	grams
$CaCO_3$	1	gram
K_2HPO_4	0.005	gram
$(NH_4)_2SO_4$	0.01	gram

The flasks were plugged with cotton and sterilized, at 15 lbs. pressure, for 15 minutes. A standard solution of ammonium sulfate was sterilized separately and added to the sterile flasks. At the various stations visited about 50–60-cc. portions of fresh sea water, brought up from the different depths, by means of sterile glass containers, were added to the flasks immediately after the samples were obtained. To determine the occurrence of nitrifying bacteria in the marine bottom of the open sea, a medium similar to the above and containing 75–cc. portions of sea water with only 5 milligrams of $(NH_4)_2SO_4$ per flask was prepared. This medium was sterilized for 1 hour in flowing steam (100° C.). The flasks were inoculated, after returning from the cruise, with small quantities (about 1 gram) of fresh mud, obtained in the Gulf of Maine under sterile conditions; this was done by removing carefully the inner part of a core of mud.

TABLE IV

Occurrence of nitrite-forming bacteria in the marine bottom of the Gulf of Maine (Stations 1329 and 1330) and Georges Bank (Station 1336).

Tr = trace of nitrite; + = positive nitrite test; ++ and +++ = extensive nitrite formation. For map showing location of stations see Rakestraw, 1933, *Biol. Bull.*, **64**: 150.

Station No.	Depth of mud	Days of incubation			
		2	6	11	15
	cm.				
1329	0–30	+	++	+++	+++
1329	30–60	0	++	++	+++
1329	60–90	0	tr	tr	tr
1330	0–30	+	+++	++	+++
1330	30–60	0	tr	tr	tr
1330	60–90	tr	tr	tr	tr
1336	Surface layer of sand bottom	0	+	+	++
Control	0	0	0	0	

The results of these experiments on the presence of nitrite-forming bacteria in the sea water and in the marine bottom of high seas, are reported in Tables III and IV. These results show that the free water in the high seas is either entirely free from nitrifying bacteria or contains only very few cells of these organisms. In the case of the marine bottoms, however, positive nitrite formation was obtained in the case of the mud bottoms (Stations 1329 and 1330). Even within two days incubation these bacteria seem to be present in the mud to quite considerable depths. The sand bottom also gave active nitrite formation, somewhat more slowly than the mud bottoms.

In addition to the cultures containing ammonium salts, other cultures containing nitrites were prepared, but all attempts to obtain nitrate-forming bacteria in these media failed. However, when old cultures of nitrite-forming bacteria, in which nitrate-formation began to take place, were transferred to fresh media containing 5 mgm. KNO_2 in 30 cc. of medium, active nitrate formation took place within 7–8 days.

The following medium was finally adopted for the purpose of demonstrating the presence of nitrifying bacteria, using large test tubes instead of flasks, due to the ease of handling. The tubes contained 30 cc. sea water, 10 gm. washed sand, 1 gm. $CaCO_3$, 5 mgm. $(NH_4)_2SO_4$, and 5 mgm. K_2HPO_4. The medium was sterilized in flowing steam; the ammonium salt was sterilized separately, then added in aliquot portions, using sterile pipettes.

The nitrite-forming organism could be readily cultivated on the above medium; on repeated transfer, active cultures were obtained. By inoculating a silica-gel medium (thoroughly dialyzed in tap water and soaked in sterile sea water), to which an ammonium salt and calcium carbonate had been added, with an active liquid culture, abundant formation of nitrite on the plate took place within 14 days. This should facilitate greatly the isolation of the organism in pure culture, although for the purpose of the above experiments this was not considered essential.

The results presented here and other data of a similar nature are quite sufficient to demonstrate definitely that free sea water, especially at the surface of the sea, has either no nitrifying bacteria at all or only very few of these organisms. On the other hand, the sea bottom, mud or sand, has an active population of nitrifying organisms. The formation and accumulation of nitrate in the sea is probably due largely to the activities of these organisms. The processes of nitrite and nitrate formation take place in the sea bottom; the nitrate then diffuses into the water. The fact that it is much easier to demonstrate in culture the formation of nitrite than that of nitrate is due largely to the specificity of the organisms and conditions of cultivation.

[*Editor's Note:* Material has been omitted at this point.]

REFERENCES

2. Bauer, E. 1902. Ueber zwei denitrifizirende Bakterien aus der Ostsee. Wiss. Meeresunters, Kiel, N.F., *6*:9.
8. Berkeley, C. 1919. A study of the marine bacteria, Straits of Georgia, B.C. Trans. Roy. Soc. Canada, *13*(V):15.
9. Brandt, K. 1899. Ueber den Stoffwechsel im Meere. Wiss. Meeresunters. Kiel, N. F., *4*:213; 1902, *6*:23; 1916–1920, *18*:185; 1919, *19*:251.
10. Brandt, K. 1904. Über die bedeutung der Stickstoffverbindungen für die Produktion im Meere. Bot. Centrbl. Beih., *16*:383.
11. Brandt, K. 1923–27. Stickstoffverbindungen im Meere. I. Wiss. Meeresunters. Kiel, *20*:201.
19. Gran, H. H. 1903. Bacteria of the ocean and their nutrition. Naturen. Bergen, *27*:33.
21. Harvey, H. W. 1925. Oxidation in sea water. J. Mar. Biol. Ass., *13*:953.
22. Harvey, H. W. 1926. Nitrate in the sea. J. Mar. Biol. Ass., *14*:71.
23. Harvey, H. W. 1928. Biological Chemistry and Physics of Sea Water. Cambridge Univ. Press.
24. Issatchenko, B. L. 1908. Zur Frage von der Nitrifikation in den Meeren. Centrbl. Bakt., II, *21*:430.
25. Issatchenko, B. L. 1914. Recherches sur les microbes de l'océan Glacial Arctique. Petrograd.
26. Issatchenko, B. L. 1926. Sur la nitrification dans les mers. Compt. Rend. Acad. Sci. Paris, *182*:185.
30. Liebert, E. 1915. Über mikrobiologische Nitrit- und Nitratebildung im Meere. Rapp. Verhandl. Rijksinst. Visschereijonderzoek, *I*(3).
31. Lipman, C. B. 1922. Does nitrification occur in sea water? Science, *56*:501.
33. Lipman, C. B. 1926. The concentration of sea water as affecting its bacterial population. J. Bact., *12*:311.
36. Natterer, K. 1892. Monatschr. *13*:873–896, 897–915; 1893, *14*:624–673; 1894, *15*:530–595; 1895, *16*:405–581; 1899, *20*:1–263; 1900, *21*:1037–1060.
37. Nathanson, A. 1906. Ueber die Bedeutung vertikaler Wasserbewegungen für die Produktin des Planktons im Meere. Abh. Matt. Phys. Kl. Kgl. Sächs. Ges. Wiss., *29*:335.
39. Rao, G. G., and N. R. Dhar. 1931. Photosensitized oxidation of ammonia and ammonium salts and the problem of nitrification in soils. Soil Sci., *31*:379.
43. Schloesing, A. 1875. Sur l'ammoniaque de l'atmotsphère. Compt. Rend. Acad. Sci., *80*:175.
44. Thomsen, P. 1907. Über das Vorkommen von Nitrobakterien im Meere. Ber. deut. bot. Gesell., *25*:16–22; 1910, Wiss. Meeresunters. Kiel, N.F., *11*:1–27.
45. Vernon, H. M. 1898. The relations between marine animal and vegetable life. Mitt. Zool. Sta. Neapel., *13*:341.
50. Waksman, S. A., H. W. Reuszer, C. L. Carey, M. Hotchkiss, and C. R. Renn. 1933. Bacteriological investigations of sea water and marine bottoms in the Gulf of Maine. Biol. Bull., *64*:183.
51. ZoBell, C. E. 1933. Photochemical nitrification in sea water. Science, *77*:27.

32

Reprinted from *Limnol. Oceanogr.*, **10**(suppl.), R274–R289 (Nov. 1965)

CHARACTERISTICS OF A MARINE NITRIFYING BACTERIUM, *NITROSOCYSTIS OCEANUS* SP. N.[1]

Stanley W. Watson
Woods Hole Oceanographic Institution, Woods Hole, Massachusetts

ABSTRACT

A new species of an ammonia-oxidizing bacterium, *Nitrosocystis oceanus*, is described. This is the first nitrifying bacterium ever isolated from open ocean waters and the first species of the genus *Nitrosocystis* observed in over 30 years. The cells are gram negative, spherical to ellipsoidal, 1.8–2.2 μ in diameter, and occur as single cells, diploids, and occasionally as tetrads. Cells at times formed cysts and zoogloea. The most consistent and outstanding generic characteristic is the ultrastructure of the cells. Cells of this genus have a cytomembrane organelle composed of a series of flattened vesicles that transverse and bisect the cell; while cells of *Nitrosomonas* sp. lack this complex membranous organelle, they do have one to two concentric membranes at the periphery of the cell adjacent to the plasma membrane. *Nitrosocystis oceanus* is an obligate autotroph using only ammonia as an energy source and carbon dioxide as a carbon source. This bacterium grows optimally at 30C and in ammonia concentrations of $5 \times 10^{3} - 2.4 \times 10^{5}$ μg-at. NH_3-N/liter. It is an obligate halophile, requiring natural seawater for growth, and it lyses when suspended in distilled water. Growing optimally in continuous cultures, individual cells produced 2×10^{-6} μM of NO_2^- per day. *Nitrosocystis oceanus* was cultured repeatedly from offshore waters of the western North and South Atlantic Ocean since 1959. It is estimated that the standing crop of these organisms is less than one bacterium per ml. This number of bacteria in the upper 100 m of tropical waters would produce approximately 0.07 μg-at. NO_2^--N liter^{-1} year^{-1}.

INTRODUCTION

In 1962, the author isolated the first known ammonia-oxidizing bacterium from the open ocean and described it briefly at the VIII International Congress of Microbiology (Watson 1962). The present report amplifies the original morphological description of this new bacterium, *Nitrosocystis oceanus* and also deals with factors controlling the growth of *N. oceanus* and draws certain conclusions on the possible rates of nitrification in the open ocean; these are based on cultural studies of the organism in the laboratory.

The author initiated studies on marine nitrifying bacteria to illuminate the process of nitrification, that is, the oxidation of ammonia to nitrite and nitrite to nitrate in the oceans. While 60% of the combined nitrogen in the deep oceans exists as nitrate (Vaccaro 1962), nothing is known about the mechanism and rate of ammonia oxidation or about the microorganisms involved in the process. It is assumed, though not experimentally verified, that nitrate is formed in the open ocean, as on land, by the microbial oxidation of ammonia to nitrite and subsequently to nitrate (ZoBell 1946). Prior to 1962, only one investigator, Thompson (1908), reported the culturing of nitrifying bacteria from the marine environment, but he failed to find them more than 1,200 m from shore.

Winogradsky (1890) discovered that two physiological groups of nitrifying bacteria exist in soil. One oxidizes ammonia to nitrite and the other nitrite to nitrate. Subsequently, he described five genera of ammonia-oxidizing bacteria and two genera of nitrite-oxidizing bacteria. *Nitrosomonas* and *Nitrobacter* are two of the best known representatives of the two bacterial groups. All of these nitrifying bacteria are obligate autotrophs deriving their sole source of energy from the oxidation of either ammonia or nitrite and their sole source of carbon from carbon dioxide.

The present report deals only with a new

[1] Contribution No. 1659 from the Woods Hole Oceanographic Institution. This investigation was supported in part by U.S. Atomic Energy Commission Contract No. AT(30-1)-1918-114, by Public Health Service Grant GM 11214-02, and by National Science Foundation Grant 861, and by the U.S. Program in Biology, International Indian Ocean Expedition.

nitrifying bacterium, *Nitrosocystis oceanus*, which oxidizes ammonia to nitrite. Since this bacterium is being classified in a much disputed genus, it is necessary to review the taxonomy of the autotrophic bacteria that oxidize ammonia to nitrite.

Taxonomic literature review

Winogradsky (1890) placed the first known nitrifying bacterium in the genus *Nitromonas* which was later changed to *Nitrosomonas* (Winogradsky 1892). He proposed the species *N. europaea* for strains isolated from Europe and Africa and the species *N. javensis* for a Javanese strain. The former was a rod-shaped organism 0.9 to 1.0 μ by 1.1 to 1.8 μ, while the latter was a small coccus 0.5 μ diameter. Both organisms, in a liquid medium, grew as single or diploid cells suspended in the medium or as aggregates, which Winogradsky first called "zoogloea" and later "cysts."

Winogradsky (1892) established a second genus, *Nitrosococcus*, for those ammonia-oxidizing bacteria isolated from South America and Australia. These were spherical, 1.5 to 2.0 μ diameter, and differed from those in the genus *Nitrosomonas* by not forming any type of an aggregate.

Numerous investigators between 1891 and 1931 cultured soil-nitrifying bacteria and invariably called them *Nitrosomonas europaea*, but no investigator reported a new species until Nelson (1931) described *Nitrosomonas monocella*, which differed chiefly from *N. europaea* by having smaller cells, 0.6 to 0.9 μ.

Following his 1904 publication, Winogradsky remained silent about the nitrifying bacteria for many years but did report on them again in 1931. At that time, he proposed splitting the organisms, formerly classified in the genus *Nitrosomonas* into two genera: *Nitrosomonas* and *Nitrosocystis*, but he failed to make an adequate explanation for his revision. He spoke of having pure cultures in 1891, while in 1931 he stated that it was not possible in his earlier studies to separate the two organisms. One can only assume that in 1892 and 1904 he had been working with mixed cultures consisting of nitrifying bacteria so strikingly different that in 1932 he was forced to recognize them as belonging to two different genera. For an additional discussion of this dilemma, reference may be made to a review by Kingma Boltjes (1935).

In this revised taxonomic scheme, Winogradsky (1931) placed 1) all spherical organisms forming cysts in the genus *Nitrosocystis*; 2) all rod-shaped organisms not forming cysts or zoogloea in the genus *Nitrosomonas*; and 3) all spiral-shaped bacteria in a new genus, *Nitrosospira*. Later, Helene Winogradsky (1935, 1937) placed both rods and cocci forming zoogloea in still another new genus, *Nitrosogloea*. After 1931, both Serge Winogradsky (1931) and Helene Winogradsky (1935, 1937) (Winogradsky and Winogradsky 1933) ignored the genus *Nitrosococcus* in their publications. But they never stated that *Nitrosococcus* was not a valid genus nor did they reassign the organisms in this genus to other genera, and *Nitrosococcus* was retained as a valid genus in the seventh edition of *Bergey's Manual of Determinative Bacteriology* (Breed, Murray, and Smith 1957).

Many workers have cultured *Nitrosomonas europaea* from the soil, but few have found species of ammonia-oxidizing bacteria belonging to the four other described genera. Besides the Winogradskys, the few investigators who have reported the culturing of nitrifying bacteria other than *Nitrosomonas* species include Bonazzi (1919), who reported isolating a *Nitrosococcus* sp.; Sims and Collins (1960), who cultured both a *Nitrosococcus* and *Nitrosogloea* species; Palleroni (1950), who found a *Nitrosospira* sp.; and Romell (1932), who was the only person other than the Winogradskys to find a *Nitrosocystis* species.

Because so few reports deal with nitrifying bacteria other than *N. europaea*, people are prone to consider this organism the most important, if not the only, ammonia-oxidizing bacterium in the soil. In fact, Grace (1951) doubted the validity of three of the described genera. Both she and Imsenecki

(1946) suggested that Winogradsky had not observed cysts of nitrifying bacteria, but had seen only fruiting bodies of myxobacteria that contaminated his culture. It seems obvious from this brief review that contradictions and confusion pervade the taxonomy of the nitrifying bacteria.

The author is grateful to Mrs. Grace Fraser, Mrs. Frederica Valois, and Mrs. Linda Graham for excellent technical assistance.

METHODS

Isolation and culture

The first strain of *N. oceanus* was found in water collected with a Van Dorn bottle at a depth of 600 m in the Atlantic Ocean approximately 370 km east of Long Island, New York. Using a sterile Nisken sampler, additional strains were cultured from both North and South Atlantic waters.

In the initial search for nitrifying bacteria, approximately 5 liters of seawater were enriched with 500 μg-at. of NH_3-N and 100 μg-at. of PO_4^{-3}-P. In later enrichment cultures, the ammonia concentration was increased tenfold. After collection and enrichment, the seawater was incubated at room temperature and examined periodically for nitrite by the method of Bendschneider and Robinson (1952) and for nitrate by the method of Mullin and Riley (1955).

When nitrites increased at an exponential rate, the organisms were inoculated into 1 liter of fresh medium in a 2-liter shake flask. The composition of this medium is as follows:

Seawater	1 liter
$(NH_4)_2SO_4$	13.2 g
$MgSO_4 \cdot 7H_2O$	200 mg
$CaCl_2$	20 mg
K_2HPO_4	114 mg
Chelated iron	
(Sequestrene 13% Fe)	130 μg Fe
$Na_2MoO_4 \cdot 2H_2O$	1 μg
$MnCl_2 \cdot 4H_2O$	2 μg
$CoCl_2 \cdot 6H_2O$	2 μg
$CuSO_4 \cdot 5H_2O$	20 μg
$ZnSO_4 \cdot 7H_2O$	100 μg

After repeated serial transfers in the described medium, the crude culture was grown in a 14-liter fermenter equipped with an automatic *p*H controller. The *p*H was kept at 7.5 by the addition of 2 M K_2CO_3. The culture continued to grow under these conditions until the cell population was 3×10^7 cells/ml and the nitrite concentration was 0.2 to 0.3 g-at. NO_2^--N/liter. Usually at this stage the ratio of heterotrophic to autotrophic cells was 1–100. Once this ratio was achieved, the culture was diluted serially and then 1 to 10 cells were inoculated into 100 ml of medium in 250-ml Erlenmeyer flasks. Following inoculation, cultures were incubated on a shaker at 25C for two to three weeks before nitrite could be detected. After 10 μg-at. NH_3-N/liter were oxidized, the entire content of the flask was inoculated into 900 ml of media in a 1-liter reaction vessel equipped with an automatic *p*H controller. The *p*H was controlled by the automatic addition of 2 M K_2CO_3. Cultures were sparged with air sterilized by passage through an HA Millipore®[2] filter and agitated by a magnetic stirring bar and stirrer. They were incubated at room temperature and were allowed to grow until there were 1×10^7 cells/ml before being checked for purity.

Cultures were checked both visually and culturally for the presence of contaminants. For visual examination, the cells were concentrated by centrifugation and examined under a phase-contrast microscope. For cultural examination, concentrated cultures were inoculated into a variety of media. The basic medium in all cases consisted of seawater plus the salt mixture used for autotrophic growth. The following organic materials were then added to this basic medium:

Nutrient broth
- full strength
- full strength—+0.1% tryptone
- full strength—+0.1% yeast extract
- ¼ strength

0.1% Tryptone
- +0.1% yeast extract
- +0.1% acetate
- +0.1% glucose
- +0.1% pyruvate
- +0.1% lactate

0.1% NO_3

[2] Registered trademark, Millipore Filter Corporation, Bedford, Massachusetts.

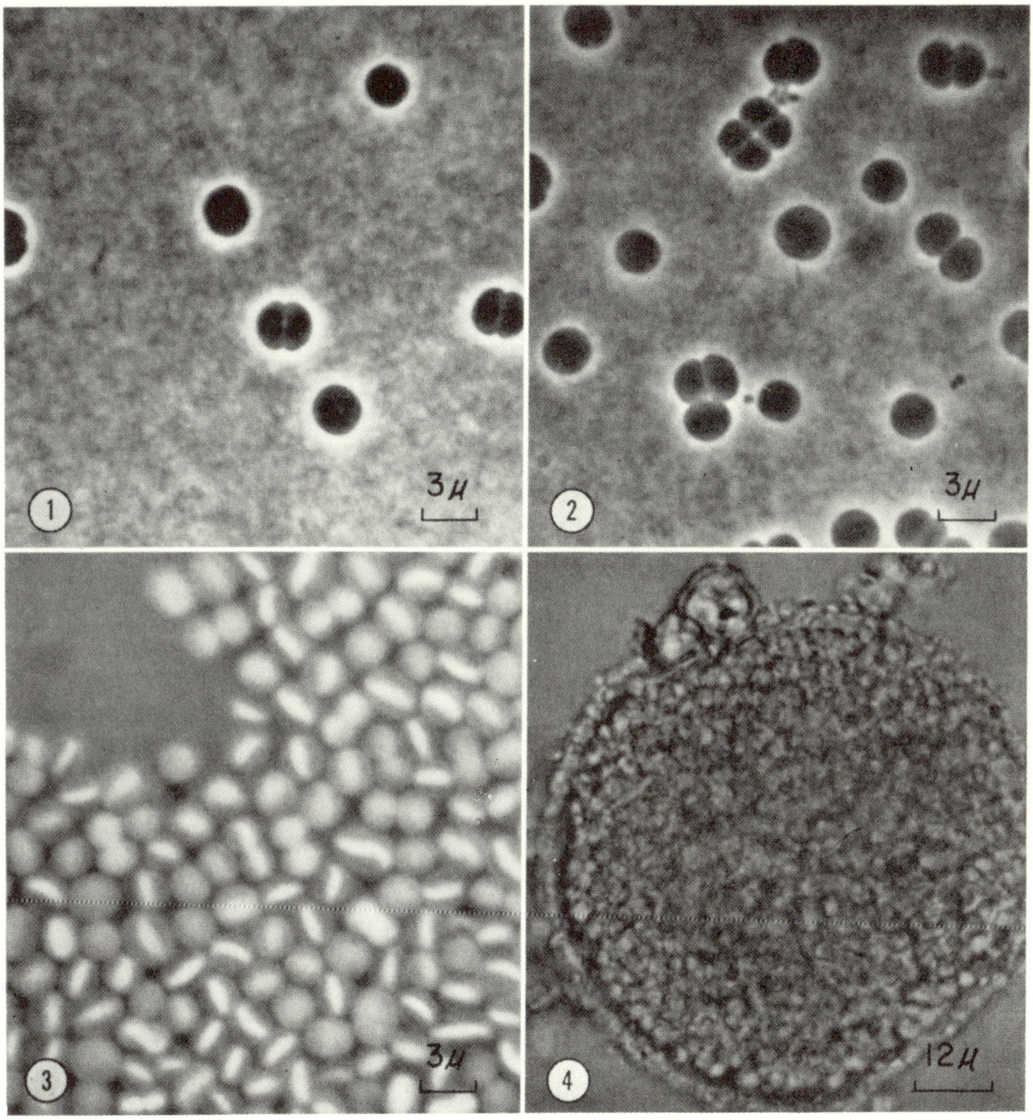

FIG. 1 AND 2. Phase photomicrographs of living cells suspended in seawater show single cells, diploids, and tetrads. FIG. 3. Phase photomicrograph of living cells suspended in seawater plus 30% gelatin shows presence of cytomembranes. FIG. 4. Photomicrograph of the thin-walled cyst composed of vegetative cells of *N. oceanus*.

+0.1% acetate
+0.1% glucose
5% Serum albumin
0.1% Yeast extract
+0.1% acetate
+0.1% glucose
+0.1% pyruvate
+0.1% lactate
+0.1% tryptone + 0.1% acetate
+0.1% tryptone + 0.1% glucose

Pure stock cultures were maintained in the logarithmic growth phase by growing them in a semicontinuous flow system in the 1-liter reaction vessel equipped with automatic *p*H control as previously described. When the *p*H of the culture dropped below 7.5, a peristaltic pump was actuated to deliver, simultaneously, sterile carbonate and fresh medium into the vessel. The volume of the culture was kept constant by means of a side arm overflow port. The density of the population and the nitrite level in these cultures were controlled by varying the tubing ratio and the molarity of the carbonate. Stock cultures were usually maintained in a cell concentration of 1×10^7 cells/ml and a nitrite level of 2×10^4 μg-at. NO_2^--N/liter. The silicone tubing used for the carbonate delivery was ⅜ × ¼ inch (0.95 × 0.64 cm) and that for the medium was 7/16 × 5/16 inch (1.1 × 0.8 cm). The K_2CO_3 was 0.075 M made up in distilled water with 3% NaCl. The dual peristaltic pump used for addition of media and carbonate was manufactured by the Harvard Apparatus Company.

Both the overflow and the culture were concentrated by centrifugation and examined daily, culturally and visually, for contaminants. *N. oceanus* was grown and kept pure in one of these semicontinuous flow systems for over three months.

Cells grown in semicontinuous cultures, as already described, were used as an inoculum. By using fewer than 10^3 cells/ml as an inoculum, a lag phase of from a few hours to more than a week occurred. No lag phase took place using a larger inoculum.

Cultural studies of factors controlling growth were carried out in 250-ml Erlenmeyer flasks containing 100 ml of media and, unless otherwise stated, were incubated on a shaker at 25C. The medium was usually inoculated with 10^5 cells and at the start of the experiment the nitrite concentration was approximately 2 μg-at. NO_2^--N/liter and the *p*H was 7.5. The ammonia concentration in the medium in all experiments (except those pertaining to the effect of ammonia concentration) was 5×10^4 μg-at. NH_3-N/liter at the initiation of the experiment. The cultures were discontinued after approximately eight days, at which time the nitrites had exceeded 200 μg-at. NO_2^--N/liter and the *p*H had fallen below 7.5.

The increase of nitrite in a culture was used as the index of growth because the cell population was directly proportional to the nitrite concentration in most cultures.

Physiological studies

Oxygen consumption was followed in a Warburg respirometer using standard techniques. Cells for these studies were washed in 0.05 M phosphate buffer with 3% NaCl until free of ammonia and nitrite and then were resuspended in the same buffer.

In cell-free studies, *N. oceanus* was grown in a 30-liter fermenter. It was then harvested by means of a Servall continuous flow centrifuge at 5C and washed with 0.05 M phosphate buffer with 3% NaCl until free of ammonia and nitrite. Finally, the cells were ruptured by sonication (Bronson Sonifier) or by pressure (Aminco French Pressure Cell) at 16,000 psi (1,120 kg/cm^2). Residual whole cells were removed by centrifugation at 5,000 rpm for 15 min. The supernatant liquid from this preparation was then recentrifuged at 144,000 × *g* (Spinco centrifuge).

Absorption of the 144,000 × *g* supernatant liquid was measured before and after reduction with dithionite in a Gilford multiple sampler absorbance recorder for the purpose of detecting cytochromes *a* and *c*.

Morphological observations

Living cells were observed through a Leitz microscope equipped with Reichert phase-contrast objectives. Cells for the observations were mounted either in seawater or in 20 to 30% gelatin (Mason and Powelson 1956). The latter method proved to be useful in making visible some internal structures of the cell.

Electron microscope studies were done in collaboration with Dr. R. G. E. Murray who kindly prepared sections and electron photomicrographs of *N. oceanus* and *N. europaea*[3] according to the method of Murray and Watson (1965).

RESULTS

Morphological observations

When *N. oceanus* was grown in a particulate-free medium in a shake flask or fermenter, single and diploid cells remained

[3] Strains of *N. europaea* obtained from Dr. David Pramer, Rutgers University.

suspended in the medium (Figs. 1 and 2). In addition, large zoogloeal growth, up to 12.7 cm diameter and 1.3 cm thick, sometimes formed on the walls of the fermenter. In a liquid medium with 1 g of $CaCO_3$/liter, *N. oceanus* occasionally formed cysts (Figs. 4, 5, and 6).

On an agar plate, *N. oceanus* formed small, warty-type microcolonies (Fig. 7) composed of a few to 500 cells. In these colonies, the cells adhered tenaciously to one another and thus it was not possible to subculture by streaking from such a colony.

The gram negative cells of *N. oceanus* were spherical to ellipsoidal and occurred singly, in pairs, and occasionally as tetrads (Figs. 1 and 2). In pure culture, the cells were fairly regular in shape but very irregular forms were observed in the enrichment cultures. They ranged in size from 1.8 to 2.2 μ, but giant cells up to 10 μ diameter were sometimes present.

No internal details of the cells were discernible when they were mounted in seawater (Figs. 1 and 2) and viewed with a phase-contrast microscope, but if mounted in 20 to 30% gelatin, a distinct bar-like structure traversing and bisecting the cell was clearly visible (Fig. 3). This, when viewed with the electron microscope, appeared as a cytomembrane organelle composed of a series of flattened vesicles extending near but usually not continuous with the plasma membrane (Fig. 8). The ultra-structure of this bacterium is reported in more detail elsewhere (Murray and Watson 1963, 1965).

The ultra-structure of *N. oceanus* was compared with that of *N. europaea* (Fig. 9), which lacked the central cytomembrane system found in *N. oceanus* but did possess one to two concentric membranes lying adjacent and internal to the plasma membrane (*see* Murray and Watson 1965 for additional details).

Cells of *N. oceanus* were frequently motile in enrichment cultures but rarely in pure cultures. These cells moved by means of a single flagellum or a small tuft of flagella (Fig. 10).

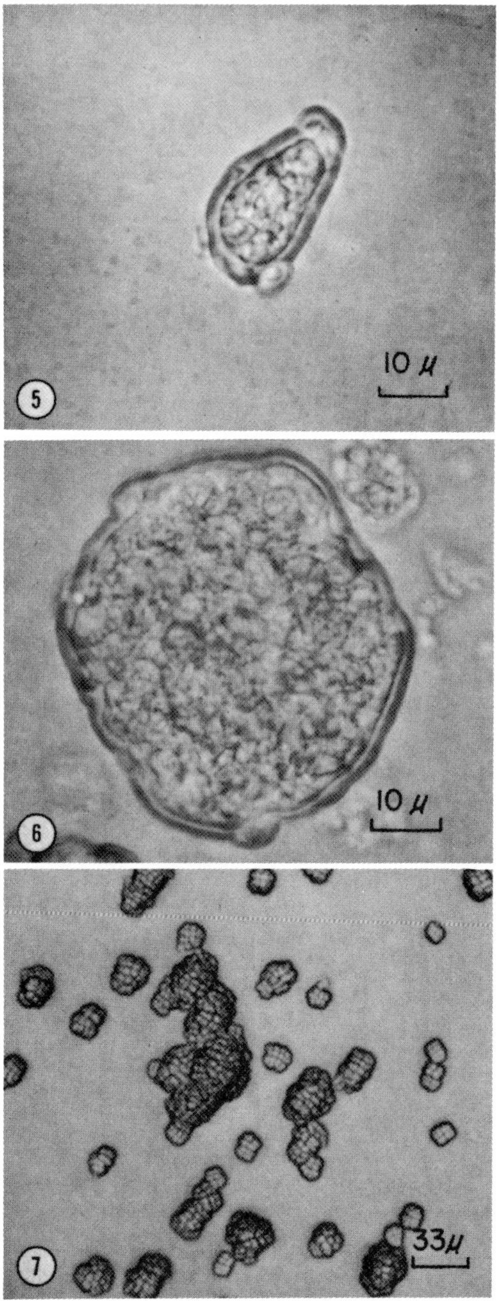

FIG. 5 AND 6. Photomicrograph of thick-walled cyst; note lack of internal detail. FIG. 7. Photomicrograph of warty-type microcolonies growing on agar.

In the presence of $CaCO_3$, two distinct types of cysts were formed. The first was a thick-walled cyst 2 to 100 μ diameter (Figs.

FIG. 8. Electron photomicrograph of a section of *N. oceanus*, shows cytomembrane organelle.

5 and 6). No internal details were resolved in this cyst and it did not stain with methylene blue, gentian violet, or iodine. Cysts could not be crushed by pressure when placed between a microscope coverslip and slide.

The second type of cyst varied in size from 10 to 100 μ and was composed of a few to 100 or more vegetative cells enveloped by a distinct cyst wall (Fig. 4). When this cyst was crushed, motile vegetative cells escaped. In this cyst, the vegeta-

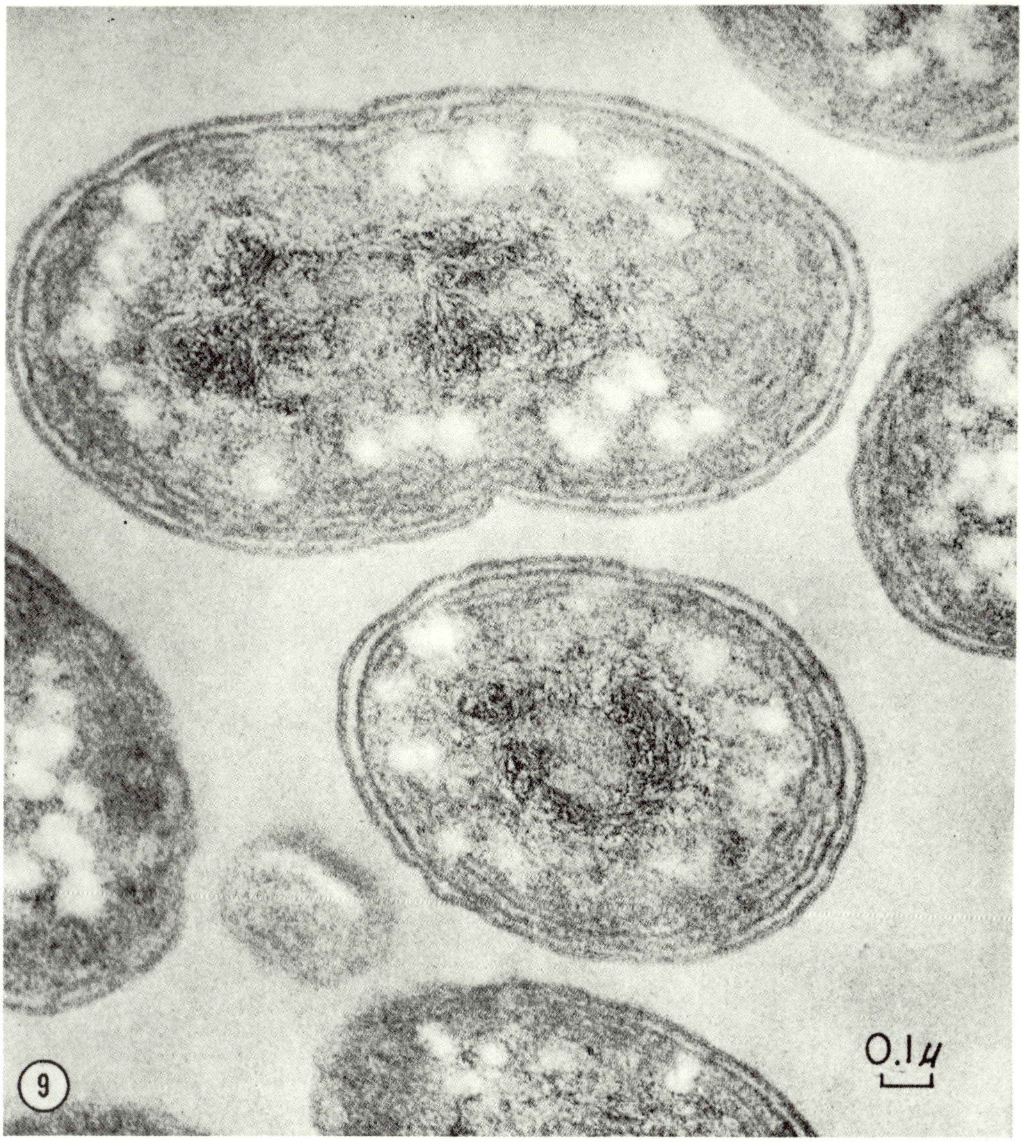

FIG. 9. Electron photomicrograph of a section of *N. europaea*, shows concentric cytomembranes located adjacent and internal to the plasma membrane.

tive cells stained a deep blue with methylene blue while the cyst wall remained a light blue.

Because of the doubts raised by Imsenecki (1946) and Grace (1951) about the ability of nitrifying bacteria to produce cysts, the author studied sections of these cysts with an electron microscope. The cysts were found to be composed chiefly of cells having the typical cytomembrane organelle (Fig. 11). Because of the uniqueness of this membrane structure, there can be no doubt that the cells within the cyst were those of *N. oceanus* and, because the cysts did not come from pure cultures, a small fraction of the cells in the cysts were obviously not cells of *N. oceanus*. It was assumed, therefore, that the cells other than *N. oceanus* were heterotrophic bacteria. Unfortunately, factors controlling cyst forma-

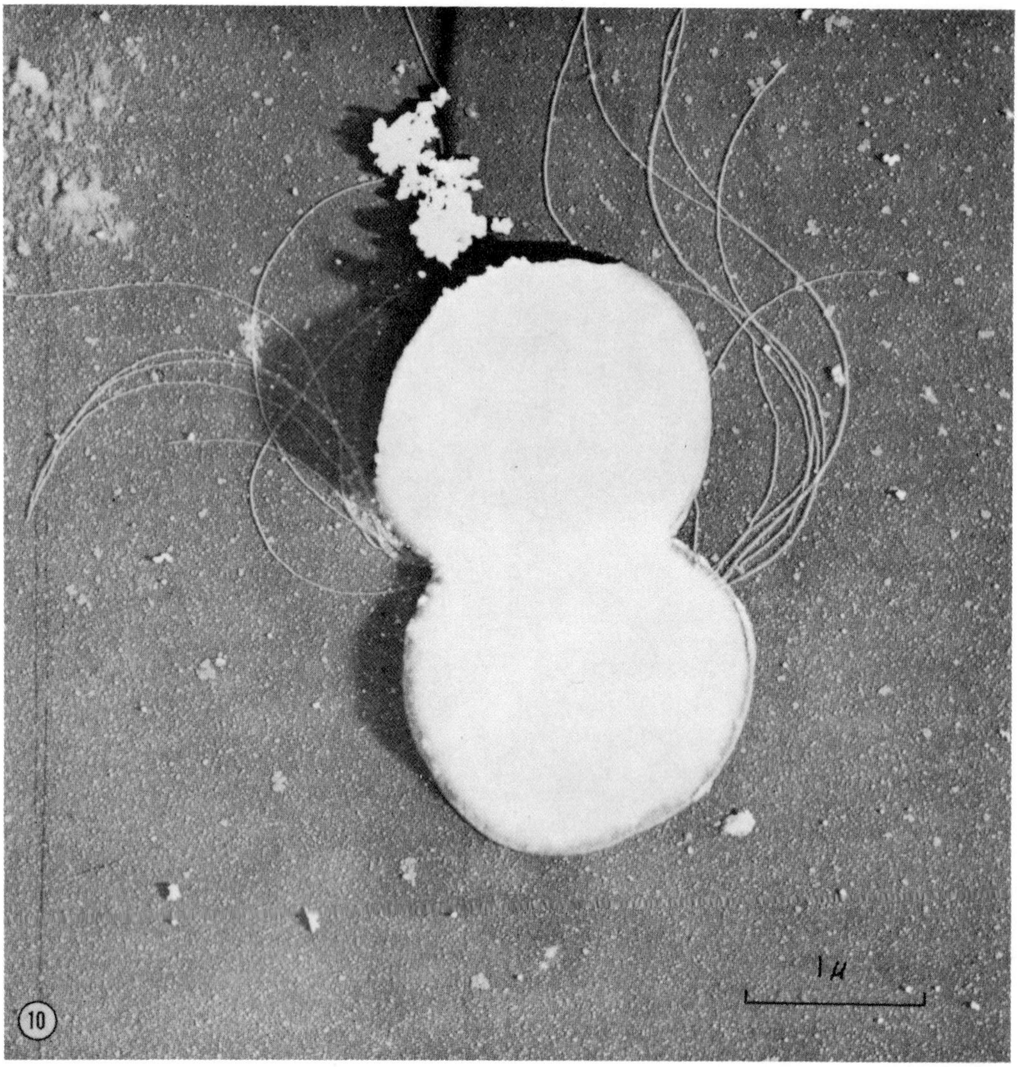

FIG. 10. Electron photomicrograph of a whole cell of *N. oceanus* in the process of division; shows tufts of flagella.

tion remain a mystery and observations of these structures were governed by chance. So far, cysts have never been observed in pure cultures.

Factors affecting rate of growth and nitrite production

The most important variables include light, temperature, *p*H, ammonia, and nitrite concentration. The effect of temperature is detailed in Table 1.

Cells incubated at 35 to 40C had a shorter generation time for the first five generations, but subsequently the generation time was prolonged to 36 hr. Therefore, 30C proved to be the optimum temperature for growth.

The optimum *p*H for growth was 7.5 to 7.8, and below *p*H 7.0 little, if any, growth occurred. The *p*H of cultures grown in shake flasks often decreased to 5.5, but if the *p*H was readjusted the cells would continue to grow. In contrast, the cells were killed immediately if the *p*H was adjusted to 8.2 or higher.

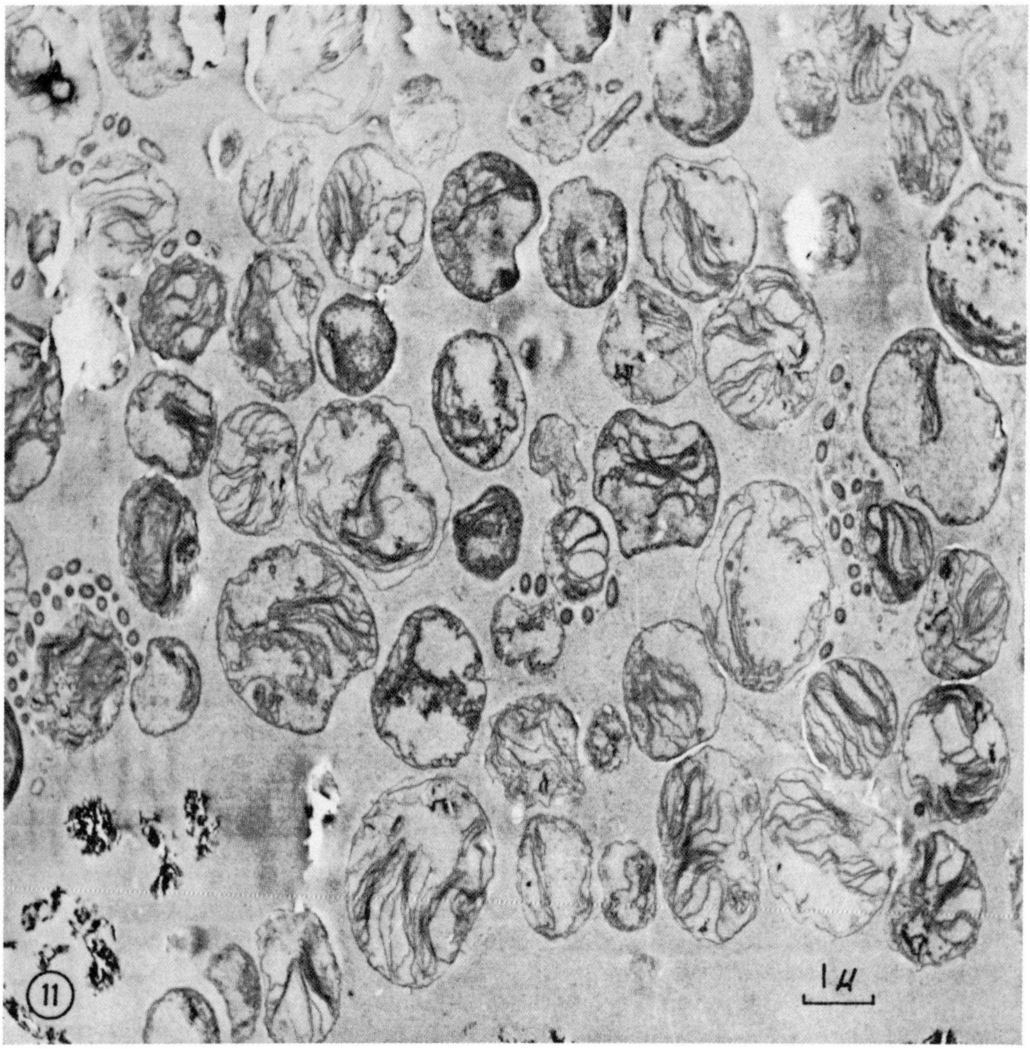

FIG. 11. Electron photomicrograph of a cyst of *N. oceanus* shows cysts to be composed of vegetative cells of *N. oceanus*.

N. oceanus grew equally well on a wide range of ammonia concentrations (Table 2). When grown in ammonia concentrations greater than 1,000 μg-at. NH_3-N/liter, the generation times were consistent from one experiment to another. Below this level, generation times varied between experiments and sometimes no growth was observed if there were less than 100 μg-at. NH_3-N/liter. While no lag phase was observed in ammonia concentrations greater than 1,000 μg-at. NH_3-N/liter, lag phases of a week or more took place in lesser concentrations. In fermenters, nitrite was not noticeably toxic to cultures until it reached a concentration of 50 to 100 mg-at. NO_2^--N/liter. Above this level, growth gradually decreased and eventually stopped when the nitrite concentration had risen to 300 mg-at. NO_2^--N/liter.

Light partially inhibited the growth of *N. oceanus*. In shake flask cultures exposed to normal laboratory light, the generation time was 84 hr while similar cultures grown in the dark had a 24-hr generation period.

N. oceanus is an obligate halophile

TABLE 1. *Effect of temperature on the growth of* N. oceanus

Temperature (C)	Generation time (hr)
40	16
35	17
30	21
25	22
20	34
12	120

whose cells lyse in distilled water. When first isolated, this bacterium grew equally well either in natural or artificial seawater but now grows only in natural seawater. Reasons for this obligate requirement for natural seawater require explanation.

Although cells failed to grow in the absence of natural seawater, they produced nitrite from ammonia in distilled water with added NaCl (Table 3).

Cells did not lyse in the absence of seawater if suspended in 0.5 M sucrose or in a 0.5 M concentration of the following cations: Ca^{++}, Mn^{++}, Mg^{++}, NH_4^+, Li^+, K^+, and Na^+. However, the cells failed to oxidize ammonia if suspended in either sucrose or any of the divalent cations. They did oxidize ammonia at equal rates in the presence of the described monovalent cations.

In semicontinuous cultures, the flow rate of which was regulated by a *p*H controller, the cell number was directly proportional to the nitrite produced when the density of cells ranged from 1×10^9 to 5×10^{10} cells/liter. When the cell populations were at these levels, the nitrite concentration was 2×10^3 to 1×10^5 μg-at. NO_2^--N/liter. Thus, for each cell formed, 2×10^{-6} μg-at. NO_2^--N/liter was produced.

N. oceanus typifies an obligate autotroph. No growth was observed when it was inoculated into a variety of organic media (*see* methods) used for checking the purity of cultures, nor were significant amounts of oxygen consumed when incubated in the presence of organic substrates (Table 4).

Organic nutrients added to the regular ammonia media never stimulated growth, but some compounds prolonged the generation time of *N. oceanus* (Table 5).

Biochemical studies

Like *N. europaea*, cell-free extracts of *N. oceanus* would oxidize hydroxylamine but not ammonia. When hydroxylamine was oxidized, cytochrome *c* was reduced. The partial purification of the enzyme responsible for the reaction, hydroxylamine cytochrome *c* reductase, has been reported by others in this laboratory (Hooper 1964).

Cells of *N. oceanus* are rich in cytochromes. A dense mass culture appears yellowish brown. When cells are concentrated by centrifugation, the pellet is brick red. The difference spectrum of cell-free extracts is given in Fig. 12.

DISCUSSION

Taxonomy

The characteristics currently used to separate genera of the ammonia-oxidizing bacteria are variable and quite likely not valid. After working with this group of organisms for the last seven years, the author is convinced that there are two distinct groups

TABLE 2. *Effect of ammonia concentrations on growth of* N. oceanus

Ammonia concentration (μg-at. NH_3-N/liter)	Generation time (hr)
5×10^5	28
2.4×10^5	24
1×10^5	24
5×10^4	24
1×10^4	24
5×10^3	24
3×10^3	32
1×10^3	156
1 – 10	500

TABLE 3. *The effect of NaCl on nitrite production by whole cells of* N. oceanus

NaCl (molarity)	μg-at. NO_2^--N/liter produced
0.0	23
0.1	134
0.2	256
0.3	398
0.4	454
0.5	430
0.75	390
1.0	400
2.0	328

of ammonia-oxidizing bacteria, a conviction based on the ultra-structure of these organisms. *N. oceanus* and *N. europaea* are representative of each group. The organisms differ chiefly on the location and complexity of their cytomembrane systems. In *N. oceanus,* the membranes occur in a membranous organelle (Fig. 8) which transverses and bisects the cells while *N. europaea* has a simpler membrane system located at the periphery of the cell (Fig. 9).

In the past, four of the five genera were separated by their ability, or lack of it, to form cysts or zoogloea. The author's observations show that the formation of zoogloea and cysts is too variable to use as taxonomic criteria. Cultural conditions alone can be responsible for their appearance or absence in any of the nitrifying bacteria including species of *Nitrosomonas.* To illustrate, the author cultured a marine nitrifying bacterium having a concentric membrane system at the periphery of the cell, and thus, its ultra-structure was similar to that found in the terrestrial bacterium *N. europaea.* However, according to the key in *Bergey's Manual of Determinative Bacteriology* (Breed, Murray, and Smith 1957), this organism could not be considered as a *Nitrosomonas* sp. because it formed both cysts and zoogloea. Furthermore, *N. oceanus* rarely forms cysts and only occasionally zoogloea. Thus, depending upon the particular cultural conditions and the specific time when a culture was examined, *N. oceanus* could be classified in any of the three genera: *Nitrosocystis, Nitrosogloea,* or *Nitrosococcus.*

This background information influenced the author's decision in classifying the new organism *Nitrosocystis oceanus.* Since the ultra-structure of the organism was so strikingly different from that of *N. europaea,* it did not seem proper to include it in the genus *Nitrosomonas.* Of the five previously described genera, *N. oceanus* could justifiably be included in the genus *Nitrosocystis, Nitrosogloea,* or *Nitrosococcus.* While the author is convinced that the organisms included in all three genera can form cysts,

TABLE 4. *Oxygen uptake and nitrite production by whole cells of* N. oceanus *incubated with organic substrates for 1 hr in the absence of ammonia*

Cell suspension	Substrate	Molarity	μM O_2 uptake	μM NO_2^- produced
1.	glutamine	10^{-1}	0.00	1.00
	glutamine	10^{-2}	0.00	0.00
	lysine	10^{-1}	0.00	0.30
	lysine	10^{-2}	0.00	0.00
	valine	10^{-1}	0.00	0.00
	control with ammonia	1.6×10^{-1}	8.70	7.80
2.	glycine	10^{-1}	0.84	0.75
	glycine	10^{-2}	0.00	0.00
	glycine	10^{-3}	0.00	0.00
	methionine	10^{-1}	0.00	0.00
	methionine	10^{-2}	0.15	0.00
	tyrosine	10^{-1}	0.14	0.00
	tyrosine	10^{-2}	0.14	0.00
	control with ammonia	1.6×10^{-1}	17.40	12.00
3.	glutamine	10^{-1}	0.00	0.00
	valine	10^{-1}	0.00	0.00
	isoleucine	10^{-1}	0.00	0.00
	lysine	10^{-1}	0.00	0.00
	threonine	10^{-1}	0.00	0.00
	control with ammonia	1.6×10^{-1}	5.60	2.30
4.	uric acid	10^{-1}	0.00	0.00
	guanine	10^{-2}	0.00	0.00
	guanine	10^{-3}	0.00	0.00
	urea	10^{-1}	0.00	0.00
	urea	10^{-2}	0.00	0.00
	control with ammonia	1.6×10^{-1}	3.40	2.88
5.	pyruvate	10^{-2}	0.00	0.00
	glutamate	10^{-1}	0.00	0.00
	control with ammonia	1.6×10^{-1}	5.10	7.10
6.	citrate	10^{-2}	0.00	0.00
	ketoglutarate	10^{-2}	0.00	0.00
	oxalacetate	10^{-2}	0.00	0.00
	succinate	10^{-2}	0.00	0.00
	fumaric	10^{-2}	0.00	0.00
	malic	10^{-2}	0.00	0.00

other investigators report that cysts are present only in the genus *Nitrosocystis.* Hence, on the evidence, this new marine-nitrifying organism should be categorized in the genus *Nitrosocystis.*

To avoid additional confusion, the author proposes the creation of a new species, *Nitrosocystis oceanus.* The descriptions of the organisms now included in the genera *Nitrosocystis, Nitrosogloea,* and *Nitroso-*

TABLE 5. *Effect of organic compounds on growth of* N. oceanus

Organic compound	Molarity	Generation time (hr)
Control	(regular ammonia medium)	24
Glucose	3×10^{-2}	24
Citrate	2×10^{-2}	24
Acetate	1×10^{-3}	24
Acetate	1×10^{-5}	24
Lactate	1×10^{-3}	24
Threonine	4×10^{-3}	110
Threonine	4×10^{-5}	26
Asparagine	4×10^{-3}	66
Asparagine	4×10^{-5}	33
Histidine	3×10^{-3}	190
Histidine	3×10^{-5}	33
Nutrient broth	(¼ strength)	130

coccus lack sufficient detail to allow a categorical statement that this new bacterium is identical to any previously described species.

N. oceanus shows similarities to *Nitrosococcus nitrosus, Nitrosocystis coccoides,* and to an unnamed strain of nitrifying bacteria

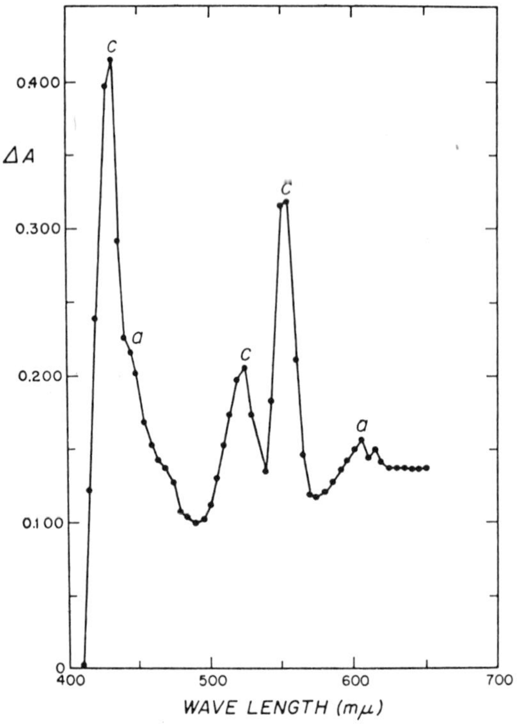

FIG. 12. Reduced minus oxidized absorption difference spectrum of a cell-free extract of *N. oceanus* shows presence of cytochrome *a* and *c.*

that Winogradsky (1904) isolated from the Petersburg area. The latter is not listed in *Bergey's Manual of Determinative Bacteriology* (Breed, Murray, and Smith 1957). All four organisms are spherical to ellipsoidal and all were approximately the same size, although *N. oceanus* is slightly larger. It is possible, though, using the criterion of ultra-structure, that one or all of the three previously described organisms could belong to the genus *Nitrosomonas.*

It seems, probably, that the organism described by Winogradsky from Petersburg is a *Nitrosocystis* sp., because he observed a central granular body within stained cells. This may be identical to the cytomembrane system seen in *N. oceanus,* but his reference is the only one in the literature, other than those by Murray and Watson, that mentions the internal morphology of a nitrifying cell.

Cyst formation

These studies should erase all doubt as expressed by Imsenecki (1946) and by Grace (1951) about the ability of nitrifying bacteria to form cysts. Although this author demonstrated cyst formation in *N. oceanus,* nothing is known about their function or conditions for formation. Likewise, no concrete evidence indicates a relationship between the two types of cysts. It is tempting to speculate, without sufficient evidence, that 1) single cells encyst; 2) cysts increase in size without cellular division; 3) cysts undergo schizogony and form daughter cells. If this is the sequence of events, it would be possible to relate the two types of cysts. Stages 1) and 2) represent the dense, nonstainable cysts while stage 3) represents the large stainable cysts containing vegetative cells.

The presence of heterotrophic bacteria in these cysts needs some explanation. The cultures were several months old and heavily contaminated with heterotrophic bacteria, some of which may have invaded the mature cyst. The presence of contaminants might also be explained if cyst formation occurred by the clumping of vegetative cells in the presence of heterotrophic bacteria prior to laying down a cyst wall.

Biochemical similarities between Nitrosocystis *and* Nitrosomonas

Although *N. oceanus* differed morphologically from *Nitrosomonas*, both were physiologically similar. Both organisms appear to be obligate autotrophs using only ammonia as an energy source and only carbon dioxide as a carbon source. Hydroxylamine, but not ammonia, was oxidized by cell-free extracts of both organisms. Neither, however, grew using hydroxylamine as an energy source. Hydroxylamine cytochrome *c* reductase has been demonstrated in both organisms (Hooper 1964), although the properties of this enzyme varied slightly between organisms.

Both organisms contain cytochrome *a, b,* and *c* (Fig. 12) (Hooper 1964). The difference spectrum of cytochrome *c* of *N. oceanus* as found in this experiment and also by Hooper (1964) was identical to that of cytochrome *c* of *N. europaea* (Hooper 1964). Extracts of both organisms had maximum absorption at 420, 523, and 553 mμ representing the α, β, and γ peaks of cytochrome *c*. The cytochrome *a* of the two organisms may differ slightly. Hooper (1964) reported maximum absorption in a difference spectrum at 445 and 605 mμ for *N. oceanus* and only at 600 mμ for *N. europaea*. The author obtained maximum absorption at 440, 605, and 615 mμ with extracts of *N. oceanus*. For cytochrome *b*, Hooper reported absorption peaks at 530 and 560 mμ, however the author did not find cytochrome *b* in extracts of *N. oceanus*. Thus, while these two organisms appear physiologically similar, subtle differences do exist.

Nitrifying bacteria in the ocean

Although ammonia-oxidizing bacteria were demonstrated in numerous samples from the North and South Atlantic and from the Indian Ocean, relatively little is known about the distribution, abundance, or types of ammonia-oxidizing bacteria in the oceans of the world.

N. oceanus was found in the majority of the enrichment cultures, but the author does not wish to imply that this is the only or even the most important ammonia-oxidizing bacterium in the ocean. In fact, *Nitrosomonas* sp. may be as or more common than *N. oceanus*, and there may be other bacteria of equal or greater importance.

Unfortunately, nitrifying bacteria in the oceans cannot be enumerated by ordinary bacteriological procedures. However, it should be possible to estimate the number of bacteria initially present in a sample by the time it takes them and their progeny to produce a given amount of nitrite in the medium. With this approach, the following assumptions have to be made: 1) no lag phase, 2) a constant amount of nitrite produced per cell division, and 3) a constant generation time. The production of nitrite in our experiments was found to be 2×10^{-6} μg-at. NO_2^--N per cell formed. As previously shown, cells did have a fairly uniform generation time of 24 hr when grown under optimal conditions. The theoretical relationship between the size of the inoculum and number of days of incubation needed for the production of 2 μg-at. of NO_2^--N, on the basis of the above assumptions, is shown in Table 6. Experimentally, when 10^2, 10^3, or 10^4 cells were inoculated into 1 liter of medium, it took about 14, 11, and 8 days, respectively, to produce 2 μg-at. of NO_2^--N/liter, partially confirming the calculations.

Using the theoretical relationship expressed in Table 6, the numbers of bacteria present in samples of seawater were estimated in Table 7. The author feels that the estimated number of bacteria present in waters from Barbados Harbor and the Indian Ocean is fairly accurate. Once detected in these samples, the concentration of nitrite doubled every 24 hr until *p*H became limiting. Thus, the assumption that these bacteria had a 24 hr generation time seemed to be true. Also, in these samples there appeared to be little if any lag phase, so the relationship presented in Table 6 seemed to hold true.

Less reliance can be placed on the estimated number of bacteria in North Atlantic Ocean waters because of an obvious prolonged lag phase. In laboratory cultures,

TABLE 6. *Days to produce 2 μg-at. NO_2^--N/liter with varying number of cells as an inoculum*

Inoculum (cells/liter)	Estimated time to produce 2 μg-at. NO_2^--N/liter (days)
1	21
10	18
10^2	14
10^3	11
10^4	8
10^5	4.25
10^6	1

TABLE 7. *Estimates of number of ammonia-oxidizing bacteria in samples of seawater based on the time it took them to produce 2 μg-at. NO_2^--N/liter*

Source of water	Time to produce 2 μg-at. NO_2^--N/liter (days)	Estimated number of ammonia-oxidizing bacteria initially present in sample
Barbados Harbor	1	10^6
Indian Ocean (near the Seychelle Islands)	7	10^4
North Atlantic Ocean	30–60	<1

when a small inoculum was used, a lag phase of a few days to a week was common, and on occasion a lag phase of two weeks was observed. If a lag phase of three weeks did occur in the North Atlantic Ocean samples, and if there were 10^3 bacteria/ml initially, then 2 μg-at. of NO_2^--N/liter would have been produced in 32 days. Thus, if these samples had no lag phase, the population would be less than one cell per liter. The maximum possible population, assuming a lag phase of three weeks, would be estimated at 1,000 bacteria per liter.

Estimated rate of nitrification in the oceans

It is impossible with present methods to measure directly the rate of nitrification in the ocean. It is possible, though, to make some crude estimates of the oxidation rate of ammonia within the ocean. These estimates are based on the following assumptions: 1) that the estimate of the number of bacteria present in seawater (Table 7) is reliable; 2) equivalent amounts of nitrite would be produced per cell division either in the ocean or in a semicontinuous culture and that this value would be, as previously stated, 2×10^{-6} μg-at. NO_2^--N/liter per cell division; 3) that the generation time will be influenced in the ocean, as in the laboratory, by the ammonia concentration and the temperature of the water; 4) the bacterial population in the ocean is in a steady state and for each cell produced one cell will die or in some other manner be removed. For the purpose of this discussion it is also assumed, since large amounts of nitrite do not accumulate in the ocean, that nitrite is oxidized to nitrate as rapidly as it is produced by ammonia-oxidizing bacteria. Therefore, the rate of nitrate production is essentially the same as nitrite production.

The theoretical amount of nitrite produced annually for a given cell population in a steady state condition is given in Table 8. Calculations are made for two temperatures (5 and 25C) and two ammonia concentrations. The lower of these, 1 μg-at./liter is typical of the sea; the other, 50,000 μg-at./liter, was the optimum concentration in our laboratory experiments. In our experiments at 5C, nitrite is produced at one-fifth the rate observed at 25C. Production at the low ammonia concentration is one-tenth that at the high concentration.

According to the estimates in Tables 7 and 8, an annual production of 70 μg-at. NO_2^--N/liter would be expected in Barbados Harbor and 0.7 μg-at. NO_2^--N/liter in the Indian Ocean. In the cold deep waters of the North Atlantic Ocean, even assuming the maximum of 1,000 bacteria per liter,

TABLE 8. *Theoretical amount of nitrite produced annually by ammonia-oxidizing bacteria growing in steady state conditions*

Bacteria/liter	5C μg-at. NH_3-N/liter 1	5C μg-at. NH_3-N/liter 50,000	25C μg-at. NH_3-N/liter 1	25C μg-at. NH_3-N/liter 50,000
1	1.4×10^{-5}	1.4×10^{-4}	7×10^{-5}	7×10^{-4}
10	1.4×10^{-4}	1.4×10^{-3}	7×10^{-4}	7×10^{-3}
10^2	1.4×10^{-3}	1.4×10^{-2}	7×10^{-3}	7×10^{-2}
10^3	1.4×10^{-2}	1.4×10^{-1}	7×10^{-2}	7×10^{-1}
10^4	1.4×10^{-1}	1.4	7×10^{-1}	7
10^5	1.4	14	7	70
10^6	14	140	70	700

only 1.4×10^{-2} μg-at. NO_2^--N/liter would be produced per year.

In the deep oceans, such as the North Atlantic, there are usually 25 to 30 μg-at. NO_3^--N/liter. Vaccaro (1962) estimated that this amount of nitrate would represent 60 years accumulation based on the yearly organic production in surface waters. Our present estimates of the numbers of bacteria and rates of nitrite production would require about 2,000 years to produce this quantity of nitrite. All estimates of the apparent age of the deep water of the North Atlantic are less than this (*cf.* Broecker 1963). Assuming an age of 250 years, and that the nitrification took place in this period of time, an annual production of 0.1 μg-at. of NO_3^--N would be required. According to Table 8, it would take 10^4 bacteria per liter to account for this amount of nitrification. This comparison suggests that the above calculations seriously underestimate the size of populations of nitrifying bacteria in the open sea. However, since no previous information is available on the rates of nitrification in the ocean, the author felt it worthwhile to make these calculations to serve as a guide for future investigations.

REFERENCES

BENDSCHNEIDER, K., AND R. J. ROBINSON. 1952. A new spectrophotometric method for the determination of nitrite in sea water. J. Marine Res., **11**: 87–96.

BONAZZI, A. 1919. On nitrification. III. The isolation and description of the nitrite ferment. Botan. Gaz., **68**: 194–207.

BREED, R. S., E. G. D. MURRAY, AND N. R. SMITH. 1957. Bergey's manual of determinative bacteriology. Seventh edition. Williams and Wilkins, Baltimore, Maryland. 1094 p.

BROECKER, W. 1963. Radioisotopes and large-scale oceanic mixing, p. 88–108. *In* M. N. Hill [ed.], The sea, v. 2. Interscience, New York.

GRACE, J. B. 1951. Myxobacteria mistaken for nitrifying bacteria. Nature, **168**: 117.

HOOPER, A. B. 1964. Oxidation of hydroxylamine to nitrite by the chemoautotrophic bacteria *Nitrosomonas* and *Nitrosocystis*. Ph.D. Thesis, Johns Hopkins Univ., Baltimore, Maryland. 66 p.

IMSENECKI, A. 1946. Symbiosis between myxobacteria and nitrifying bacteria. Nature, **157**: 877.

KINGMA BOLTJES, T. V. 1935. Untersuchungen über die nitrifizierenden Bakterien. Arch. Mikrobiol., **6**: 79–138.

MASON, D. J., AND D. M. POWELSON. 1956. Nuclear division as observed in live bacteria by a new technique. J. Bacteriol., **71**: 474–479.

MULLIN, J. B., AND J. P. RILEY. 1955. The spectrophotometric determination of nitrate in natural waters, with particular reference to sea water. Anal. Chim. Acta, **12**: 464–480.

MURRAY, R. G. E., AND S. W. WATSON. 1963. An organelle confined within the cell wall of *Nitrosocystis oceanus* (Watson). Nature, **197**: 211–212.

———, AND ———. 1965. The structure of *Nitrosocystis oceanus* and comparison with *Nitrosomonas* and *Nitrobacter*. J. Bacteriol., **89**: 1594–1609.

NELSON, D. H. 1931. Isolation and characterization of *Nitrosomonas* and *Nitrobacter*. Zentr. Bakteriol. Parasitenk. Abt. II, **83**: 280–311.

PALLERONI, N. J. 1950. Sobre la presecia de *Nitrosospira* en tierras de la Antártida Argentina. Rev. Fac. Cienc. Agrar., Min. Educ. Univ. Nac. Cuyo (Mendoza), **2**: 46.

ROMELL, L. G. 1932. A *Nitrosocystis* from American forest soil. Svensk Botan. Tidskr., **26**: 303–312.

SIMS, C. M., AND F. M. COLLINS. 1960. The numbers and distribution of ammonia-oxidizing bacteria in some northern territory and South Australian soils. Australian J. Agr. Res., **11**: 505–512.

THOMPSON, P. 1908. Über da Vorkommen von Nitrobaketerien im Meere. Ph.D. Thesis, Univ. Kiel. 27 p.

VACCARO, R. F. 1962. The oxidation of ammonia in sea water. J. Conseil, Conseil Perm. Intern. Exploration Mer, **27**: 3–14.

WATSON, S. W. 1962. *Nitrosocystis oceanus* Sp. Nov. Abstract, Intern. Congr. Microbiol., 8th, Montreal, 1962.

WINOGRADSKY, H. 1935. Sur la microflore nitrificatrice des boues activées de Paris. Compt. Rend., **200**: 1886–1888.

———. 1937. Contribution a l'étude de la microflore nitrificatrice des boues activées de Paris. Ann. Inst. Pasteur, **58**: 326–340.

WINOGRADSKY, S. 1890. Sur les organismes de la nitrofication. Compt. Rend., **110**: 1013–1016.

———. 1892. Contributions à la morphologie des organismes de la nitrification. Arch. Sci. Biol., **1**: 86–137.

———. 1904. Die Nitrifikation, p. 132–181. *In* Handbuch der technischen Mykologie. Lafar, Jena.

———. 1931. Nouvelles recherches sur les microbes de la nitrification. Compt. Rend., **192**: 1000–1004.

———, AND H. WINOGRADSKY. 1933. Nouvelles recherches sur les organismes de la nitrification. Ann. Inst. Pasteur, **50**: 350–432.

ZOBELL, C. E. 1946. Marine microbiology. Chronica Botanica Co., Waltham, Mass. 240 p.

33

Reprinted from *J. Bacteriol.*, **106**(2), 356–361 (1971)

Separate Nitrite, Nitric Oxide, and Nitrous Oxide Reducing Fractions from *Pseudomonas perfectomarinus*

W. J. PAYNE, P. S. RILEY,[1] AND C. D. COX, JR.

Department of Microbiology, University of Georgia, Athens, Georgia 30601

Received for publication 23 December 1970

Pseudomonas perfectomarinus was found to grow anaerobically at the expense of nitrate, nitrite, or nitrous oxide but not chlorate or nitric oxide. In several repetitive experiments, anaerobic incubation in culture media containing nitrate revealed that an average of 82% of the cells in aerobically grown populations were converted to the capacity for respiration of nitrate. Although they did not form colonies under these conditions, the bacteria synthesized the denitrifying enzymes within 3 hr in the absence of oxygen or another acceptable inorganic oxidant. This was demonstrated by the ability, after anaerobic incubation, of cells and of extracts to reduce nitrite, nitric oxide, and nitrous oxide to nitrogen. From crude extracts of cells grown on nitrate, nitrite, or nitrous oxide, separate complex fractions were obtained that utilized reduced nicotinamide adenine dinucleotide as the source of electrons for the reduction of (i) nitrite to nitric oxide, (ii) nitric oxide to nitrous oxide, and (iii) nitrous oxide to nitrogen. Gas chromatographic analyses revealed that each of these fractions reduced only one of the nitrogenous oxides.

Nitrite and nitric and nitrous oxides were identified as intermediate products of complete respiratory nitrate reduction (denitrification) by a variety of bacteria with nitrogen released as the terminal product of reduction (4, 8, 17). Other intermediates may occur as well but are yet unidentified. In earlier studies of events characterizing the flow of electrons in denitrification by resting cells and extracts of various bacteria (1, 9, 15), investigators made use of manometric assays of gas release. The identity of the gases was thus at times uncertain. Mass spectrometric assays for products of reduction of ^{15}N-labeled nitrate were also employed (4) to provide a greater degree of certainty in the identification of products, but this procedure is not technically well suited for the necessarily repetitive analyses of a detailed investigation.

Gas chromatographic procedures are now known to be reliable for identifying components of prepared mixtures of the gases likely to result from denitrification (6). Barbaree and Payne (2) and Payne and Riley (11) recently demonstrated that this method can be employed to separate and identify the products of various steps in the denitrification carried out by cells and extracts of *Pseudomonas perfectomarinus*. Moreover, Matsubara and Mori (8) used gas chromatography to establish the identity of nitrous oxide as a transient product of nitrite reduction by *P. denitrificans;* more recently, Renner and Becker (14) employed this method to demonstrate that nitrous oxide is both the terminal product of nitrate respiration and the specific result of nitric oxide reduction by resting cells of *Corynebacterium nephridii*.

Equipped with this more effective analytical capability, it has now been possible to undertake studies of the separate subterminal and terminal events in denitrification carried out by enzymatic fractions from cell-free extracts of *P. perfectomarinus*. In an earlier report, we indicated that a complex fraction containing reducing activity for nitric oxide (but not for nitrate, nitrite, or nitrous oxide) was obtained by gel chromatography (11). The current paper describes additional procedures for separation, from crude extracts, of enzymatic components that reduce nitrite to nitric oxide and nitrous oxide to nitrogen—each free of the other activities. Moreover, we have observed that these enzymes are synthesized by the bacteria under anaerobic conditions even if none of the nitrogenous oxides is present during

[1] Present address: Department of Biochemistry, Medical College of Georgia, Augusta, Ga. 30904.

incubation.

MATERIALS AND METHODS

Culture of the bacteria. *P. perfectomarinus* (18) was maintained and cultured either aerobically or at the expense of nitrous oxide on a liquid tryptone-yeast extract-sea salt medium (TYS) as previously described (3, 11). When cells were grown at the expense of nitrate or nitrite (one served as well as the other), TYS was supplemented with 0.1% potassium nitrate or nitrite and designated TYSN. An asparagine-minimal salts medium (15) was employed in one experiment. The possibility of chlorate (12) or nitric oxide serving as terminal oxidant was investigated by supplementing anaerobic cultures in TYS with various concentrations of potassium chlorate or bubbling nitric oxide through the cultures (in a hood to avoid poisoning).

Conversion of populations from aerobic to denitrifying metabolism. Quantitative estimates of the fraction of aerobic populations of *P. perfectomarinus* that may be transformed to denitrifying growth were obtained by diluting samples from 24-hr cultures in TYS into small vials of tempered TYS and TYSN agar supplemented with 2.4% thioglycolate and incubating anaerobically at 30 C. Colonies formed within 72 to 96 hr in TYSN agar were counted and compared with the viable aerobic population in identical samples as determined by the ordinary pour-plate method for preparing TYS agar cultures that were incubated at 30 C.

These bacteria do not grow anaerobically without an acceptable inorganic terminal oxidant (11) but will survive an indefinite period of anaerobiosis. To determine whether anaerobiosis in the absence of an appropriate oxidant would result in the formation of the denitrifying enzymes, the flow of air was stopped near the end of the exponential phase of cultures growing with strong aeration in tryptone-sea salt or asparagine-minimal medium. The effects of sudden change from air to helium flow were also examined. After incubating for 1 to 3 hr, during which time the oxygen content of the atmosphere over the cultures was observed by gas chromatographic analyses to decrease sharply, cells were harvested by centrifugation, suspended under helium in homologous medium containing 100 μg chloramphenicol per ml (11), and assayed for the ability to reduce nitrite and nitrous oxide (the first and last intermediates) to nitrogen. Samples of these cells were frozen at the time of harvest, and crude cell-free extracts were prepared soon after the whole-cell assays were completed. The crude extracts were then assayed for the ability to reduce nitrite, nitric oxide, and nitrous oxide to nitrogen.

Preparation and assay of reductive activity in extracts and fractions. *P. perfectomarinus* was cultured at the expense of nitrate, nitrite, or nitrous oxide in TYS. Cells were harvested, washed, ruptured, and centrifuged, and crude extracts were assayed for the capacity for denitrification as previously described (11). Reduced nicotinamide adenine dinucleotide (NADH) was employed as electron donor. The fraction containing the nitric oxide-reducing complex (Fr II) was obtained in best yield from cells grown on nitrous oxide. This preparation was free of enzymes capable of reducing nitrate, nitrite, or nitrous oxide (11).

Separation of the nitrite-reducing complex (Fr I) was accomplished as follows. (i) Nitrate or nitrite grown cells suspended in 0.2 M potassium phosphate buffer (*p*H 7.4) were passed through the French pressure cell, clarified by centrifugation, dialyzed against 0.02 M potassium phosphate buffer (*p*H 7.4) for 4 to 6 hr with three changes, and treated with enough ammonium sulfate (Mann, enzyme grade) to provide 60% saturation; (ii) the precipitate was removed leaving the nitrite-reducing enzymes, free of other activities, "soluble" in the supernatant fluid; (iii) excess ammonium sulfate was then removed by passage of this soluble fraction through a refrigerated Sephadex G-75 column (0.5 by 10 cm). Activity was determined as previously described (11) with nitrate, nitrite, nitric oxide, and nitrous oxide supplied in separate assays as potential electron acceptors.

The nitrous oxide-reducing fraction (Fr III) was obtained from cells that had been grown with nitrous oxide as the terminal oxidant. Crude extracts of cells that were passed through the French pressure cell contained only low levels of nitrate- and nitrite-reducing activities. There was significant reduction of nitric oxide by the crude extract, but separation and 14-fold purification of the capacity for nitrous oxide reduction from the other activities were obtained by the following procedures. (i) The precipitate resulting from 0 to 40% ammonium sulfate saturation of crude extract was removed; (ii) material precipitated between 40 and 70% saturation was suspended in 0.2 M potassium phosphate buffer (*p*H 7.4), placed on refrigerated Sephadex G-100 in columns (2.5 by 90 cm), and eluted with 0.02 M potassium phosphate buffer (*p*H 7.4) that was 0.001 M with respect to 2-mercaptoethanol; this yielded (iii) three protein-rich fractions which were assayed separately and in combinations for the ability to reduce each of the electron acceptors. The first eluted had no activity or influence on activity, and the other two were inactive singly. However, a combination (which we designated Fr III) of the second and third protein-rich components did reduce nitrous oxide but not nitrate, nitrite, or nitric oxide.

No attempt was made to separate and retain a nitrate-reducing fraction.

RESULTS

Assays for conversion of aerobic populations to denitrification. *P. perfectomarinus* did not grow anaerobically at the expense of chlorate (12) or nitric oxide but responded only to nitrate, nitrite, and nitrous oxide.

When samples comprising 4×10^9 to 5×10^9 viable cells of *P. perfectomarinus* per ml, subcultured more than 10 times aerobically in the absence of nitrate, nitrite, or nitrous oxide, were incubated anaerobically in agar cultures containing nitrate, an average of 82% of the viable cells in 20 repetitive experiments formed colonies. Of this fraction, an average of 58% produced visible gas bubbles around the submerged colonies, and nitrogen was the only gas detected in samples taken by needle and syringe from several of the bubbles. There was no anaerobic growth in TYS agar lacking nitrate. Convertibility to the capacity for denitrification is thus a property of a very large fraction of aerobically growing populations of *P. perfectomarinus*.

Since assay of these experiments depended on colony formation, the omission of nitrate from the agar revealed no more than a requirement for a terminal oxidant for growth. However, we were interested in determining if the presence of nitrate was necessary for conversion of cells to the capability for nitrate respiration, whether or not they grew. We found, accordingly, that when the aerobically growing cultures were deprived of air near the end of the exponential phase of growth, the capacity for reducing the oxides to nitrogen was derepressed even in the absence of nitrate or any other of the nitrogenous oxides during conversion. Capacity to reduce nitrite at 40% of the rate exhibited by the cells grown on nitrate was observed after anaerobic incubation for 3 hr (Fig. 1A). Aerobic cells and those incubated anaerobically for only 1 hr did not reduce nitrite to nitrogen. It can be seen that capacity to reduce nitrous oxide was partially derepressed in 1 hr and fully derepressed after anaerobic incubation for 3 hr (Fig. 1B). Aerobically grown cells reduced nitrous oxide at a low rate. This suggests that the aerobic respiratory mechanism may function to some degree with nitrous oxide substituting for oxygen.

What fraction of the total, aerobically viable population produced the denitrifying enzymes anaerobically was not investigated. However, sufficient quantities were produced to permit

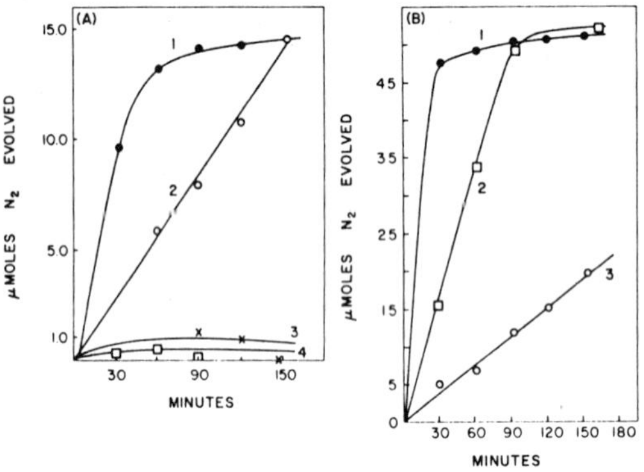

FIG. 1. *Reduction of nitrite and nitrous oxide by resting cells after anaerobic incubation in asparagine-minimal salts medium in the absence of an inorganic oxidant. After anaerobic incubation, cells were harvested, suspended in the same volume of fresh medium containing chloramphenicol (100 μg/ml), and dispensed into Warburg vessels. Each vessel, containing 3 ml of cell suspension (139 mg, dry weight), was flushed for 2 min with helium. The side arms were then stoppered, and the vessel mouth was closed with a rubber septum. Either a solution containing 30 μmoles of KNO_2 or 2 ml (approximately 49 μmoles) of nitrous oxide gas was injected through the cap into the liquid, and the reaction mixtures were incubated with shaking at 30 C. Samples of the atmosphere over the reactions were obtained by hypodermic needle and gas-tight syringe and assayed as previously described (11). Both cells which were grown aerobically and not incubated anaerobically and cells that were grown anaerobically at the expense of nitrate and then treated identically in every other way were assayed as controls. (A) Nitrite provided: curve 1, cells grown on nitrate; curve 2, cells incubated anaerobically for 3 hr; curve 3, cells incubated anaerobically for 1 hr; curve 4, aerobic cells. (B) Nitrous oxide provided: curve 1, cells grown on nitrate or cells incubated anaerobically for 3 hr; curve 2, cells incubated anaerobically for 1 hr; curve 3, aerobic cells.*

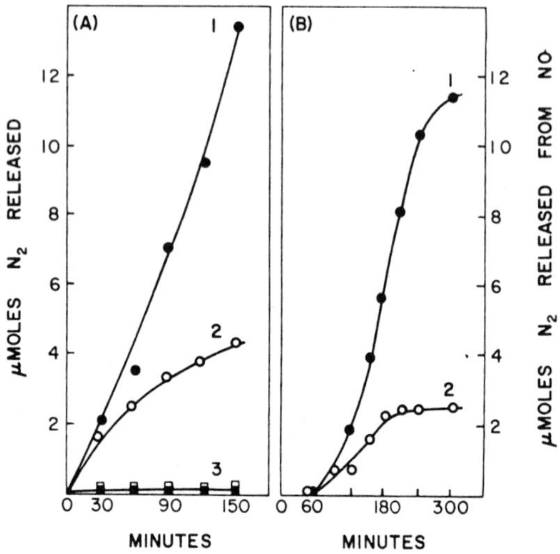

FIG. 2. *Reduction of nitrite, nitric oxide, and nitrous oxide by extracts of anaerobically incubated cells. After incubation for 3 hr after sparging ceased, the cells were harvested and ruptured by passage through a French pressure cell; the crude extract was clarified by centrifugation. Fully aerobically grown cells were harvested and extracted in the same manner to provide controls. Reaction mixtures contained: NADH, 1 μmole; flavine adenine dinucleotide, 0.5 μmole; flavine mononucleotide, 0.5 μmole; potassium phosphate buffer (pH 7.4), 300 μmoles; 30 mg of extract protein. Final liquid volume, 3 ml. Reaction vessels were flushed with helium for 2 min, the side arms were stoppered, and the vessels were capped. The oxidants were injected, and the reaction mixtures were incubated with shaking at 30 C. (A) Curve 1, nitrous oxide, approximately 50 μmoles; curve 2, KNO_2, 10 μmoles; curve 3, no acceptor or extracts of aerobic cells with either. (B) Curve 1, extracts from anaerobically incubated cells; curve 2, extracts from aerobic cells, both with nitric oxide, approximately 26 μmoles. Omission of NADH delayed reactions for 30 min.*

their demonstration in cell-free extracts of these cells. Nitrite and nitrous oxide were reduced without a lag (Fig. 2A). When nitric oxide was supplied, there was a lag before nitrogen was released (Fig. 2B). Nitrous oxide accumulated during the first hour but was rapidly depleted as nitrogen began to appear. We are currently examining the possibility that, when it is present in significant excess, nitric oxide suppresses reduction of nitrous oxide, when the latter is present in small quantity, as these results would suggest.

Replacement of air with helium as the sparging gas for the conversion of cultures from aerobic to anaerobic conditions did not result in synthesis of the denitrifying enzymes. Natural depletion of residual oxygen appeared to be necessary for this conversion.

Separation of reducing fractions. Crude extracts of cells grown on nitrate, nitrite, or nitrous oxide reduced each of the nitrogenous oxides with NADH as electron donor. But, nitrite-reducing activity was particularly rich in extracts from cells grown in nitrate or nitrite and nitric- and nitrous oxide-reducing activity in extracts of those grown on nitrous oxide.

The freedom of Fr I from nitrate, nitric oxide, and nitrous oxide reducing activities was established by demonstration (i) that nitric oxide was the only gaseous product detected from reduction of nitrite (Fig. 3) and (ii) that the fraction failed to reduce nitrate, nitric oxide, or nitrous oxide. The addition of nitrate to the reaction mixtures did not suppress the rate of release of nitric oxide from nitrite.

The separateness of Fr II was demonstrated by gas chromatographic analysis (Fig. 4) of the activity of preparations obtained by discontinuous gradient elution of extract protein from diethylaminoethyl cellulose columns as previously described (11). Nitrous oxide was the only product detected when nitric oxide was reduced. Suppres-

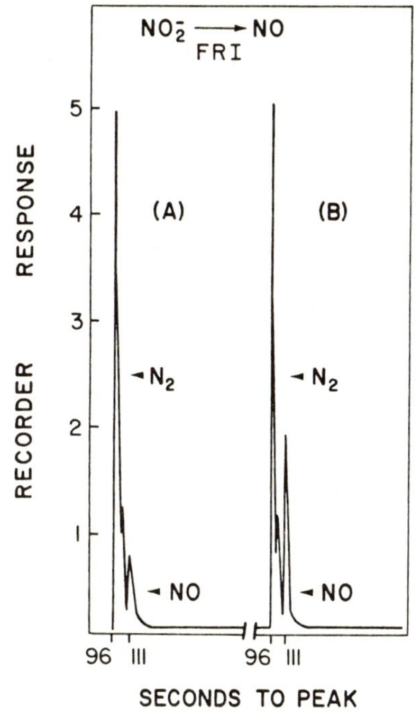

FIG. 3. *Reduction of nitrite to nitric oxide by Fr I. Reaction mixture was prepared and assayed as described in legend for Fig. 2 with 1 mg of Fr I protein provided as enzyme source and 30 μmoles of KNO_2, supplied as electron acceptor. Final volume, 2 ml. (A) Sample taken at 30 min, (B) at 75 min. In identical experiments, KNO_3, 30 μmoles; nitric oxide gas, approximately 26 μmoles; or nitrous oxide gas, approximately 49 μmoles, was supplied instead of KNO_2 as potential electron acceptor. None of these was reduced.*

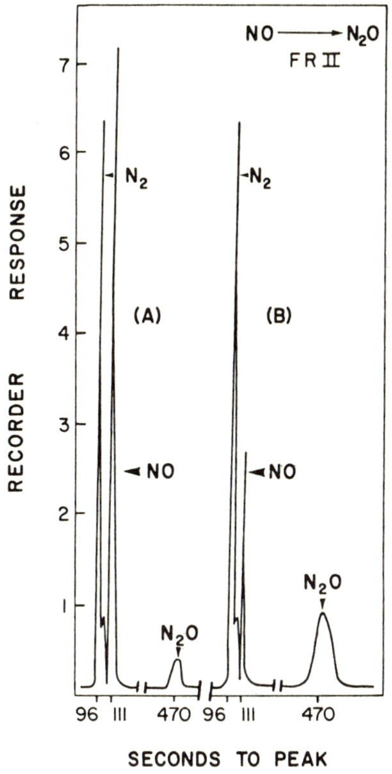

FIG. 4. *Reduction of nitric oxide to nitrous oxide by Fr II. Reaction mixture was prepared and assayed as described in legend for Fig. 2 with 0.5 mg of Fr II protein provided as enzyme source and approximately 26 μmoles of nitric oxide gas supplied as electron donor. Final volume, 2 ml. (A) Sample taken at 30 min, (B) at 60 min. In identical experiments, KNO_2 or KNO_3 (30 μmoles) or nitrous oxide gas (approximately 49 μmoles) was supplied instead of nitric oxide as potential electron acceptor. None of these was reduced.*

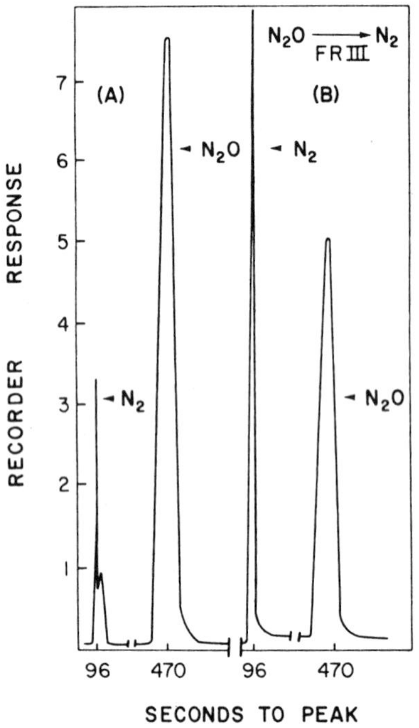

FIG. 5. *Reduction of nitrous oxide to nitrogen by Fr III. Reaction mixture was prepared and assayed as described in legend for Fig. 2 with 5 mg of Fr III provided as enzyme source and nitrous oxide gas (approximately 49 μmoles) supplied as electron acceptor. Final volume, 2 ml. (A) Sample taken at 10 min, (B) at 60 min. The nitrogen peak went off the recorder scale at 10. In identical experiments, KNO_2 or KNO_3 (30 μmoles) or nitric oxide gas (approximately 26 μmoles) was supplied instead of nitrous oxide as potential electron acceptor. None of these was reduced.*

sion of the activity of this fraction by nitrate was previously described (11). We can now report that nitrite also suppressed the nitric oxide-reducing activity of Fr II, but four to five times as much nitrite as nitrate was required under identical conditions to obtain the maximum (60%) suppression.

The parting of nitrate-, nitrite-, and nitric oxide-reducing activities from Fr III was demonstrated (Fig. 5) by incubating the latter fraction and electron transfer cofactors with each of the nitrogenous oxidants. Only nitrous oxide was reduced by Fr III. Nitrogen was the product detected.

DISCUSSION

Reduction of nitrite, nitric oxide, and nitrous oxide by enzymes in crude extracts or complex fractions of bacteria was previously observed in other laboratories (1, 4, 10, 13, 16, 17) and our own (3, 11). In addition, a number of workers investigated the repressive effects of oxygen on the synthesis of denitrifying enzymes (5, 7), but the stimulating effects of anoxia on their synthesis were overlooked. In previous studies, nitrate was included in the anaerobic culture medium because an oxidant is required for growth. The possibility that onset of anaerobiosis in the absence of a reducible oxide may serve unaided in aerobically cultured populations to initiate synthesis of the denitrifying enzymes was not previously investigated. From the current work, it seems necessary for the cells to be able to use up the residual quantity of oxygen present during synthesis of the denitrifying enzymes, perhaps as a means of getting energy for the conversion. It would thus be reasonable that immediate replacement of air with helium did not permit the initiation of synthesis, for apparently these pseudomonads have no truly fermentative energy-yielding mechanisms.

Since the enzymes required for reduction of each of the known intermediates of denitrification can be separated, it may now be possible to design more definitive studies of cytochromes and possibly nonheme metals involved specifically in the electron transport carried out in each of the three fractions. *Pseudomonas* species were reported to produce a variety of cytochromes concomitant with nitrite reduction. Walker and Nicholas (16) associated *c*-type and other cytochromes with nitrite reductase from *P. aeruginosa*, and Yamanaka (17) reported nitrite reduction by a *c*-551 pigment and an oxidase from *P. aeruginosa*. According to Matsubara and Mori (8), *Pseudomonas denitrificans* produces a cytochrome *c*-552 which accepts electrons from tetramethyl-*p*-phenylenediamine and reduces nitrite. A two-heme nitrite reductase, first discerned in *Micrococcus denitrificans*, was also found in *P. aeruginosa* (10). Cytochromes that carry out reduction of nitric and nitrous oxides, however, were not determined.

ACKNOWLEDGMENTS

This study was supported by grant Nonr 3677(01) from the Office of Naval Research and by stipends from Public Health Service training grants 1 T01 GM01968-01 and 02, from the National Institute of General Medical Sciences (to P.S.R. and C.D.C.).

LITERATURE CITED

1. Allen, M. G., and C. B. van Niel. 1952. Experiments on bacterial denitrification. J. Bacteriol. **64**:397–412.
2. Barbaree, J. M., and W. J. Payne. 1967. Products of denitrification by a marine bacterium as revealed by gas chromatography. Mar. Biol. **1**:136–139.
3. Best, A. N., and W. J. Payne. 1965. Preliminary enzymatic events in asparagine-dependent denitrification by *Pseudomonas perfectomarinus*. J. Bacteriol. **89**:1051–1054.
4. Delwiche, C. C. 1959. Production and utilization of nitrous oxide by *Pseudomonas denitrificans*. J. Bacteriol. **77**:55–59.
5. Downey, R. J., D. F. Kiszkiss, and J. H. Nuner. Influence

of oxygen on development of nitrate respiration in *Bacillus stearothermophilus*. J. Bacteriol. **9**:1056–1062.
6. Hollis, O. L. 1966. Separation of gaseous mixtures using porous polyaromatic beads. Anal. Chem. **38**:309–316.
7. Lam, Y., and D. J. D. Nicholas. 1969. A nitrite reductase with cytochrome oxidase activity from *Micrococcus denitrificans*. Biochim. Biophys. Acta **180**:459–472.
8. Matsubara, T., and T. Mori. 1968. Studies on denitrification. IX. Nitrous oxide, its production and reduction to nitrogen. J. Biochem. **64**:863–871.
9. Najjar, V. A., and C. W. Chung. 1956. Enzymatic steps in denitrification, p. 260–291. *In* W. D. McElroy and B. H. Glass (ed.), Symposium on inorganic nitrogen metabolism. Johns Hopkins Press, Inc., Baltimore.
10. Newton, N. 1969. The two-haem nitrite reductase of *Micrococcus denitrificans*. Biochim. Biophys. Acta **185**:316–331.
11. Payne, W. J., and P. S. Riley. 1969. Suppression by nitrate of enzymatic reduction of nitric oxide. Proc. Soc. Exp. Biol. Med. **132**:258–260.
12. Pichinoty, F., and M. Piechaud. 1968. Recherche des nitrate-reductases bacteriennes A et B: methodes. Ann. Inst. Pasteur **114**:77–98.
13. Radcliffe, B. C., and D. J. D. Nicholas. 1968. Some properties of a nitrite reductase from *Pseudomonas denitrificans*. Biochim. Biophys. Acta **153**:454–554.
14. Renner, E. D., and G. L. Becker. 1970. Production of nitric oxide and nitrous oxide during denitrification by *Corynebacterium nephridii*. J. Bacteriol. **101**:821–826.
15. Rhodes, M. E., A. N. Best, and W. J. Payne. 1963. Electron donors and cofactors for denitrification by *Pseudomonas perfectomarinus*. Canad. J. Microbiol. **9**:799–807.
16. Walker, G. D., and D. J. D. Nicholas. 1961. Nitrite reductase from *Pseudomonas aeruginosa*. Biochim. Biophys. Acta **49**:350–360.
17. Yamanaka, T. 1964. Identity of *Pseudomonas* cytochrome oxidase with *Pseudomonas* nitrite reductase. Nature (London) **204**:253–245.
18. Zobell, C. E., and H. C. Upham. 1944. A list of marine bacteria including descriptions of sixty new species. Bull. Scripps Inst. Oceanogr. **5**:239–292.

Part VI

SECONDARY ACTIVITIES OF MARINE BACTERIA

Editor's Comments on Papers 34 Through 37

34 DiSALVO
Isolation of Bacteria from the Corallum of Porites lobata *(DANA) and Its Possible Significance*

35 EHRLICH
Bacteriology of Manganese Nodules: II. Manganese Oxidation by Cell-Free Extract from a Manganese Nodule Bacterium

36 MARSHALL, STOUT, and MITCHELL
Mechanism of the Initial Events in the Sorption of Marine Bacteria to Surfaces

37 SOROKIN, PETIPA, and PAVLOVA
Quantitative Estimate of Marine Bacterioplankton as a Source of Food

The obvious degradative abilities of bacteria toward nitrogen and carbon compounds naturally led most investigators to undertake extensive examinations of these aspects of microbial function in the seas. However, other workers, either from chance observation or an intuitive feeling, have sought out additional, sometimes secondary, roles for the bacteria in the ocean. The five papers presented in Part VI present descriptions of some of these interesting secondary activities: bacterial involvement in the precipitation and decomposition of calcium carbonate, the sorption to solid surfaces of these microorganisms, and their entrance into the food chain as food sources for zooplankton and protozoans.

Since Drew (9) presented his hypothesis of the mechanism of calcium carbonate formation in tropical waters, there have been several attempts to find evidence to prove or disprove his theory (27, 46). As mentioned previously in this volume, both Waksman and Lipman were highly skeptical of Drew's explanation of such activity. However, many microbiologists have observed on agar surfaces pure cultures that produced a white halo or even sharp, short projections into the medium. Most frequently these formations have been ignored or blamed on fungal contaminants. In 1947, Hewitt (21) reported the formation of bacterial calculi and cautioned against confusing their formation with bacterial colonial dissociation. The conflict was still not resolved, however, because Nadson (32) had proposed that ammonia production, not nitrate reduction, was responsible for the formation of loci of calcium carbonate. He also interjected another requisite condition for such an activity, that sulfate reduction occurs at the same time. Because sulfate reduction to hydrogen sulfide is generally an anaerobic process (33, 49), obviously his concept could not explain the bacterial role in calcium carbonate precipitation in the aerobic water column.

Lalou (25), working with marine sediments supplemented with glucose, obtained "at will" various quantities of calcium carbonate crystals following fermentation and anaerobic sulfate reduction, with the release of hydrogen sulfide. This H_2S then complexed with iron oxide, the pH of the medium was raised, and $CaCO_3$ was precipitated. These successful experiments tended to keep the controversy alive. Lalou does not state whether the crystals formed were calcite or aragonite, but he mentions that this demonstration in an aquarium of $CaCO_3$ precipitation does not rule out physiochemical processes. A report by Hutchison (22) in 1961 on *Staphylococcus aureus* calculus formation indicated that not all such crystals are $CaCO_3$, but that they can also be a complex of calcium phosphate. Monaghan and Lytle (30) and Oppenheimer (34) both noted that the formation of aragonite crystals in association with bacteria, especially from the Bahamas Bank, took up to one month for their appearance.

This led Greenfield (16) into a study of both the magnesium and calcium metabolism of marine bacteria. Working with a pure culture of *Pseudomonas piscicida,* he verified his previous studies with MB-1, which showed crystallization occurring with the bacterial cell as a nucleus, and extended that observation to demonstrate that the cell envelope is the primary site for this process. Thus, the early hypothesis by Nadson (32) that ammonia production from protein breakdown resulted in a higher pH and, thereby, carbonate precipitation as aragonite was disproven.

The possibility of the localized increase in CO_2 resulting from bacterial metabolism has also been considered to play an important part in this sedimentation process. That this is not an isolated phenomenon relating only to tropical waters has been amply demonstrated by Shinano and Sakai (36), who isolated crystal-forming bacteria from the North Pacific. These authors found both calcium carbonate and ammonium phosphate hexahydrate crystals associated with marine bacterial growth. Shinano then proceeded to isolate and identify over 600 strains of organisms participating in this process (37). He noted that gram-positive cocci (*Micrococcus*) (38), gram-negative rods (*Flavobacterium* and *Achromobacterium* genera) (39), and yeasts (*Rhodotorula, Candida,* and *Debaryomyces*) (40) were the predominant microbes involved in calcium carbonate precipitation. Lately, Krumbein (24) has reported that the formation of these crystals results in the death of the bacterium, due to crystal pressure, and this could be observed with scanning electron microscopy within 36 to 90 hours of cultivation.

Because we are concerned here with calcium carbonate precipitation in tropical waters, it seems logical to ask if this action has any effect on the corals and their exoskeletal formation. Actually, at this point it is not known if any bacteria are directly involved in assisting in the formation of coral reefs. DiSalvo (5, 6) has contributed most of the information on our knowledge of bacteria in or on corals. Starting in 1963, he began investigating the microbial flora of dead corals that might be contributing to their conversion to sedimentary particles. During the course of these studies, he noted that bacteria could invade at least one type of live coral, *Porites lobata* (DANA), and decompose the internal organic matter. This might then lead to further breakdown of the carbonate skeleton, either from acid production resulting from microbial metabolic by-products or through structured weakening due to the loss of the organic matrix (7, 8). Paper 34 is his first major report on this subject.

Another secondary activity of marine microorganisms that has also stirred some controversy (15, 18, 19, 45) concerns microbial involvement in the formation of manganese nodules. In 1963, Ehrlich (10) reported that bacteria found on the surfaces of manganese nodules increased the rate of sorption from seawater of the Mn^{2+} form, especially if glucose or peptone was present. The organism used in these studies was a strain of *Arthrobacter*. In addition, he noted that some bacteria were also capable of releasing manganese from the nodules, thereby presenting the possibility of controlling the rate of manganese nodule growth (4, 11, 43, 44). By 1968, Ehrlich presented in Paper 35 data showing the enzymatic nature of the conversion of the soluble manganous to insoluble manganic oxide. The inhibition by sulfhydryl complexing agents, the optimum temperature of 17.5°C with total inhibition resulting from heating to 100°C, and finally separation of the active fraction on Sephadex G-150 columns all favor an enzymatic explanation for the observation over simply a pH or redox change in the environment (42, 43). Ehrlich (13) also isolated a gram-negative bacterium that would enzymatically reduce manganic oxides to manganous ions. Further studies have revealed that the presence of bacteria on nodules is quite common, and frequently these numbers are very high (12, 41). Current studies undertaken by LaRock include scanning electron microscopy of the nodules and ATP analysis of the total biomass (unpublished data).*

Obviously, microbial involvement in the precipitation or solution of carbonates, iron and manganese oxides, and probably other untested geochemical processes does not completely explain the entire story. There are physical and chemical factors involved, as well as macroscopic biological involvement [boring sponges, for example (14)]. Whether any of these factors are accentuated by microbiologically induced changes in the pH, Eh, or chemical balance of the microenvironment is certainly too early to tell. The point that needs to be stressed here is that, in any consideration of models for geochemical processes in the oceans, the microbial flora undoubtedly play a role and must be considered if these models are to have any reality to nature. At the same time, the obvious challenge for the marine microbiologist is to provide hard data on the rate and extent of these activities. The isolation of one organism capable of performing a transformation does not mean that the transformation is an ecologically important function. The burden of proof is on us.

Long before man knew about bacteria or fungi as organisms in the oceans, he had learned to recognize one of their more deleterious functions, the fouling of ship bottoms and the rotting of wood. Because of the obvious safety and economic importance to man, there is an extensive literature on the subject, including a recent book that covers macroscopic as well as microscopic forms of fouling organisms (23). This book deals especially with the fungi, their identification, biology, and ecology. A review of bacterial attachment and fouling was published by Corpe (2) in which he described many of the types of bacteria involved in such activities, as well as some of the mechanisms for attachment: slime, holdfasts, stalks, and so on. Despite demonstrations of the reversible and irreversible types of sorption by ZoBell (50, 51), Wood (47),

*Since these comments were written, this information has been published: P. A. LaRock and H. L. Erlich, "Observations of bacterial microcolonies on the surface of ferromanganese nodules from Blake Plateau by scanning electron microscopy," *Microbial Ecol.* 2:84–96 (1975).

and Rubentschik et al. (35), the mechanism by which nonappendaged bacteria were held onto solid surfaces was still largely unknown.

Then, in Paper 36, Marshall, Stout, and Mitchell described the reversible and irreversible sorption processes and thus provided a reasonable explanation for this phenomenon. The authors considered both the biological and physical factors that are important in this process. They particularly noted the effects of electrolytes, the dependence on time, and the effect of the age of the cultures on this sorption–desorption phenomenon. This study was followed by their investigation of the selective sorption of marine bacteria (29), in which they confirmed Corpe's earlier hypothesis that polymeric fibrils were more important than holdfasts or pili (2). A widely held theory that the fouling of surfaces in the sea is in some way dependent on primary film development by bacteria is still awaiting verification. At least, now, though, we have begun to understand the first steps in the fouling process—the attachment of bacteria to these solid surfaces.

Fortunately, the growth of bacteria in oceans, sediments, or on solid surfaces is limited. Besides the obvious controlling factors of nutrient depletion, biological crowding with attached forms, and possible autolytic factors, another mechanism exists which is perhaps more effective than any of these in controlling bacterial populations in the marine environment. Although included here under "secondary activities," the protozoans, nematodes, copepods, and invertebrates that feed upon bacteria might take exception to this denial of primacy. The scarcity of definitive studies on the nutritional requirements of planktonic animals until the last 20 years or so has resulted in the comparative ignorance of most marine microbiologists about the role these higher forms might play in controlling natural aquatic bacterial populations (49).

It has been recognized for some time that bacteria can serve as a major food source for mussels, sand crabs, and worms as a result of the studies reported by ZoBell and Feltham in 1938 (52). Much later, Wood (48) summarized some of the information accumulating on oyster and larval ingestion of bacteria, but studies on nematodes, copepods (20), and protozoan (11) requirements for bacterial sources of food had not received as thorough attention until recently. An early summary of the importance of bacteria to protozoan nutrition is to be found in a review by Luck et al. (28), in which they discuss the studies completed by German workers until 1931 and then describe their studies regarding the culturing of marine protozoa on various bacteria. The curious aspect of this paper is that they did not include any marine bacteria in their feeding studies, and concentrated instead on what proved to be *Pseudomonas aeruginosa* and *Escherichia coli*. They did note that spent culture filtrate and extracts and dialysates of the cells would not support the growth of their marine protozoan (28).

A revival in interest in the possibility that bacteria might serve *in situ* as food sources has resulted in several recent publications. Lighthart (26) observed that bacteriovorous protozoans were more common in shallow water bodies than in the deep ocean water of the Pacific. About the same time, Muller and Lee (31) reported that foraminifera required bacteria for growth, and they described means for producing axenic cultures of foraminifera.

Shortly after this, Paper 37, of Sorokin, Petipa, and Pavlova, was published. In this study, reporting on water samples from the Black Sea and the tropical Pacific, the authors used a ^{14}C-labeled marine pseudomonad grown in pure culture, and also radioactive natural populations obtained *in situ* and labeled using ^{14}C-algal hydrolysate. These two nutrient sources were then added to the test samples. Planktonic organisms were grouped according to fine, coarse, or mixed filter feeders, and the authors concluded that the minimum level of bacteria required for "recognition" by the filter feeders did indeed approach that found in the ocean layers where the zooplankton are most concentrated. However, they noted that the copepods will assimilate algae more efficiently and effectively than planktonic bacteria. Sorokin (42) has continued these studies, again using radioactively labeled microbial populations, to trace the uptake of bacteria by some scleractinian corals. More recently, in his excellent review of meiofaunal trophic relationships, Coull (3) discussed the importance of obtaining more definitive data on such factors as meiofaunal predation of bacterial sedimentary populations. Hamilton (17) has cultured protozoans in chemostats and tested their selectiveness toward bacteria; Chua and Brinkhurst (1) have postulated that the bacterial food source may influence tubificid worm distribution.

Thus, from this brief review and the selected papers, we see that bacterial "secondary activities" may be quite as important to the local microenvironment as the more generally recognized primary activities of organic carbon and nitrogen turnover.

REFERENCES

1. Chua, K. E., and Brinkhurst, R. O. 1973. Bacteria as potential nutritional resources for three sympatric species of tubified oligochaetes, pp. 513–517. In: H. L. Stevenson and R. R. Colwell (eds.), *Estuarine*

Microbial Ecology. University of South Carolina Press, Columbia, S.C., 536 pp.

2. Corpe, W. A. 1970. Attachment of marine bacteria to solid surfaces, pp. 73–87. In: R. S. Manley (ed.), *Adhesion in Biological Systems*. Academic Press, New York, 302 pp.
3. Coull, B. C. 1973. Estuarine meiofauna: a review: trophic relationships and microbial interactions, pp. 499–512. In: H. L. Stevenson and R. R. Colwell (eds.), *Estuarine Microbial Ecology*. University of South Carolina Press, Columbia, S.C., 536 pp.
4. DeCastro, A. F., and Ehrlich, H. L. 1970. Reduction of iron oxide minerals by a marine *Bacillus*. *Antonie van Leeuwenhoek J. Microbiol. Serol. 36*:317–327.
5. DiSalvo, L. H. 1969. Isolation of bacteria from the corallum of *Porites lobata* (DANA) and its possible significance. *Am. Zool.* 9:735–740.
6. DiSalvo, L. H. 1969. On the existence of a coral reef regenerative sediment. *Pac. Sci. 23*:129.
7. DiSalvo, L. H. 1971. Regenerative functions and microbial ecology of coral reefs. II. Oxygen metabolism in the regenerative system. *Can. J. Microbiol. 17*:1091–1100.
8. DiSalvo, L., and Gundersen, K. 1971. Regenerative functions and microbial ecology of coral reefs. I. Assays for microbial population. *Can. J. Microbiol. 17*:1081–1089.
9. Drew, G. H. 1913. On the precipitation of calcium carbonate in the sea by marine bacteria, and on the action of denitrifying bacteria in tropical and temperate seas. Papers from the Marine Biological Laboratory at Tortugas. *Carnegie Inst. Wash. Publ. 182*:2–45.
10. Ehrlich, H. L. 1963. Bacteriology of manganese nodules. I. Bacterial action on manganese in nodule enrichments. *Appl. Microbiol. 11*: 15–19.
11. Ehrlich, H. L. 1966. Reactions with manganese by bacteria from ferromanganese nodules. *Dev. Ind. Microbiol. 7*:279–286.
12. Ehrlich, H. L., Ghiorse, W.C., and Johnson, G. L., II. 1972. Distribution of microbes on manganes nodules from the Atlantic and Pacific oceans. *Dev. Ind. Microbiol. 13*:57–65.
13. Ehrlich, H. L., Yang, S. H., and Mainwaring, J. D. 1973. Bacteriology of manganese nodules. VI. Fate of copper, nickel, cobalt and iron during bacterial and chemical reduction of manganese(IV) oxide matrix of nodules. *Z. Allg. Mikrobiol. 13*:39–48.
14. Fütterer, D. K. 1974. Significance of the boring sponge *Cliona* for the origin of fine grained material of carbonate sediments. *J. Sediment. Petrol. 44*:79–84.
15. Glasby, G. P. 1972. The mineralogy of manganese nodules from a range of marine environments. *Mar. Geol. 13*:57–72.
16. Greenfield, L. J. 1963. Metabolism and concentration of calcium and magnesium and precipitation of calcium carbonate by a marine bacterium. *Ann. N.Y. Acad. Sci. 109*:23–45.
17. Hamilton, R.D. 1973. Interrelationships between bacteria and protozoa, pp. 491–498. In: H. L. Stevenson and R. R. Colwell (eds.), *Estuarine Microbial Ecology*. University of South Carolina Press, Columbia, S.C., 536 pp.
18. Hammond, A. L. 1974. Manganese nodules. I. Mineral resources on the deep seabed. *Science 183*:502–503.

19. Hammond, A. L. 1974. Manganese nodules. II. Prospects for deep sea mining. *Science 183*:644–646.
20. Harding, G. C. H. 1974. The food of deep-sea copepods. *J. Mar. Biol. Assoc. U.K. 54*:141–155.
21. Hewitt, H. B. 1947. Bacterial "calculi." *J. Pathol. Bacteriol. 59*:657–664.
22. Hutchison, J. G. P. 1961. Crystals in colonies of *Staphylococcus aureus. J. Pathol. Bacteriol. 82*:214–217.
23. Jones, E. B. G., and Eltringham, S. K. 1971. *Marine Borers, Fungi and Fouling Organisms of Wood.* Organization for Economic Co-operation and Development, Paris and Washington, D.C.
24. Krumbein, W. E. 1974. On the precipitation of aragonite on the surface of marine bacteria. *Naturwissenschaften 61*:167–169.
25. Lalou, C. 1957. Studies on bacterial precipitation of carbonates in sea water. *J. Sediment. Petrol. 27*:190–195.
26. Lighthart, B. 1969. Planktonic and benthic bacteriovorous protozoa at eleven stations in Puget Sound and adjacent Pacific Ocean. *J. Fish. Res. Board Can. 26*:299–304.
27. Lipman, C. B. 1929. Further studies on marine bacteria with special reference to the Drew hypothesis on $CaCO_3$ precipitation in the sea. *Carnegie Inst. Wash. Publ. 26*:231.
28. Luck, J. M., Sheets, G., and Thomas, J. O. 1931. The role of bacteria in the nutrition of protozoa. *Quart. Rev. Biol. 6*:46–58.
29. Marshall, K. C., Stout, R., and Mitchell, R. 1971. Selective sorption of bacteria from seawater. *Can. J. Microbiol. 17*:1413–1416.
30. Monaghan, P. H., and Lytle, M. L. 1956. The origin of calcareous ooliths. *J. Sediment. Petrol. 26*:111–118.
31. Muller, W. A., and Lee, J. J. 1969. Apparent indispensability of bacteria in foraminiferan nutrition *J. Protozool. 16*:471–478.
32. Nadson, G. A. 1928. Beitrag zur Kenntis der bakteriogen Kalkabla gerunzen. *Arch. Hydrobiol. 19*:154–164.
33. Oppenheimer, C. H. 1960. Bacterial activity in sediments of shallow marine bays. *Geochim. Cosmochim. Acta 19*:244–260.
34. Oppenheimer, C. H. 1961. Note on the formation of spherical aragonitic bodies in the presence of bacteria from the Bahama Bank. *Geochim. Cosmochim. Acta 23*:295–299.
35. Rubentschik, L., Roisin, M. B., and Bieljansky, F. M. 1936. Adsorption of bacteria in salt lakes. *J. Bacteriol. 32*:11–31.
36. Shinano, H., and Sakai, M. 1969. Studies of marine bacteria taking part in the precipitation of calcium carbonate. I. Calcium carbonate deposited in peptone medium prepared with natural sea water and artificial sea water. *Bull. Jap. Soc. Sci. Fish. 35*:1001–1005.
37. Shinano, H. 1972. Studies of marine microorganisms taking part in the precipitation of calcium carbonate. II. Detection and grouping of the microorganisms taking part in the precipitation of calcium carbonate. *Bull. Jap. Soc. Sci. Fish. 38*:717–725.
38. Shinano, H. 1972. Studies of marine microorganisms taking part in the precipitation of calcium carbonate. III. Taxonomic study of marine bacteria taking part in the precipitation of calcium carbonate. *Bull. Jap. Soc. Sci. Fish. 38*:727–732.
39. Shinano, H. 1973. Studies of marine bacteria taking part in the precipitation of calcium carbonate. VI. A taxonomic study of marine

bacteria taking part in the precipitation of calcium carbonate. *Bull. Jap. Soc. Sci. Fish. 39*:85–90.

40. Shinano, H. 1973. Studies of marine bacteria taking part in the precipitation of calcium carbonate. VII. A taxonomic study of the yeasts of marine origin taking part in the precipitation of calcium carbonate. *Bull. Jap. Soc. Sci. Fish. 39*:91–95.
41. Sorokin, Yu. I. 1971. Microflora of iron–manganese concretions from the ocean floor. *Microbiology USSR 40*:493–495.
42. Sorokin, Yu. I. 1973. On the feeding of some scleractinian corals with bacteria and dissolved organic matter. *Limnol. Oceanogr. 18*:380–385.
43. Trimble, R. B., and Ehrlich, H. L. 1968. Bacteriology of manganese nodules. III. Reduction of MnO_2 by two strains of nodule bacteria. *Appl. Microbiol. 16*:695–702.
44. Trimble, R. B., and Ehrlich, H. L. 1970. Bacteriology of manganese nodules. IV. Induction of an MnO_2-reductase system in a marine *Bacillus. Appl. Microbiol. 19*:966–972.
45. Turekian, K. K. 1968. *Oceans*. Prentice-Hall, Inc., Englewood Cliffs, N.J., 120 pp.
46. Waksman, S. A., Hotchkiss, M., and Carey, C. L. 1933. Marine bacteria and their role in the cycle of life in the sea. II. Bacteria concerned in the cycle of nitrogen in the sea. *Biol. Bull. 65*:137–167.
47. Wood, E. J. F. 1950. Investigations on underwater fouling. I. The role of bacteria in the early stages of fouling. *Aust. J. Mar. Fresh. Res. 1*: 85–91.
48. Wood, E. J. F. 1953. Heterotrophic bacteria in marine environments of Eastern Australia. *Aust. J. Mar. Fresh. Res. 4*:160–200.
49. Wood, E. J. F. 1965. *Marine Microbial Ecology*. Chapman & Hall Ltd., London, and Reinhold, New York, 243 pp.
50. ZoBell, C. E. 1943. The effect of solid surfaces upon bacterial activity. *J. Bacteriol. 46*:39–56.
51. ZoBell, C. E., and Allen, E. C. 1935. The significance of marine bacteria in the fouling of submerged surfaces. *J. Bacteriol. 29*:239–251.
52. ZoBell, C. E., and Feltham, C. B. 1938. Bacteria as food for certain marine invertebrates. *J. Mar. Res. 1*:312–327.

34

Reprinted from *Am. Zool.*, **9**, 735–740 (1969)

Isolation of Bacteria from the Corallum of *Porites lobata* (Dana) and Its Possible Significance

Louis H. DiSalvo

Department of Zoology, University of North Carolina, Chapel Hill, N. C. 27514

Synopsis. Bacteria were recovered from loci within skeletal regions of the glomerate coral, *Porites lobata*. The origin of these bacteria is unknown, although areas of discoloration suggest invasion from the substratum in the region of basal attachment. Weakened areas of internal corallum contained from 10^4 to 10^5 bacteria per gram dry weight as determined by plate counts on a peptone-agar medium. Some of the isolated bacteria were capable of digesting chitin *in vitro*. This finding suggested that the mechanism for skeletal weakening might be bacterial breakdown of the organic matrix. Absence of change from aragonitic to calcitic crystals from a discolored region supported the contention that skeletal weakening was due to the breakdown of organic matrix rather than dissolution of carbonate.

Results were obtained as part of a survey to determine the number and distribution of bacteria in some coral reef environments.

Coral reefs are entire ecosystems whose existence and perpetuity depend on the buildup and re-arrangement of calcified structures. In these diversified communities specialized organisms biochemically construct skeletons from calcium, carbonate, and small amounts of other ions removed from the surrounding seawater. Concurrent with skeletal buildup, physical and chemical forces, as well as the specialized activities of boring organisms, fragment and perhaps dissolve portions of secreted skeletons. Two major groups of microscopically active organisms have been implicated in penetration and destruction of calcareous reef substrates. These include the clionid sponges (Goreau and Hartman, 1963; Neumann, 1966) and a diverse taxonomic assemblage of "boring algae" which penetrate limestone substrates (Purdy and Kornicker, 1958).

The activities of bacteria have sometimes been implicated in coral reef carbonate dynamics. Hypotheses were proposed by Purdy (1963) and Swinchatt (1965) concerning bacteria-mediated breakdown and buildup of calcareous substrates. Additional hypotheses were reviewed and augmented by Wood (1960), who admitted to the almost total lack of knowledge about bacteria on reefs.

This work was supported by the U.S. Public Health Service (PHS5701 Es 0006104), the Hawaii Institute of Marine Biology, and the U.S. Atomic Energy Commission through the Eniwetok Marine Biological Laboratory.

BACTERIA ON CORAL REEFS

This report includes data collected during studies on the number, distribution, and activities of bacteria in selected coral reef environments. During exploratory research it was apparent that bacteria invaded coral skeletons. Casual splitting of glomerate coral heads during studies of Eniwetok Atoll in 1964 revealed discolored (light brown) internal regions. These regions did not have the appearance of the typical boring algal bands, and the discolored skeleton was more friable than the adjacent white skeleton. Peptone-agar cultures of scrapings from the discolored regions yielded many bacterial colonies. Similar cultures of scrapings taken from adjacent intact skeletal areas of the same head gave negative results. An additional observation was that some penetration paths of boring sponges appeared to follow pre-existing streaks of discoloration into deep regions of the corallum.

More recently I carried out research at the Hawaii Institute of Marine Biology at Kaneohe Bay, Oahu, during which I at-

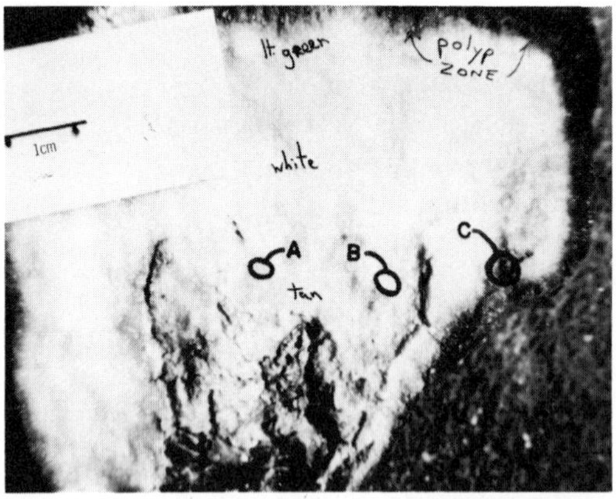

FIG. 1. Corallum of *Porites lobata* split apart to show internal altered region. Sites A, B, and C represent zones sampled in the initial qualitative bacterial recovery (Table 2). Areas A and B are light tan, C is gray-black.

tempted to verify and enlarge upon the exploratory results using a single species of coral. The only glomerate coral species commonly found in Kaneohe Bay is *Porites lobata* (Vaughn). These colonies proved excellent for the studies because different coralla when split apart showed internal disturbances of varied intensity and appearance. Figure 1 illustrates the areas sampled in an internal decalcified region. Most commonly these areas were brown, although occasionally a gray-black zone was found. The gray-black regions were probably sites of sulfate reduction, smelling weakly of H_2S.

Bacterial invasion of coral skeletons undoubtedly depends on the quality and quantity of nutrient material in the corallum. Wainwright (1963) described some organic materials which might occur in coral skeletons. He found a fibrillar chitinous matrix in the coral, *Pocillopora damicornis* as well as remnants of blue-green boring algae, and a group of unidentified organic filaments. Histochemical tests showed that protein, polysaccharide, and lipid moieties were present.

My initial experiment was conducted to determine if the *Porites* skeleton contained alcohol-extractable substances which would support the growth of bacteria *in vitro*. The overall procedure included extraction of powdered corallum, recovery of extracts, and preparation of an organic matter-free liquid medium into which the extract was placed as an organic enrichment. A bacterial inoculum was then injected into control and extract-enriched media, and bacterial multiplication in both media was periodically monitored by bacteriological plating. Procedural details are as follows. A 10-kg head of *Porites lobata* was obtained from the reef and split apart with a hammer and chisel. Superficial pieces of the head were successively split off until fragments were obtained from the interior which were at least 10 cm away from any head surface and free of macroscopically visible borers. No algal bands were visible in the sample. In all subsequent handling procedures chromic acid cleaned labware was used to prevent organic contamination of the skeletal material. The material was ground in a mortar and pestle and dried at 60°C yielding a weight of 173 g. The powdered skeleton was then extracted in 100 ml absolute ethyl alcohol for 2 hr at 60°C in a sealed flask which was agitated at 0.5-hr intervals. The alcohol was filtered through Whatman brand #1 paper and evaporated to dryness in the oven at 60°C. The recovered greenish-white material weighed 0.319 g (0.184% of corallum extracted), and was kept desiccated at room temperature until used. An inorganic (control) medium was prepared from seawater which had been aged in the dark for six weeks and then passed through a 0.45 μ (pore size) Millipore filter. To this water I added 0.05% $(NH_4)_2SO_4$. Experimental medium was prepared by enriching control medium with 0.4% of the coral skeletal extract (4 g l^{-1}). Five-ml aliquots of each of the two media were placed in 75 $\times$ 15 mm test tubes, closed with cotton, and sterilized by autoclaving. Each set of media was inoculated with approximately 1×10^7 viable bacteria from a seawater suspension of coral reef regenerative sediment (DiSalvo, 1969). All tubes were incubated in the dark at 26°C and agitated by hand about once every three days. One control tube

TABLE 1. *Stimulation of bacterial growth by adding coral skeletal extract to an inorganic liquid medium.*

Day of incubation	Replicate plate	Colony count	
		No skeletal extract	Skeletal extract
5	1	50	*ca.* 400
12	1	60	*ca.* 1000
	2	50	*ca.* 1000
18	1	30	*ca.* 600
	2	16	*ca.* 600
32	1	40	150
	2	35	180

and one enriched tube were harvested at intervals recorded in Table 1. Duplicate bacteriological platings were made using ZoBell 2216e medium which contained 0.5% peptone, 0.1% yeast extract, 0.01% $FePO_4$, 1.5% agar, and aged seawater *qs* (final pH = 7.5-7.8). A 0.1 ml aliquot of a 1:100 dilution of each culture was inoculated into Petri dishes (9 cm diameter) and pour-plates were made using about 10 ml molten agar which had been cooled to 40-45°C.

Table 1 demonstrates that the inoculated bacteria were able to utilize the coral skeletal extract for growth, producing 5 to 20 times more colonies over a 4-week period.

Concurrently, I made several attempts to isolate bacteria from selected areas within *Porites lobata* heads. Specific attempts were made to culture chitinoclastic bacteria which might have been utilizing the coral matrix as a nutrient source. In the following experiments, coral heads weighing 2-4 kg were collected from the reef and returned alive to the laboratory in seawater.

Each head was drained, dried of superficial water, and cleaved with a hammer and chisel for sampling. Sampling zones within the head were chosen by arbitrary visual inspection. The split coral face was immediately placed in a working chamber illuminated with a UV light prior to sampling to reduce the possibility of aerial contamination. Sample scrapings were taken with a flame-sterilized microspatula and rinsed into sterile seawater dilution blanks. All dilution tubes were vigorously agitated by hand for several minutes before plating or further dilution in order to suspend as many bacteria as possible. After the rapid aseptic sampling had been completed, additional sample scrapings were obtained and observed microscopically for presence of boring sponge tissue.

Qualitative measurements. In the first sampling, about 0.05 g skeletal scrapings from each of three discolored areas were deposited in test tubes containing 1 ml of sterile seawater. Well agitated 0.1-ml sample aliquots were pour-plated for total counts in the peptone medium. Clearance of particulate chitin suspended in a mineral agar medium was the criterion used to ascertain the presence of chitinoclastic bacteria in the samples. The chitin-agar medium contained 1.5% agar, 0.05% $(NH_4)_2SO_4$, 0.5% precipitated chitin particles, and aged seawater *qs* (Lear, 1963). The head areas sampled were designated "A", "B", and "C" (Fig. 1). Areas A and B appeared light brown, while C was gray-black. The scrapings from areas A and B were cultured under aerobic conditions, and C was cultured anaerobically by using a vacuum desiccator for a culture chamber. The chamber was flushed repeatedly with inert gas ("Q" gas, Nuclear Chicago Corp.) after the inoculated plates were installed. Pyrogallol was included in the bottom of the desiccator to absorb residual O_2. All cultures were incubated in the dark at room temperature (26°C). The results are given in Table 2.

A second qualitative sampling was carried out as suggested by Odum and Odum (1956), in which a bacteriological profile of a *Porites* head was obtained. About 0.05 g of skeletal scrapings were taken from each of seven sampling areas as listed in Table 3. The scrapings were each initially placed in 4.5 ml sterile seawater diluents. Following procedures outlined above, 0.1 ml aliquots of each dilution were plated for aerobic and anaerobic total counts as well as for aerobic chitinoclast counts. The results are listed in Table 3.

The above results show that bacteria are present in discolored regions. No bor-

TABLE 2. *Culture of aseptically sampled internal skeletal scrapings of* Porites lobata *to ascertain presence of bacteria.*

Sampling zone	Appearance of corallum	Replicate plate	Colony counts: Peptone agar (3 days)	Colony counts: Chitin agar (12 days)
A	white-brown	1	*ca.* 600	12
		2	*ca.* 600	6
B	white-brown	1	*ca.* 600	80
		2	*ca.* 800	75
C*	gray-black	1	*ca.* 800	0
		2	*ca.* 800	0

* Cultured under anaerobic conditions.

ing-sponge tissue was microscopically visible in the scrapings from these areas, suggesting that the bacteria were the sole agents responsible for the degradative effects on the corallum. The growth of chitinoclasts suggested a possible role in the breakdown of the chitinous coral matrix.

Quantitative measurements. An estimate of bacteria per gram of skeleton was obtained for an internal *Porites* region affected by bacteria. An amount of skeleton was recovered under aseptic conditions, and was ground in a heat-sterilized mortar and pestle with 4.5 ml sterile seawater. A 1-ml aliquot of this *ca.* 10 ml of slurry was diluted 1:100 and 1:1000 in sterile seawater, and the skeletal remains were recovered by filtering onto Whatman #1 paper. The skeletal material was dried in the oven at 60°C and weighed, yielding a dry weight of 8.0 grams. One-tenth ml aliquots of the 1:100 and 1:1000 dilutions were plated in peptone and chitin agar as described above. The results of this determination are presented in Table 4.

A similar determination was performed on *Porites* skeleton which showed incipient invasion by boring sponge. A sample of sponge-invaded skeleton was obtained using a sterile microspatula as described above. The scrapings were placed in 4.5 ml

TABLE 3. *Qualitative profile of a* Porites lobata *head with regard to the presence of total aerobic, anaerobic, and chitinoclastic bacteria.*

Sample no.	Description of zone	Replicate plate	Peptone (aerobic) 3	Peptone (aerobic) 7	Peptone (anaerobic) 3	Peptone (anaerobic) 7	Chitin (aerobic) 3	Chitin (aerobic) 7
			Colony count; medium / days incubated					
1	polyp zone	1	2	5	0	1	nd	nd
		2	0	25	0	0	0	0
2	sub-polyp green band	1	0	25	0	0	0	0
		2	0	10	0	0	1 m	1 m
3	green-brown band 10 mm sub-surface	1	0	1	0	0	0	0
		2	0	2	0	0	0	0
4	lt. brown area 20 mm sub-surface	1	0	4	0	0	0	0
		2	0	3	0	0	0	0
5	center white area	1	0	0	0	0	nd	nd
		2	0	1 m	0	0	nd	nd
6	lt. brown band 20 mm from base	1	1	8	0	0	0	0
		2	1	10	0	0	1	1
7	discolored zone, 5 mm from base	1	14	100	2	5	0	0
		2	17	85	0	2	1	3
Blank plates for sterility tests		1	0	0	0	0	0	0
		2	0	1 m	0	0	0	0

m = mycelium

TABLE 4. *Relation between bacterial numbers and weight of corallum,* Porites lobata, *internal discolored region.*

Medium (incubation time)	Replicate plate	Colony count Dilution 1.25×10^3	1.25×10^4
Peptone (48 hr)	1	*ca.* 300	10
	2	*ca.* 300	15
Chitin (100 hr)	1	17	2
	2	20	1

Calculation:

	Mean plate count	·	Dilution factor	÷	Grams skeleton	=	Colonies/gram skeleton
(peptone)	12.5		1.25×10^4		8.0		1.95×10^4
(chitin)	18.5		1.25×10^3		8.0		0.29×10^4

sterile seawater and agitated, and 0.1 ml aliquots were plated in duplicate in peptone agar. The skeletal scrapings were recovered by filtration onto a small-tared Whatman #1 paper. The dry weight of the scrapings was 0.06 g. After 24 hr incubation the two replicate plates showed counts of 30 and 40 colonies, representing a minimum of 2.7×10^4 viable cells per gram of sponge-infiltrated corallum.

Three replicate sample scrapings from a light brown, interior, weakened zone of *Porites* were assayed for crystal type by X-ray crystallography (kindly performed by Steven Smith, Univ. of Hawaii, Department of Oceanography). The results showed all calcium carbonate to be in the form of aragonite.

DISCUSSION

The present results provide the first evidence that bacteria are important in the breakdown of coral skeletons, although implied evidence is found in the statement of Sorby (1879, p. 71) that: ". . . powdered coral, kept for weeks in water, loses organic matter and gives rise to such minute particles that the water is like dilute milk." My most important finding was that bacteria isolated from the coral skeletons were able to digest chitin suspended in otherwise nutrient-free agar. The agar was seldom attacked. Results of the crystal analysis showed no crystalline changes in the weakened areas, indicating absence of chemical decalcification. These results suggest that weakening of the corallum is due to bacterial breakdown of the organic matrix. No X-ray crystallography was performed on gray-black regions of weakened skeleton where chemical reactions accompanying sulfate reduction might be expected to convert aragonite to calcite (Revelle and Fairbridge, 1957).

A survey which included the results presented above showed that sediments in proximity to the bases of coral skeletons contained minimum counts of 10^7-10^8 bacteria per gram dry weight, of which an estimated 10-20% were chitin-digesting species. Thus, there exists a large pool of potential infectivity in contact with the basal regions of most coral skeletons. Studies on the environmental breakdown of coral skeletons are complicated by the several types of boring organisms which are typically present and contributing to acute skeletal defects. Further studies must determine rates of bacterial activity, conditions which favor the sole presence of bacteria in certain skeletal areas, and the roles played by bacteria in the overall pattern of succession through which healthy coralla are reduced to reef sediments.

REFERENCES

DiSalvo, L. H. 1969. On the existence of a coral reef regenerative sediment. Pacific Sci. 23:129.

Goreau, T. F., and W. D. Hartman. 1963. Boring sponges as controlling factors in the formation and maintenance of coral reefs, p. 25-54. *In* R. F. Sognnaes, [ed.], Mechanisms of hard tissue destruction. AAAS Publ. 75. Washington, D. C.

Lear, D. W. 1963. Occurrence and significance of chitinoclastic bacteria in pelagic waters and zooplankton, p. 594-610. *In* C. H. Oppenheimer, [ed.], Symposium on marine microbiology. C. C. Thomas, Springfield, Ill.

Neumann, A. C. 1966. Observations on coastal erosion in Bermuda and measurements of the boring rate of the sponge *Cliona lampa*. Limnol. Oceanogr. 11:92-109.

Odum, H. T., and E. P. Odum. 1956. Corals as producers, herbivores, carnivores, and possibly decomposers. Ecology 37:385.

Purdy, E. G., and L. S. Kornicker. 1958. Algal distintegration of Bahamian limestone coasts. J. Geol. 71:472-497.

Purdy, E. G. 1963. Recent calcium carbonate facies of the Great Bahama Bank. Parts I, II. J. Geol. 71:334-355, 472-497.

Revelle, R., and R. Fairbridge. 1957. Carbonates and carbon dioxide, p. 239-295. *In* J. W. Hedgepeth, [ed.], Treatise on marine ecology and paleoecology. Vol. I, Mem. 67, Geol. Soc. Amer.

Sorby, H. C. 1879. The structure and origin of lime stones. Proc. Geol. Soc. London 25:56-95.

Swinchatt, J. P. 1965. Significance of constituent composition and texture of skeletal breakdown in some recent carbonate sediments. J. Sed. Petr. 35:71-90.

Wainwright, S. A. 1963. Skeletal organization of the coral *Pocillopora damicornis*. Quart. J. Microsc. Sci. 104:169-183.

Wood, E. J. F. 1960. Microbiology of coral reefs. Proc. 9th Pacific Sci. Congr. 4:171-172.

35

Reprinted from *Appl. Microbiol.*, **16**(2), 197–202 (1968)

Bacteriology of Manganese Nodules

II. Manganese Oxidation by Cell-free Extract from a Manganese Nodule Bacterium

H. L. EHRLICH

Department of Biology, Rensselaer Polytechnic Institute, Troy, New York 12181

Received for publication 14 July 1967

A cell-free extract from *Arthrobacter* 37, isolated from a manganese nodule from the Atlantic Ocean, exhibited enzymatic activity which accelerated manganese accretion to synthetic Mn-Fe oxide as well as to crushed manganese nodule. The reaction required oxygen and was inhibited by $HgCl_2$ and *p*-chloromercuribenzoate but not by Atebrine dihydrochloride. The rate of enzymatic action depended on the concentration of cell-free extract used. The enzymatic activity had a temperature optimum around 17.5 C and was destroyed by heating at 100 C. The amount of heat required for inactivation depended on the amount of nucleic acid in the preparation. In the cell-free extract, unlike the whole-cell preparation, peptone could not substitute for $NaHCO_3$ in the reaction mixture. An enzyme-containing protein fraction and a nucleic acid fraction could be separated from cell extract by gel filtration, when prepared in 3% NaCl but not in seawater. The nucleic acid fraction was not required for enzymatic activity.

Previous reports have indicated that some bacteria associated with manganese nodules from the Atlantic Ocean accelerate manganese accretion to such nodules and to other manganese oxides (1, 2). It was inferred from laboratory experiments that, in this process, Mn^{2+} is first adsorbed on the surface of some preexistent manganese oxide and that bacteria on the oxide then oxidize the adsorbed Mn^{2+} to additional manganese oxide. This reaction creates new adsorption sites for Mn^{2+}, which may then be bacterially oxidized in turn (1, 2). This process may be repeated many times. In the laboratory, bacterial acceleration of manganese accretion may be estimated quantitatively by measuring the loss of Mn^{2+} from solution to insoluble manganese oxide adsorbent in the presence and absence of active bacteria. In such experiments, more Mn^{2+} is lost to the adsorbent with active bacteria than without active bacteria. Reactions describing the process of manganese accretion brought about by *Arthobacter* 37, and probably by other marine manganese-oxidizers as well, may be written as follows:

$$H_2MnO_3 + Mn^{2+} \rightarrow MnMnO_3 + 2H^+ \quad (1)$$

$$MnMnO_3 + 2H_2O + \tfrac{1}{2}O_2 \xrightarrow[\text{peptone or } NaHCO_3]{\text{bacteria}} (H_2MnO_3)_2 \quad (2)$$

Reaction 1 illustrates the nonbiological adsorption step, and reaction 2 illustrates the biological oxidation step. The presence of a complexing agent, such as peptone or $NaHCO_3$, is essential in resting-cell reactions (2; Ehrlich, *unpublished data*). The concentration of peptone or $NaHCO_3$ is critical (2; Ehrlich, *unpublished data*). Equations 1 and 2 are undoubtedly rough approximations of manganese accretion to nodules or other manganese oxides. They do not predict that manganese-iron oxides, such as those found in manganese nodules or synthetic preparations, are better adsorbent substrates than simple hydrous manganese oxides or pyrolusite, MnO_2 (H. L. Ehrlich, Bacteriol. Proc., p. 42, 1964). The equations however, do not show how iron is incorporated into manganese nodules.

Although all of the earlier evidence supports the idea that bacteria effect an enzymatic oxidation of adsorbed Mn^{2+} in manganese accretion to nodules and other manganese oxides, this investigation provides the first direct evidence of such enzymatic activity.

Materials and Methods

Organism. The organism employed in these experiments is *Arthrobacter* 37, isolated from an Atlantic manganese nodule (1). The culture was carried in test tube slants of seawater-nutrient agar containing Peptone, 5 g; Beef Extract, 3 g; agar, 15 g; and filtered aged seawater, 1 liter.

To extract enzyme activity, cultures were grown on

an appropriate number of slants of seawater-nutrient agar in Roux or Kolle flasks. The cells were harvested after about 17 hr of incubation; older cells yielded no enzyme activity. To obtain 4 ml of active extract, the harvest from 3 Roux flasks was sufficient, whereas the harvest from 28 to 30 Kolle flasks was required to obtain 10 to 12 ml of active extract.

Preparation of cell extract. Cell crops were harvested in 5 ml of filtered natural seawater or 3% NaCl solutions and were centrifuged in 50-ml Nalgene PR centrifuge tubes at 9,700 or 17,300 × *g* for 10 min at 4 C. After discarding the supernatant fluid, the cells were washed three times in 10 ml of the same solution in which they were harvested, by centrifugation at 9,700 or 17,300 × *g* for 10 min at 4 C. After the last centrifugation, the supernatant fluid from a large cell harvest was removed, as completely as possible, by pipette. The cells from a small harvest were transferred in 4 ml of appropriate fresh solution to a 56 × 14 mm glass centrifuge tube and were recentrifuged at 6,300 × *g* for 10 min at 4 C. After respinning, the supernatant fluid was removed, as completely as possible, by pipette.

To extract enzyme activity from the washed cells, an extraction method was adopted from the work of Person and Zipper (6). About 10 mg of Zeolite 3A (Linde Division of Union Carbide Corp., Buffalo, N.Y.), preheated in a muffle furnace at 400 F for 30 min and cooled to room temperature in a desiccator immediately before use, was added to the cell crop in a centrifuge tube and was blended with a glass rod. When the blending was completed, an additional 10 mg of Zeolite was added and blended with the cells. Zeolite additions and blending were continued until a thick paste was formed, and this paste thinned somewhat after additional stirring. Throughout the extraction procedure, the centrifuge tube was kept in an ice bath. The whole Zeolite treatment took about 6 min for a small cell harvest and 12 min for a large cell harvest. After blending was completed, a prechilled solution of the same composition as that used for harvesting was added to the cell-Zeolite mixture. A 4.5 ml amount was added to extracts from small cell harvests and a 10 to 12 ml amount was added to extracts from large cell harvests. The contents of the tubes were then stirred until all solid matter was completely resuspended, and the solutions were centrifuged at 9,700 or 17,300 × *g* for 10 min at 4 C. The supernatant fluid was transferred to another centrifuge tube and was recentrifuged for 10 min at the above-mentioned speed and temperature. The final supernatant fluid contained the active cell extract. Microscopic examination with a phase-contrast microscope revealed no residual cells or cell debris. The extract could be stored in a deep freeze at −4 C for several weeks if first quick frozen in a dry ice-acetone bath.

Enzyme assay. The enzyme assay was carried out in sets of duplicate 50-ml Erlenmeyer flasks containing 0.1 g of synthetic Mn-Fe oxide overlaid with 10 ml of filtered seawater or 10 ml of 3% NaCl solution containing 0.001 M $MnSO_4$, and 1.14×10^{-2} M $NaHCO_3$, *p*H 7.0. Cell extract (1 ml or another quantity as specified) was included in the 10-ml volume of overlying solution. Control flasks, similarly set up, with the exception that the cell extract was replaced by an equivalent volume of appropriate suspension medium, were also prepared. Incubation was carried out for 3 hr at room temperature (or at another temperature when specified). After incubation 3.0 ml of the supernatant fluid was removed from appropriate flasks and was centrifuged at 1,000 × *g* for 10 min. The *p*H of each centrifuged sample was checked with Alkacid paper (Fisher Scientific, Inc., Pittsburgh, Pa.) and was always found to be 7.0. Duplicate 1.0-ml portions of each centrifuged sample were then assayed by the persulfate method (1). The excess manganese (ΔMn), removed from the supernatant fluid by enzyme in the cell extract, was calculated from the difference in Mn^{2+} concentration in the supernatant fluids of control and experimental flasks after 3 hr of incubation.

Preparation of Mn-Fe oxide. The procedure used was described previously (2). Particles of mesh size −60 +80, or −80 (Tyler scale), were selected as adsorbents. After air-drying, some of these preparations released significant amounts of Mn^{2+} when first brought in contact with solution. These preparations subsequently readsorbed some or all of this Mn^{2+}. The manganese was probably sequestered as a salt from the oxide micelles during drying. Dried manganese-nodule substance was never observed to release any measurable Mn^{2+} on first contact with solution.

Heat inactivation of cell extracts. To test the heat lability of enzyme activity in the cell extracts, 2.1-ml portions, from 12 ml of cell extract in either seawater or a 3% NaCl solution, were heated in a boiling-water bath (100 C) for 0, 1, 5, 10, and 15 min and were then chilled in a cold-water bath. Two 1-ml samples from each heat-treated extract were then tested by the enzyme assay.

Gel filtration of enzyme extracts. Partial purification of the cell extract was attempted by passing 5 ml of enzyme through 7 to 8 g of Sephadex G-150 (Pharmacia Fine Chemicals, Inc., Piscataway, New Market, N.J.) in a K-25/45 Laboratory Column (Pharmacia Fine Chemicals, Inc.) at 4 C. The solution in which the cell extract was contained was also employed as the eluant but it was autoclaved before use. The flow rate was 0.75 ml per min for the NaCl extract and 1.3 ml per min for the seawater extract; the effluents were collected in 3-ml fractions for the NaCl extract and in 5-ml fractions for seawater extract. The fractions were tested for nucleotide content by measuring optical density (OD) of 10-fold diluted samples at 260 mμ with a DU spectrophotometer (Beckman Instruments, Inc., Fullerton, Calif.). The presence of protein was determined by measuring the OD at 680 mμ with a junior spectrophotometer (Coleman Instruments Corp., Maywood, Ill.), after 0.4 ml of undiluted sample had been reacted by the method of Lowry et al. (4). A salt precipitate which formed in the protein assay of seawater extracts was removed by centrifugation before reading the OD. Enzyme activity was measured as previously described.

TABLE 1. *Effect of crude cell extract on Mn^{2+} adsorption by two different Mn-Fe oxide preparations*

Mn-Fe oxide prepn	Enzyme[a]	Time (min)	Mn in supernatant fluid (mμmoles/ml)[b]	Δ Mn in supernatant fluid (mμmoles/ml)[c]
A	−	0[d]	1,040	
	+	0[d]	1,040	
	−	0.5[e]	2,990	
	+	0.5[e]	2,950	
	−	180	2,480	
	+	180	2,250	−230
B	−	0[d]	1,030	
	+	0[d]	1,030	
	−	0.5[e]	1,120	
	+	0.5[e]	1,110	
	−	180	600	
	+	180	500	−100

[a] A different cell extract was used with each preparation. All reactions were conducted in seawater at room temperature.

[b] All results are expressed as averages of duplicates.

[c] The difference in Mn^{2+} removed from the supernatant fluids of control flasks and flasks containing enzyme. A minus sign indicates that more Mn^{2+} was removed from the supernatant fluid containing enzyme than from the supernatant fluid lacking this substance; a plus sign denotes the opposite.

[d] Based on the measurement of Mn^{2+} in 10 ml of reaction mixture in the absence of any Mn-Fe oxide.

[e] This time is 0.5 min after addition of cell extract or 4 min after addition of first reagent solution.

RESULTS

Demonstration of activity in crude extract. In repeated trials, Mn^{2+} disappeared faster from the supernatant fluid of flasks with cell extract than from flasks without extract (Table 1). The last column in Table 1 lists the difference in Mn^{2+} removed with and without cell extract after 180 min. The final *p*H in all flasks was 7.0. No catalytic activity could be recovered from Zeolite 3A that was not used to extract cells.

When 0.1 g of crushed Atlantic manganese nodule was substituted for synthetic Mn-Fe oxide, and 1.14×10^{-3} M $NaHCO_3$ was substituted for 1.14×10^{-2} M $NaHCO_3$ (*p*H 7.0), in 3 hr, 30 mμmoles more of Mn^{2+} per ml disappeared from the supernatant fluid of flasks containing cell extract than from flasks lacking extract.

It is evident from Table 1 that, when Mn-Fe oxide preparation A was used, the concentration of Mn^{2+}, 0.5 min after enzyme addition, was significantly in excess of the 1,040 mμmoles of $MnSO_4$ per ml of reaction mixture at 0 min. The Mn^{2+} concentration decreased again 180 min after enzyme addition. This decrease was greater in preparations containing cell extract than in preparations lacking this substance. Some of the Mn-Fe oxide preparations released large amounts of Mn^{2+} 5 to 10 min after first contact with solution but then readsorbed some of this Mn^{2+} in the following hours (*see* Materials and Methods). An example of the occurrence of this reaction in the absence of enzyme is shown in Fig. 1. Each point in this figure represents the Mn^{2+} contained in 1 ml of supernatant fluid in a separate reaction vessel at a specified time. The titer at 0 min represents the measured Mn^{2+} concentration in the solution before any Mn^{2+} was released by the Mn-Fe oxide. Mn-Fe oxide preparation B (Table 1) released only small amounts of Mn^{2+} 0.5 min after enzyme addition; after 180 min, preparation B adsorbed enough Mn^{2+} to lower the titer well below the concentration at 0 min. The adsorption of Mn^{2+} in this case was also accelerated by cell extract. It is, therefore, concluded that the cell extract has the same qualitative effect on Mn^{2+} adsorption, whether a type A or a type B Mn-Fe oxide preparation is used.

Oxygen requirement. To show that the reaction depends on the presence of oxygen, an enzyme assay was performed at room temperature in a vacuum desiccator in a nitrogen atmosphere containing less than 0.1% oxygen. Displacement of the air by nitrogen gas was achieved by evacuating the desiccator with an aspirator to a residual pressure of 75 mm Hg, introducing high purity nitrogen (Union Carbide Corp., Linde Div., New York, N.Y.) to atmospheric pressure, and repeat-

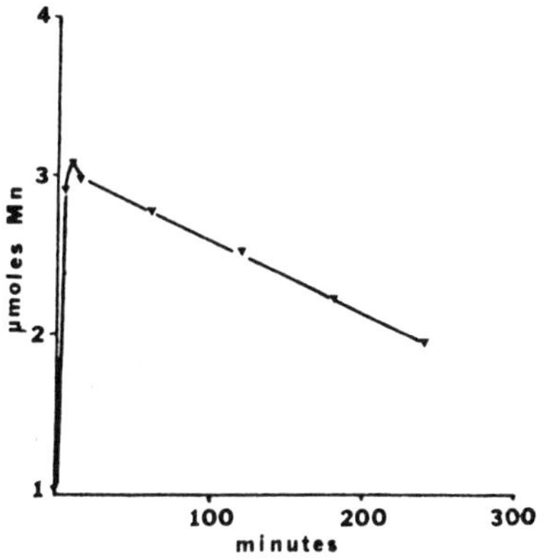

FIG. 1. *Manganese release and adsorption by a Mn-Fe oxide preparation in the absence of cell-free extract. Each reaction vessel contained all reagents used in an enzyme activity assay, except for cell extract.*

TABLE 2. *Chemical effects on enzymatic manganese oxidation*

Treatment[a]	Enzyme[b]	Mn in supernatant fluid after 3 hr[c] (mμmoles/ml)	Δ Mn in supernatant fluid after 3 hr[d] (mμmoles/ml)
Untreated	−	1,380	
	+	1,300	−80
N_2 atm	−	1,390	
	+	1,380	−10
0.1% peptone (no $NaHCO_3$)	−	2,110	
	+	2,120	+10
10^{-6} M $HgCl_2$	−	1,650	
	+	1,630	−20

[a] All reactions were conducted in seawater at room temperature.

[b] The same cell extract and the same Mn-Fe oxide preparation were used for all experiments in this table.

[c] All results are expressed as averages of duplicates.

[d] See footnote (c) in Table 1.

ing this procedure two more times. Table 2 shows that most of the enzyme action was eliminated under nitrogen. The residual activity was probably the result of enzyme action before the original air was sufficiently replaced by nitrogen. Complete displacement of air took about 15 to 20 min. In a similar experiment with a different extract, no activity was exhibited under nitrogen.

Substitution of peptone for $NaHCO_3$. Unlike the experiments using whole cells, 0.1% peptone could not substitute for 1.14×10^{-2} M $NaHCO_3$ when cell extract was used (Table 2). Indeed, 10 mμmoles more of Mn^{2+} per ml was left in the flasks containing extract than in the flasks without extract. This finding may be attributed to the combined complexing power of peptone and nonspecific protein in the extract. The more extensive complexing power of peptone, as compared to $NaHCO_3$, is shown in Table 2. In the absence of cell extract, peptone held back more Mn^{2+} in the supernatant fluids after 3 hr than did $NaHCO_3$ (untreated setup; N_2-atmosphere setup).

Effect of enzyme inhibitors. The catalytic activity of the extract in seawater was strongly inhibited by 10^{-6} M $HgCl_2$ (Table 2). When other enzyme preparations in 3% NaCl were used, activity was 85, 90, and 100% inhibited by 10^{-5} M *p*-chloromercuribenzoate. With still other enzyme preparations in 3% NaCl, no inhibition was observed with 10^{-3} M Atebrine dihydrochloride. In the experiments with $HgCl_2$, it was important to add the inhibitor to both control and experimental flasks because the inhibitor displaced some exchangeable Mn^{2+} on the Mn-Fe oxide adsorbent (Table 2). The other two inhibitors did not have this effect.

Enzyme-concentration effect with crude extract. Enzyme assays with increasing amounts of cell extract showed increasing amounts of activity (Table 3). The fact that the increase in activity over the range of extract concentrations tested was not completely uniform may be the result of the crude nature of the extract.

Temperature effects on enzyme activity. When activity tests with 1-ml portions of the same extract were conducted in water baths for 4 hr at four different temperatures, the results shown in Table 4 were obtained. Greatest activity occurred at 17.5 C. Examination of the data from the control flasks without added extract shows that the extent of adsorption of Mn^{2+} in 4 hr decreased progressively with a drop in temperature.

Heat inactivation of extract. Heating the crude cell extract in a boiling-water bath (100 C) for various lengths of time showed different degrees of heat lability of the enzyme activity, depending on whether the extract was prepared in seawater or in 3% NaCl (Table 5). In repeated experiments, greater lability of the activity was observed in NaCl-extract than in seawater. This was attributed to the presence of nucleic acid in the crude ex-

TABLE 3. *Effect of enzyme concentration on activity*

Extract concn[a] (ml)	Δ Mn in supernatant fluid after 3 hr[b] (mμmoles/ml)
1	−80
2	−110
3	−140
4	−210

[a] Reactions performed in seawater at room temperature. The same Mn-Fe oxide preparation was used in all reaction vessels.

[b] See footnote (c) in Table 1.

TABLE 4. *Effect of temperature on enzyme activity*

Temp (C)	Enzyme[a]	Mn in supernatant fluid after 4 hr (mμmoles/ml)	Δ Mn in supernatant fluid after 4 hr[b] (mμmoles/ml)
4	−	2,600	
	+	2,600	0
12	−	1,940	
	+	1,870	−70
17.5	−	1,870	
	+	1,670	−200
25	−	1,800	
	+	1,690	−110

[a] Reactions performed in seawater. The same cell extract and Mn-Fe oxide preparation were used in all reaction vessels.

[b] See footnote (c) in Table 1.

TABLE 5. *Heat inactivation of crude enzyme extract*

Heating (min)	Δ Mn in supernatant after 3 hr[a] (mμmoles/ml)	
	3% NaCl	Seawater
0	−60	−80
1	−40	−70
5	−10	−50
10	+10	−50
15	+40	−20

[a] See footnote (c) in Table 1. The same Mn-Fe oxide preparation was used in all reaction vessels.

tracts. The nucleic acid somehow protected the enzyme system in seawater of approximately 30 ‰ salinity but did so less efficiently, or not at all in 3% NaCl. Subsequent purification experiments provided evidence that the nucleic acid may be partly combined with the enzyme when extracted in seawater. Removal of nucleic acid from cell extract in 3% NaCl by gel filtration, as described in the next experiment, allowed complete heat inactivation in 1 min at 100 C.

Partial purification of enzyme by gel filtration. Crude cell extract could be shown to contain nucleotides, as well as protein, by measuring the OD at 260 and 280 mμ and by performing a protein determination by the method of Lowry et al. (4). When cell-free extract was passed through Sephadex G-150 (*see* Materials and Methods), different degrees of separation of nucleotides from protein were achieved, depending on whether the extract was prepared in seawater or in 3% NaCl (Fig. 2). Extraction and gel filtration in 3% NaCl gave better separation. In seawater, the nucleotide fraction seemed to carry much of the protein with it (Fig. 2A). The delivery of nucleotide material from the column immediately after the void-volume, regardless of whether 3% NaCl or seawater was used, indicates that the nucleotide material had a high molecular weight and thus was nucleic acid. A test for enzyme activity in fractions 8 through 15 (from the extract in seawater) revealed catalytic activity in fractions 8 through 12, with an absolute maximum in fraction 9 (Fig. 2A). A test for enzyme activity in fractions 10 through 19 and 22 through 31 (from the 3% NaCl extract) revealed activity in fractions 10 through 27, with an absolute maximum in fraction 15 (Fig. 2B). The peaks of enzyme activity in the series of fractions, especially from the NaCl extract, may indicate an association of the enzyme system with different-sized particles or cell fragments in the extract. Since intimate contact is required between whole cells and Mn-Fe oxide for adsorbed Mn^{2+} to be oxidized (2), it is not unreasonable to assume that the enzyme system is related to the cell wall-membrane structure of the cell. Hence, the different-sized particles of fragments in the extract may constitute pieces of cell wall-membrane complex produced by the Zeolite treatment.

The results of gel filtration of 3% NaCl extract make it clear that the enzyme activity is not stimulated by the nucleic acid fraction, since the greatest activity was found in a fraction almost free of nucleic acid. This is an important difference from the cell-free iron-oxidase system of *Ferrobacillus ferrooxidans,* in which iron-oxidizing activity can be stimulated by nucleic acid or nucleotides (10).

A comparison of the enzyme activities of the most active fractions from the crude 3% NaCl extract with the activity of the crude extract before gel filtration reveals that the specific activity of the enzyme was significantly increased by gel filtration No exact quantitative estimate of the change in specific activity of the enzyme is possible at this time because of the complexity of the reaction system.

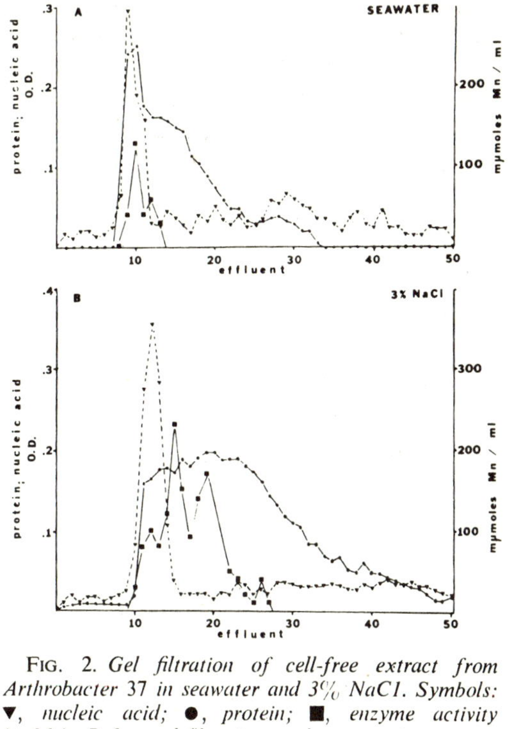

FIG. 2. *Gel filtration of cell-free extract from Arthrobacter 37 in seawater and 3% NaCl. Symbols:* ▼, *nucleic acid;* ●, *protein;* ■, *enzyme activity* (Δ *Mn*). *Before gel filtration crude extract in seawater contained 1.500 OD units of protein and in 3 hr caused a* Δ*Mn of 110 mμmoles per ml of supernatant fluid. Similarly, crude extract in 3% NaCl contained 1.300 OD units of protein before gel filtration and in 3 hr caused a* Δ*Mn of 100 mμmoles per ml of supernatant fluid. All enzyme assays were performed with the same Mn-Fe oxide preparation.*

DISCUSSION

The results obtained are consistent with the notion that acceleration of manganese accretion to nodules or other manganese oxides by cell-free extract involves an enzymatic step. The finding that oxygen is required for the enzymatic reaction indicates than an oxidation of manganese is involved. Since it can be shown that, in the absence of an adsorbent such as Mn-Fe oxide, oxidation of Mn^{2+} by the extract is not detectable, the adsorption of Mn^{2+} before oxidation must be essential. The reason for this adsorption requirement can only be speculated upon. It may be related to a lower activaticn-energy requirement for oxidizing adsorbed Mn^{2+} than for oxidizing freely dissolved Mn^{2+}.

Bacterial manganese oxidation has been reported (3, 5, 7–9, 11), and, in most of these cases, the cause of the observed oxidation was not directly determined. The oxidation may be the result of enzymatic catalysis, biodegradation of manganese chelate followed by auto-oxidation of the freed Mn^{2+} or auto-oxidation of Mn^{2+} because of more favorable conditions of *p*H and E_h created by the bacteria. The observations made by Johnson and Stokes on *Sphaerotilus discophorus* (3) come closest to demonstrating enzymatic oxidation. If *S. discophorus* is enzymatically active on Mn^{2+} (which seems probable), it differs from *Arthrobacter* 37 in its ability to catalyze the oxidation of free Mn^{2+}.

Work to characterize the manganese-oxidizing enzyme system of *Arthrobacter* 37 is being continued in this laboratory.

ACKNOWLEDGMENTS

The expert technical assistance of Alice R. Ellett is gratefully acknowledged. This investigation was supported by contract 591 (22), work unit number NR 103-665, between the Office of Naval Research, Department of the Navy and Rensselaer Polytechnic Institute. Thanks are also extended to the Linde Division of Union Carbide Corp., Buffalo, N.Y., for a gift of Zeolite 3A.

LITERATURE CITED

1. EHRLICH, H. L. 1963. Bacteriology of manganese nodules. I. Bacterial action on manganese in nodule enrichments. Appl. Microbiol. **11**:15–19.
2. EHRLICH, H. L. 1966. Reactions with manganese by bacteria from marine ferromanganese nodules. Develop. Ind. Microbiol. **7**:279–286.
3. JOHNSON, A. H., AND J. L. STOKES. 1966. Manganese oxidation by *Sphaerotilus discophorus*. J. Bacteriol. **91**:1543–1547.
4. LOWRY, O. H., N. J. ROSEBROUGH, A. L. FARR, AND R. J. RANDALL. 1951. Protein measurement with the Folin phenol reagent. J. Biol. Chem. **193**:265–275.
5. MOESE, J. R., AND H. BRAUTNER. 1966. Mikrobiologische Studien an manganoxydierenden Bakterien. Zentr. Bakteriol. Parasitenk. Abt. II **120**:480–495.
6. PERSON, P., AND H. ZIPPER. 1964. Disruption of mitochondria and solubilization of cytochrome oxidase by a synthetic zeolite. Biochem. Biophys. Res. Commun. **17**:225–230.
7. SCHWEISFURTH, R., AND R. MERTES. 1962. Mikrobiologische und chemische Untersuchungen ueber Bildung und Bekaempfung von Manganschlammablagerung in einer Druckleitung fuer Talsperrwasser. Arch. Hyg. Bakteriol. **146**:401–417.
8. SILVERMAN, M. P., AND H. L. EHRLICH. 1964. Microbial formation and degradation of minerals. Advan. Appl. Microbiol. **6**:153–206.
9. TYLER, P. A., AND K. C. MARSHALL. 1967. Microbial oxidation of manganese in hydro-electric pipelines. Antonie van Leeuwenhoek J. Microbiol. Serol. **33**:171–183.
10. YATES, M. G., AND A. NASON. 1966. Enhancing effect of nucleic acids and their derivatives in the reduction of cytochrome c by ferrous ions. J. Biol. Chem. **241**:4861–4871.
11. ZAVARZIN, G. A. 1964. The mechanism of manganese deposition on mollusk shells. Dokl. Akad. Nauk SSSR **154**:944–945.

36

Reprinted from *J. Gen. Microbiol.*, **68**, 337–348 (1971)

Mechanism of the Initial Events in the Sorption of Marine Bacteria to Surfaces

By K. C. MARSHALL*, RUBY STOUT AND R MITCHELL

Laboratory of Applied Microbiology, Division of Engineering and Applied Physics Harvard University, Cambridge, Massachusetts, U.S.A.

(*Accepted for publication* 3 *August* 1971)

SUMMARY

The sorption of two marine bacteria to surfaces involved an instantaneous *reversible* phase, and a time-dependent *irreversible* phase. Reversible sorption of the non-motile *Achromobacter* strain R8 decreased to zero as the electrolyte concentration decreased, or as the thickness of the electrical double-layer increased. The electrolyte concentration at which all bacteria were repelled from the glass surface depended on the valency of the cation. The reversible phase is interpreted in terms of the balance between the electrical double-layer repulsion energies at different electrolyte concentrations and the van der Waals attractive energies. Even at the electrolyte concentration of seawater, the bacteria probably are held at a small distance from the glass surface by a repulsion barrier. Reversible sorption often led to rotational motion of the motile *Pseudomonas* sp. strain R3 at a liquid–glass interface.

Pseudomonas R3 produced polymeric fibrils in artificial seawater; these may be concerned in the irreversible sorption of the bacteria to surfaces. Sorption and polymer production were stimulated by 7 mg./l. glucose but higher levels inhibited irreversible sorption. Omission of Ca^{2+} and Mg^{2+} from the artificial seawater prevented growth, polymer production, and sorption to surfaces by Pseudomonas R3.

INTRODUCTION

The sorption of bacteria to surfaces is a general phenomenon encountered in natural environments with important ecological implications (Wood, 1967; Marshall, 1971). Primary microbial film formation on surfaces immersed in seawater is considered by some investigators to be a prerequisite to fouling by larger organisms such as barnacles (Wood, 1967). The mechanism whereby marine bacteria sorb to surfaces has received scant attention. ZoBell (1943) suggested that, once bacteria are attracted to a surface, firm attachment requires incubation for several hours. He attributed this delay to the need for the synthesis of extracellular adhesive materials. Recently, Corpe (1970*a*) has reported the production of an extracellular acid polysaccharide by a primary film-forming bacterium, *Pseudomonas atlantica*. Glass slides coated with this polymer became fouled with micro-organisms more rapidly than uncoated slides. Corpe (1970*b*) has reviewed the literature on attachment of bacteria to surfaces immersed in marine environments.

The present investigation combines a study of some of the colloidal and biological properties of pure cultures of marine bacteria to obtain information on processes involved in the sorption to surfaces.

* Permanent address: Department of Agricultural Science, University of Tasmania, Hobart, Tasmania, Australia.

Definition of phases of sorption

Our studies have confirmed ZoBell's (1943) suggestion that sorption consists of two phases. The bacteria were first attracted to a surface, and, after several hours, some became firmly attached. These phases are defined as follows.

Reversible sorption is an essentially instantaneous attraction of bacteria to a surface. Such bacteria are held weakly near the surface; they still exhibit Brownian motion and are readily removed by washing the surface with 2·5 % NaCl.

Irreversible sorption involves the firm adhesion of bacteria to the surface; they no longer exhibit Brownian motion and are not removed by washing with 2·5 % NaCl.

METHODS

Organisms. The bacteria used in this investigation were a motile *Pseudomonas* sp. strain R 3 and a non-motile *Achromobacter* sp. strain R 8. Both organisms were isolated from a glass coverslip that had been immersed in natural seawater for 1 h., rinsed several times in sterile 2·5 % NaCl, plated by smearing the coverslip on an artificial seawater agar, and incubated 24 h. at 25°. Pseudomonas R 3 required 2.5 % NaCl in the medium, while Achromobacter R 8 grew equally well in both high and low salt media. Logarithmic and stationary phase cultures of Pseudomonas R 3 refer to cultures incubated for 4 and 24 h., respectively.

Studies of reversible sorption. The behaviour of the motile Pseudomonas R 3 at liquid–glass interfaces was examined by preparing films (Kodak 4-X reversal film type 7277) of the bacteria viewed at the plane of a coverslip by phase-contrast microscopy. Detailed examination of these films made possible a reasonable interpretation of the curious gyratory motions observed with motile bacteria at such interfaces. The maximum velocity of Pseudomonas R 3 was determined by the motility tracking method of Vaituzis & Doetsch (1969).

The effect of electrolyte concentration on reversible sorption was investigated using Achromobacter R 8. This non-motile organism grew in low salt media and very low electrolyte concentrations did not cause lysis. Achromobacter R 8 was grown on nutrient agar (Difco) at 25° for 24 h. After washing twice in distilled water, portions of the suspension were mixed with equal volumes of a range of concentrations of NaCl or $MgSO_4$. Drops of these suspensions were run under coverslips supported above slides by broken coverslip pieces. The preparations were stood 30 min. to allow those organisms not sorbed at the liquid–glass interface to fall from view by gravity, so that these cells attracted to the coverslip surface could be counted. The same technique was employed to investigate the effect of divalent cations on the reversible sorption of Pseudomonas R 3.

Studies of irreversible sorption. Glass slides were cleaned in chromic acid, rinsed in distilled water, and then supported in slots in a sheet of foam polystyrene over an evaporating basin (1 l. capacity) containing an appropriate medium inoculated with Pseudomonas R 3. Duplicate slides were removed at intervals, rinsed thoroughly with 2·5 % NaCl, and the previously immersed area of the slide covered with a coverslip. Organisms firmly sorbed (not exhibiting Brownian motion) were counted in at least ten fields on both slides. All counts are expressed as numbers sorbed/cm^2 of glass surface.

The media used included either 2·5 % NaCl or an artificial seawater (ASW), suggested by T. Waite, which contained (g./l.): NH_4Cl, 0·0007; NaCl, 24; KCl, 0·6; $MgSO_4.7H_2O$, 5; $MgCl_2.6H_2O$, 3·6; $CaCl_2$, 0·3; $NaNO_3$, 0·1; KH_2PO_4, 0·01; $FeCl_3$, 0·001; tris-HCl, 5·32; tris Base, 1·97; P_{11} trace metals (below), 10 ml.; pH 7·8. The P_{11} trace metal solution contained (g./l.): di-sodium ethylene-diaminotetracetic acid (EDTA), 1; $FeCl_3$, 0·01; H_3BO_3,

0·2; $MnCl_2$, 0·04; $ZnCl_2$, 0·005; $CoCl_2$, 0·001. Modifications of these media are detailed in the Results section. The concentration of glucose added to artificial seawater is indicated by the number after ASWG, e.g. ASWG-30 indicates 30 mg. l. glucose.

Pseudomonas R3 was inoculated on large slopes in 100 ml. bottles of nutrient agar (Difco) containing 2·5 % NaCl, and normally incubated 24 h at 25°. The bacteria were washed once in 2·5 % NaCl before being inoculated into the test medium to give a final concentration of between 1 and 10×10^7 bacteria/ml.

For electron-microscope studies of irreversibly sorbed bacteria, nickel grids (formvar films) were immersed directly in the medium containing Pseudomonas R3; removed after 1 h., fixed in 2·5 % formaldehyde in 2·5 % NaCl (Hodgkiss & Shewan, 1968) for 30 min., dried, rinsed in distilled water, and dried prior to shadowing with gold–palladium alloy (60 % gold–40 % palladium).

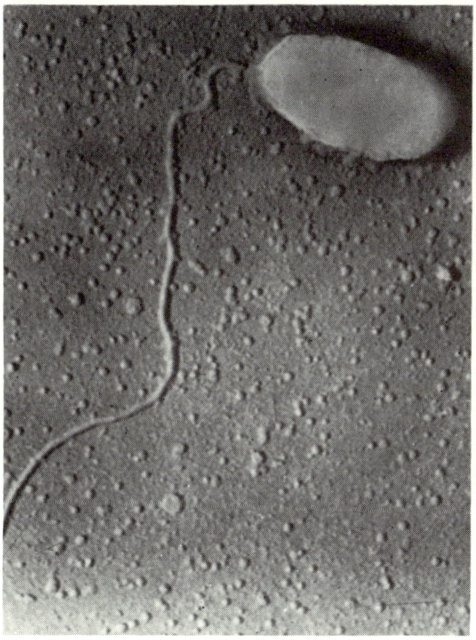

Fig. 1. Single polar flagellum of *Pseudomonas* strain R3.

RESULTS

Motility and reversible sorption of Pseudomonas R3. Pseudomonas R3 was a motile rod possessing a single polar flagellum (Fig. 1). In common with most motile bacteria, this organism exhibited a peculiar rotational behaviour at a liquid–glass interface. Cine films indicated that such bacteria sorbed at the pole (an edge-to-face association) and, by virtue of the motive force of the flagellum, rotated violently around the fixed pole (Fig. 2). If sorbed in a face-to-face position (Fig. 2), they rotated in either direction in a propeller-like fashion. The bacteria occasionally broke away from the surface and often sorbed at another point. Any non-motile cells were sorbed in a face-to-face position (Fig. 2) and showed only Brownian motion. Both motile and non-motile cells were readily desorbed by washing the glass surfaces in 2·5 % NaCl.

The kinetic energy of motile Pseudomonas R 3 cells was estimated: the maximum velocity was 33 μm./sec.; assuming an average mass of 10^{-12} g., then the kinetic energy of a motile organism was $5{\cdot}45 \times 10^{-18}$ ergs.

Effect of medium composition on irreversible sorption of Pseudomonas R 3. As reported below, the irreversible sorption of Pseudomonas R 3 was affected by divalent cations in ASW. However, the results in Table 1 show that deletion of divalent cations from ASW + glucose at 7 mg./l. (ASWG-7) did not influence the numbers of bacteria initially attracted to the glass surface.

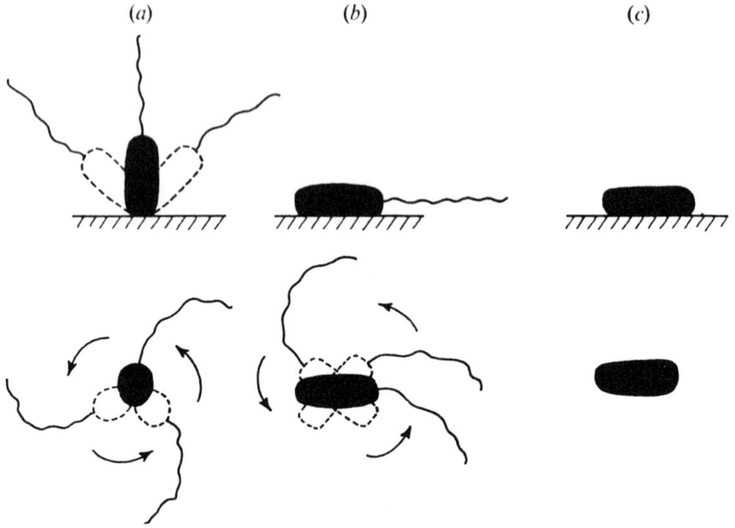

Fig. 2. A diagrammatic interpretation of the reversible sorption of Pseudomonas R 3 to a glass surface: (*a*) and (*b*) illustrate the rotational movements of motile bacteria in an edge-to-face and a face-to-face manner, respectively; (*c*) face-to-face sorption of non-motile bacteria.

Table 1. *Effect of divalent cations in artificial seawater on the reversible sorption of Pseudomonas* R 3

Medium	No. bacteria sorbed/cm²
ASWG	$5{\cdot}35 \times 10^3$
ASWG-Ca^{2+}	$5{\cdot}27 \times 10^3$
ASWG-Mg^{2+}	$5{\cdot}58 \times 10^3$
ASWG-Ca^{2+}, -Mg^{2+}	$5{\cdot}42 \times 10^3$

Effect of electrolyte concentration on reversible sorption of Achromobacter R 8. The number of achromobacteria reversibly sorbed increased with increasing electrolyte concentration or as the thickness of the electrical double layer decreased (Fig. 3). The theoretical thickness of the diffuse double layer ($1/K$) was calculated from the expression

$$K = 0{\cdot}327 \times 10^8 Z\sqrt{c},$$

where Z = valency and c = molar concentration of electrolyte. This expression holds for aqueous solutions of symmetrical electrolytes at 25° (Shaw, 1966).

The bacteria were reversibly sorbed at lower concentrations of a divalent electrolyte ($MgSO_4$) than of a monovalent electrolyte (NaCl), an effect clearly related to the greater compression of the double-layer in the divalent system at comparable concentrations. In both electrolyte systems, all the bacteria were repelled from the surface when the value of $1/K$ exceeded about 200 Å. This result suggests that the initial, reversible attraction of bacteria to a surface depends on the magnitude of the double-layer repulsion energy at different electrolyte concentrations compared with the van der Waals attraction energy.

Effect of glucose on the irreversible sorption of Pseudomonas R3. Sorption was negligible from NaCl, but appreciable from ASW (Fig. 4). Glucose (7 mg./l.) stimulated sorption from

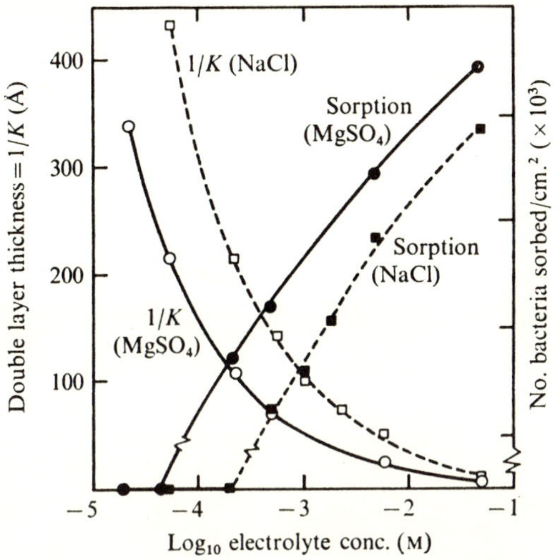

Fig. 3. Reversible sorption of Achromobacter R8 and the theoretical double-layer thickness (1/K) in relation to electrolyte concentration and valency. □—□, 1/K for NaCl; ○—○, 1/K for $MgSO_4$; ■—■, sorption of Achromobacter R8 from NaCl; ●—●, sorption of Achromobacter R8 from $MgSO_4$. The ionic strength of natural seawater is approximately 0·65 M.

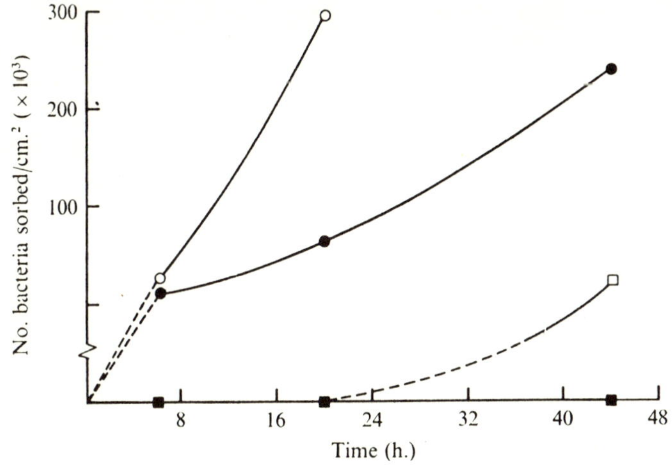

Fig. 4. Irreversible sorption of Pseudomonas R3 from NaCl and artificial seawater (ASW). ■—■, 2·5% NaCl; □—□, 2·5% NaCl+glucose (7 mg./l.); ○—○, ASW+glucose (7 mg./l.); ●—●, ASW.

both media. Electron micrographs of grids suspended in ASW or ASWG-7 showed very fine extracellular polymeric fibrils on bacteria sorbed from ASW (Fig. 5); bacteria sorbed from ASWG-7 had many more of such fibrils (Fig. 6).

Although the growth of Pseudomonas R3 was stimulated by higher levels of glucose irreversible sorption of the bacteria from ASWG-30 and ASWG-70 was inhibited completely (Fig. 7). Sorption was rapid from ASWG-7 and moderate from ASW. The irreversible sorption of Pseudomonas R3 was lowered dramatically even at glucose levels of 14 and 21 mg./l. (Table 2). Flocculation of bacteria in the bulk suspension was observed in conditions in which sorption was greatest.

Electron micrographs of grids immersed in ASWG-7 revealed large numbers of bacteria all with extracellular polymeric fibrils. Sometimes it appeared that some sorbed bacteria had been sheared from the grid surface by washing, leaving polymer 'footprints' on the grid surface (Fig. 8). Very few bacteria were found on grids from ASWG-14 and ASWG-21. Some bacteria showed little evidence of polymer production (Fig. 9), while others appeared to produce abundant polymer (Fig. 10).

Although ASW contained both NH_4Cl and $NaNO_3$ as nitrogen sources, the individual salts at equivalent nitrogen levels did not influence growth or the irreversible sorption of

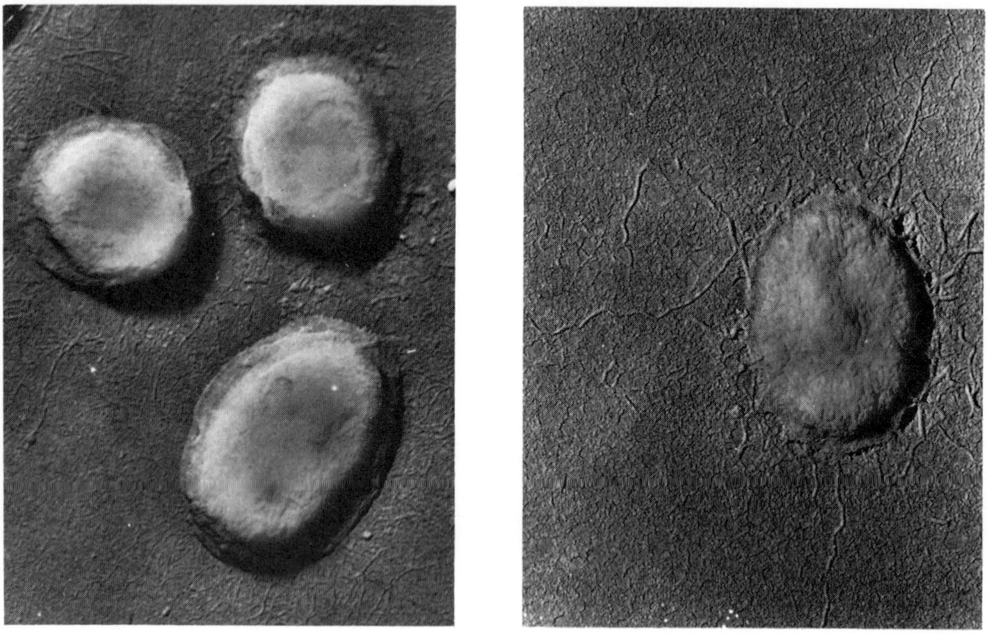

Fig. 5 Fig. 6

Fig. 5. Extracellular polymeric fibrils on bacteria sorbed on grid immersed in artificial seawater.
Fig. 6. As in Fig. 5, but grid immersed in artificial seawater + glucose (7 mg./l.).

Table 2. *Effect of glucose levels in artificial seawater on growth and irreversible sorption of Pseudomonas* R3

Glucose (mg./l.)	No. bacteria/ml. after 24 h*	No. bacteria sorbed/cm.^{2} ($\times 10^3$) 20 h	4 days
0	$14{\cdot}8 \times 10^7$†	44·6	488
7	$26{\cdot}1 \times 10^7$†	65·8	514
14	$33{\cdot}2 \times 10^7$	2·3	78
21	$44{\cdot}1 \times 10^7$	2·1	31

* Initial inoculum = 10.0×10^7 bacteria/ml.
† Flocculation evident in the bulk suspension.

Pseudomonas R3 to glass surfaces. Since all the nitrogen in the basic ASW medium was available to Pseudomonas R3, then the C/N ratios ranged from 0·17 in ASWG-7 to 1·70 in ASWG-70. Consequently, all of these media were carbon deficient, so *N*-limitation was not the reason for production of extracellular polymeric materials in the media.

Effect of divalent cations on irreversible sorption of Pseudomonas R3. The omission of Ca^{2+} and Mg^{2+} (with Na_2SO_4 added to provide SO_4^{2-}) from ASWG-7 prevented the irreversible sorption of Pseudomonas R3 (Fig. 11). Mere addition of Ca^{2+} and Mg^{2+} to 2·5 % NaCl + glucose did not stimulate the sorption of Pseudomonas R3 from such a medium. Total numbers of Pseudomonas R3 in the suspension remained constant in all but ASWG-7, where rapid sorption appeared to be related to the limited growth of the bacteria. Few bacteria adhered to electron microscope grids immersed in ASWG-7 lacking Ca^{2+} and Mg^{2+}, and those that did showed no evidence of polymer production (Fig. 12).

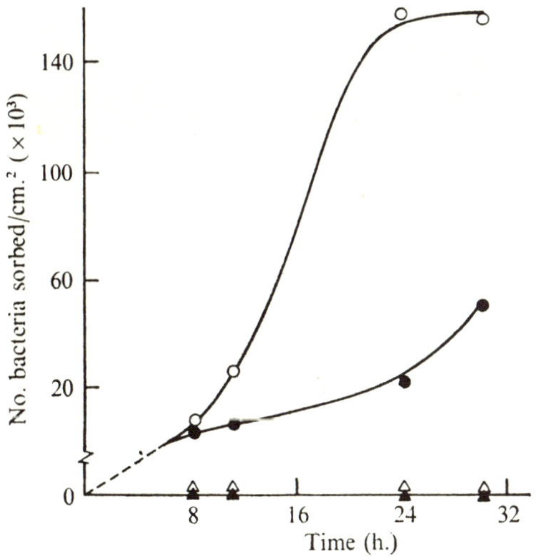

Fig. 7

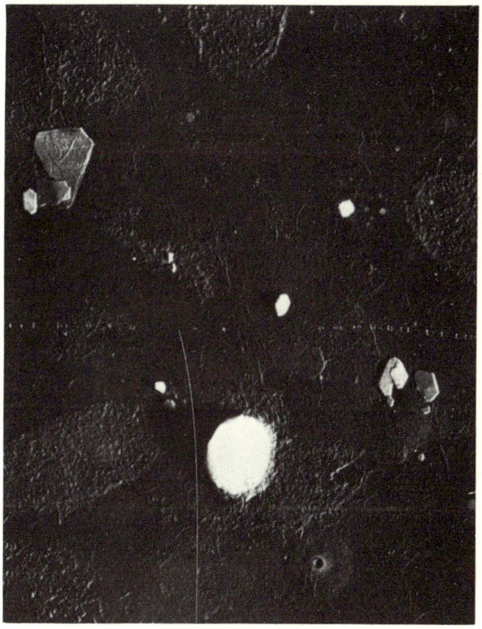

Fig. 8

Fig. 7. Effect of glucose concentration on the irreversible sorption of Pseudomonas R3 from ASW. ●—●, ASW; ○—○, ASW + glucose (7 mg./l.); △—△, ASW + glucose (30 mg./l.); ▲—▲, ASW + glucose (70 mg./l.).

Fig. 8. Polymer 'footprints' remaining after bacteria were sheared from the grid surface.

Table 3. *Effect of divalent cations in artificial seawater on the irreversible sorption of Pseudomonas* R3

Medium	No. bacteria sorbed/cm² 24 h.	48 h.
ASWG	242×10^3*	464×10^3*
ASWG-Ca^{2+}	155×10^3	400×10^3
ASWG-Mg^{2+}	126×10^3*	280×10^3*
ASWG-Ca^{2+}, -Mg^{2+}	0	0

* Flocculation evident in the bulk suspension.

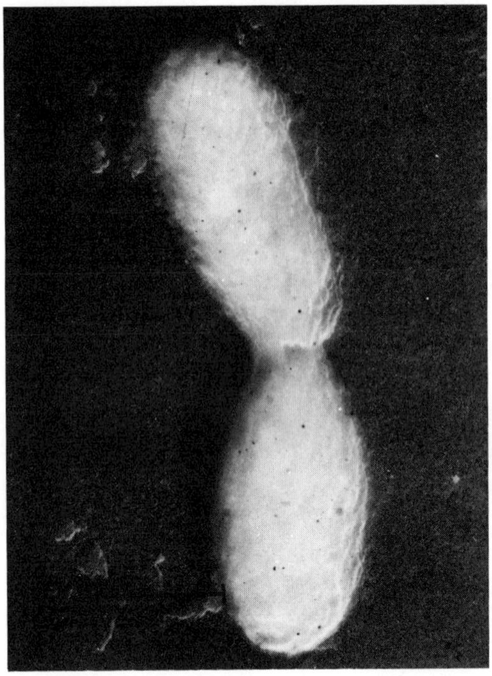

Fig. 9

Fig. 10

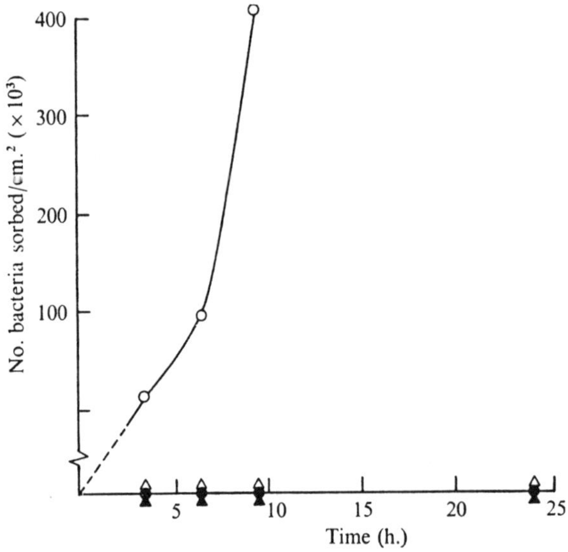

Fig. 11

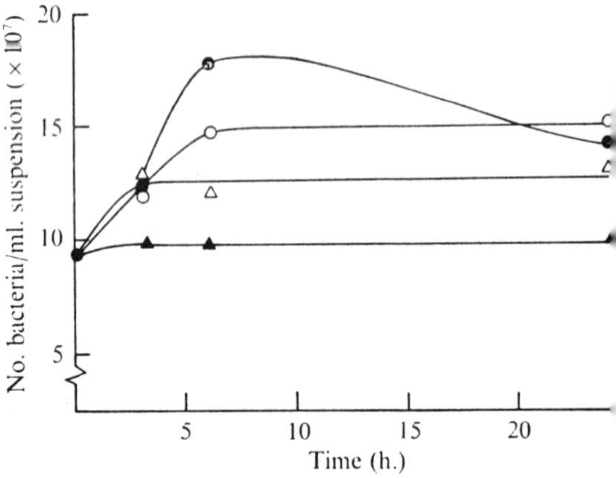

Fig. 12

Fig. 9. Bacterium lacking extracellular polymeric fibrils. From ASWG without Ca^{2+} and Mg^{2+}.

Fig. 10. Grid immersed in artificial seawater+glucose (14 mg./l.). Bacterium showing abundant polymer production.

Fig. 11. Effect of divalent cations on the irreversible sorption of Pseudomonas R 3 from NaCl and ASW. ▲—▲, 2·5% NaCl+ glucose; △—△, 2·5% NaCl+glucose+Ca^{2+}+Mg^{2+}; ○—○, ASW+ glucose; ●—●, ASW+glucose -Ca^{2+}, -Mg^{2+}.

Fig. 12. Effect of divalent cations on growth of Pseudomonas R 3 in ASWG (glucose at 7 mg./l.). ●—●, ASWG; ○—○, ASWG -Ca^{2+}; △—△, ASWG -Mg^{2+}; ▲—▲, ASWG -Ca^{2+}, -Mg^{2+}.

Although irreversible sorption was inhibited when both Ca^{2+} and Mg^{2+} were omitted from ASWG-7, sorption did occur when either was omitted (Table 3). The degree of sorption from the various media (Table 3) appeared to be related to the growth of Pseudomonas R 3 in the media (Fig. 12). Flocculation was observed consistently in the ASWG and ASWG-7 without Mg^{2+}, but not in ASWG without Ca^{2+} or ASWG-7 with neither Ca^{2+} nor Mg^{2+}. Thus, flocculation in the bulk suspension did not necessarily involve the same mechanisms as sorption.

Effect of age of inoculum on irreversible sorption of Pseudomonas R 3. Sorption of young (log phase) bacteria from ASWG was much faster than that of older bacteria (Fig. 13), but within 24 h. the originally turbid bacterial suspension had completely lysed. Neither growth, lysis, nor sorption were observed with young bacteria in ASWG without Ca^{2+} and Mg^{2+}.

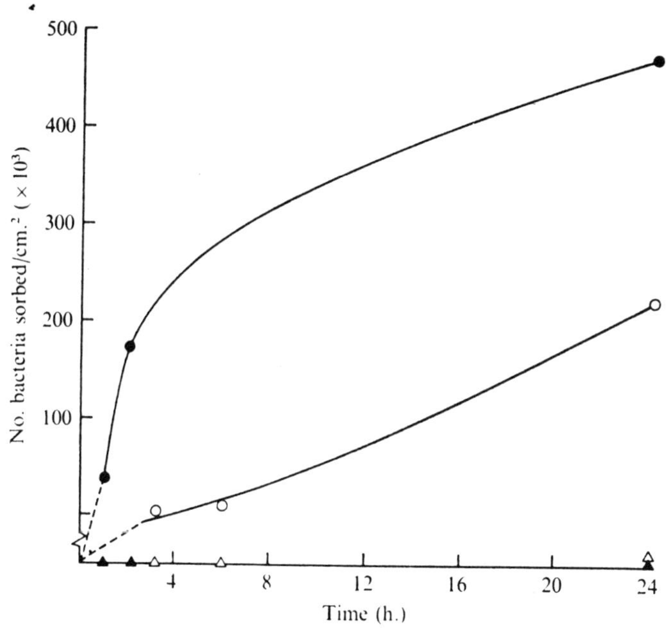

Fig. 13. Effect of age of inoculum on reversible sorption of Pseudomonas R 3 from ASWG. ●—●, Log phase bacteria in ASWG, all bacteria lysed; ▲—▲, log phase bacteria in ASWG -Ca^{2+}, -Mg^{2+}, no lysis; ○—○, stationary phase bacteria in ASWG, no lysis; △—△, stationary phase bacteria in ASWG -Ca^{2+}, Mg^{2+}, no lysis.

DISCUSSION

In the sorption of marine bacteria to surfaces, different mechanisms must be involved in the essentially instantaneous reversible phase and in the time-dependent irreversible phase of sorption.

Reversible sorption. The effects of different concentrations of monovalent and divalent electrolytes on the reversible sorption of Achromobacter R 8 suggest that this phenomenon may be explained in terms of the Derjaguin-Landau and Verwey-Overbeek theory (Shaw, 1966; Weiss, 1968). This theory involves an estimation of the magnitude, and variation with interparticle distance, of the London–van der Waals attractive energies between two surfaces and the electrical repulsive energies resulting from the overlapping ionic atmospheres (diffuse double-layers) around the surfaces.

The energies of interaction between Pseudomonas R3 and a glass surface at different values for the double-layer thickness have been computed (Weiss & Harlos, 1971) with a programme based on a rearranged version of the formula derived by Hogg, Healy & Fuerstenau (1966):

$$V_T = V_R + V_A,$$

where

$$V_R = \text{repulsion energy} = \frac{\epsilon}{4}\left(\frac{a_1 a_2}{a_1+a_2}\right)\left[(\psi_1+\psi_2)^2 \ln(1+e^{-KH}) + (\psi_1-\psi_2)^2 \ln(1-e^{-KH})\right],$$

$$V_A = \text{attraction energy} = -\frac{A}{6}\frac{a_1 a_2}{a_1+a_2}\frac{1}{H},$$

where a is the radius of curvature of the particle, $a_1 = 10^4$ nm. (glass) and $a_2 = 0{\cdot}4\ \mu$m. (bacterium); ψ is the surface potential of the particle, $\psi_1 = -15$ mV (glass), $\psi_2 = -25$ mV (bacterium); K is the inverse Debye-Huckel length, $1/K$ varies from 0·645 nm. at 2×10^{-1} M-NaCl to 20 nm. at 2×10^{-4} M-NaCl; H is the distance of closest approach of the 2 particles; ϵ is the dielectric constant (of water); A is Hamaker's constant (5×10^{-15} ergs).

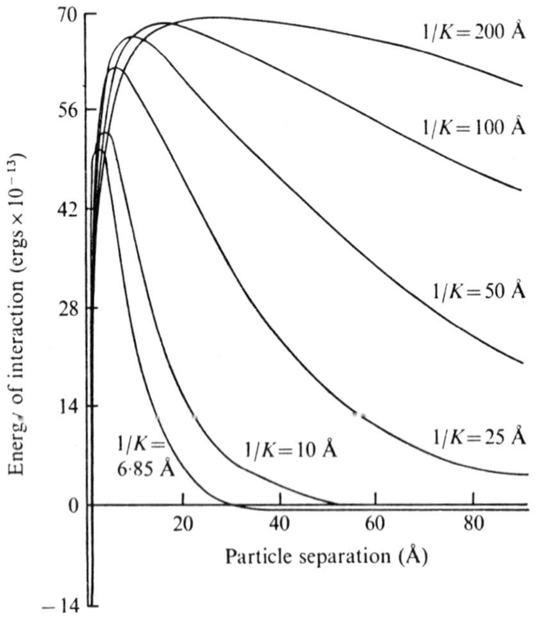

Fig. 14. Computed curves of the energy of interaction between glass and bacterial (Achromobacter R8) surfaces at varying electrolyte concentrations (varying double-layer repulsion) using a value of the Hamaker constant (A) of 5×10^{-15} ergs.

The series of curves in Fig. 14 demonstrate the increase in magnitude of the resultant repulsion curves and the increase in particle separation with decreasing electrolyte concentration (increasing values of $1/K$). At high electrolyte concentrations, a secondary minimum ('attractive trough') is apparent. The progressive increase in repulsion energy with decreasing electrolyte concentration closely parallels the observed decrease in reversible sorption of Achromobacter R8 (Fig. 3).

In the reversible phase of sorption in seawater or 2·5% NaCl it is likely that Pseudomonas R3 organisms are attracted to the point of the secondary minimum, a small but finite distance from the glass surface. It is unlikely that thermal motion would provide sufficient energy for the bacteria to overcome the repulsion barrier. In fact, the kinetic energy of motile Pseudomonas R3 ($5 \cdot 45 \times 10^{-18}$ ergs) is not sufficient to overcome the repulsion barrier at any value for $1/K$ shown in Fig. 14. The magnitude of the attractive energy at the secondary minimum may not be sufficient to hold the bacteria against the shearing effect of the rinsing process. The cine films clearly showed that motile Pseudomonas R3 could break away from the surface, although the attractive energy held them against violent rotational movements resulting from flagellar activity.

Irreversible sorption. The irreversible phase of sorption implies a firmer adhesion of bacteria to a surface. Polymeric bridging between the bacterial surface and that of the test surface might overcome the repulsion barrier between such surfaces. Such a mechanism has been proposed for the adhesion of tissue cells (Moscona, 1962) and sponge cells (Humphreys, 1965), and for the flocculation of microbial cells (Tenney & Stumm, 1965; Busch & Stumm, 1968). The electron micrographs presented provide evidence for the production of polymeric fibrils by Pseudomonas R3. In ASWG without Ca^{2+} and Mg^{2+}, polymer was not detected and irreversible sorption did not occur. The observation that polymer 'footprints' were left behind when bacteria sheared from the surface suggests that this polymer bound more tightly to the formvar coated surface than to the bacteria.

The fact that extremely low levels of available carbon stimulated irreversible sorption while higher levels inhibited this process may be relevant to microbial ecology. In natural seawaters the available carbon levels are usually very low, and such conditions probably favour the firm adhesion of micro-organisms to surfaces immersed in such environments.

Other factors are probably involved in both phases of sorption of bacteria to surfaces. The sorption to a surface of monolayers of various macromolecules present in seawater or even in bacterial cultures might drastically alter such physical characteristics of the surface as wettability and surface charge properties (Baier, Shafrin & Zisman, 1968).

This project was supported by a contract no. N00014-67-A-0298-0026 from the Office of Naval Research. The authors gratefully acknowledge the help of Dr J. Harlos and Dr L. Weiss (Roswell Park Memorial Institute, Buffalo, N.Y.) for the computations, Dr L. Spielman for helpful suggestions, and Mr G. Pearce for assistance with the electron microscopy.

REFERENCES

BAIER, R. E., SHAFRIN, E. G. & ZISMAN, W. A. (1968). Adhesion: mechanisms that assist or impede it. *Science, New York* **172**, 1360–1368.

BUSCH, P. L. & STUMM, W. (1968). Chemical interactions in the aggregation of bacteria bioflocculation in waste treatment. *Environmental Science and Technology* **2**, 49–53.

CORPE, W. A. (1970*a*). An acid polysaccharide produced by a primary film-forming marine bacterium. *Developments in Industrial Microbiology* **11**, 402–412.

CORPE, W. A. (1970*b*). Attachment of marine bacteria to solid surfaces. In *Adhesions in Biological Systems.* Edited by R. S. Manly. New York: Academic Press.

HODGKISS, W. & SHEWAN, J. M. (1968). Problems and modern principles in the taxonomy of marine bacteria. In *Advances in Microbiology of the Sea*, vol. 1, 127–166. Edited by M. R. Drood & E. J. F. Wood. London and New York: Academic Press.

HOGG, R., HEALY, T. W. & FUERSTENAU, D. W. (1966). Mutual coagulation of colloidal dispersions. *Transactions of the Faraday Society* **62**, 1938–1951.

HUMPHREYS, T. (1965). Cell surface components participating in aggregation: evidence for a new cell particulate. *Experimental Cell Research* **40**, 539–543.

MARSHALL, K. C. (1971). Sorptive interactions between soil particles and microorganisms. In *Soil Biochemistry*, vol. II, ch. 14. Edited by A. D. McLaren and J. J. Skujins. New York: Marcel Dekker.

MOSCONA, A. A. (1962). Analysis of cell recombinations in experimental synthesis of tissues *in vitro. Journal of Cellular and Comparative Physiology* **60**, 65–80.

SHAW, D. J. (1966). *Introduction to Colloid and Surface Chemistry.* London: Butterworth.

TENNEY, M. W. & STUMM, W. (1965). Chemical flocculation of micro-organisms in biological waste treatment. *Journal of the Water Pollution Control Federation* **37**, 1370–1388.

VAITUZIS, A. & DOETSCH, R. N. (1969). Motility tracks: technique for quantitative study of bacterial movement. *Applied Microbiology* **17**, 584–588.

WEISS, L. (1968). Studies on cellular adhesion in tissue-culture X. An experimental and theoretical approach to interaction forces between cells and glass. *Experimental Cell Research* **53**, 603–618.

WEISS, L. & HARLOS, J. P. (1971). Short term interactions between cell surfaces. *Progress in Surface Science (in the Press).*

WOOD, E. J. F. (1967). *Microbiology of Oceans and Estuaries.* Amsterdam: Elsevier.

ZOBELL, C. E. (1943). The effect of solid surfaces upon bacterial activity. *Journal of Bacteriology* **46**, 39–59.

37

Reprinted from *Okeanologiya*, **10**, 253–260 (1970)

QUANTITATIVE ESTIMATE OF MARINE BACTERIOPLANKTON AS A SOURCE OF FOOD

Yu. I. Sorokin, T.S. Petipa and Ye. V. Pavlova

The radiocarbon method was used to evaluate the role of bacteria as a source of food for the mass forms of planktonic animals from the Black Sea and the tropical Pacific. The natural bacterioplankton of which 30 to 40 per cent are formed by aggregate bacterial cells was found to be consumed as intensively as phytoplankton by thin and rough filtrators and to a lesser degree by capturing carnivores.

Optimum concentrations of the natural bacterioplankton at which it is intensively consumed by filtrators are 0.3 to 0.7 g/m^3. Similar concentrations were found in the gradient layers of the oligotrophic waters of the ocean where zooplankton is concentrated.

The important trophic role of the bacterial population in fresh waters has been confirmed by work carried out by limnologists [2, 3, 11]. It has been shown that the production of bacteria in these waters is frequently comparable with the production of phytoplankton [13]. The bacterioplankton can satisfy the nutritional requirements of invertebrates living in water, filter feeders at their natural concentrations in fresh water reservoirs, while the assimilability of bacterial food is not lower than that of plankton algae [14]. The nutrient role of bacteria in water reservoirs increases considerably in the case of a large supply of allochthonous organic material and on mass development of large inedible forms of phytoplankton. The inclusion of this organic material into the production process takes place via the bacterial link with considerable participation of Protozoa as an intermediate link between the bacterioplankton and the copepod zooplankton [15]. The bacterial flora takes part in the formation of nutritional properties of detritus [12, 16] and in the utilization of dissolved organic matter [17, 21, 22].

Information on the trophic role of marine microflora is extremely scanty, particularly as regards the pelagic region of the ocean [1, 19, 23]. Here coarse filter feeders of the Calanidae type, which at first sight appear to be unable to filter out such fine particles as bacterial cells, predominate. It has also been shown that in tropical oligotrophic surface waters of the ocean the production of bacterioplankton is comparable with the production of phytoplankton, which here is extremely low.

It has been reported in a series of papers that the production of phytoplankton in oligotrophic waters is not sufficient to supply the existing animal population with food [20]. As an additional source of food it has been suggested to use aggregated organic matter that is formed in seawater by the action of mechanical and biological factors [18, 22].

It follows that the quantitative evaluation of the trophic role of the micro-flora is a necessary stage in the study of the productivity of marine plankton communities. Below are presented some basic results of investigation into the nutritional role of bacterial plankton in the Black Sea and Pacific Ocean. The investigations were carried out at the Karadagsk Department of the Institute of the Biology of Southern Seas, USSR Academy of Sciences in 1967 and during the 44th voyage of the R/V Vityaz' in 1969.

METHODS

The nutritional value of bacteria for marine animals was studied using the radiocarbon method [14]. The C^{14}-labelled bacterial food was prepared in the form of a cell suspension or in the form of natural bacterioplankton. In the first case the bacterial culture of the genus Pseudomonas isolated from seawater was grown on a mineral medium containing labelled glucose. The bacterial cells were washed free of the medium by centrifugation and a thick suspension was prepared from them. In the suspension measurements were made of the specific activity of carbon in bacterial bodies (C_r) and their number by direct microscopy on membrane filters. This suspension was used in the experiments. For the preparation of the required concentration of bacteria in the experiment into the seawater, freed of bacteria by filtration through membrane filter No. 3, was introduced the appropriate volume of the suspension of labelled bacteria. In order to obtain labelled natural bacterioplankton, into the seawater freed of plankton by filtration through a double plankton gauze No. 76 was introduced a portion of C^{14}-labelled dissolved organic material in the form of an algae hydrolysate. The latter was prepared by means of acid hydrolysis of C^{14}-labelled Chlorella. The radio-activity of the hydrolysate was approximately $2 \cdot 10^6$ counts/ml at a organic carbon content of approximately 1 mg/ml. From 0.2 to 0.5 ml/liter of this solution was added to the sample of water. Thus into one liter of water were introduced only 0.2–0.5 mg of organic carbon, which on the whole changed only to a small extent the natural background of the content of dissolved organic material in the water, this consisting of approximately 2 mg/liter. On absorbing the labelled organic matter the natural bacterioplankton acquires after one or two days a certain label (300–500 counts/ml). The water with the labelled bacterioplankton was used in the experiments. For the preparation of the required concentration of bacterioplankton in experiments on the study of feeding, the water with the labelled bacterioplankton after counting of the number of bacteria in it was dissolved in the appropriate proportion of bacteria-free seawater, freed of bacteria by filtration through membrane filter No. 3.

The specific activity of natural labelled bacterioplankton was calculated using formula

$$C_r = B/10 \cdot r \ \mu g \ C/count,$$

where C_r is the inverse specific activity of the bacterial materisl; B the raw biomass of bacteria in mg/liter determined by direct microscopy using the number of bacteria and their sizes on membrane filters; r the radioactivity of bacteria in the same volume.

The animals for the experiment were caught from samples taken with a plankton net. The volume of water used in the experiments was from 100 ml to 2 liter. The number of animals in the experiments varied from 3 to 200, depending on their dimensions. Each test was usually set up in duplicate or triplicate.

The ability of animals to feed on bacteria was evaluated by means of determination of value C_y/C (daily index of assimilation) and its comparison with the analogous value on its being fed on phytoplankton [14]. The daily index of assimilation corresponds to the percentage ratio of the amount of C^{14}-labelled carbon food consumed and assimilated by the animal calculated per day (a) to its content in the body of the animal (W):

$$C_y/C = \frac{a \cdot 100}{W} \ \%.$$

Value a was determined in the experiments. The animals were kept in aquaria in the presence of the labelled food at the appropriate concentration for 2–24 hours. After this they were washed free of the labelled food, from their bodies were made preparations of dried homogenate and the radioactivity (r) was determined. Then $a = \frac{r \cdot C_r \cdot 24}{t}$ μg C/specimen per day, where t is the duration of the experiment, in hours, and C_r is the inverse specific activity of the labelled food, in μg C/count.

The assimilation of food on feeding with bacterioplankton was determined in balance experiments with labelled food [7, 9, 10].

RESULTS OF DETERMINATIONS

a. Consumption of distributed bacteria by fine and coarse filter feeders. Penilia avirostris (Fam. Sididae) is a normal component of the summer plankton in the Black Sea. Like the majority of plant-eating Cladocera, it belongs to the number of fine filter feeders. In experiments with Penilia it was established to what degree and at which concentrations bacteria are able to satisfy the nutritional requirements of this crustacean [7]. With this aim at the start was determined the comparable intensity of feeding (value C_y/C) on feeding Penilia on bacteria and fine plankton algae at optimum concentration of both types of food (approximately 1-2 g/m^2 of raw substance). In experiments with Penilia a suspension of single bacterial cells grown in a culture was used as food. Typical results of experiments with Penilia are presented in Table 1. They show that for the fine filter feeder Penilia dispersed bacteria are just as accessible and valuable food as are phytoplankton. Analogous experiments were carried out with two other crustaceons from the order Copepoda-Paracalanus parvus and Acartia clausi (9).

Table 1

Comparative Intensity of Feeding by Planktonic Crustaceans on Algae and Bacteria in the Black Sea

Species of crustaceans	Type of food	Concentration of food g/m^3	C_r of food	Carbon content in the body of Crustacean mgc/spec.	Duration of test	Radioactivity of crustacean bodies at end of test, counts/spec.	Daily index of assimilation
Penilia	Bacteria	0.7	0.0030	0.0011	7	25	24
	Pedinella	0.7	0.0003		7	300	28
Paracalanus	Bacteria	1.5	0.003	0.00115	5	7	5
	Exuviaella	3.0	0.0015		2	34	54
Acartia	Bacteria	1.5	0.003	0.0019	5	2	1.6
	Glenodinium	2.6	0.002		24	256	28

The results of the experiments are also presented in Table 1. Paracalanus which is regarded to be one of the finest filter feeders among the Calanidae [8] is able to a certain degree to filter out dispersely distributed bacteria. The "coarse filter feeder" Acartia, or more correctly prehensile, virtually does not possess this property. The experiments elucidating the dependence of feeding of filter feeders on the concentration of bacteria yielded very interesting results. It was shown that Penilia begins to consume bacteria at negligibly low concentrations ($5 \cdot 10^3$ cells/ml or 0.004 g/m^3). Already at a concentration of 3.10^5 cells/ml or 0.3 g/m^3 the curve showing the dependence of the intensity of feeding on the concentration of bacteria becomes horizontal (Fig. 1). For freshwater crustaceans among the fine filter feeders this optimum concentration is nearly 10 times higher, $2\text{-}4 \cdot 10^6$ cells/ml. It is characteristic that concentrations of bacteria found to be optimum for these crustaceans are normal for that natural medium which they inhabit. The total number of bacteria, of the order of 100,000-300,000 cells/ml is normal for the neritic zone of the Black Sea [4]. In mesotrophic freshwater basins which are inhabited by the fine filter feeder Daphnia longispina the total number of bacteria is $1\text{-}2 \cdot 10^6$ cells/ml [6]. For Paracalanus the concentration optimum is somewhat higher than that for Penilia. It is important that the optima of concentrations of bacterial and algal food proved to be close (from 0.5 to 3-4 g/m^3 raw substance). This confirms that the bacterioplankton is equal in value to the phytoplankton as a food source for fine filter feeders.

It follows from experiments carried out with the Black Sea Acartia and tropical copepods [9, 10] that for coarse filter feeders and prehensiles, being mass forms of the pelagic zones of seas and oceans, dispersely distributed bacterial cells are not high quality food. However, this does not prove their inability to feed on natural bacterioplankton which, according to much data, can be represented as accumulated cell aggregates. Indeed, experiments carried out with natural labelled bacterioplankton showed its accessibility for consumption by coarse filter feeders and even prehensiles.

b. Experiments with natural bacterioplankton. Experiments with labelled natural bacterioplankton were carried out with some mass species of plankton animals in the tropical

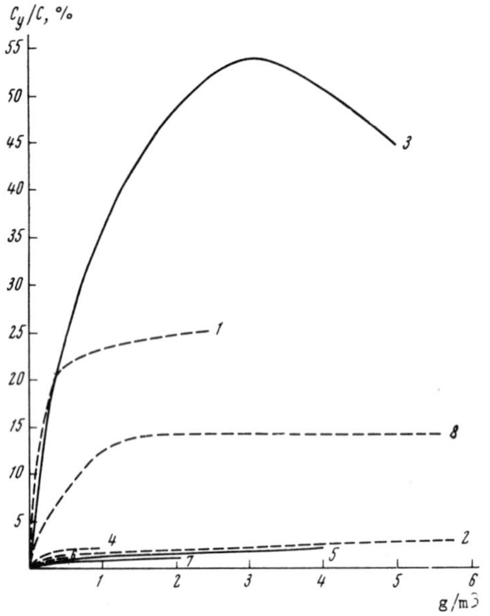

Fig. 1. Dependence of the "index of assimilation" on the concentration of bacteria and algae.

1) Penilia awirostris (bacteria); 2) Paracalanus parvus (bacteria); 3) P. parvus (Exuviaella); 4) Undinula and other Calanidae (bacteria); 5) Undinula and other Calanidae (Amphidinium); 6) Eucalanus attenuatus (bacteria); 7) E. attenuatus (Biddulphia); 8) Daphnia longispina (bacteria)

waters of the Pacific Ocean: of Copepoda, Pontellidae, a mixture of small Calanidae, Acartia, Euchaeta, Eucalanus, Haloptilus, Rhincalanus, Copilia, of Tunicata, Salpa, Oikopleura, after this Ostracoda and larvae of flying fish (Exocoetidae). Among these only the representatives of Tunicata can be regarded as belonging to fine filter feeders. The remaining ones belong to coarse filter feeders and prehensiles. Typical results of determination of comparative intensity of feeding by animals (C_y/C) on phytoplankton and bacterioplankton at their optimum con centration (0.5-2 g/m^3) are given in Table 2. The highest values of C_y/C on feeding on bacterioplankton were obtained for the fine filter feeder Oikopleura (up to 50%). In coarse filter feeders, Calanidae, the C_y/C values are lower for feeding on bacterioplankton, from 1 to 3%. These values, however, are close to those that were obtained for these crustaceans on their feeding on plant food, phytoplankton. Experiments with animals with mixed type of feeding, Acartia, Eucalanus, have shown that they are also able to feed on bacterioplankton. Furthermore, even predators-prehensiles such as Rhincalanus, Euchaeta, Haloptilus, Pleuromamma are able to consume bacterioplankton although to a lesser degree ($C_y/C \sim 0.1\%$). The bacterioplankton proved to be relatively satisfactory food also for euphausiids [5].

The accessibility of natural bacterioplankton for coarse filter feeders, and prehensiles even, can undoubtedly be explained by the fact that it consists to a considerable degree of rather large aggregates of bacterial cells and not of dispersely distributed bacteria. Indeed, the analysis of labelled natural bacterioplankton has shown that it consists to the extent of 30-40% of aggregated particles with a diameter of more than 4 microns, retained by the membrane filter No. 6. Special experiments on feeding filter feeders on natural bacterioplankton, freed from aggregates by filtering through filter No. 6, have shown that this type of dispersely distributed bacterioplankton is consumed 10 times less by coarse filter feeders (Fig. 2).

For the evaluation of the true trophic role of bacterioplankton under natural conditions

Table 2

Intensity of Feeding of Plankton Animals on Algae and Natural Bacterioplankton in the Pacific Ocean

Species of consumer	Type of food	Concentration of food, g/m³	C_y/C, %
Eucalanus attenuatus	Bacterioplankton	1.4—2.0	0.95
	Amphidinium	1.0	0.28
	Gymnodinium	0.84	0.67
	Blue-green	1.0	5.54
Euchaeta marina	Bacterioplankton	1.4—2.0	0.11
	Biddulphia	0.75	0.70
	Blue-green	1.0	0.31
Euchaeta wolfendeni	Bacterioplankton	1.4—2.0	0.09
	Gymnodinium	1.25	0.01
	Biddulphia	0.75	0.39
Rhincalanus cornutus	Bacterioplankton	1.4—2.0	0.07
	Gymnodinium	0.84	0.07
	Biddulphia	0.75	2.47
Mixture of small Calanidae	Bacterioplankton	1.4—2	0.57—3.22
	Biddulphia	0.75	2.89
	Blue-green	1.0	1.11
Ostracoda	Bacterioplankton	1.4	0.003
	Biddulphia	0.3—0.5	0.2—1.7
Appendicularia (Oikopleura)	Bacterioplankton	1.4—2.0	11.0—50.0
Salpa maxima	Bacterioplankton	1.5	0.8
	Amphidinium	1.0	0.18

it is necessary to have available the characteristics of the dependence of the intensity of feeding on the concentration and data on the concentration of the bacterioplankton in the ocean. Figure 1 shows typical curves of the dependence of the intensity of feeding of a series of mass species of animals on the bacterioplankton concentration.

Measurements of the biomass of bacterioplankton in tropical waters of the ocean give its average values as 10-30 mg/m^3, which is slightly lower than the experimental optimum found.

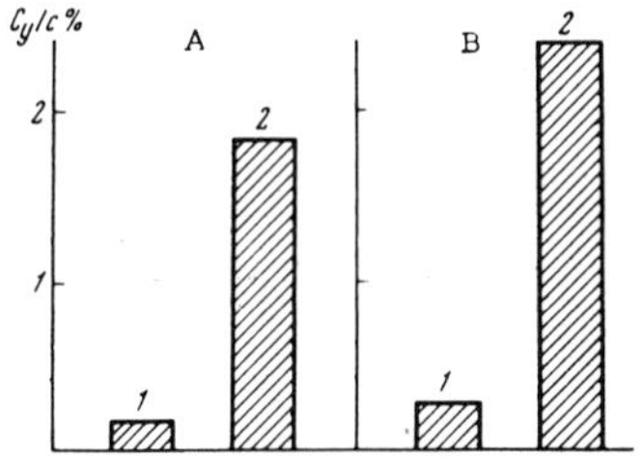

Fig. 2

Fig. 2. Comparative intensity of feeding of animals on dispersely distributed bacteria and natural bacterioplankton containing aggregates:

A) Undinula and other Calanidae; B) Mixture of small Calanidae; 1) Dispersely distributed bacteria; 2) Bacterioplankton (concentration of bacteria in experiments 0.5-1.15 g/m^3.

In evaluation of the trophic role of bacterioplankton, however, apparently one should concentrate not on average values of its concentration but on maximum values, as the consumers are mobile and are able to accumulate in those water layers where there is a larger amount of food available, in particular bacterioplankton. Such layers with increased concentration and a higher bacterioplankton production were found in tropical waters of the ocean. By probing the intensity of bioluminescence down through the water it was established that phosphorescent organisms, predominantly zooplankton, form a permanent layer at depths of 60-80 m. It was found that in this layer the concentration of bacterioplankton exceeds by 5-10 times the mean concentration for the surface water of the ocean (Fig. 3) and reaches optimum values necessary for the feeding of the zooplankton (100-300 mg/m^3). Thus even in oligotrophic tropical waters there exist conditions ensuring the bacterioplankton an important place as a food link in the community of plankton organisms.

c. Assimilation of bacterioplankton. The results of determination in some marine animals of elements of the food equilibrium characterizing the consumption and utilization of bacterial food in comparison with algae and also the calculations of assimilability are presented in Table 3. They show that for fine filter feeders bacteria are a satisfactory food, although the assimilability for this food is usually lower than that for algal food. If one compares the values for the intake of fine filter feeders, expressed as a percentage of body weight, and the assimilability for bacterial and algal food it can be observed that at a low assimilability in the first case the animals have a higher intake than in the second; the amount of the assimilated food in both cases, i.e., the amount of the assimilated bacterial and plant food, is close, however (Table 3).

Table 3

Elements of Equilibrium and Assimilability in Some Plankton Crustaceans at Optimum Concentration of Bacteria and Algae

Type of food	Assimilated		Not assimilated		Intake		Assimilability
	cal/spec	% body weight	cal/spec	% body weight	cal/spec	% body weight	% of intake
Penilia							
Bacteria	0.0067	37.2	0.0123	67.8	0.0189	105	35
Algae	0.0088	49.0	0.0031	17.7	0.0119	67	75
Paracalanus							
Bacteria	0.00180	16.2	0.00085	7.4	0.00265	23.8	66
Algae	0.00046	4.0	0.00012	1.0	0.00058	5.0	80
Eucalanus							
Bacteria	0.0159	3.8	0.0223	5.4	0.0382	9.3	41.3
Algae	0.0304	8.1	0.0276	7.4	0.0580	15.5	52.9

Thus the data presented is sufficiently significant evidence of the important nutritional role of bacterioplankton in the ocean. The intensity of consumption and assimilation of bacterioplankton by filter feeder plankton invertebrates is comparable with the intensity and effectiveness of phytoplankton consumed by them. Active consumption of bacterioplankton by animals takes place at its concentrations close to those observed in gradient layers.

CONCLUSIONS

1. Using the radiocarbon method the consumption and assimilation of bacteria contained in dispersely distributed suspension and in natural bacterioplankton by some mass species of aqueous invertebrates in the Black Sea and tropical waters of the Pacific Ocean have been studied.

2. It has been determined that find filter feeders of the Penilia and Paracalanus type are able to feed on dispersely distributed bacteria. The optimum concentrations of bacteria for the feeding of Penilia is about $3 \cdot 10^5$ cells/ml, which is 5-10 times lower than the value for freshwater Cladocera.

3. The natural bacterioplankton is represented by 30-40% of aggregates of bacterial cells whose diameter exceeds 4 microns. It is consumed with the same intensity as phytoplankton not only by fine but also by coarse filter feeders, and to a certain degree even by prehensiles, animals with a mixed type of feeding and predators.

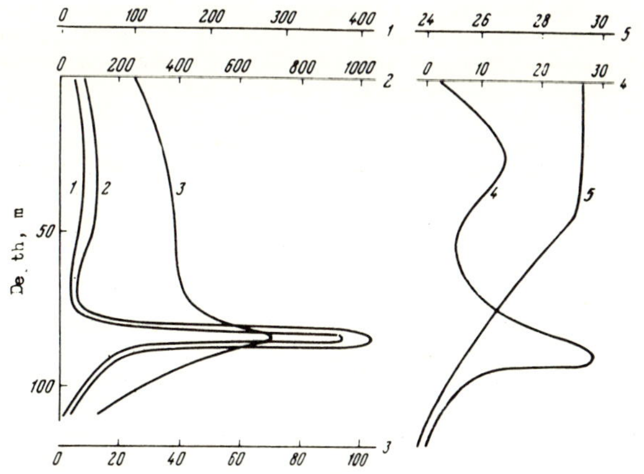

Fig. 3. Distribution of bacterial biomass, mg/m^3 (1); total number, thousand cells/ml (2); daily production of bacteria, mg/m^3 (3); relative intensity of bioluminescence, % (4); and temperature (5) at Station 6052 (Solomon Sea).

4. Optimum concentrations of natural bacterioplankton at which it is intensively consumed by filter feeders is 0.3-0.7 g/m^3. The same concentrations of bacterioplankton were found in narrow layers in oligotrophic waters of the ocean where the zooplankton is also concentrated.

5. Assimilability for bacterial food in fine filter feeders was 35 and 66% and was lower than that for plant food which was 75-80%. The amount of the assimilated bacterial and algal food, however, in these animals did not appreciably differ, or the assimilation of bacteria was even higher.

In coarse filter feeders and prehensiles, to which belongs the majority of copepods, the assimilability for bacterial food and the amount of assimilated bacteria is usually lower than the corresponding values obtained for feeding on algae.

REFERENCES

1. Vinogradov, M.Ye. Vertical distribution of ocean zooplankton, Nauka, 1968.
2. Gorbunov, K.V. Cellulose bacteria as link in the food chain of freshwater reservoirs. Mikrobiologiya, 15, No. 2, 1946.
3. Ivanov, M.V. Method of determination of bacterial biomass in water. Mikrobiologiya, 24, No. 1, 1955.
4. Lebedeva, M.N. Ecological principles of microorganism distribution in the Black Sea. Tr. Sevastop. biol st., 10, 1968.
5. Lukanina, Ye.A. and L.A. Ponomareva. Feeding of Euphausiacea in the tropical region of the Pacific Ocean. In: Sb. rabot, povedennykh v 44-m reyse i/s "Vityaz' " (Reports on work carried out during the 44th voyage of the Vityaz'). Nauka (in press).
6. Novoshilova, M.I. Generation time of bacteria and production of bacterial biomass in the water of Rybinsk reservoir. Mikrobiologiya, 26, No. 2, 1957.
7. Pavlova, Ye.B., and Yu.I. Sorokin, Bacterial feeding of the plankton copepod Penilia avirostris Dana from the Black Sea. In: Produktsiya i pishchevyye svyazi v soobshchestvakh planktonnykh organizmov (Production and feeding relationships in plankton organism communities). Naukova Dumka, Kiev, 1970.
8. Petipa, T.S. Feeding of plankton crustaceans in the Black Sea. In: Biologicheskiye issledovaniya Chernogo marya i ego resursov (Biological investigations in the Black Sea and its resources). Okeanogr.komis. AN SSSR, Nauka, 1968.
9. Petipa, T.S., Yu.I. Sorokin and L.A. Lanskaya. Investigations on feeding of Acartia clausi Giesbr. by the radiocarbon method. In: Produktsiya i pishchevyye svyazi v soobshchestve planktonnykh organizmov (Production and feeding relationships in plankton organism communities). Kiev, Naukova Dumka, 1970.

10. Petipa, T.S., Ye.V. Pavlova and Yu.I. Sorokin. Study of feeding of mass forms of plankton in the tropical region of the Pacific Ocean by the radiocarbon method. In: Sb. rabot, provedennykh v 44-m reyse i/s "Vityaz' " (Reports on work carried out during the 44th voyage of Vityaz'). Nauka (in press).
11. Rodina, A.G. Role of individual groups of bacteria in the productivity of water reservoirs. Tr. probl. tem. soveshch. Zool. in-ta AN SSSR, No. 1, 1951.
12. Rodina, A.G. Bacterial content in detritus of lakes in the Ladoga region. Mikrobiologiya, 32, No. 6, 1963.
13. Romanenko, V.I. Characterization of microbiological processes of formation and consumption of organic matter in the Rybinsk water reservoir. Tr. Inst. biol. vnutr. vod. AN SSSR, No. 13/16, 1966.
14. Sorokin, Yu.I. Application of C^{14} for the study of aqueous animals. In: Plankton i bentos vnytrennikh vodoyemov (Plankton and benthos of inland bodies of water). Nauka, 1966.
15. Sorokin, Yu.I. Characterization of productivity of the coastal zone in the Volga reach of the Rybinsk reservoir. Byul. Inst. biol. vnutr. vod AN SSSR, No. 5, 1969.
16. Syshchenya, L.M. Detritus and its use in the production process in water reservoirs. Gidrobiologicheskiy zhurnal, 4, No. 2, 1968.
17. Khaylov, K.M. and Yu.A. Gorbenko. External metabolic regulation in the system: community of periphyton microorganisms — dissolved organic matter of ocean waters. Dokl. AN SSSR, 173, No. 6, 1967.
18. Baylor, E.R. and W.H. Sutcliff. Dissolved organic matter in seawater as a source of particulate food. Limnol. and Oceanogr., 8, 1963.
19. Jorgensen, C.B. The food of filter feeding organisms. Rapp. Proc. Verb. Réun., 153, 1962.
20. Marshall, S.M. and A.P. Orr. The biology of a marine copepod. London, Oliver and Boyd, 1955.
21. Parsons, T.R. and J.H. Strickland. On the production of particulate organic carbon by heterotrophic processes in seawater. Deep-Sea Res., 8, 1962.
22. Sieburth, G.M. Observations on bacteria planktonic in Narraganset Bay, Rhode Island. Proc. U.S.-Japan seminar on Marine Microbiology (Kyoto), 1968.
23. Wood, E.J.F. Microbiology of oceans and estuaries. Amsterdam-London-W.G., 1967.

Institute of the Biology of Inland Waters,
USSR Academy of Sciences
Institute of the Biology of Southern Seas
Academy of Sciences of the Ukrainian SSR

Received September 9, 1969

Part VII

MICROBES IN THE SEA

Editor's Comments on Papers 38 Through 45

38 **BARGHOORN and LINDER**
Marine Fungi: Their Taxonomy and Biology

39 **FELL, AHEARN, MEYERS, and ROTH**
Isolation of Yeasts from Biscayne Bay, Florida, and Adjacent Benthic Areas

40 **WEYLAND**
Actinomycetes in North Sea and Atlantic Ocean Sediments

41 **SPENCER**
Indigenous Marine Bacteriophages

42 **LISTON**
Distribution, Taxonomy and Function of Heterotrophic Bacteria on the Sea Floor

43 **KRISS, ABYZOV, LEBEDEVA, MISHUSTINA, and MITSKEVICH**
Geographic Regularities in Microbe Population (Heterotroph) Distribution in the World Ocean

44 **SIEBURTH**
Distribution and Activity of Oceanic Bacteria

45 **TAGA**
Some Ecological Aspects of Marine Bacteria in the Kuroshio Current

THE FILAMENTOUS FUNGI

During the preceding sections, it may have seemed as though bacteria were the only heterotrophic microorganisms that could be found in the oceans. Part of this apparent imbalance stems from the fact that bacteriologists who have been concerned with marine microbes have generally believed that the appearance of fungal colonies on their plates, following inoculation with seawater or especially sediments, was the result of poor technique and/or air contamination. Also, mycology has traditionally been more concerned with soil forms and pathogenic species than with aquatic fungi. Until recently, comparatively few mycologists examined the fate, distribution, and ecological role of fungi in the oceans. Another factor accounting for the lag in marine mycological studies has been the fact that the genera isolated from the sea were frequently the same genera as those found in soil and freshwater. This led many workers to conclude that true marine fungi simply did not exist. In fact, as recently as 1961, Johnson and Sparrow (50) emphasized that forms which could be found only in the sea with a specific requirement for seawater or its inorganic ions had not been reported. They, therefore, proposed the working definition of a marine fungus as one that could grow and sporulate on seawater-containing media.

The historical background to marine fungal studies dating back to the nineteenth century has been previously described by ZoBell (114), Johnson and Meyers (48), and Johnson and Sparrow (50). Part of the resurgence of interest in marine fungi no doubt stemmed from the general awakening to the importance of the oceans to man. In addition, the time was ripe for expanding marine mycological studies because the basic groundwork had been laid in 1944 by Barghoorn and Linder. Excerpts from their classic paper are reprinted here (Paper 38). They demonstrate that marine fungi indeed have a tolerance to higher salt concentrations, but, more importantly, the authors found that their marine isolates were more adapted to a slightly alkaline pH, such as found in the ocean, than are terrestrial or freshwater forms. Thus, the basis for a biochemical-physiological explanation of fungal adaptation to the sea was established.

Since that time, we have seen the publication of a book by Johnson and Sparrow (50), and an extensive body of literature has developed on the cellulolytic and lignin-degrading abilities of marine fungi (7, 43, 48, 49, 66). The economic importance of this aspect has been internationally recognized by the publication of the proceedings of a symposium on marine fungi as contributors to the deterioration of wood and fouling processes (45). Even earlier, in 1934, the economic importance of aquatic phycomycetes to marine ecosystems was recognized by Renn (75), who found that labryinthula-like organisms were responsible for the destruction of the eelgrass beds along the Atlantic coast. Since that time, numerous investigators have implicated this heterogeneous group of fungi as the causative agents in major diseases of clams and oysters and in various plant infestations (4, 18, 20, 56, 97).

One aspect of marine mycological research that is under intensive study today concerns the role of fungi in the marine environment. These studies are typified by investigations of fungal succession during the degradation of algae, leaves, and other woody materials in the seas (32, 41, 44). Descriptions of the isolates, their identification, and consequently sequential determinations of succession during the decay of mangrove leaves and mangrove seedlings have been published by Fell and Master (25) and Newell (70), respectively. In addition, Jones (44) has noted similar types of fungal succession on test wood blocks placed into various northern waters. Scanning electron microscopic evidence for orderly fungal succession on algae, *Spartina alterniflora* leaves, and wood have been beautifully illustrated by Sieburth and co-workers (10, 29, 87) and more recently in his monograph, *Microbial Seascapes* (88). Furthermore, Kohlmeyer (55) has successfully isolated ascomycetes that are capable of degrading the chitinous and perhaps keratin-like tubes of hydrozoans and polychetes.

As concepts regarding distribution patterns and successional sequences of the fungi have developed, interest has naturally increased in the physiology of these organisms. Frequently, as a direct outgrowth of these experiments, investigators have sought biochemical answers to explain the timing and order of fungal attack on submerged objects. In fact, it is widely recognized now that true marine fungi exist; that is, there are species which are found only in the seas and which have specific requirements for seawater for growth and reproduction (3). Other fungi, especially the phycomycetous *Thraustochytrium roseum* (89) and *Dermocystidium* sp. (6) require seawater for phosphate uptake. Meanwhile, *Pythium marinum* has been isolated from the alga *Porphyra miniata* and shown not only to require cations, at the concentrations found in seawater, but to be facultatively psychrophilic as well (52).

Sguros and co-workers have been investigating the nutrition and physiology of several cellulolytic marine ascomycetes, and have noted that thiamine is required and biotin is stimulatory for at least one organism in this group, *Halosphaeria mediosetigera* (83). They have also demonstrated the importance of inorganic ions at seawater concentrations for other ascomycetes (84) and a preferential utilization of cellulosic by-products in several other isolates (85). Earlier, Vishniac (106, 107) discovered a requirement in *Labyrinthula* M, P, and A strains for all the major cations in seawater and an additional steroid requirement for growth in the P strain. Fungal bioassays for thiamine and cobalamine in seawater have also been developed with a fungus requiring these vitamins (108, 109). The organism was later identified as *Thraustochytrium* sp. (8). Thus, at least in the Labyrinthulales, the physiological complexity in relation to the requirement for seawater is similar to that described earlier for the bacteria.

An attempt was made in 1968 by Meyers (62) to coordinate the successional patterns, nutritional needs, and isolation data of various marine fungi from selected geographical and ecological niches, particularly *Thalassia* leaves. A wide variety of fungal genera were isolated at different stages in the growth of the turtle grass, and many of these fungi either required seawater or had optimal growth under simulated marine environmental conditions. However, the nutritional patterns exhibited by the selected isolates failed to display any general trend. That may be more a reflection of our ignorance of the changes in the biochemistry of maturation of *Thalassia* sp. than our ignorance about fungal biochemistry.

Comparative biochemical studies between marine and terrestrial fungal isolates are rather sparse, and when *Leptosphaeria* from various ecological niches were analyzed, the hyphal cell walls showed no chemical modification despite the differences in environmental sources (100).

It is now generally recognized that the appearance of fungal colonies on bacteriological media is not necessarily an indication of poor technique. In fact, because of the studies initiated by Barghoorn and Linder (4) and the work of others since that time, fungi should be considered a major portion of the microbial flora and certainly major contributors to the turnover of macromolecules such as cellulose and lignin.

YEASTS IN THE SEA

Another group of fungi that often appear on standard marine bacteriological plates is the yeasts. Because they form smooth, rounded, often pigmented, and more often glisteny colonies, they are often counted as bacteria. That the yeasts are ubiquitous and frequently a significant proportion of the heterotrophic population has been amply demonstrated by Fell and co-workers (26). The first paper by Fell, Ahearn, Meyers, and Roth on distribution patterns of yeasts is reprinted here (Paper 39). Although yeasts had been recognized as occurring in the oceans as early as 1894

(27, 28), the first extensive listings of the species found in the oceans were coordinated and published by Kriss (56) and Kriss et al. (58). This paper by Fell et al., however, seemed to alert marine microbiologists to the possibility that all along they had been ignoring a significant group of microbes. Certainly, since this paper was published, it has been demonstrated that truly marine yeasts exist (26) and that these are frequently adapted to their environment; for example, obligate psychrophiles have been isolated from the Antarctic (24) and temperature tolerant forms from the Indian Ocean (23).

The distribution of yeasts has been correlated with the amount of organic matter in the environment. In areas with high organic matter, such as eelgrass beds, estuaries, and algal beds, there is a substantial number of yeast species and high total numbers (26, 105), whereas in the open ocean this value can drop to an average of two to three colony-forming units per liter of seawater (1). Although yeasts are known to degrade oil and the numbers present increase immediately after an oil spill, there is a diminution in these numbers, for unknown reasons, even when oil is still present (1). One explanation that has not been adequately tested by field measurements of the *in situ* chemistry of the seawater is whether this limitation in growth is due to a decrease in available nitrogen, vitamins, phosphate, oxygen, or even pH.

Aside from the general axiom that more yeasts are to be found in estuarine sediments than in the overlying waters (63, 65, 105), these microorganisms have been isolated from marine birds and fish (11, 105), from the stomachs of the little toothed whale (67), and other marine mammals (105), crustaceans, and invertebrates (79). The predominant types of marine yeasts appear to be species of *Candida, Rhodotorula, Torulopsis, Debaryomyces, Metschnikowiella,* and *Sporobolomyces*. Many of the investigations on yeasts have reported that nutritionally some yeasts require the vitamins biotin and thiamine; the growth of other strains is stimulated by these vitamins, whereas the majority of isolates display no such nutritional requirement (2, 79). The oxidative metabolism of *Rhodotorula* species from terrestrial and marine sources has been compared, and it was found that the marine isolates were generally less active than their terrestrial counterparts; but they were able to degrade and/or assimilate, among other carbon sources, galactose, xylose, and frequently cellobiose (60). Similarly, diverse nutritional patterns have been reported by Kriss and his co-workers (56, 58), who noted that most inorganic sources of nitrogen were utilized, and that many of the carbohydrates contained in the macromolecules of other organisms could be assimilated by the yeasts. Other than the previously mentioned involvement in oil degradation, little is known specifically about the contributions of yeasts to the marine environment. Were they introduced through various pollution routes and just happened to survive, and in some cases have adapted to growth in a saline world, or are they an integral part of the microflora? Answers to these questions await further studies.

BETWEEN FUNGI AND EUBACTERIA: THE ACTINOMYCETES

In his compilation of microorganisms in the oceans, ZoBell (114) mentioned that the genera *Mycobacterium, Actinomyces, Micromonospora,* and *Nocardia,* but not *Streptomyces,* had been isolated from marine waters and sediments. These organisms were generally able to attack aliphatic, aromatic, and naphthenic hydrocarbons (114). Later, Kriss (56) reported the specific cultural and biochemical characteristics of isolates from the genera *Mycobacterium* and *Actinomyces* that he had obtained. He especially commented on their infrequent occurrence in the Black Sea and noted that their isolation was dependent on the use of a mineral salts medium and that they could not be cultured on the usual enriched medium. Although the actual numbers found were few, Kriss et al. (58) reported the isolation, primarily of *Mycobacterium* species, from the Indian and Pacific Oceans, but not from the Atlantic Ocean. However, Weyland, as reported in Paper 40, found members of this group to be widely distributed when the proper isolation medium was used. He obtained isolates in 102 out of 107 sediment samples, but he had to use six different media to achieve these isolations. This need for multiple isolation media indicates a wide diversity in physiological types among the marine actinomycetes.

When, or if, actinomycetes are isolated during routine bacteriological sampling, they are usually not specifically listed; thus, the impression is left that they are only rarely found in the sea (9), and then are probably contaminants from land (33, 72). Singal et al. (90), though, found them in high numbers in Indian Ocean sediments; in some cases the actinomycetes comprised up to 60 percent of the bacterial flora. Again, the most commonly isolated genera were *Actinomyces,* followed by *Micromonospora.* Earlier, Grein and Meyers (33) had studied sediments from the

New Jersey coast and the Bahamas and isolated several *Streptomyces* species, as well as representatives of other genera in this group. Based upon salinity tolerance studies, they hypothesized a terrestrial origin for their "marine" actinomycetes. From numerous algae, muds, sand, and soil, a total of 21 strains of actinomycetes were isolated by Chesters et al. (16). Thirteen of these could hydrolyze laminarin and three could attack laminarin plus alginic acid. These isolates were tentatively identified as *Streptomyces* sp. These authors also studied the effects of seawater salts and temperature on the expression of the enzyme(s) from *Streptomyces* XIV, noting maximum concentrations of the laminarin lytic factor at 37°C and 0 to 1 percent NaCl. All these observations indicate a terrestrial origin for the "marine" actinomycetes.

In an attempt to demonstrate the mechanism by which terrestrial forms could enter the marine environment, Okami and Okazaki (71) examined the spore structure of *Actinomyces* and various types of soil from which they could be isolated. The authors were particularly interested in seeking a correlation between the elution of spores from soils and the particular spore structure and appendages that might influence whether these bacteria could be washed into the sea. They found that many spores were precipitated upon contact with seawater, and that many of their isolates were unable to grow, while some were killed at salinities and temperatures approaching *in situ* marine levels. This inquiry lends further credence to the supposition that actinomycetes found in coastal areas, at least, are most likely of terrestrial origin. However, the known cellulolytic and agar-digesting abilities of these microbes (16, 42) would seem to make them logical normal components of the marine microflora.

THE ROLE OF SULFUR AND PHOTOSYNTHETIC BACTERIA IN THE SEA

During the descriptive phase of marine microbiology, it has been amply demonstrated that, with the proper enrichment technique, almost any type of microorganism can be isolated from the ocean environment. One can look for specific degradative abilities, such as cellulose or chitin decomposition (114), or for specific types of bacteria, such as the autotrophic sulfur bacteria (30, 103) or the photosynthetic bacteria (61, 102) or the rosette and exotic pleomorphic designs of the hyphomicrobia and budding bacteria (39). The exact contribution these and many other types of bacteria make to the overall balance in the microbial ecosystem is still not known.

Unfortunately, aside from scattered reports of their isolation, not much else is known about the photosynthetic bacteria, as normally no attempt is made to separate bacterial photosynthetic processes from algal activity. Eimhjellen (21), however, examined the possible contribution of the photosynthetic bacteria to the carotenoid composition of a marine sponge. He isolated species of *Chlorobium,* a purple sulfur bacterium, and *Thiocystis,* a sulfur bacterium, from the spongy material of *Halichondrium panicea.* However, he concluded that the compositional differences between the sponge and the bacteria were too great to permit bacteria to be considered as the source of sponge carotenoids.

Indications that the photosynthetic bacteria may, however, play a major role in the oceans can be found in the report by Matheron and Baulaigue (61), who isolated 15 strains from marine environments and studied their nutritional and physiological needs. Most of the photosynthetic isolates required a minimum of 1 percent NaCl for survival and could photoheterotrophically utilize several organic substrates: pyruvate, lactate, propionate, butyrate, crotonate, acetate, malate, succinate, and formate. Although Brisou (9) lists several isolates from the photosynthetic bacterial genera as occurring in the sea, there is no mention of their role or any of their physiological or nutritional requirements. The thorough study on the culture and isolation of the photosynthetic sulfur bacteria by Trüper (102) should eventually result in further studies of this group of bacteria with great potential importance to the sea.

Kriss (56), on the other hand, considers these photosynthetic microorganisms major contributors to the sulfur cycle inasmuch as they can frequently grow in environments containing high levels of H_2S. Such an environment is anaerobic, frequently high in organic matter, and thus it provides an ideal system for photoheterotrophic growth of the purple sulfur bacteria.

The first of the nonphotosynthetic sulfur bacteria to be isolated from seawater, *Thiobacillus thioparus,* was found by Nathanson in the well-studied Gulf of Naples (68). Marine thiobacilli have since been isolated from seawater in the Caribbean and Carioca Trench (101) and numerous stations in the Atlantic Ocean (101, 103), the Black Sea (46, 56, 103), and nearshore waters by Woods Hole, Massachusetts (103). In the latter investigation, the authors reported that, although isolates could oxidize reduced sulfur compounds, 95 percent could also grow on organic-containing media. Tuttle and Jannasch (103) have further compared the sulfur-oxidizing bacteria from the Black Sea and the Carioca Trench and found them to be fairly similar, although the Black Sea isolates were obtained from slightly greater depths. In both cases, the majority of isolates were obtained from below the oxygen maximum zone.

Research at many laboratories during the last 30 years has demonstrated the occurrence of species of *Desulfovibrio* (35, 56, 58, 91, 114), which is capable of reducing sulfates to sulfides in seawater and sediments. The sulfides are then either precipitated by the heavy metal components of the system or dissolved and reoxidized (46, 51). Several workers have indicated that sulfide production can result from the decomposition of organic matter (69, 74), as well as from the reduction of sulfates. Skyring and Trudinger (92) have analyzed the adenosine-5′-triphosphate sulfurylase, the adenylyl sulfate reductase, and the sulfite reductase from various species of *Desulfovibrio* and *Desulfomaculum,* and shown a rather heterogeneous group of enzymes to be present when analyzed by polyacrylamide gel electrophoresis. There is also one report of "free" arylsulfatase activity in marine sediments with the enzymatic activity presumably originating from bacterial sources (15). From all of these reports, therefore, the bacteria of the sulfur cycle undoubtedly constitute a major fraction of the geochemically active bacteria in the sea.

As a model of the sulfur cycle, the Black Sea has been under intensive investigation by Russian microbiologists for almost 80 years. The literature on the role of marine bacteria in the reduction and reoxidation of sulfurous compounds has been described by Kriss in his chapter "Biochemical Activities of Marine Microorganisms" in *Marine Microbiology (Deep Sea).* All these studies can perhaps best be summarized by the following quotation from that chapter: "the Black Sea represents a kind of gigantic biochemical laboratory which determines the existing hydrochemical regime of this sea" (56). Certainly, the Black Sea does provide a unique environment in the marine world for the study of the interactions of aerobic, facultatively anaerobic, and obligately anaerobic bacteria.

SUBMICROSCOPIC MICROBES

Although at the smaller end of the size scale in the microbial world and, hence, more difficult to culture and visualize, the rickettsia and viruses must not be forgotten. The first of this group, the rickettsia, have been almost totally ignored as contributors to fish and plant diseases. To date, there is no literature on "marine" rickettsia.

Viruses have fared a little better because of the widespread public health interest in the survival of polio and hepatitis viruses in seawaters. Additionally, though, it has been recognized for some time that specific bacteriophage survive and have the potential to invade their host cells in marine waters and sediments. The first report of the isolation of bacteriophage specifically from the sea was by Hauduroy (36) in 1923, who isolated a phage that lysed *Bacillus typhi (Salmonella typhosa).* He obtained water from such diverse sources as offshore, from the Monaco aquarium, the town sewage out-fall, and from around various aquatic plants and animals, for a total of 15 different water samples. Following the successful isolation, he postulated that with further improvements in the technique one should be able to isolate such phages active against bacteria pathogenic to marine animals. In a succeeding paper (37), he reported that water samples held since 1907 were found to be bacteriologically sterile and no bacteriophage could be isolated from them, thus showing a certain lability in seawater.

Three years later Fejgin (22) confirmed the isolation of bacteriophage from seawater, but from his description of the bacterium lysed it is unclear as to whether he was dealing with a marine or human intestinal bacterial isolate. The agglutination reactions with various antisera seem to imply that the organism was *Bacillus typhi* or Shiga's bacillus. ZoBell (114) and many others during the ensuing 20 years tried with only moderate success to isolate bacteriophages. It was a prevailing opinion of the time that the toxicity of seawater to freshwater bacteria, especially coliform types, was the result of bacteriophage activity in the sea. No doubt this resulted from these early studies by Hauduroy and Fejgin.

Despite the early lead in the investigation of marine bacteriophage by the French workers, not much interest was expressed after the 1920 period in this area of research. In fact, in his 1955 book, Brisou (9) does not even mention any later work on the isolation of viruses from the sea.

The first report of the specific isolation of bacteriophage against specific marine bacteria is rather recent. The short but influential paper by Spencer describing the modifications in existing techniques necessary for the successful isolation of marine phages is reprinted here as Paper 41. Phages active against *Photobacterium phosphoreum,* three species of *Pseudomonas,* one *Flavobac-*

terium sp., and one *Cytophaga* sp. were obtained by Spencer. This report was shortly followed by another (98) on some characteristics of the *Photobacterium* phage. A more detailed comparison of isolation methods and the effects of temperature and salinity for adsorption, penetration, and lysis of the host bacterium were also presented by Spencer (99). In general, he found that temperatures above 25°C prevented lysis, and that Na^+ and Mg^{2+} were required for optimal lysis (99). Johnson (47), working with a marine virbriophage system, noted 98 percent adsorption in 15 minutes at 6°C, but none at 25°C (47). In further studies along this line, Wiebe and Liston (111) have examined the combined action of temperature and pressure on phage attachment and lysis. They noted that a temperature range of 5 to 12°C and pressures below 200 atm were required for lysis; Chaina (14) has reported that phage isolated from the Indian Ocean had greater temperature tolerance. An ionic strength comparable to that of seawater has also been found optimal for adsorption by a pseudomonas-specific phage (17).

Kriss (56, 57) has published a comparison of the various techniques for the isolation of marine bacteriophage from the waters and muds of the Black Sea. Although phages were frequently found, his description of the distribution and isolation methods indicate why the reports of successful isolations had been so few until Spencer's paper. In over 2500 water samples examined from the Pacific, Atlantic, and Indian oceans, Kriss and co-workers (58) were able to find zones of clearing in only 603 of the cultures, or 24 percent of the time (range of 2 to 55 percent). Hidaka (38), in extensive studies on marine bacteriophages, has found three predominant morphological types, while noting distinct differences in the types of plaques formed.

Generally, it appears that the marine bacteriophages are also less specific in regards to host attachment and lysis, inasmuch as most of the isolates lyse pseudomonads, vibrios, and/or aeromonads (14, 31, 38, 99), although some viruses do show a degree of specificity (53, 113). In one of the most recent reports, a transducing phage was isolated. This virus has a specificity for *Photobacterium harveyi* [*Lucibacterium harveyi* (12)] and is reported to carry several of the genes of the tryptophan pathway (53). This is a significant finding for at least two reasons: (1) theoretically, bacteriophage can act as regulators of bacterial population levels, and (2) since phages may also act as agents for the transmission of specific physiological characteristics, such as the one just mentioned, they can serve to aid in the survival of the lysogenic host cells. It will certainly be of interest to see to what extent this phenomenon can occur in the oceans and perhaps affect our current concepts of bacterial distributions and survival in the sea.

SURVEY OF SEDIMENTARY MICROBIAL POPULATIONS

Several of the previous papers have been concerned with the isolation and characterization of the microbial flora in the sediments. Lloyd (Paper 7) was among the first to make quantitative analyses of bacteria in sediments, and ZoBell and Morita (Paper 24) not only examined bacteria that could grow under conditions of elevated hydrostatic pressures, but they also noted the different physiological types to be found there: ammonifiers were more frequently enumerated by their methods than nitrate reducers, which were more numerous than starch hydrolyzers or sulfate reducers. In addition, they tested for these activities under various pressures approximating those found *in situ*, and also examined the sediments for anaerobes (119). At least those bacteria which survived the journey from 10,000 m were generally facultative anaerobes, although the authors do report that some clostridia were found (119). Earlier, Rittenberg (78, 116) had investigated the total bacterial flora of some long cores (up to 355 cm long), during which studies he used the roll tube method for the isolation of anaerobes. In all the cases studied, though, he had to expose the cells to molten agar; thus his total numbers are undoubtedly low, although the relative proportion of aerobic forms to anaerobic forms is probably about the same. Surprisingly, he found aerobes were 10 to 100 times more numerous than anaerobes (77), thus confirming the earlier work of ZoBell and Anderson (115), who examined sediments from off the California coast and noted that proteolysis was the most frequently encountered characteristic, and that aerobic bacteria greatly outnumbered the anaerobic bacteria. However, most of the aerobic isolates were really facultatively anaerobic and only at great core depths did anaerobes ever become predominant. Despite colder overlying water, even thermophiles have been isolated from deep ocean sediments (5).

It has been repeatedly shown that pH, Eh, and sediment composition affect the types and total numbers of bacteria found (19, 59, 60, 73, 112, 117). Thus, the early studies of ZoBell and Feltham (117) and Waksman and co-workers (110) have become the standards for the study of marine sediments because of their emphasis on investigating the sediment ecosystem as well as the bacterial flora. Liston, in Paper 42, has amply demonstrated this integrated approach by examining not only open ocean sediments, but also nearshore and Puget Sound sediments. Moreover, in this work, he studied the interrelationship between cellulose addition to model sediments and the appearance of cellulose-degrading organisms. He also considered the dependency of the protozoan population on the bacterial flora. From these observations, he concluded that the role of sedimentary bacteria was twofold: (1) they function as mineralizers of organic matter, and (2) they act as primary converters for the benthic food chain, serving as a major food source.

DISTRIBUTION OF BACTERIA IN THE WATER COLUMN

During the last 90 years great effort has been expended to determine which bacteria are present in the sediments and ocean waters. The results of all these descriptive studies can best be summarized by a rereading of Russell (80), who noted in 1892 that the numbers of bacteria decrease with increasing distance from land and that the numbers also decrease with increasing depth. What Russell was unable to determine, though, was at what rates microbial activity was occurring. From all these studies, the concept of a baseline level of bacteria in the various marine environments is emerging. In recognition of this background concentration of bacteria, many investigators have sought indicator bacteria or biochemical properties which might typify these organisms. For example, Kriss (58) has postulated the use of *Bacterium agile* as one such indicator bacterium for different water masses.

In Paper 43, Kriss, Abyzov, Lebedeva, Mishustina, and Mitskevich examined over 4000 water samples from throughout the oceans of the world. The authors concluded that greater numbers of microorganisms could be found in tropical waters than in colder water masses. However, it must be noted that the incubation temperatures they employed were 18 to 35°C, which would certainly prove lethal to the obligate psychrophiles one might reasonably expect to find in the colder waters of the higher latitudes. The authors also attempted to correlate these distribution patterns with the presumed level of "semidigested" organic matter, stating that the allochthonous organic material is higher in the tropical regions and, therefore, more bacteria must occur there. This finding has not been confirmed by other investigators (34, 40, 86, 94).

In an attempt to define more precisely the bacterial distribution patterns in the oceans by using the best techniques available, such as aseptic sampling with nontoxic containers, incubation at *in situ* temperatures, and biochemical testing of the isolates, Sieburth examined more than 380 water samples from both the Pacific and Atlantic oceans and the Caribbean Sea. This extensive and thorough study resulted in Paper 44. In summary, Sieburth was unable to substantiate the findings of Kriss regarding the decrease in numbers of bacteria at higher latitudes or a decrease in the biochemical activity of the isolates. From a comparison of the two papers, it is obvious that great attention to detail and a thorough study of the isolated organisms are essential if one is to draw conclusions about the microbial flora of different water masses. This basic concept had been amply stated in 1939 when Butkevich and Bogdanova (13) emphatically stated, "Correct results in this field may be obtained only upon investigating fresh material or material kept under conditions close to those of the natural medium."

In previous papers dealing with the bacterial flora of the water column, especially Papers 4, 7, 31, and 37, the authors were primarily concerned with the total numbers of bacterial heterotrophs present. By now it should be clear that the chemical composition and the hydrography of the water must be considered before any meaningful conclusions can be drawn about the bacterial role in the system. Also, as we see from Paper 44, these types of studies must include identification of, at least, the major physiological types of bacteria. An example of the types of baseline survey that are currently underway is to be found in a study by Taga, Paper 45. In this paper, which is reprinted here except for pages on the statistical treatment of the data and photographs of the cruise track and aseptic sampler, Taga has integrated all these parameters. He has also examined four depths every 4½ hours at an oceanic station to obtain some idea of the diurnal variation in the bacterial count. Diurnal fluctuations do exist and seem to be greatest in the surface and 50-m samples; the total numbers of colony-forming units remain relatively constant below this depth.

Other investigators are beginning to follow this pattern of intensive study and even include such additional details as the ATP content and/or the mineralization or heterotrophic potential of the system (34, 81). Other measurements that are important because they frequently have a positive correlation with traditional microbial measurements are dissolved organic carbon, particulate organic carbon, nitrate-nitrite-ammonium, phosphate, and the organic nitrogen content of the water. Besides the above measurements, in sedimentary analyses it is important to include the pH, redox potential, percent moisture of the system, and, perhaps, some information about the type of sediment (sandy, clay, silt, etc.).

Steps toward this total system analysis have been made by Hobbie et al. (40) in their analysis of two stations in the Atlantic Ocean where they measured ATP, heterotrophic uptake of three labeled compounds, bacterial and phytoplankton biomass, the respiratory electron-transport system of the active organism, oxygen uptake, CO_2, DNA, chlorophyll and pheophytin pigments, salinity, oxygen, and temperature. Although only tentative relationships could be drawn because of the uniqueness of this cruise, it is obvious that the day of the lone microbiologist on a cruise is over. It now takes many people and much effort to obtain meaningful information. Thus, the modern marine microbiologist must be knowledgeable about the environment of the microorganisms as well as about which microbes are there if he expects to understand the function and rate of activity of marine microorganisms in the ocean.

REFERENCES

1. Ahearn, D. G. 1973. Effects of environmental stress on aquatic yeast populations, pp. 433–440. In: L. H. Stevenson and R. R. Colwell (eds.), *Estuarine Microbial Ecology*. University of South Carolina Press, Columbia, S.C., 536 pp.
2. Ahearn, D. G., and Roth, F. J., Jr. 1962. Vitamin requirements of marine-occurring yeasts. *Dev. Ind. Microbiol. 3*:163 173.
3. Alderman, D. J., and Jones, E. B. G. 1971. Physiological requirements of two marine Phycomycetes, *Althornia crouchii* and *Ostracoblabe implexa. Trans. Br. Mycol. Soc. 57*:213–225.
4. Barghoorn, E. S. and Linder, D. H. 1944. Marine fungi: their taxonomy and biology. *Farlowia 1*:395–467.
5. Bartholomew, J. W., and Rittenberg, S. C. 1949. Thermophilic bacteria from deep ocean bottom cores. *J. Bacteriol. 57*:658.
6. Belsky, M. M., Goldstein, S., and Menna, M. 1970. Factors affecting phosphate uptake in the marine fungus *Dermocystidium* sp. *J. Gen. Microbiol. 62*:399–402.
7. Borut, S. Y., and Johnson, T. W., Jr. 1962. Some biological observations on fungi in estuarine sediments. *Mycologia 54*:181–193.
8. Breen, G., and Goldstein, S. 1975. Identification of the marine phycomycete isolate "S-3." *Mycologia 67*:177–180.
9. Brisou, J. 1955. *La Microbiologie du milieu marin*. Éditions Médicales Flammarion, Paris, 272 pp.
10. Brooks, R. D., Goos, R. D., and Sieburth, J. McN. 1972. Fungal infestation of the surface and interior vessels of freshly collected driftwood. *Mar. Biol. 16*:274–278.

11. Bruce, J., and Morris, E. O. 1973. Psychrophilic yeasts isolated from marine fish. *Antonie van Leeuwenhoek J. Microbiol. Serol. 39*:331–339.
12. Buchanan, R. E., and Gibbons, N. E. 1974. *Bergey's Manual of Determinative Bacteriology*, 8th ed. The Williams & Wilkins Co., Baltimore, Md., 1268 pp.
13. Butkevich, V. S., and Bogdanova, I. V. 1939. Some specificities of the bacterial population of Arctic Seas. *Microbiology USSR 8*:1073–1095.
14. Chaina, P. N. 1965. Some recent studies on marine bacteriophages. *J. Gen. Microbiol. 41*:XXV.
15. Chandramohan, D., Devendran, K., and Natarajan, R. 1974. Arylsulfatase activity in marine sediments. *Mar. Biol 27*:89–92.
16. Chesters, C. G. C., Apinis, A., and Turner, M. 1956. Studies of the decomposition of seaweeds and seaweed products by microorganisms. *Proc. Linn. Soc. Lond. 166*:87–97.
17. Cota-Robles, E. H., Gregory, J. P., and Rucinsky, T. E. 1974. Adsorption of the marine bacteriophage PM2 to its host bacterium, pp. 457–466. In: R. R. Colwell and R. Y. Morita (eds.), *Effect of the Ocean Environment on Microbial Activities*. University Park Press, Baltimore, Md., 587 pp.
18. Couch, J. N. 1942. A new fungus on crab eggs. *J. Elisha Mitchell Sci. Sco. 58*:158–162.
19. Dale, N. G. 1974. Bacteria in intertidal sediments: factors related to their distribution. *Limnol. Oceanogr. 19*:509–518.
20. Davis, H. C., Loosanoff, V. L., Weston, W. H., and Martin, C. 1954. A fungus disease in clam and oyster larvae. *Science 120*:36–38.
21. Eimhjellen, K. E. 1967. Photosynthetic bacteria and carotenoids from a sea sponge *Halichondrium panicea*. *Acta Chem. Scand. 21*:2280–2281.
22. Fejgin, B. 1926. Études sur les microbe marins. 2. Étude sur la forme imperceptible des bactéries dans l'eau de mer. *Bull. Inst. Oceanogr. No. 484*:1–7.
23. Fell, J. W. 1967. Distribution of yeasts in the Indian Ocean. *Bull. Mar. Sci. 17*:454–470.
24. Fell, J. W. 1974. Distributions of yeasts in water masses of the southern oceans, pp. 510–523. In: R. R. Colwell and R. Y. Morita (eds.), *Effects of the Ocean Environment on Microbial Activities*. University Park Press, Baltimore, Md., 587 pp.
25. Fell, J. W., and Master, I. M. 1973. Fungi associated with the degradation of mangrove (*Rhizophora mangle* L.) leaves in south Florida, pp. 455–465. In: L. H. Stevenson and R. R. Colwell (eds.), *Estuarine Microbial Ecology*. University of South Carolina Press, Columbia, S.C., 536 pp.
26. Fell, J. W., and Van Uden, N. 1963. Yeasts in marine environments, pp. 329–340. In: C. H. Oppenheimer (ed.), *Symposium on Marine Microbiology*. Charles C Thomas, Publisher, Springfield, Ill., 769 pp.
27. Fischer, B. 1894. Die Bakterien des Meeres nach den Untersuchungen der Plankton-Expedition. *Ergeb. Plankton-Exped. Humbolt-Stiftung 4*:1–83.

28. Fischer, B., and Brebeck, C. 1894. *Zur Morphologie, Biologie, und Systematik der Kahmpilze, der* Monitia candida. Hansen und des Soorerregus. G. Fischer, Jena.
29. Gessner, R. V., Goos, R. D., and Sieburth, J. McN. 1972. The fungal microcosm of the internodes of *Spartina alterniflora. Mar. Biol. 16*: 269–273.
30. Gietzen, J. 1931. Untersuchungen über marine Thiorhodaceen. *Zentralbl. Bakteriol., Abt. II, 83*:183–217.
31. Gill, M. L., and Nealson, K. 1972. Isolation and host range studies of marine bacteriophage. *Biol. Bull. 143*:463–464.
32. Gold, H. S. 1959. Distribution of some lignicolous Ascomycetes and Fungi Imperfecti in an estuary. *J. Elisha Mitchell Sci. Soc. 75*:26–28.
33. Grein, A., and Meyers, S. P. 1958. Growth characteristics and antibiotic production of Actinomycetes isolated from littoral sediments and materials suspended in sea water. *J. Bacteriol 76*:457–463.
34. Gunderson, K., Mountain, C. W., Taylor, D., Ohye, R., and Shen, J. 1972. Some chemical and microbiological observations in the Pacific Ocean off the Hawaiian Islands. *Limnol. Oceanogr. 17*:524–531.
35. Hata, Y. 1960. Influence of heavy metals upon the growth and the activity of marine sulfate-reducing bacteria. *J. Shimonoseki College Fish. 9*:363–375.
36. Hauduroy, P. 1923. I. Recherches sur le bactériophage de D'Hérelle (Présence du principe dans l'eau de mer). *Bull. Inst. Oceanogr. No. 433*:1–12.
37. Hauduroy, P. 1923. Recherches du bactériophage de D'Hérelle dans de l'eau de mer conservée seize ans au laboratorie (IIe Note). *Bull. Inst. Oceanogr. No. 434*:1–2.
38. Hidaka, T. 1971. Isolation of marine bacteriophages from sea water. *Bull. Jap. Soc. Sci. Fish. 37*:1199–1206.
39. Hirsch, P., and Rheinheimer, G. 1968. Biology of budding bacteria. V. Budding bacteria in aquatic habitats: occurrence, enrichment, and isolation. *Arch. Mikrobiol. 62*:289–306.
40. Hobbie, J. E., Holm-Hansen, O., Packard, T. T., Pomeroy, L. R., Sheldon, R. W., Thomas, J. P., and Wiebe, W. J. 1972. A study of the distribution and activity of microorganisms in ocean water. *Limnol. Oceanogr. 17*:544–555.
41. Hughes, G. C. 1960. Ecological aspects of some lignicolous fungi in estuarine waters. *Diss. Abst. 60-05499*:2086.
42. Humm, H. J., and Shepard, K. S. 1946. Three new agar-digesting actinomycetes. *Duke Univ. Mar. Sta. Bull. 3*:76–80.
43. Jones, E. B. G. 1962. Marine fungi. *Trans. Br. Mycol. Soc. 45*:93–114.
44. Jones, E. B. G. 1963. Observations on the fungal succession on wood test blocks submerged in the sea. *J. Inst. Wood Sci. 11*:14–23.
45. Jones, E. B. G., and Eltringham, S. K. 1971. *Marine Borers, Fungi and Fouling Organisms of Wood,* No. 27-923. Organization for Economic Co-operation and Development, Paris, 367 pp.
46. Jones, G. E., and Starkey, R. L. 1957. Fractionation of stable isotopes of sulfur by microorganisms and their role in deposition of native sulfur. *Appl. Microbiol. 5*:111–118.
47. Johnson, R. M. 1968. Characteristics of a marine vibrio-bacteriophage system. *J. Arizona Acad. Sci. 5*:28–33.

48. Johnson, T. W., Jr. and Meyers, S. P. 1957. Literature on halophilous and halolimnic fungi. *Bull. Mar. Sci. Gulf Caribbean 7*:330–359.
49. Johnson, T. W., Jr., Ferchau, H. A., and Gold, H. S. 1959. Isolation, culture, growth and nutrition of some lignicolous marine fungi. *Phyton 12*:65–80.
50. Johnson, T. W., Jr., and Sparrow, F. K., Jr. 1961. *Fungi in Oceans and Estuaries*. Hafner Publishing Co., New York, 685 pp.
51. Kaplan, I. R., and Rittenberg, S. C. 1962. The microbiological fractionation of sulfur isotopes, pp. 80–93. In: M. L. Jensen (ed.), *Symposium on Biogeochemistry of Sulfur Isotopes*.
52. Kazama, F. Y., and Fuller, M. S. 1973. Mineral nutrition of *Pythium marinum*, a marine facultative parasite. *Can. J. Bot. 51*:693–699.
53. Keynan, A., Nealson, K., Sideropoulos, H., and Hastings, J. W. 1974. Marine transducing bacteriophage attacking a luminous bacterium. *J. Virol. 14*:333–340.
54. Kohlmeyer, J. 1963. The importance of fungi in the sea, pp. 300–314. In: C. H. Oppenheimer (ed.), *Marine Microbiology*. Charles C Thomas, Publisher, Springfield, Ill., 769 pp.
55. Kohlmeyer, J. 1972. Marine fungi deteriorating chitin of hydrozoa and keratin-like annelid tubes. *Mar. Biol. 12*:277–284.
56. Kriss, A. E. 1962. *Marine Microbiology (Deep Sea)*, J. M. Shewan and Z. Kabata (trans.). Wiley-Interscience, New York, 536 pp.
57. Kriss, A. E., and Rukina, E. A. 1947. Bacteriophage in the sea. *Akad. Nauk. SSSR Dokl. 57*:833–836.
58. Kriss, A. E., Mishustina, I. E., Mitskevich, I. N., and Zemtsova, E. V. 1967. In: K. Syers (trans.) and G. E. Fogg (ed.), *Microbial Population of Oceans and Seas*. St. Martin's Press, New York, 287 pp.
59. Lindblom, G. P., and Lupton, M. D. 1961. Microbiological aspects of organic geochemistry. *Dev. Ind. Microbiol. 2*:9–22.
60. Litchfield, J. H., and Roppel, R. M. 1961. The oxidative metabolism of *Rhodotorula* and *Cryptococcus* species isolated from a marine environment. *Dev. Ind. Microbiol. 2*:283–290.
61. Matheron, R., and Baulaigue, R. 1972. Bactéries photosynthétiques sulfureuses marines. *Arch. Mikrobiol. 86*:291–304.
62. Meyers, S. P. 1968. Observations on the physiological ecology of marine fungi. *Bull. Misaki Mar. Biol. Inst., Kyoto Univ. 12*:207–225.
63. Meyers, S. P., Ahearn, D. G., Gunkel, W., and Roth, F. J., Jr. 1967. Yeasts from the North Sea. *Mar. Biol. 1*:118–123.
64. Meyers, S. P., Ahearn, D. G., and Roth, F. J., Jr. 1967. Mycological investigations of the Black Sea. *Bull. Mar. Sci. 17*:576–596.
65. Meyers, S. P., Nicholson, M. L., Rhee, J., Miles, P., and Ahearn, D. G. 1970. Mycological studies in Barataria Bay, Louisiana, and biodegradation of oyster grass, *Spartina alterniflora*. *Coastal Stud. Bull. 5*:111–124.
66. Meyers, S. P., and Reynolds, E. S. 1959. Cellulolytic activity in lignicolous marine ascomycetes. *Bull. Mar. Sci. Gulf Caribbean 9*:441–455.
67. Morii, H. 1973. Yeasts predominating in the stomach of marine little toothed whales. *Bull. Jap. Soc. Sci. Fish. 39*:333.
68. Nathanson, A. 1902. Ueber eine neue Gruppe von Schwelfebacterien und ihren Stoffwechsel. *Mitt. Zool. Sta. Neapel 15*:655–680.

69. Nedwell, D. B., and Floodgate, G. F. 1972. Temperature-induced changes in the formation of sulphide in a marine sediment. *Mar. Biol. 14*:18–24.
70. Newell, S. Y. 1973. Succession and role of fungi in the degradation of red mangrove seedlings, pp. 467–480. In: L. H. Stevenson and R. R. Colwell (eds.), *Estuarine Microbial Ecology*. University of South Carolina Press, Columbia, S.C., 536 pp.
71. Okami, Y., and Okazaki, T. 1974. Studies on marine microorganisms. III. Transport of spores of actinomycetes into shallow sea mud and the effect of salt and temperature on their survival. *J. Antibiot. 27*: 240–247.
72. Okazaki, T., and Okami, Y. 1972. Studies of marine microorganisms. II. Actinomycetes in Sagami Bay and their antibiotic substances. *J. Antibiot. 25*:461–466.
73. Oppenheimer, C. H. 1960. Bacterial activity in sediments of shallow marine bays. *Geochim. Cosmochim. Acta 19*:244–260.
74. Ramm, A. E., and Bella, D. A. 1974. Sulfide production in anaerobic microcosms. *Limnol. Oceanogr. 19*:110–118.
75. Renn, C. E. 1936. The wasting disease of *Zostera marina*. I. A phytological investigation of the diseased plant. *Biol. Bull. 70*:148–158.
76. Ritchie, D. 1957. Salinity optima for marine fungi affected by temperature. *Am. J. Bot. 44*:870–874.
77. Rittenberg, S. C. 1940. Bacterial analysis of some long cores of marine sediments. *J. Mar. Res. 3*:191–201.
78. Rittenberg, S. C., Anderson, D. Q., and ZoBell, C. E. 1937. Studies on the enumeration of marine anaerobic bacteria. *Proc. Soc. Exp. Biol. Med. 35*:652–653.
79. Roth, F. J., Jr., Ahearn, D. G., Fell, J. W., Meyers, S. P., and Meyer, S. A. 1962. Ecology and taxonomy of yeasts isolated from various marine substrates. *Limnol. Oceanogr. 7*:178–185.
80. Russell, H. L. 1892. Untersuchungen über im Golf von Neapel lebende Bacterien. *Zeit. Hyg. Infectionskr. 11*:165–206.
81. Seki, H., Wada, E., Koike, I., and Hattori, A. 1974. Evidence of high organotrophic potentiality of bacteria in the deep sea. *Mar. Biol. 26*:1–4.
82. Sguros, P. L., Meyers, S. P., and Simms, J. 1962. Role of marine fungi in the biochemistry of the oceans. I. Establishment of quantitative technique for cultivation, growth measurement and production of inocula. *Mycologia 54*:521–535.
83. Sguros, P. L., and Simms, J. 1963. Role of marine fungi in the biochemistry of the oceans. III. Growth factor requirements of the ascomycete *Halosphaeria mediosetigera*. *Can. J. Microbiol.* 9:585–591.
84. Sguros, P. L., and Simms, J. 1964. Role of marine fungi in the biochemistry of the oceans. IV. Growth responses to seawater inorganic macroconstituents. *J. Bacteriol. 88*:346–355.
85. Sguros, P. L., Rodrigues, J., and Simms, J. 1973. Role of marine fungi in the biochemistry of the oceans. V. Patterns of constitutive nutritional growth responses. *Mycologia 65*:161–174.
86. Sieburth, J. McN. 1971. Distribution and activity of oceanic bacteria. *Deep-Sea Res. Oceanogr. Abst. 18*:1111–1121.

87. Sieburth, J. McN., Brooks, R. D., Gessner, R. V., Thomas, C. D., and Tootle, J. L. 1974. Microbial colonization of marine plant surfaces as observed by scanning electron microscopy, pp. 418–432. In: R. R. Colwell and R. Y. Morita (eds.), *Effect of the Ocean Environment on Microbial Activity*. University Park Press, Baltimore, Md., 587 pp.
88. Sieburth, J. McN. 1975. *Microbial Seascapes*. University Park Press, Baltimore, Md.
89. Siegenthaler, P. A., Belsky, M. M., and Goldstein, S. 1967. Phosphate uptake in an obligately marine fungus: a specific requirement for sodium. *Science 155*:93–94.
90. Singal, E. M., Mishustina, I. E., Rudaya, S. M., and Solovieva, N. K. 1973. Actinomycete populations in some soils of equatorial tropical zones of the Indian Ocean. *Antibiotiki (Mosc.) 18*:605–608.
91. Sisler, F. D., and ZoBell, C. E. 1950. Hydrogen-utilizing, sulfate-reducing bacteria in marine sediments. *J. Bacteriol. 60*:747–756.
92. Skyring, G. W., and Trudinger, P. A. 1973. A comparison of the electrophoretic properties of the ATP-sulfurylases, APS-reductases, and sulfite reductases from cultures of dissimilatory sulfate-reducing bacteria. *Can. J. Microbiol. 19*:375–380.
93. Smith, L. S., and Krueger, A. P. 1954. Characteristics of a new vibrio-bacteriophage system. *J. Gen. Physiol. 38*:161–173.
94. Sorokin, Yu. I. 1964. On the primary production and bacterial activities in the Black Sea. *J. Cons. Perm. Int. Explor. Mer 29*:41–60.
95. Sorokin, Yu. I. 1970. Experimental investigation of the rate and mechanism of oxidation of hydrogen sulfide in the Black Sea using ^{35}S. *Okeanologiya 10*:51–61.
96. Sorokin, Yu. I. 1971. On the role of bacteria in the productivity of tropical ocean waters. *Int. Rev. Gesamten Hydrobiol. 56*:1–48.
97. Sparrow, F. K., Jr. 1934. Observations on marine Phycomycetes collected in Denmark. *Dan. Bot. Ark. 8*:1–33.
98. Spencer, R. 1955. A marine bacteriophage. *Nature 175*:690–691.
99. Spencer, R. 1962. Bacterial viruses in the sea, pp. 350–365. In: C. H. Oppenheimer (ed.), *Marine Microbiology*. Charles C Thomas, Publisher, Springfield, Ill., 769 pp.
100. Szaniszlo, P. J., and Mitchell, R. 1971. Hyphal wall compositions of marine and terrestrial fungi of the genus *Leptosphaeria*. *J. Bacteriol. 106*:640–645.
101. Tilton, R. C., Cobet, A. B., and Jones, G. E. 1967. Marine thiobacilli. I. Isolation and distribution. *Can. J. Microbiol. 13*:1521–1528.
102. Trüper, H. G. 1970. Culture and isolation of phototrophic sulfur bacteria from the marine environment. *Helgoländer Wiss. Meeresunters. 20*:6–16.
103. Tuttle, J. H., and Jannasch, H. W. 1972. Occurrence and types of thiobacillus-like bacteria in the sea. *Limnol. Oceanogr. 17*:532–543.
104. Tuttle, J. H., and Jannasch, H. W. 1973. Sulfide and thiosulfate-oxidizing bacteria in anoxic marine basins. *Mar. Biol. 20*:64–70.
105. Van Uden, N., and Fell, J. W. 1968. Marine yeasts, pp. 167–202. In: M. R. Droop and E. J. F. Wood (eds.), *Advances in Microbiology of the Sea*, Vol. 1. Academic Press, New York, 239 pp.
106. Vishniac, H. S. 1955. Marine mycology. *Trans. N.Y. Acad. Sci. 17*: 352–360.

107. Vishniac, H. S. 1956. On the ecology of the lower marine fungi. *Biol. Bull. 111*:410–414.
108. Vishniac, H. S. 1961. A biological assay for thiamine in sea water. *Limnol. Oceanogr. 6*:31–35.
109. Vishniac, H. S., and Riley, G. A. 1961. Cobalamin and thiamine in Long Island Sound: patterns of distribution and ecological significance. *Limnol. Oceanogr. 6*:36–41.
110. Waksman, S. A., Reuszer, H. W., Carey, C. L., Hotchkiss, M., and Renn, C. E. 1932. Studies on the biology and chemistry of the Gulf of Maine. III. Bacteriological investigations of the seawater and marine bottoms. *Biol. Bull. 64*:183–205.
111. Wiebe, W. J., and Liston, J. 1968. Isolation and characterization of a marine bacteriophage. *Mar. Biol. 1*:244–249.
112. Wood, E. J. F. 1953. Heterotrophic bacteria in marine environments of Eastern Australia. *Aust. J. Mar. Fresh. Res. 4*:160–200.
113. Zachary, A. 1974. Isolation of bacteriophages of the marine bacterium *Beneckea natriegens* from coastal salt marshes. *Appl. Microbiol. 27*:980–982.
114. ZoBell, C. E. 1946. *Marine Microbiology*. Chronica Botanica, Waltham, Mass., 240 pp.
115. ZoBell, C. E., and Anderson, D. Q. 1936. Vertical distribution of bacteria in marine sediments. *Bull. Am. Assoc. Petrol. Geol. 20*:258–269.
116. ZoBell, C. E., and Anderson, D. Q. 1940. Observations on marine anaerobes in oval tubes. *J. Bacteriol. 36*:253.
117. ZoBell, C. E., and Feltham, C. B. 1942. The bacterial flora of a marine mud flat as an ecological factor. *Ecology 23*:69–78.
118. ZoBell, C. E., and Morita, R. Y. 1957. Barophilic bacteria in some deep sea sediments. *J. Bacteriol. 73*:563–568.
119. ZoBell, C. E., and Morita, R. Y. 1959. Deep-sea bacteria. *Galathea Rep. Copenhagen 1*:139–154.

38

Reprinted from pp. 395–401, 436–440, 452, 456, 463–465 of *Farlowia*, **1**(3), 395–467 (1944)

MARINE FUNGI: THEIR TAXONOMY AND BIOLOGY [1]

E. S. Barghoorn and D. H. Linder

INTRODUCTION

Although thousands of species of fungi are known from what may be considered terrestrial habitats, and a much lesser number from freshwater habitats, very few have been described from marine environments, although a number of the latter do exist. Cotton (1), in 1907, in summarizing the situation, noted that fifteen species had been described as marine forms, and to these he added a sixteenth. Of these species, six were members of the phycomycetous order Chytridiales and ten of the ascomycetous sub-class Pyrenomycetes. Since Cotton's paper, some of the species have been relegated to synonymy, but in 1915 Sutherland (14, 15, 16) described fourteen new species of Pyrenomycetes, two of which were the types of new genera. A year later, he (17) added a number of new species of Fungi Imperfecti. Petersen (6) made a study of the chytridiaceous forms parasitic on algae, and more recently Sparrow (11, 12, 13) has added to our knowledge of that group. The only marine fungi that have been published to date as occurring on plant remains, other than algae, are three species of *Ophiobolus* described by Ellis (2), Saccardo (10), and Mounce and Diehl (5). The last named authors described *Ophiobolus halimus* which apparently caused considerable damage to the eelgrass, *Zostera marina*. Subsequently Renn (7, 8, 9) demonstrated that a species of *Labyrinthula* (Myxomycete), later determined by Young (19) as *L. macrocystis* Cienk., actually was the agent that caused the destruction of eelgrass over a considerable geographic range. In view of these later findings, there arises the thought that *Ophiobolus halimus* may be a secondary parasite or indeed might even be saprophytic, thus placing the species in the same category as those which are to be considered in the present paper.

With the exception of the papers of Saccardo, Ellis, and Mounce and Diehl, there is, as far as the authors are aware, only one published reference to the occurrence of marine fungi on phanerogamic plant remains. This is a brief note by Johnson and McNeil (4) indicating the presence

[1] This study is a joint endeavor arising from a project undertaken by E. S. Barghoorn on the microbiological factors responsible for the decomposition of submerged plant remains in the sea. D. H. Linder has made all contributions to the taxonomy and relationships of the fungi described. Problems encountered in the isolation and culturing of the various species have been worked out together. E. S. Barghoorn has undertaken the field work, has gathered together most of the specimens and pertinent collection data, has made the isolations of pure cultures and has carried out the experimental studies.

The present paper is one of a contemplated series of investigations on the various microbiological, chemical and physical factors involved in the decomposition or preservation of submerged plant materials.

of a seawater fungus in the decomposition of hardwood timbers in the port of Sydney, Australia. No description of the fungus is made beyond mention of the fact that it was referred to the class Ascomycetes and that "some timbers are weakened by a softening of the surface fibres to a depth of one-quarter of an inch of truewood in a period of seven years immersion."

The fact that a score or more of species have been described as occurring in the sea is of importance since it shows that fungi not only tolerate salt water, but indeed that marine conditions furnish a normal habitat for the relatively small number of fungi that have become adapted to it.

The fungi considered in this study differ from the majority of those already mentioned in that they do not occur on algae, but on wood and plant remains in the sea. All of the species were obtained from specimens of wood that had been either submerged continually from five months to one year and were below low tide level for the entire period, or were exposed in the inter-tidal zone where they were subjected twice a day to immersion in seawater. Under these conditions, it is highly improbable that the fungi are terrestrial forms which penetrated the wood after it had been removed from the sea, or before it had been immersed. This fact becomes evident when it is realized that the majority of the samples were examined within one to three days of their removal from the marine location, a time much too short to permit the formation and fruiting of the ascigerous stage, and even too brief a time for the establishment of any extensive colonies of the conidial phases. Another convincing line of evidence that these fungi are normally marine in their occurrence is the fact that the majority of them are very tolerant of sea water, even to the extent of growing on artificial media made with sea water of three times the normal saline concentration, as will be shown later in this study. For the present, the collection data given below may serve to give some insight into the conditions under which the various samples of wood and cordage have become inoculated by the fungi.

SPECIMEN COLLECTION DATA

Specimen	*Location*	*Submerged*	*Removed*	*Conditions*	*Type of Specimen*
1	Old Concord Wharf, Portsmouth, New Hampshire.	Sept. 23/41	May 19/42	4 ft. below low water.	Test block.
2	Searsport, Maine.	Sept. 30/41	May 26/42	5 ft. below low water.	Test block.
3	Woods Hole, Mass.	Oct. 13/41	June 13/42	7–8 ft. below low water. 5 ft. above mud.	Test block.
4	Baylies Wharf, Fall River, Mass.	Feb. 1942	July 18/42	8 ft. below low water.	Test board.
5	Tide flats, No. Truro. 200 ft. from shore at low tide.	?	June, 1942	Covered by 8 ft. of water at high tide.	Driftwood embedded in sand.
6	Chelsea, Mass.	1918	May, 1942	Near high tide line.	Pile chip.

SPECIMEN COLLECTION DATA—*Continued*

Specimen	*Location*	*Submerged*	*Removed*	*Conditions*	*Type of Specimen*
7	Same as 5.	?	June, 1942	Same as 5.	Driftwood.
8	Same as 5.	?	June, 1942	Same as 5.	Driftwood.
9	Same as 5.	?	June, 1942	Same as 5.	Driftwood.
10	Woods Hole, Mass	March 6/42	Aug. 5/42	Below low water at all times.	Test board.
11	B. & M. Bridge, Portsmouth, N. H.	Dec. 29/41	Aug. 25/42	Below low water at all times.	Test block.
13	Bucksport, Maine.	Dec. 19/41	Aug. 25/42	Below low water at all times.	Test block.
14	Provincetown, Mass.	1933?	July 24/42	2 ft. above low water.	Pile chip. Cold Storage Wharf.
15	Woods Hole, Mass.	April 6/42	Sept. 5/42	Below low water at all times.	Test board.
16	Salem, Mass.	May, 1942	Sept. 18/42	Below low water at all times.	Test board.
17	Saybrook, Conn.	May 1/42	Oct. 1/42	Below low water at all times.	Test board.
18	Provincetown, Mass	?	July 24/42	Tideflats. Covered 8 ft. at high water.	Timber embedded in sand
19	Newburyport, Mass.	May 25/42	Oct. 16/42	Below low water at all times. Salinity low.	Test board.
20	Woods Hole, Mass.	?	Aug. 24/42	Intertidal zone.	Manila rope.
21	Rockland, Maine.	March 10/42	Nov. 18/42	Below low water at all times.	Test block.
22	Thomaston, Maine.	Sept., 1938	Nov. 27/42	Below low water at all times.	Back board for test blocks.
23	Provincetown, Mass.	?	Sept. 24/42	3 ft. above low water.	Submerged stump.
24	Provincetown, Mass.	?	Sept. 24/42	3 ft. above low water.	Submerged stump (bark).
25	Provincetown, Mass.	?	Sept. 24/42	3 ft. above low water.	Submerged stump.
26	Provincetown, Mass.	?	Sept. 24/42	3 ft. above low water.	Submerged stump.
27	Provincetown, Mass.	1900 or 1901	Sept. 24/42	4–5 ft. above low water.	Jetty bulkhead E. Breakwall, Provincetown.
28	Provincetown, Mass.	?	Sept. 24/42	2 ft. above low water.	Driftwood.
29	Provincetown, Mass.	?	Sept. 24/42	2 ft. above low water.	Driftwood.
30	Bridgeport, Conn	Dec. 8/41	Dec. 21/42	Below low water.	Test board.

Since these fungi appear to be terrestrial forms that have become adapted to marine conditions, it may not be amiss at this time to point out briefly some of the modifications that have occurred which appear to make these forms better adjusted to their present environment. One of these changes which, with few exceptions, appears to hold for all species belonging in the Pyrenomycetes, is that the ascus wall breaks down at an early stage, very shortly after the ascospores have been delimited. In some cases, the ascus wall can be observed to be deeply eroded by the time that the spores have been delimited. Exactly why dissolution of the ascus wall is so common is difficult to conjecture, although it may be presumed that with the breakdown of the wall, there is an increase in the amount of hygroscopic material which, in addition to that already formed within the ascus, aids in discharging the spores into the aquatic environment. When it is realized that the greater number of terrestrial species are dependent for dispersal upon the forceful discharge of the spores into the air where they are carried by air currents, it is apparent that a substitute method would be more effective for the dispersal of spores of the marine species. The deliquescence of the ascus wall, then, would seem to answer this need. In addition to this almost general characteristic, there is another that immediately draws attention, viz., the presence of processes on the spores that serve to keep the spores suspended for a greater length of time in the water. Thus, *Peritrichospora*, which appears to be closely allied to the terrestrial genus *Ceriospora*, is characterized by numerous cilium-like processes that are produced around the middle septa of the spores (Pl. V) and formed after the spores have been freed from the ascus. In addition to the cilium-like processes, the spores have an appendage at either end which soon becomes modified and eventually deliquesces. These appendages appear to serve for the attachment of the spores to the new substratum since when the spores are mounted in water, the softened ends often cling to the cover glass or slide and can be removed from their point of attachment only after considerable agitation of the mount. Similarly, the spores of *Remispora*, while not having the cilium-like processes, are equipped at each end with a pair of appendages which appear to serve both for the attachment of the spores to the substratum, as well as to assist in suspending the spores in water and facilitating their dispersal. Unlike the appendages of the spores of *Peritrichospora*, those of *Remispora* appear to be in some way connected with the astral rays of the nucleus since they are already formed within the ascus, and, while the ascospores are immature, nearly surround them (Pl. III, figs. 7–10). As another example of adaptation in the pyrenomycetous genera, the species of *Halophiobolus* furnish excellent illustration of their suitability for aquatic dissemination. In these species, the spores (Pl. VI) are filamentous and slender and sink less readily than would the rounded or ellipsoid spores. While such filamentous spores are found in the discomycetous genus *Vibrissea* which grows on decaying wood or leaves in freshwater habitats, and are very common in the hypocreaceous genus *Cordyceps* which, with few exceptions, parasitizes insects and is thus terrestrial, the two ends of the spores of the species of *Halophiobolus* are equipped with modified tips which are somewhat gelatinous and thus are capable of attaching the spores to any substratum with which they come in contact. Indeed, it is very difficult because of the adhesiveness of the tips, to discharge the spores from a glass pipette.

A similar difficulty is encountered in handling the spores of several other species for the following reason. Surrounding the spore there is a jelly-like sheath or wing-like structure (Pl. III, fig. 1–4) which becomes more and more diffluent upon the release of the spores into water. Owing to the tenacity of the jelly, the spores tend to cling to any object with which they come in contact, even to such smooth surfaces as glass. A spore modification closely resembling this is reported in the case of an aquatic (freshwater) Ascomycete described by Weston (18) in which he showed that the jelly firmly attaches the spore to the substratum or to any surface upon which it settled. Thus it can be seen that there are special modifications of the ascospores which not only aid in keeping the spores suspended in water, but which also assist in holding them in contact with any solid substratum which they encounter in the water. Among the conidial forms, perhaps the most outstanding is *Orbimyces* (Pl. I, fig. 5-6), the conidia of which are dark and spherical or subspherical, and equipped with finger-like processes which serve as a sort of parachute to keep the conidia from sinking too rapidly. This modification of the conidium parallels that recently described by Ingold (3) for a number of fresh-water Fungi Imperfecti which also are characterized by their form and the number of appendages they bear. The morphological response of *Orbimyces* to its aquaeous habitat is so similar to that of the several species described and delineated by Ingold that there seems no reason to deny that *Orbimyces* is a form morphologically adapted to marine conditions, just as are the other species to their freshwater habitat.

The morphological modifications of the ascospores and conidia lead one to speculate as to the ancestry of these forms and the rôle that they play in the scheme of things. Of the former, there is so little evidence on which to base theories that it does not seem worth while entering into a discussion. The most that can be said is that these forms belong in the same genera or in genera closely related to those that occur in terrestrial habitats. As regards the place that these forms take in the natural cycle of events, it would seem that they act more or less in the same capacity as wood-rotting fungi in forests. The action on plant remains does not appear to be as rapid as that brought about on land, but unquestionably they play an important rôle since many of the forms are strong destroyers of cellulose, pectic substances, and other carbohydrates, and to a lesser extent of lignin. Their economic significance to man is at present being further studied, but from the evidence now available there is not the least doubt that these forms are instrumental in speeding the rate of decay of wooden piling and particularly of cordage by the successive and continuous decaying of the outer parts that are submerged intermittently or all the time in salt or brackish water. Since rope is essentially cellulose, certain of these fungi play a very important part in its deterioration. One of our collections, already listed, demonstrates this fact most clearly for the samples of rope not only were penetrated in all directions by fungous hyphae, but the fungi were producing large numbers of fruiting bodies all over the surface of the substratum and producing spores in great numbers, as evidenced by the number of mature asci still present in a goodly proportion of the fruiting bodies. But sufficient has been written at this stage to indicate the potential importance of at least a fair proportion of these fungi. More extensive studies on their physiology are being made, the results of which will appear later in this and in subsequent papers.

ACKNOWLEDGEMENTS

The writers are deeply indebted to Dr. William F. Clapp and Miss Ruth Lindquist, formerly of the Clapp Laboratories at Duxbury, Mass., for their keen interest in this problem as well as for their continued coöperation in sending test boards for examination as soon as possible.

To Dr. Frederick S. Hammett, Director of the Marine Experimental Station of the Lankenau Hospital Research Institute, appreciation and thanks are extended for his hearty coöperation and interest in these investigations and for making available the facilities of the Laboratory at North Truro, Mass., where the use of laboratory space and equipment greatly aided field collection and preparation of material for later study.

The writers wish to thank Dr. Alfred C. Redfield and Dr. Bostwick Ketchum of the Woods Hole Oceanographic Institute for supplying specimens of cordage of known conditions of exposure to sea water.

E. S. Barghoorn is indebted to the American Academy of Arts and Sciences at Boston for grants-in-aid which have enabled him to carry on extensive field work and other activities in the prosecution of this work.

BIBLIOGRAPHY

1. **Cotton, A. D.** Notes on marine Pyrenomycetes. Trans. Brit. Myc. Soc. **3:** 92–99. 1907.
2. **Ellis, J. B. & B. M. Everhart.** New fungi. Journ. Myc. **1:** 148–[150]–154. 1885.
3. **Ingold, C. T.** Aquatic Hyphomycetes of decaying alder trees. Trans. Brit. Myc. Soc. **25:** 339–417. *pl. 12–17. 48 text figs.* 1942.
4. **Johnson, R. A. & F. A. McNeil.** Supplementary Rept. No. 2. Maritime Services Board of N. S. Wales. p. 36. 1941.
5. **Mounce, I. & W. W. Diehl.** A new Ophiobolus on eelgrass. Canad. Journ. Res. **11:** 242–246. *1 pl.* 1934.
6. **Petersen, H. E.** Contributions à la connaisance des Phycomycètes marins (Chytridineae Fischer). Oversigt Kgl. Danske Vidensk. Selskabs. Forhandl. **1905:** 439–488. 11 figs. 1905.
7. **Renn, C. E.** A mycetozoan parasite of Zostera marina. Nature **135:** 544–545. 1935.
8. ———. Persistence of the eel-grass disease and parasite on the American Atlantic Coast. Nature **138:** 507–508. 1936.
9. ———. The wasting disease of *Zostera marina.* I. A phytological investigation of the diseased plant. Biol. Bull. **70:** 148–159. *6 figs.* 1936.
10. **Saccardo, P. A.** Sylloge Fungorum **2:** 350. 1883.

11. **Sparrow, F. K.** Observations on marine Phycomycetes collected in Denmark. Dansk. Bot. Arkiv **8**: 1–24. *4 pls.* 1934.
12. ———. Biological observations on the marine fungi of Woods Hole waters. Biol. Bull. **70**: 236–263. *3 pl. 35 text figs.* 1936.
13. ———. Aquatic Phycomycetes exclusive of the Saprolegniaceae and Pythium. Univ. Mich. Stud. Sci. ser. **15**: 1–785. *1 pl. 68 figs.* 1943.
14. **Sutherland, G. K.** New marine Pyrenomycetes. Trans. Brit. Myc. Soc. **5**: 147–154. *pl. 3.* 1915.
15. ———. New marine fungi on Pelvetia. New Phytol. **14**: 33–43. *4 figs.* 1915.
16. ———. Additional notes on marine Pyrenomycetes. New Phytol. **14**: 183–193. *5 text figs.* 1915.
17. ———. Marine Fungi Imperfecti. New Phytol. **15**: 35–38. *5 text figs.* 1916.
18. **Weston, W. H.** Observations on Loramyces, an undescribed aquatic Ascomycete. Mycologia **21**: 55–76. *pl. 8–9.* 1929.
19. **Young, E. L., III.** *Labyrinthula* on Pacific Coast eel-grass. Canad. Journ. Res. **16C**: 115–117. 1938.

[*Editor's Note:* Material has been omitted at this point.]

MATERIALS AND METHODS

The fungi with which this study is concerned were obtained from specimens of wood or cordage exposed for varying lengths of time to the full salinity of sea water or to the brackish waters of estuaries. In general, two types of materials were used, viz., specimens of wood or cordage of known conditions and duration of submergence, and specimens of wood and cordage of undetermined length of exposure. In all cases the fungi were isolated from a natural substratum which was either continuously submerged in water or covered for the greater part of the interval elapsing between successive high tides. A summary of the specimen collection data is included in the introduction.

The "test" boards and blocks referred to in this paper were supplied through the courtesy of the William Clapp Laboratories in Duxbury, Mass. These blocks, chiefly of fresh-sawed hard pine, had been exposed for six to twelve months at a depth of two to eight feet below low water and had not been exposed to the air from the time of their submergence until the date of removal. Abundant perithecia on the surface of or embedded in many of the blocks provided an excellent source of material for spore isolations. The majority of the blocks were examined and used as sources of inoculum within three or four days after their removal from the water. This brief period of time precludes the possibility of perithecium production by any contaminating terrestrial fungi. Further proof that perithecium production is commonly vigorous under sea water was obtained by sectioning specimens of wood freshly collected from Cape Cod Bay and Provincetown Harbor. Sections of such material, bearing perithecia with viable spores, also show the presence of a very extensive and vigorous mycelium in the wood. Comparable material has recently been obtained from Chesapeake Bay, Maryland, and Biscayne Bay, Florida.

In addition to wood, specimens of submerged rope and cordage were examined for attack by fungi. All rope samples which have been studied whether of treated or untreated material show varying degrees of infection and deterioration by fungi. In cases in which perithecia occurred on or embedded in the rope the species have been found identical with those obtained from wood blocks. Although studies are now in progress on the mode and rate of attack of the marine fungi on both wood and cordage it is desirable to point out here that rope, because of its fibrous texture, appears to be more readily attacked than wood by the cellulose destroying members of the group.

In addition to the collection data tabulated in the introduction the following locations along the Atlantic Coast have yielded specimens of wood (or cordage) bearing perithecial stages of marine fungi:

- Newfoundland
  - Corner Brook
- Maine
  - Eastport
  - Machias
  - Portland
  - Rockland
  - Wiscasset
- Massachusetts
  - Beverley
  - Boston
  - Charlestown
  - Edgartown
  - Gloucester
  - Marshfield
  - Neponset
  - Quincy
  - Vineyard Haven
  - Weymouth
- Connecticut
  - South Groton
- New York
  - City Island
  - Rosebank, L. I.
- Maryland
  - Gibson Island, Chesapeake Bay
- Florida
  - Biscayne Bay

In isolating certain of the fungi from the natural substratum great difficulties were encountered in freeing the germinating spores from bacteria. The poured plate method using nutrient agar made up in sea water usually proved useless owing to the very rapid spread of motile marine bacteria which frequently destroyed the spores even before their germination occurred. Modifications of the sprayed-spore technique (Wehmeyer 1923) were likewise of little value. The acidification of the spore suspension to retard bacterial activity completely inhibited germination of the fungal spores. The best results were obtained with stab cultures made by detaching a small mass of spores from an exuding or crushed perithecium and thrusting it with a streaking motion deep into agar. The spores, germinating beneath the surface of the agar, rapidly developed a bacteria-free mycelium. In certain cases, however, the rapid spread of bacteria on the surface of the agar made it necessary to invert the stab-culture plates, free the agar from the dish and cut blocks from the lower surface in order ultimately to free the fungus from bacteria. The excessive motility of certain of the bacteria, in conjunction with a definite agar-liquefying activity, permitted their development along agar-glass interfaces, although they were unable to penetrate deeply into free agar. In making the stab cultures, therefore, the needle was not allowed to penetrate completely through the agar layer. This precaution prevented the spread of bacteria on the lower surface of the culture. The medium most useful for spore germination with a minimum of bacterial growth was a three per cent crude unwashed agar made up in sea water.

In determining the reaction of the fungi to the substratum, to temperature, and to hydrogen-ion concentration various agars were used. All of the media, with the exception of the fresh water and concentrated sea water, were solidified with three per cent Bacto agar made up in filtered sea water with added nutrient.

Stock cultures were maintained on a four per cent agar containing 0.5 per cent malt extract. Other agars were made up as follows:

Cellulose agar. One per cent regenerated cotton cellulose prepared by the method of Northrup (1919), pH 7.4[1].

Maltose, Xylose and Galactose agars. Sugar concentration two per cent, pH 7.4.

Starch agar. Two per cent purified corn starch, pH 7.5.

Asparagine agar. Four-tenths per cent asparagine Merck, pH 7.2.

Pectin agar. Five-tenths per cent lemon pectin purified by the method of Baxter (1925) pH 6.5. The pH of the pectin agar was adjusted to 7.2 before use by the addition of dilute sodium hydroxide. This medium doubtless contains many degradation products of pectin hydrolysis.

Wood flour agar. 0.75 per cent finely powdered white pine filings. The wood dust was extracted with boiling chloroform, and repeated treatments with benzol and absolute alcohol to remove, insofar as possible, the resins and terpenes. The treated wood was washed repeatedly with boiling distilled water and dried on a Buchner funnel. pH 7.4.

Fresh water agar. Glass distilled water four parts, tap water one part. pH 7.2. Tap water contained traces of organic matter in addition to traces of inorganic salts.

Sea water agar. Filtered sea water with no added nutrient, pH 7.4.

Concentrated sea water agar. Prepared from sea water evaporated to exactly one-third its original volume, with no added nutrient. A portion of the calcium precipitated out during evaporation. pH 8.2.

The agar used for the study of the influence of temperature was made with four-tenths per cent malt extract and no additional nutrient. The experiments on the influence of hydrogen-ion concentration were run with five per cent malt extract agar; the medium was acidified by the addition of hydrochloric acid and made alkaline by addition of sodium hydroxide.

In studying the rate of radial growth of tho fungi in culture a more or less standardized procedure was followed. Pure cultures of the various species were maintained by repeated transfers on a stock culture medium in Petri dishes. Small blocks of this medium, one to two millimeters square were used as inoculum in the various growth studies. The blocks were cut from the outer third of the spreading mycelium. Rate of radial growth was determined by measuring the spread of the mycelium on Petri dish cultures containing the different media. After inoculating the plates the colonies were allowed to grow for three to six days and then measured. Measurements were made radially in three directions from the edges of the inoculum to the outer extreme of the mycelial mat. The plates were then incubated for a varying length of time, in most cases six to ten days, and measured again. The rate of radial growth was taken as the increase in radius of the colony in millimeters per day between the first and second measurements. The value for any one measurement was obtained by averaging the three readings.

[1] pH values are for media after sterilization.

In eliminating from measurement the growth for the first three to six days, the influence of nutrient carried over in the inoculum was greatly reduced.

Although radial increase of a colony is a rather arbitrary index of the amount of growth of a mycelium it is by and large a more practical means of measurement than determinations of dry weight increase. As far as was practicable in this study the radial increase of the colony was roughly correlated with the apparent visible density of the fungal mat. In all but a few cases extensive radial spread was quite closely associated with abundant branching and production of hyphae. The cases in which this correlation did not obtain, were those of "starvation media" such as the fresh water and pure sea water cultures. Certain species on these two media produced extremely sparse but far-ranging mycelia. These facts must be taken into account in considering the rate of growth on certain of the media deficient in available nutrient.

In preliminary experiments to determine the availability of different carbohydrates it was found that the depth of medium in the culture dish exerted considerable influence on the rate of spread of the mycelium. Certain species spread more rapidly on deep cultures while others spread more rapidly on shallow cultures. In order to counteract this effect which is particularly significant in the study of the influence of temperature and hydrogen-ion concentration, the plates were poured to as uniform depth as possible and then sorted before inoculation so that a given species was grown on media of the same depth.

The growth studies in this investigation were made with fourteen selected species of fungi obtained either by spore isolation or direct inoculation from infected material. Inasmuch as the latter source of cultures always presents the uncertainty of possible contaminating organisms not originally present in the material, only species obtained from spore isolations are considered in detail. Of these, seven species were selected as representative of the marine fungi as a group. The seven species cover fairly well the morphological range of the forms known at present and probably represent fairly well the physiological diversity of this interesting group of fungi.

RELATION TO SALINITY

As indicated by the collection data summarized in the introduction, the marine fungi occur naturally in both brackish water and sea water of normal salinity. In view of their undoubted derivation from terrestrial forms it seemed interesting to determine their reaction in culture to extremes of salinity as some measure of their adjustment to the marine environment. For this purpose five species were chosen and grown on three different media: fresh water, normal filtered sea water and sea water evaporated to one-third its original volume. The three series of experiments were run in duplicate, simultaneously and at the same temperature.

TABLE 1. EFFECT OF SALINITY ON RATE OF RADIAL GROWTH

	Radial Growth in Millimeters Per Day		
	Fresh water pH 7.2	Sea water pH 7.4	Concentrated Sea water pH 8.2
Amphisphaeria maritima	.55	.30	.12
Helicoma salinum	.68	1.14	.31
Peritrichospora integra	.79	.96	.38
Ceriosporopsis halima	1.33	1.17	.54
Halophiobolus opacus	.52	3.10	.07
Halophiobolus salinus	2.74	3.16	.07

In Table 1 it will be noted that two species, *Amphisphaeria maritima* and *Ceriosporopsis halima* develop more rapidly in fresh water than in sea water, while the remainder grow radially at a more rapid rate in normal sea water. In concentrated sea water, the growth rate for all is considerably decreased although *Ceriosporopsis halima* and *Peritrichospora integra* grow at nearly one-half the rate determined in fresh water. An important factor in the experiment which is not expressed in the table is the relative denseness of the mycelia. In all cases the hyphal production per colony area was conspicuously greater in normal sea water than in either fresh water or concentrated sea water. In fresh water agar the hyphal mats were more hyaline and the hyphae less branched, particularly in the outer third of the colony. In concentrated sea water the hyphae branched abnormally, the secondary hyphae being crooked and much shortened. In this latter medium it might be presumed that the higher pH value would adversely affect growth. However, as will be shown later, a pH of above 8.0 is more favorable for growth in culture than a pH below 8.0 in the case of at least three species.

Although *Amphisphaeria maritima* and *Ceriosporopsis halima* in the above experiments showed a more rapid radial growth in fresh water than sea water, neither of these species shows preference for such conditions in nature as far as collection data and field study indicate. Indeed, their occurrence in fresh water is to be doubted, although further collections may demonstrate their presence in the non-saline portions of estuarine rivers. Of the five species studied in respect to salinity reactions all but *Halophiobolus opacus* and *Peritrichospora integra* have been collected from brackish waters as well as the open sea.

Two species of the marine fungi, *Halophiobolus opacus* and *Remispora maritima* were collected in conditions in which a fairly high salt content presumably occurred in the natural substratum. Both of these forms were found, in the perfect stage, on wooden pilings in such a situation as to be exposed to the drying action of wind and sun for many hours at low tide, although completely submerged at high tide. *Remispora maritima* in particular was collected on wood exposed to the air for six to seven hours between tides. Such conditions would suggest a high tolerance to salinity as well as to sudden changes in salinity such as would occur with recurring incoming tides.

[*Editor's Note:* Material has been omitted at this point.]

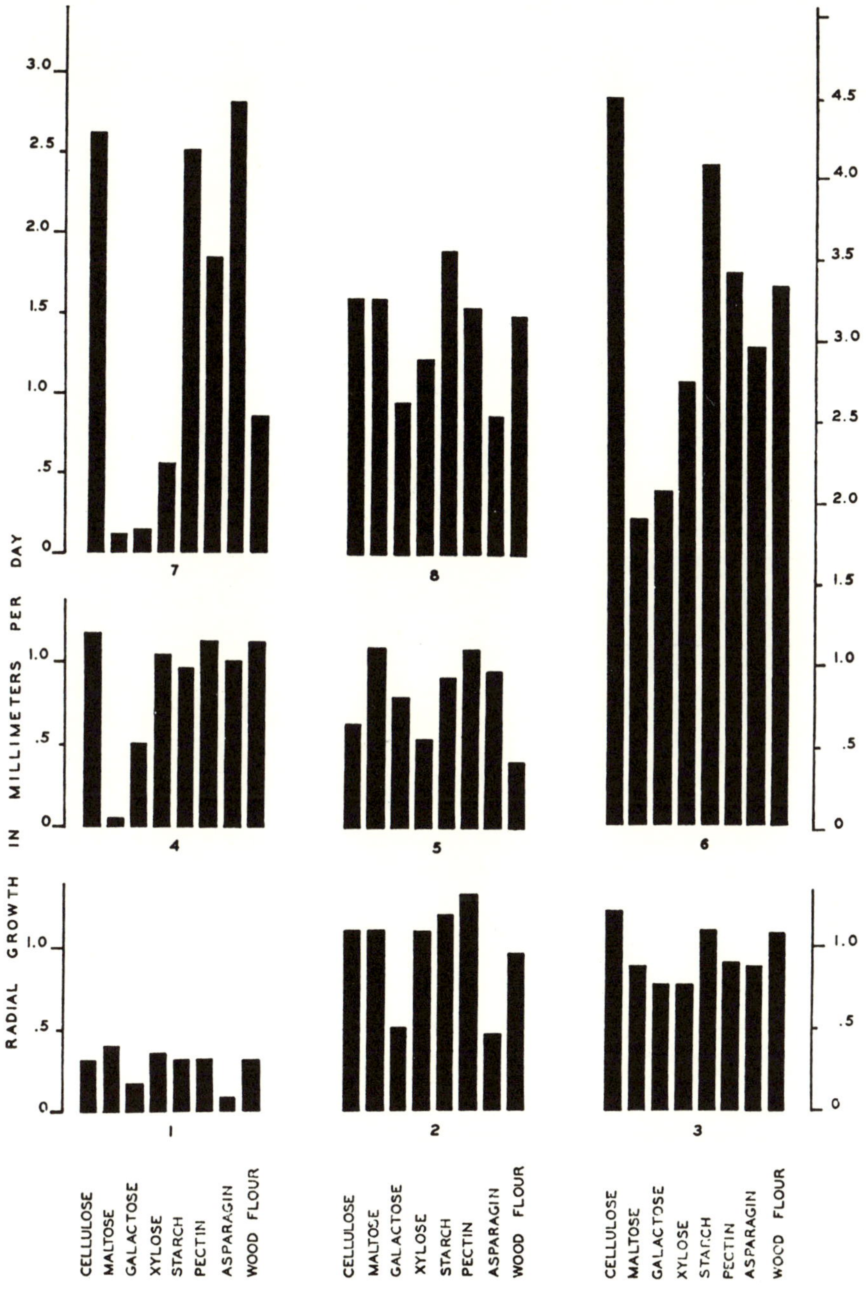

Fig. 1. Rate of growth per day of the different fungi on various substrata. 1. *Amphisphaeria maritima.* 2. *Ceriosporopsis halima.* 3. *Halophiobolus cylindricus.* 4. *Peritrichospora integra.* 5. *Phialophorophoma litoralis.* 6. *Halophiobolus salinus.* 7. *Halophiobolus opacus.* 8. *Helicoma salinum.*

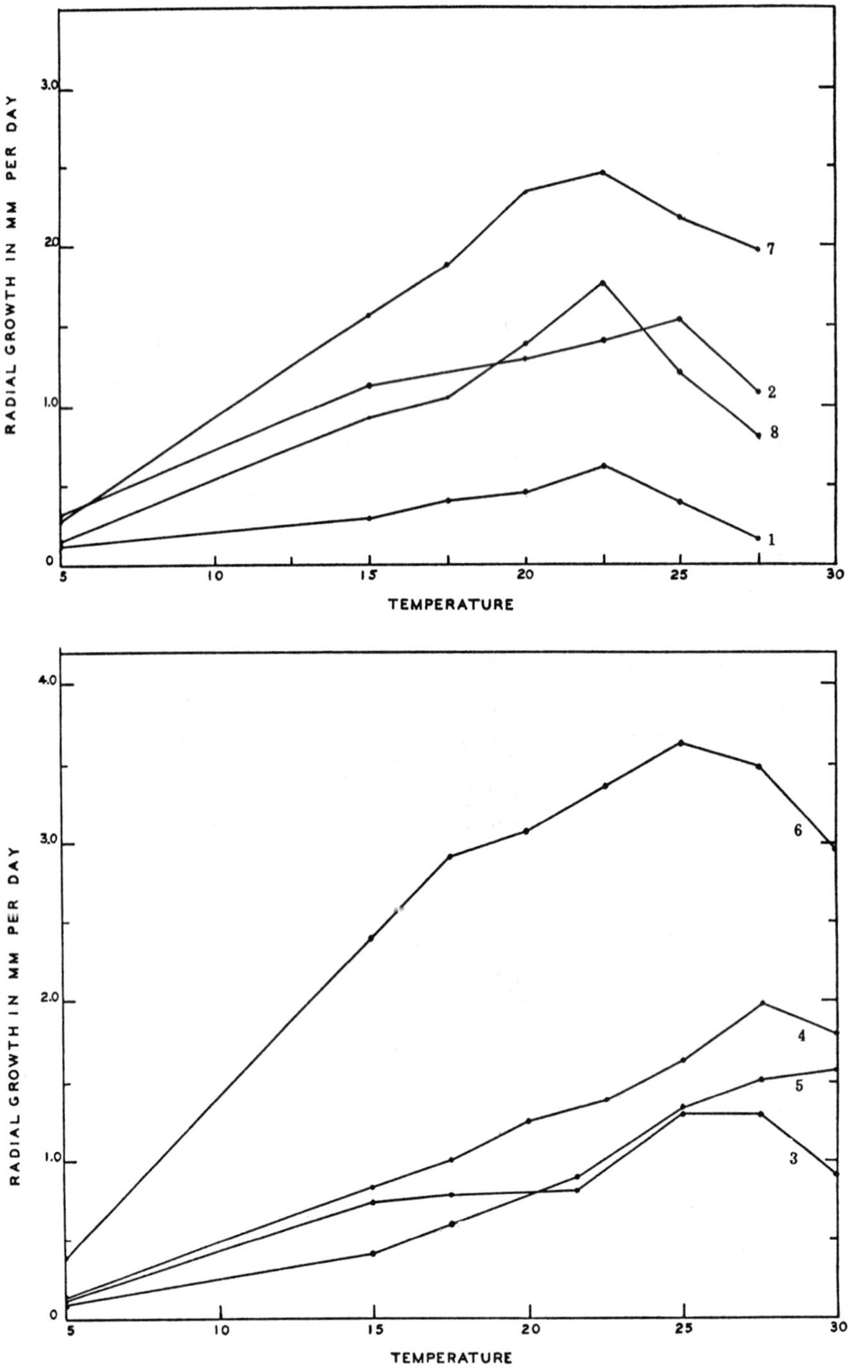

Fig. 2. The effect of temperature on the rate of radial growth. 1. *Amphisphaeria maritima.* 2. *Ceriosporopsis halima.* 3. *Halophiobolus cylindricus.* 4. *Peritrichospora integra.* 5. *Phialophorophoma litoralis.* 6. *Halophiobolus salinus.* 7. *Halophiobolus opacus.* 8. *Helicoma salinum.*

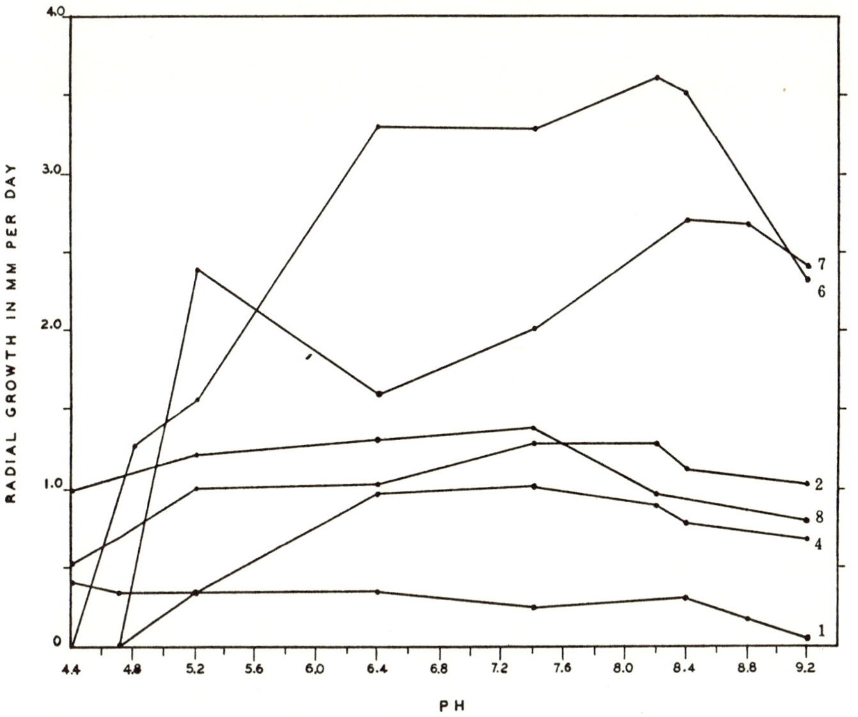

Fig. 3. The effect of hydrogen-ion concentration on the rate of radial growth. 1. *Amphisphaeria maritima.* 2. *Ceriosporopsis halima.* 4. *Peritrichospora integra.* 6. *Halophiobolus salinus.* 7. *Halophiobolus opacus.* 8. *Helicoma salinum.*

[*Editor's Note:* Material has been omitted at this point.]

Although the relative significance of bacteria and fungi in the degradation of wood and cordage in sea water has not been quantitatively evaluated, it seems clear to the writer from the evidence presented in this study that the marine cellulose destroying fungi are the more significant organisms, particularly in the early stages of the decomposition processes.

SUMMARY

1. The physiological behavior of seven species of marine fungi, obtained from ascospore isolations, has been studied particularly in regard to their response to temperature, hydrogen-ion concentration and rate of growth on artificial substrata.

2. Growth of the fungi on media of varying salinity shows that these forms are adapted to the saline environment of the sea. Tolerance of relatively high salt concentrations is demonstrated by their growth in sea water of three times normal salinity.

3. When grown on sea water agar containing separately cellulose, maltose, galactose, xylose, starch, pectin, asparagine and wood flour, growth as measured by radial increase in the colonies was most rapid on cellulose, pectin and starch, respectively. Growth was vigorous and fairly rapid on wood flour agar. Galactose induced the formation of abnormalities in both the hyphae and spores of certain forms. Asparagine supported considerable radial growth but the mycelia produced on this sub-

stratum were very sparse and nearly devoid of aerial hyphae. Rate of growth on the various substrata was measured in terms of radial increase; the extent of production of hyphae by the colonies, however, was also considered in evaluating the availability of the substratum.

4. Forms which normally produce conidiospores in culture sporulated most heavily on cellulose, wood flour and pectin.

5. Color of the colonies varied widely in response to the substratum; the greatest frequency of pigmentation in the group as a whole, however, occurred on pectin.

6. The most favorable temperature for growth on complex synthetic media was between 22.5° and 27.5°. Several species grew most rapidly at 27.5° C. and one form at 30° C. All the species cultured made an appreciable growth at 5° C. The relatively high optimum temperatures for growth in culture, contrast strongly with the temperature of the natural environment of the sea.

7. With one exception the entire group developed best in media with an initial pH above 7.6 and growth was definitely unfavorable in acid media. Two species failed to develop at pH 4.6 or below. The response to hydrogen-ion concentration may be logically interpreted as a physiological modification to marine conditions.

8. All the marine fungi described in these investigations were isolated from specimens of wood or rope exposed for varying periods to partial or complete submergence in salt or brackish tide waters. Histological examination of their natural substrata shows that they penetrate and ramify in the cell walls of wood and cordage fibers inducing decay by enzymatic hydrolysis of the cellulose and other constituents of the cell wall. Their mode of action in the enzymatic hydrolysis of woody tissues is quite comparable to that of terrestrial wood destroying fungi. Laboratory experiments under controlled conditions demonstrate their ability to attack the various constituents of wood in artificial culture.

Extensive studies are now in progress to determine the enzyme formation by these fungi and to measure their cellulose and lignin resolving power.

AMHERST COLLEGE
AMHERST, MASS.

LITERATURE CITED

Baxter, D. V. The biology and pathology of some hardwood rotting fungi. Amer. Jour. Bot. **12**: 522–554. 1925.

Northrup, Z. A new method for preparing cellulose for cellulose agar. Abstr. Bact. **3**: 7. 1919.

Wehmeyer, L. E. The imperfect stage of some higher Pyrenomycetes obtained in culture. Papers Mich. Acad. Sci. Arts and Letters, **3**: 245–266. 1923.

39

Reprinted from *Limnol. Oceanogr.*, 5(4), 366–371 (1960)

ISOLATION OF YEASTS FROM BISCAYNE BAY, FLORIDA AND ADJACENT BENTHIC AREAS[1,2]

Jack W. Fell, Donald G. Ahearn, Samuel P. Meyers and Frank J. Roth, Jr.[3]

The Marine Laboratory, University of Miami, Miami, Florida

ABSTRACT

Investigations of the yeasts present in Biscayne Bay, Florida, have indicated the occurrence of various yeast taxa, representing 179 isolates, including species of *Saccharomyces, Hansenula, Debaryomyces, Candida, Cryptococcus, Rhodotorula, Trichosporon* and *Torulopsis*. Two asporogenous species, *Candida tropicalis* and *Rhodotorula mucilaginosa*, were the most abundant and widely distributed species in the Bay. A yeast biota, with many species similar to those isolated from Biscayne Bay, has been collected in deep sea sediments from The Bahamas. A selective enrichment culture technique has been developed, greatly facilitating the collection of yeasts from the marine environment. In general, the limited deep sea collections showed a predominance of oxidative yeasts as compared to collections made in Biscayne Bay.

The biology of yeasts in nonmarine environments has received considerable attention, but our knowledge of the yeast population present in the ocean is exceedingly meager (Fisher 1894, ZoBell and Feltham 1934, ZoBell 1946). Most of these reports, especially the earliest work, fail to describe the yeasts or yeast-like organisms so that identification is not possible, and other studies are incidental to the primary investigation of the bacterial population. Only recently have several papers appeared that deal specifically with investigations of the marine yeast biota (Bhat and Kachwalla 1955, Bhat *et al.* 1955, Novozhilova 1955).

In 1958 studies of the yeasts present in Biscayne Bay, Florida, and adjacent benthic areas were started in The Marine Laboratory of the University of Miami. This paper notes the genera and species of yeasts collected together with a brief discussion of their distribution. Further work in progress involves investigations of the possible ecological significance of yeasts and salient physiological features of selected dominant species.

MATERIALS AND METHODS

Collections in Biscayne Bay were made at 45 sites, encompassing an area from approximately 2,000 yd north of the Venetian Causeway to the southern point of Key Biscayne (Fig. 1). In general, Biscayne Bay is a shallow estuary characterized by the development of an offshore bar on a shore line of low relief.

Both sedimentary material and submerged banana stalk sections were examined. The Biscayne Bay sediments were taken with a shallow water coring device (Grein and Meyers 1958) from areas totally submerged even at low spring tide. This coring apparatus was provided with a tubular extension to permit sampling at depths to 20 ft. The banana stalk sections were submerged in area C (Fig. 1) from the end of The Marine Laboratory pier. These sections, approximately 2 in. long, were cut

[1] Contribution No. 274 from The Marine Laboratory, University of Miami.

[2] This work was supported by Grant G-5004 from the National Science Foundation and project NR 103-305 from the Microbiology Branch, Office of Naval Research. The authors express their gratitude to Dr. L. J. Wickerham, Northern Utilization Research and Development Division, Agricultural Research Service, Peoria, Illinois, for his extensive cooperation and invaluable counsel throughout the development of this work. We appreciate the suggestions of Drs. H. J. Phaff, Dept. of Food Science and Technology, University of California, Davis, Calif., and M. E. diMenna, Dept. of Scientific and Industrial Research, Lower Hutt, New Zealand, in various taxonomic aspects. Mr. J. K. McNulty of The Marine Laboratory gave generously of his time on consultations on the hydrography of Biscayne Bay, Florida. Mrs. C. Edith Marks and Mr. Steve Ebert, The Marine Laboratory, assisted immeasureably in numerous technical procedures.

[3] Department of Microbiology, University of Miami.

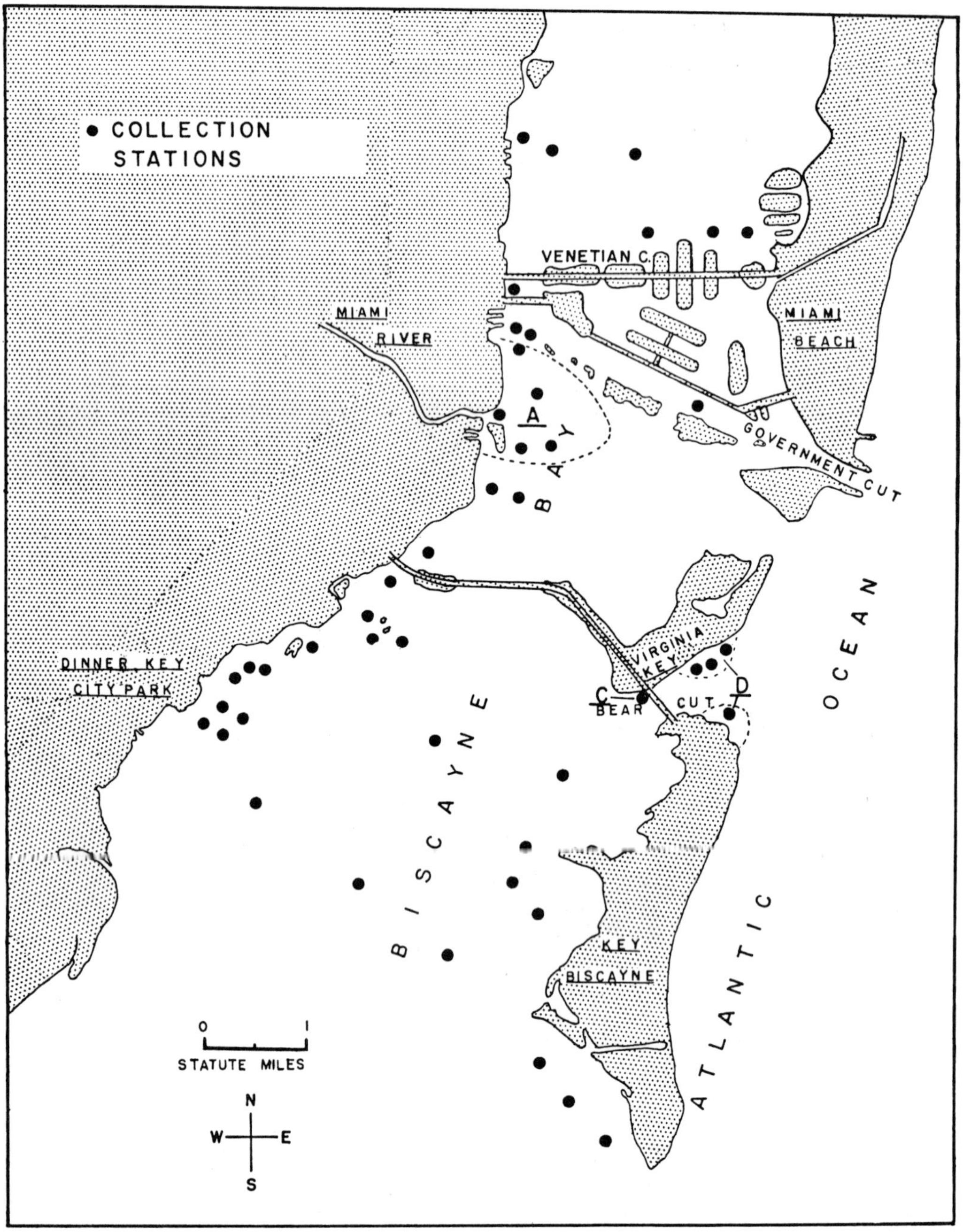

FIG. 1. Map of Biscayne Bay, Florida, with collection stations.

from nearly ripe banana stalks with a ¾ in. diameter cork borer, were sterilized (121°C for 15 min) in petri dishes, and immediately before submergence, were transferred aseptically to alcohol-sterilized plastic vials. Four to 6 vials, each with several small holes drilled at both ends to allow circulation around the banana core, were anchored along a Kordite line to ensure complete submergence at low tide. The banana stalk sections were collected after 3 to 10 days' exposure. All sediment samples from Biscayne Bay were transported to the laboratory under refrigeration (10–15°C) and processed within 24 hr.

Deep sea sediments were collected using a gravity corer, provided with a 3-ft plastic tube liner, previously alcohol-sterilized. During transit to the collection site the tubes were sealed with sterile plugs which were removed at the time of insertion of the tube into the coring device. Immediately following collection, the tubes with the sediment core were sealed aseptically and stored in the refrigerator (6–10°C) of the vessel until subsequent examination in the laboratory. Five cores were examined from the following localities. 1) High Cay, Andros Island, The Bahamas; 500 fathoms. 2) Ten to 15 miles west of Bimini, The Bahamas; 450–455 fathoms. 3) Gun Cay Light, The Bahamas; 377 fathoms.

Three cores were taken in collection 2) and were treated as a composite sample. In collections 1) and 2), only the uppermost portion of the sediment core was examined. From collection 3), the core was divided into successive sections, each section approximately 1 cm thick. The uppermost 12 sections were used in inoculation experiments.

Sectioning was accomplished by pushing a sterile plunger through the plastic corer liner to extrude the sediment. In order to prevent possible contamination from the sides of the core immediately adjacent to the plastic liner, the material used for the inoculum was taken from the central portion of the sediment. Some of the liquid present in the tube also was used as inoculum.

Approximately 2.0 g of each separate sediment sample were used as inoculum. In tests where a submerged banana stalk section was used, the infested section was collected and aseptically transferred to a sterile blendor and ground thoroughly in 20 ml of sterile sea water. The inoculum comprised 1 ml of this suspension.

For the selective isolation of yeasts by enrichment incubation procedures, the culture medium was treated with the filter sterilized antibiotic mixture to give a final concentration of 10 mg % chlortetracycline. HCl (Aureomycin), 2 mg % chloramphenicol, and 2 mg % streptomycin sulfate. The two enrichment broths used consisted of a medium containing 2.0% glucose (Medium G), and a medium composed of 2.0% glucose, 0.1% yeast extract, and 0.5% peptone (Medium GYP). The incubation vessels consisted of 125-ml Erlenmeyer flasks, each containing 20 ml of the nutrient medium. Culture media were prepared with natural sea water as well as with distilled water. However, unless noted otherwise, all media used in the incubation and isolation procedures were prepared with sea water.

After inoculation the culture vessels were placed on a rotatory shaker (55 rpm) at the laboratory temperature of 25°C. Each flask was sampled at 24-hr intervals during the 144-hr incubation period. Full strength broth and suitable dilutions were inoculated on the surface of the primary isolation medium consisting of 2.0% glucose, 1.0% peptone, 0.1% yeast extract and 1.7% agar, prepared both with distilled water and with sea water. Replicate platings were made at each sampling. To obtain data on the approximate concentration of yeasts present in the natural environment, sediments from selected areas were inoculated directly without prior incubation enrichment, to the primary isolation medium containing the antibiotic combination noted previously. After incubation, individual yeast colonies exhibiting distinctive colonial characteristics on this medium were transferred to slants of the stock culture sea water medium of nutrient agar (Bacto) supplemented with 2.0% glucose and 0.1% yeast extract. These cultures are maintained in the microbiology culture collection of The Marine Laboratory.

To identify the organisms, the isolates were tested for carbohydrate assimilation, fermentation capacities and nitrate utilization. These tests and other diagnostic procedures used are those described by Wickerham (1951) and Lodder and Kreger-van Rij (1952). Dark pigmented species have been grouped provisionally under the gen-

TABLE 1. *Yeasts collected in Biscayne Bay, Florida*

Taxon	No. of isolates	No. of stations
Sporogenous species		
Debaryomyces kloeckeri	8	7
Hansenula anomala	9	1
Saccharomyces fructuum	2	1
Saccharomyces sp.	2	1
Asporogenous species		
Candida tropicalis	42	19
C. parapsilosis	18	10
C. guilliermondii	4	2
C. intermedia	4	1
C. boidinii	1	1
C. melinii	1	1
Cryptococcus laurentii	6	4
C. albidus	1	1
Rhodotorula mucilaginosa	24	14
R. glutinis	13	9
R. texensis	5	5
R. minuta	4	4
R. graminis	1	1
Rhodotorula sp.	1	1
Trichosporon cutaneum	10	3
Torulopsis sp.	4	1
Pullularia pullulans	2	2
"Black yeasts"	16	10

TABLE 2. *Yeasts isolated from deep sea sediments in the Bahamas*

Cruise No. or Collection No.	Taxon	No. of isolates
(1)	*Rhodotorula mucilaginosa*	3
	R. glutinis	2
	R. texensis	1
	R. marina	1
	Debaryomyces kloeckeri	4
	Torulopsis famata	1
	Cryptococcus diffluens	1
(2)	*Torulopsis famata*	1
	Candida parapsilosis	2
	C. tenuis	2
(3)	*Cryptococcus albidus*	3
	C. neoformans var. *uniguttulatus*	2
	Candida parapsilosis	3
	C. curvata	5
	C. guilliermondii	1

eral designation "black yeasts." The media used for identification were prepared with Difco products in distilled water according to Wickerham's formulae. Media affected by autoclaving, such as those used in the assimilation tests, were sterilized by filtration through a No. 03 Selas filter candle. All other media were sterilized at 121°C for 15 min.

RESULTS

A total of 179 isolates representing 4 sporogenous and more than 18 asporogenous species were obtained in Biscayne Bay. The species collected together with the number of isolates and stations are given in Table 1. The two most commonly isolated genera were *Candida* and *Rhodotorula*, the predominant species being *C. tropicalis*, *C. parapsilosis*, *R. mucilaginosa*, and *R. glutinis*. Except for an unidentified black yeast, only *C. tropicalis* and *R. mucilaginosa* were isolated from the vicinity of the mouth of the Miami River (Area A), a locality characterized by a rather limited yeast biota. These two species found in samples taken at stations in Biscayne Bay other than in areas A, C and D were not obtained from collections in the shallow marine coastal environment of area D.

The largest variety of species was collected in area C from the banana stalk sections. This area is located adjacent to The Marine Laboratory in Bear Cut, a region subject to considerable tidal currents. A total of 17 species were isolated from the banana sections, while at the most, only 4 different species were found in sediment samples at other individual collection stations. An examination of the sediment beneath the banana traps revealed only two species, *Cryptococcus laurentii* and *Candida tropicalis*.

Samples taken from the center of the Bay showed lower numbers of yeasts and fewer species than samples collected near the shore. In general, silty muds exhibited more yeasts than sandy sediments. Periodic sampling of sediment from certain bay stations demonstrated discontinuous yeast populations.

The species of yeasts found in the deep sea sediments are listed in Table 2. While large numbers of yeasts were not found in the deep sea sediments, a considerable variety of isolates were observed. Differences were apparent in the dominant species of individual collections. In the mycological examination of the upper 12 cm of the core from Cruise No. 3, yeasts were found only in the uppermost 2-cm layers.

The direct inoculation of selected Biscayne Bay sediments on the primary isolation medium yielded only comparatively few yeasts. Higher colony counts were obtained on this medium prepared with sea water than on a corresponding medium prepared with distilled water. In an examination of sediments from 7 different bay localities using direct plating, a maximum number of 20 yeast colonies per gram of wet sediment was noted. Frequently, further development of different species occurred upon incubation of other portions of the same inoculum in the enrichment culture broths. The antibiotics added to the latter media, as well as the antibiotics present in the primary isolation medium, permitted yeasts to develop but inhibited the growth of marine bacteria.

The 2.0% glucose medium (G) supported development of a variety of yeasts. However, selected pure cultures of actively metabolizing yeast cells, inoculated into this medium, displayed only negligible growth, although the cells remained viable over a 10-day period. It is probable that the growth of yeasts in Medium G, following inoculation of sediments or banana suspensions, is due in large part to the various metabolites in the natural substrate used as the inoculum. The addition of yeast extract and peptone to Medium G greatly increased the growth rates of individual species as well as the total yeast growth in the enrichment flasks. A more or less general succession or sequence in the development of different types of yeasts occurred during the 6-day incubation period especially in Medium G. The white or hyaline yeasts developed initially followed by the rhodotorulas. The latter reached a maximal population after approximately 96 hr. Species of black yeasts, when present, were not isolated generally until after the fifth day of incubation. The growth of species of *Rhodotorula*, especially *R. mucilaginosa*, was particularly favorable in Medium GYP. Black yeasts, found in certain samples incubated in Medium G, were not isolated along with other yeasts from the same sample in the GYP medium.

DISCUSSION

Most of the species of yeasts collected in Biscayne Bay are taxa known from nonmarine localities, and the majority of the isolates of individual species are quite similar physiologically to their terrestrial counterparts. However, the isolation of microorganisms considered terrestrial in origin from marine and brackish environments is not uncommon (Aaronson 1956, Grein and Meyers 1958, Meyers and Reynolds 1959). Of course, the mere occurrence of fungi in marine areas does not prove that the organisms are of marine origin, or conversely, that they are adventitious halotolerant terrestrial species. The criteria for determining affinity of fungi to the marine environment require much further study and evaluation before the significance of the occurrence in the sea of fungal representatives, usually considered "typically" terrestrial or nonmarine, is known.

While the occurrence of indigenous marine species of yeasts, as such, has not been demonstrated in this work, the possible existence of marine strains of particular species warrants consideration. Certain of the species, especially *Candida parapsilosis*, examined on a large number of carbon compounds have shown distinctive physiological characteristics different from those of isolates of the same species of known terrestrial origin. Further studies of these physiological differences are in progress in this laboratory.

In general, the limited deep sea collections showed a predominance of oxidative yeasts as compared to collections made in Biscayne Bay. Possibly this may be due, in large part, to the low organic content of the sediments of the open sea.

The abundance of *Candida tropicalis* in the Biscayne Bay collections agrees with the observations of Bhat and Kachwalla (1955) that this species is quite common in marine localities. However, the examinations of deep sea sediments (as well as plankton) from the subtropical Atlantic Ocean have not revealed the presence of *C. tropicalis*, although other yeasts, including *Rhodotorula mucilaginosa*, prevalent in Biscayne Bay, have been found in deep sea material. The latter species, as well as *Cryptococcus laurentii, Candida guilliermondii, Debaryomyces kloeckeri,* and *Saccharomyces fructuum*, also were isolated by Bhat and Kachwalla in the Indian Ocean.

The large number of species isolated from the banana stalk traps may be attributed to a complexity of factors. The hydrography of the Bear Cut channel is such that tidal currents subject the banana stalk substrates to alternate flushing with bay and ocean waters, thereby permitting the establishment of fungal representatives from the respective environments. The plastic vials, containing the banana stalk cores, were heavily encrusted with fouling organisms such as the barnacle, *Balanus amphritrite*, the amphipods, *Podocercus brasiliensis* and *Erichthonius brasiliensis* and various types of algae, including *Sargassum*, entangled with the lines. It is conceivable that these communities serve as a reservoir of, or stimulate the growth of, yeast and yeast-like organisms. The association of species of *Rhodotorula* with various marine invertebrates observed in these studies suggests that this group of pigmented yeasts is well adapted for halophilic metabolic activities.

A larger spectrum of cultural techniques than those used here is necessary to obtain a more complete analysis of the different types and species of yeasts in a particular sample. The different growth rates of the various species, especially during the initial 24 hr of incubation, may account in a large part for the frequency of isolation of individual species and the patterns of succession observed. Present knowledge of the biology of marine yeasts is still too meager for a definitive evaluation of the ecological observations noted in this paper, particularly in view of the variabilities in the enrichment culture method. Work in progress in this laboratory on the metabolism of various of the marine isolates should facilitate a better understanding of the role of yeasts in the sea.

REFERENCES

AARONSON, S. 1956. A biochemical-taxonomic study of a marine micrococcus, *Gaffkya homari*, and a terrestrial counterpart. J. Gen. Microbiol., **15**: 478–484.

BHAT, J. V., AND N. KACHWALLA. 1955. Marine yeasts off the Indian coast. Proc. Indian Acad. Sci., Sec. B, **41**: 9–15.

———, ———, AND B. N. MOODY. 1955. Some aspects of the nutrition of marine yeasts and their growth. J. Sci. Ind. Res., Sec. C, **14**: 24–27.

FISHER, B. 1894. Die Bakterien des Meers nach den Untersuchungen der Plankton-Expedition unter gleichzeitiger Berücksichtigung einiger älterer und neuerer Untersuchungen. Ergebnisse der Plankton-Expedition der Humboldt-Stiftung, **4**: 1–83.

GREIN, A., AND S. P. MEYERS. 1958. Growth characteristics and antibiotic production of actinomycetes isolated from littoral sediments and materials suspended in sea water. J. Bacteriol., **76**: 457–463.

LODDER, J., AND N. J. W. KREGER VAN RIJ. 1952. *The yeasts. A taxonomic study.* North Holland Publ. Co., Amsterdam, Holland.

———, ———. 1955. Classification and identification of yeasts, part III. Lab. Pract., **4**: 53–57.

MEYERS, S. P., AND E. S. REYNOLDS. 1959. Growth and cellulolytic activity of lignicolous deuteromycetes from marine localities. Canad. J. Microbiol., **5**: 493–503.

NOVOZHILOVA, M. I. 1955. Kolichestvennaya kharakteristika, vidovoi sostav i rasprostranenie drozhzhevykh organizmov v Chernom, Okhototskom moryrakh i v Tikhom okeane. Trudy Inst. Mikrobiol., **4**: 155–195.

PHAFF, H. J., E. M. MRAK, AND O. B. WILLIAMS. 1952. Yeasts isolated from shrimp. Mycologia, **44**: 431–451.

WICKERHAM, L. J. 1951. Taxonomy of yeasts. U. S. Dept. Agr. Tech. Bull. No. 1029, pp. 1–55.

ZOBELL, C. E., AND C. B. FELTHAM. 1934. Preliminary studies on the distribution and characteristics of marine bacteria. Bull. Scripps Inst. Oceanog. Tech. Ser., **3**: 279–296.

———. 1946. *Marine microbiology.* Chronica Botanica Co., Waltham, Mass.

40

Reprinted from *Nature*, **223**(5208), 858 (1969)

Actinomycetes in North Sea and Atlantic Ocean Sediments

THERE have been few reports of actinomycetes in the sea, and then usually only in the littoral zone and inshore localities[1]. While making extensive microbiological surveys Kriss *et al.*[2] found actinomycetes only occasionally and so it was assumed that these microorganisms are confined to the terrestrial habitat, and that those isolated from ocean sediments are forms not indigenous to the sea.

In 1966 we investigated the population density of heterotrophic bacteria in the water and the top layer of sediment of the Weser estuary and the German Bight. Actinomycetes were rarely observed on agar plates inoculated with water samples, but they were seen more often if bottom sediment was used, when plates were incubated for 4–6 weeks at 18° C. We later made a survey for the selective detection of actinomycetes in the bottom sediments of the North Sea.

During five cruises of the fisheries research vessel Anton Dohrn, between 1967 and February 1969, sediments were collected at 107 stations in various parts of the North Sea using van Veen and Shipek grabs. Material from the top 2 cm of sediment was immediately inoculated on pour plates with six of twelve previously tested culture media (Table 1).

Table 1. COMPOSITION OF MEDIA

	Medium designation					
Components	SWA	DWA	CA	CHA	STNA	CZD
Peptone (g)	5	5	—	0·5	—	—
Peptone from casein (g)	—	—	1	—	—	—
Yeast extract (g)	1	1	—	0·1	—	—
KNO_3 (g)	—	—	—	—	3	—
$NaNO_3$ (g)	—	—	—	—	—	2
Saccharose (g)	—	—	—	—	—	15
Starch (g)	—	—	10	—	10	—
Chitin, hydrolysed and precipitated (g)	—	—	—	10	—	—
$FePO_4 \cdot H_2O$ (g)	0·01	0·01	0·01	0·01	0·01	0·01
Magnesium glycerophosphate (g)	—	—	—	—	—	0·5
Seawater (ml.)	750	—	750	750	750	750
Distilled water (ml.)	250	1,000	250	250	250	250
Agar (g)	15	15	15	15	15	15
*p*H	7·6 to 7·8	7·6 to 7·8	7·6 to 7·8	7·5	7·6 to 7·8	7·6 to 7·8

Of the 107 stations representing various types of sediment, 102 yielded colonies of actinomycetes in one or more types of culture medium. Actinomycete colonies derived from different samples often grew best on different culture media, suggesting that there is a variety of physiological strains or species.

There were 23–2,909 actinomycetes per cm^3. In coarse sand 30–69 m deep from the English Channel between Dover and the Isle of Wight were 92–1,485 per cm^3 (mean 510); in silt 435–690 m deep from the Skagerrak were 23–1,458 per cm^3 (mean 764); in various sediments 48–235 m deep from the central North Sea, including Devils Hole, were 23–115 per cm^3 (mean 54); in various sediments 76–164 m deep from the northern North Sea (transect Haugesund—Orkney Islands) were 23–230 per cm^3 (mean 128).

During a cruise of the research vessel Meteor in May 1968 similar results were obtained in the Atlantic Ocean off West Africa from 19° 00′ to 20° 40′ N. Sediment samples were taken from depths between 25 and 3,362 m and up to 175 nautical miles offshore. Nine out of twelve samples yielded actinomycetes although fewer colonies developed on the pour plates than when samples werefrom the North Sea. There were between twenty-three actinomycetes per cm^3 175 miles offshore at a depth of 3,362 m and 136 actinomycetes per cm^3 40 miles offshore at a depth of 299 m.

So far we have isolated 1,348 strains of actinomycetes from marine sediments. All pure cultures are being preserved in previously sterilized marine sediment for taxonomic investigation. The isolates include species of *Nocardia*, *Micromonospora*, *Microbispora* and *Streptomyces*. About half the strains develop aerial mycelia in pure culture.

In view of the high percentage of sediment samples that yielded actinomycetes and the fact that millilitre samples could be used for the detection of actinomycetes, these microorganisms in the sea seem to be neither random individuals nor temporary survivors of terrestrial run-off, but part of the marine ecosystem.

H. Weyland

Institut für Meeresforschung,
Bremerhaven.

Received May 5, 1969.

[1] Grein, A., and Meyers, S. P., *J. Bact.*, **76**, 457 (1958).

[2] Kriss, A. E., Mishustina, J. E., Mitskevich, J. N., and Zemtsova, E. V., *Microbial Population of Oceans and Seas* (Arnold, 1967).

41

Reprinted from *J. Bacteriol.*, **79**(4), 614 (1960)

INDIGENOUS MARINE BACTERIOPHAGES[1]

R. SPENCER

Humber Laboratory, Department of Scientific and Industrial Research, Hull, England

Received for publication November 24, 1959

The presence in sea water beyond the littoral zone of bacteriophages active against marine bacteria has not been conclusively demonstrated despite their possible importance in marine microbiology. Although ZoBell (*Marine Microbiology*, 1946) isolated phages from sea water of the littoral zone, he was unsuccessful with water taken beyond. The phages isolated by Kriss and Rukina (Rept. U. S. S. R. Acad. Sci., **57,** 833, 1947) from the Black Sea were active against such typically terrestrial species as *Bacillus subtilis* and *Micrococcus albus* (Kriss *et al.*, Trans. biol. Sta. Sebastopol, **7,** 50, 1949, and Kriss, *personal communication*) and may thus have been adventitious, whereas the phage isolated by Smith and Krueger (J. Gen. Physiol., **38,** 161, 1954) against a marine vibrio was not strictly marine in origin. Consequently, attempts were made to isolate phages from sea water taken well beyond the littoral zone active against typically marine bacteria.

The sea water samples were taken from the North Sea, some 10 miles off Aberdeen, Scotland, and the marine bacteria consisted of two groups, strains of *Photobacterium phosphoreum* isolated from marine fish, and strains of several species isolated from a further sample of sea water. After preliminary experiments, two methods of isolation were used in parallel, one direct and one indirect, each with incubation at both 20 and 0 C.

The direct method consisted of layering on a nutrient agar base a mixture of 10 ml of 1.5 per cent nutrient agar, 10 ml of sea water previously membrane filtered to remove interfering bacteria, and 2 ml of a dense suspension of the appropriate bacterial culture. After incubation, the presence of any phage active against that particular culture was detectable by plaque formation. Samples (600 ml) of sea water at both 20 and 0 C were normally examined at a time against a total of 40 strains of bacteria.

In the indirect method, 300 ml of sea water were mixed with 100 ml of quadruple strength nutrient broth and 8-ml amounts of broth culture of four different bacterial strains. After incubation, the bacteria were removed by either filtration or by an adaption of the chloroform technique of Fredericq (Compt. rend. soc. biol., **144,** 295, 1950) involving carbon tetrachloride, and the bacteria-free preparation tested for lytic action against the appropriate bacterial strains by the above layer technique. Samples (3 L) of sea water at both 20 and 0 C were normally examined at a time, again with 40 strains of bacteria.

In all, 6 L of sea water were examined by the direct method and almost 40 L by the indirect method, and seven phages were isolated. Four of the phages were detected by the parallel direct and indirect methods at both 20 and 0 C; three were present in samples in a concentration of 1 to 5 particles per 10 ml, and one in a concentration of approximately 100 particles per 10 ml. The remaining three phages were detected by the indirect method.

One phage was active against a strain of *Photobacterium phosphoreum*, three against unidentified nonpigmented *Pseudomonas* species, two against an unidentified Flavobacterium, and one against an organism provisionally classified as a *Cytophaga* species.

[1] The work described in this paper was carried out as part of the program of the Department of Scientific and Industrial Research.

42

Reprinted from *Bull. Misaki Mar. Biol. Inst. Kyoto Univ.*, **12**, 97–104 (Feb. 1968)

DISTRIBUTION, TAXONOMY AND FUNCTION OF HETEROTROPHIC BACTERIA ON THE SEA FLOOR*

J. Liston

College of Fisheries, University of Washington Seattle, Washington 98105, *U.S.A.*

The bacterial content of the water column in the sea shows wide fluctuations from area to area and from time to time. Thus in Puget Sound counts have been found to vary between <10 bacteria/ml to $>10^4$ bacteria/ml, and though counts in the open ocean are usually lower than this they show equally wide variations. Populations of bacteria in surface sediments on the continental shelf and slope are numerically much more stable (Hayes, 1964).

We have found consistently that counts of bacteria in shelf and slope sediment off the coasts of Washington and Oregon are *ca* 10^4 bacteria per gram of wet sediment (± 0.5 log) in samples taken from depths beyond the zone of mixing with fresh water (i.e., >50 meters). Populations in surface sediments in estuarine areas such as the Columbia River have been found to vary between 10^2 and 10^6 bacteria per gram and this has also been found in comparable areas of Puget Sound. Multiple samples taken from both anchored and drifting vessels in the same general area confirmed the quantitative uniformity of sediment bacterial populations (Table 1).

Table 1. Examination of (Intrastation Sample Variability) Mean and Standard Deviation for Logarithmically Transformed Counts of Multiple Sampled Stations

Region	Cruise	Station depth in fathoms	No. of Samples	Mean (log)	SD (log)
Puget Sound	5	100	10	4.326	0.297
	10	112	11	4.683	0.469
Washington Coast	9a	50	5	4.748	0.458
		850	5	3.702	0.406
Pacific Ocean off the Columbia River	10	50	3	4.485	0.264
		350	3	4.625	0.332
		850	4	3.954	0.265
Puget Sound		13	4	5.003	0.168

The small quantitative variation in bacterial populations on sediments in deeper areas as compared with those in shallow inshore areas of the sea undoubtedly reflects the stability and variability of the two environments. This is also

---

* Contribution No. 276, College of Fisheries, University of Washington, Seattle, Washington 98105. The results reported in this paper were derived mostly from the work of Dr. W. J. Wiebe, Dr. Bruce Lighthart; Mr. John Chan and Mr. John Baross in investigations supported by NSF Grant G19434 and ONR Contract Nonr 477(41).

apparent in the qualitative differences between inshore and offshore bacteria.

Bacteria isolated from inshore sediments show marked physiological differentiation into psychrophilic, mesophilic, halophobic, and halophilic groups though some are eurythermic and euryhalotolerant. Remarkably, most bacteria from deeper shelf and slope sediments are quite undifferentiated in this respect, being predominantly eurythermic and euryhaline (Tables 2, 3). However, the majority of these bacteria yield more rapid and abundant growth in media containing sea water. The response of sea-water-exacting sediment bacteria to elevated levels of various cations is complex and associated with physical factors such as temperature, but synthetic media made with distilled water containing Na^+ and Mg^{++} added at levels commonly found in sea water, provided for normal metabolism in most (but not all) cases.

Table 2. Salt and Temperature Relations of Marine Benthic Bacteria from Columbia River Area of Pacific Ocean

Isolation Zone:	Up River	Mixing (54 M)	Shelf (135 M)	Slope (120/810 M)
		% Incidence of Physiological Types		
Obligate SW	50	23	6	2
Euryhaline	50	77	94	98
Psychrotrophic	55	61	23	12
Psychrophilic	(1)*	(1)*	(2)*	(0)*
Eurythermic	41	32	68	77
No. in sample	30	45	63	41

* Actual numbers of organisms

Table 3. Temperature and Media Effects on 432 Bacteria from the Columbia River Trackline

	ca 1°C		28°C		37°C	
	SWA	FWA	SWA	FWA	SWA	FWA
	percent showing growth					
All isolates	90	68	96	78	61	39
Estuarine and inshore*	97	60	98	43	31	11
Offshore**	87	72	96	86	76	50

* 124 Organisms
** 308 Organisms

The physiological differentiation of the estuarine and inshore bacterial populations is believed to arise from the unstable nature of the environment in such areas. Salinity, temperature and nutrient composition of the overlying water will change with tide, fresh-water run off, seasonal and even periodic air temperature changes, etc. In addition, there is much greater exchange of bacteria between terrestrial sources and these inshore environments. This is apparent

Table 4. Salt Responses of Benthic Bacteria Isolated on Fresh Water Agar (0.5% NaCl)

Isolation Zone:	Up River	Mixing (90 M)	Shelf (135 M)	Slope (540 M)	No. of Isolates
		(Percent showing response)			
Halophobic	84	27	0	0	28
Euryhaline	26	83	100	100	81

in the occurrence of halophobic bacteria in such areas (Table 4). The lack of physiological differentiation among bacteria taken from deeper sediments probably reflects the stability of the offshore environment. We can only infer that the versatility shown by most strains in their response to varying salt concentrations and temperatures of growth is a reflection of their broad based capability of dealing with the nutritional and other problems of life in the deeper waters since no obvious advantage could accrue to a deep benthos bacterium from an ability to grow at 37°C or in the virtual absence of sea salts.

Enrichment procedures have confirmed that sediments contain microorganisms capable of dealing with a wide variety of substrates, including simple and complex carbohydrates, amino acids, proteins, etc. Chitin digestion is shown, usually at a low level, by all sediments, but cellulose hydrolysis and agar digestion appear to be common only in inshore sediments (e.g., Columbia River estuary and Puget Sound). Estimates of unit incidence of particular characteristics by a Most Probable Number Count procedure indicate that most degradative characteristics are present at a level corresponding closely with the count of heterotrophic bacteria (Table 5). This indicates a universally biochemically versatile population in sediments rather than a composite of individually specialized groups of organisms.

Table 5. Biochemical capabilities of sediment samples expressed as a ratio of the total plate count

	Alanine[1]	Glucose	Gelatin	PSW	Mean SWA Count × 100
Puget Sound*					
Depth 30M (H)	14.1	100.8	67.8	100	341.0
40M (I)	31.9	387	36.6	100	255.0
100M (C)	1.6	92.4	99.1	100	100.0
135M (D)	4.3	691.8	0.4	100	167.0
Offshore**					
90M	0.25	0.25	1.3	1.5	260
640M	0.30	0.25	0.9	2.3	73
1550M	0.60	0.40	0.55	4.3	51.5

* Averaged values for 3 cruises

** Averaged values for 2 cruises

1 Ability to use alanine as sole source of C, N, and energy

The total biochemical capability of sediment and overlying water column is similar in Puget Sound, where fairly extensive tests have been run. Comparison of the dilution count/total count ratios for various characteristics for Puget Sound and offshore sediment samples reveals that there is much closer concordance between the various offshore values than among Puget Sound samples. This is further evidence of the diversified nature of the inshore flora as compared with the offshore benthos bacteria. Confirmation of this interpretation is provided by a tabulation of characters of high frequency in offshore sediment with characters of high frequency in organisms isolated from sediments which reveals that most of the sediment characters examined occur concomitantly in a large proportion of the organisms (Table 6).

Table 6. Incidence of Characters in Sediment and in Isolated Bacteria

	55M		270M		165M	
Character	S*	B**	S	B	S	B
Gelatin	+	+	+	+	+	+
Alanine	+	+	+	+	+	+
NH_3	+	+	+	+	+	−
NO_3	+	+	+	+	−	−
Glucose	+	+	+	+	+	+
Galactose	+	+	+	+	−	−
Mannitol	+	−	+	+	+	+
Celluobiose	+	+	+	+	+	+
Citrate	−	−	+	+	−	−
Cysteine	−	+	+	+	+	+
No. of strains	52		52		30	

* Character present in sediment at dilution equivalent to total count

** Characters present as a group in 40% of the bacteria tested

The heterotrophic bacteria derived from sediments are mainly gram negative rod-shaped organisms. In our studies, over 70% of sediment bacteria examined have proved to be aerobic Pseudomonas. Achromobacter and Aeromonas and the gram positive coryneforms, Bacillus and Micrococcus make up the remaining 30% or so. It is notable that Vibrio and Flavobacterium groups which are quite common in the water column appear to be much rarer in sediment samples. Taxonometric procedures have proved useful in separating the bacteria into sub-generic groups, and the median organism concept provides a simple method for clustering similar organisms and at the same time delineating their principal characteristics. Bacteria derived from inshore samples are more easily classified than offshore organisms due to the biochemical homogeneity of the deeper sediment isolates.

Generally, the total biochemical capabilities of organisms from inshore, shelf and slope samples are similar. However, the bacteria from shallower regions show a degree of biochemical specialization which is not apparent in bacteria from deeper samples. These deeper sediment bacteria share a common wide range of characteristics and thus exhibit a high degree of metabolic versatility concomitantly with a wide physiological range noted earlier.

The rather small number of bacteria from the water columns in Puget Sound which have been examined indicate a higher incidence of fermentative and sea water requiring bacteria in this environment.

Particular studies have been conducted in our laboratory on the breakdown of refractory substances such as cellulose and chitin by benthos bacteria.

Cellulose digesting bacteria isolated from enrichments of sea water and sediment have proved to be Cytophaga (most commonly) or aerobic Pseudomonads. Chitinoclastic bacteria have proved to be mainly facultative Vibrio types. Agar digesting bacteria include Cytophaga and Pseudomonas and Vibrio types. Commonly, of course, the cellulose digesting Cytophaga will also break down agar (Table 7).

In experiments using our artificial sea bed system, we have studied the degradation of cellulose and chitin by the natural populations of sea sediments. Mostly aerobic Gram negative bacteria have been isolated from these systems. The population changes in cellulose breakdown are quite well marked with cellulose digesting bacteria dominating early and cellobiose utilizers arising

Table 7. Characteristics of Cellulose Digesting Bacteria

Properties	Organism Number 1	2	3	4	5	6
Gram stain	NEG	NEG	NEG	NEG	NEG	NEG
Temp. range, °C	8-22	8-30	8-30	8-30	8-37	8-37
Motility	GLID	GLID	+	+	−	−
Agar digestion	+	+	−	+	−	+
Cellulose digestion	+	+	+	+	+	+
Starch hydrolysis	+	+	ND	ND	ND	ND
NO_3 reduction						
a)→NO_2	+	+	−	+	−	−
b)→NH_3	−	−	−	−	−	−
Glucose	−	−	−	A+G−	−	−
Galactose	−	−	−	A+G−	−	−
Gelatin	+	+	+	−	+	+
Anaerobic growth	±	−	−	+	+	−
Pigment	Yellow	Yellow	White	White	White	White
Chitinase	−	−	−	−	−	−
SW requirement	−	−	+	−	+	+

ND=No data

NOTE: 1) Organisms 1 and 2 are cytophaga

2) Organisms 3, 4, 5, and 6 displayed esoteric morphological variations, thereby, prohibiting their classification into families.

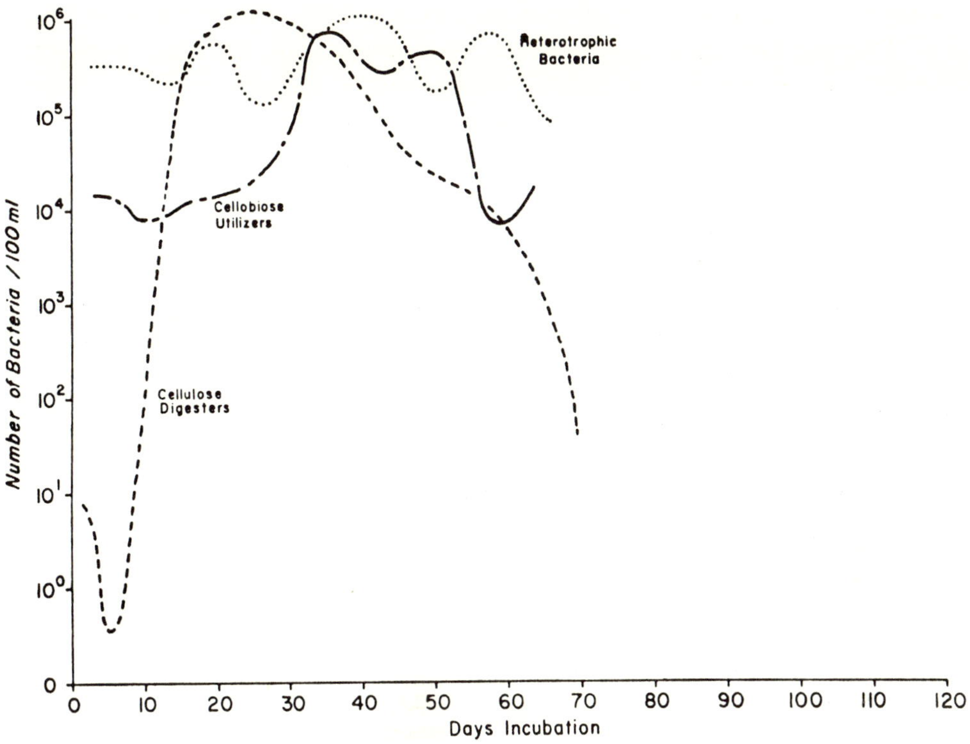

Fig. 1. Plot of the number of indicated bacteria 1-5 mm. above the sediment in an artificial seabed 0-80 days after cellulose substrate addition.

later (Fig. 1). Chitin digestion has been shown to result in a net conversion of chitin C to biomass C. It is noteworthy that the efficiency of conversion is much higher for offshore sediment than for Puget Sound sediment (Table 8). This is assumed to be another reflection of the difference between the microfloras of the two areas. Activity is greater in the case of inshore mud, but more C is oxidized to CO_2 or otherwise lost to the ecosystem.

The versatility of function shown by the bacteria from deeper waters' sediment is believed to be related to their successful survival and growth in the marine environment. Such organisms are well equipped to take advantage of the constant low level supply of organic materials of widely differing chemical com-

Table 8. Chitin-Biomass Conversion in Model Sea Bed

Sediment	Chitin C digested	Biomass produced	i.e. Conversion
Pacific Ocean (Slope)	47.2	22.8	48.4%
Puget Sound (100F)	118.6	12.64	10.7%

position which might be expected to be made available to deep sea bacteria.

Protozoa are commonly associated with bacteria in the natural environment. For lack of a better term, we call these bacteriovorous protozoa, though clearly not all of them are actually ingesting bacteria *per se*. We have observed such protozoa in more or less large numbers in sea bed experiments. Indeed, since they were more numerous in the Puget Sound mud system during the chitin experiments referred to earlier, they may have been partly responsible for the reduced efficiency of C conversion in that system.

Assays of the incidence of protozoa in water and sediment have been made using two methods: the Singh enrichment plate technique, and a direct dilution procedure. Protozoa are much more abundant in inshore waters and sediment than in offshore samples (Tables 9, 10).

Table 9. Protozoa and Bacterial Counts in Sediment from the Pacific Ocean

Depth	Bacteria SWA Plate	Protozoa Singh Plate	Dilution Estimate
90M	3×10^3	<5	240
800M	2.8×10^3	<5	<0.5
1520M	4×10^3	<5	<0.5

Table 10. Protozoa and Bacterial Counts in Sediment from Puget Sound

Station	Depth	Bacteria SWA plate	Protozoa Singh plate
A	30M	2.85×10^4	62
B	230M	5.1	62
C	100M	10.0	620
D	110M	2.75	13
F	25M	0.64	.6
G	27M	8.0	21
H	30M	7.8	240
I	40M	6.1	240

In terms of the ecology of the benthos, the sediment bacteria serve a number of functions. Quite clearly, they are responsible to some extent for the degradation of organic materials. But the absence of large numbers of chitin digesting types or other bacteria specialized to deal with refractory substance such as cellulose in deeper sediments suggests that mineralization may not be the primary ecological function of these bacteria. Experiments using model sea bed systems have indicated that sediment bacteria *in situ* can extract very low levels of dissolved organic material from sea water and convert it into biomass which enters the benthic food chain via bacteriovorous protozoa (Table 11). Since there is an enormous quantity of organic material dissolved in sea water (at low con-

Table 11. Conversion of Dissolved C and Sediment C in Model Sea Bed

Sediment	Initial C	Biomass C produced	Water flow
Pacific Ocean	N.A.	15.5 mg	39.23 l
Puget Sound	3.4 mg	18.78 mg	18.60 l

centration), the amount of organic conversion to biomass which proceeds by this route is only limited by the rate of assimilation and biosynthesis of the benthic bacteria and the cropping efficiency of the protozoa. Thus, sediment bacteria might be considered to have two important functions in the sea, as mineralizers and as primary converters in the benthic food chain.

References

Hayes, F. R. 1964. The Mud-water Interface. *Oceanograph. Mar. Bid. Ann. Rev.*, **2:** 121-145.
Wiebe, W. J. 1965. Ph.D. Thesis, University of Washington, Seattle, Wn.

43

Reprinted from *J. Bacteriol.*, **80**(6), 731–736 (1960)

GEOGRAPHIC REGULARITIES IN MICROBE POPULATION (HETEROTROPH) DISTRIBUTION IN THE WORLD OCEAN

A. E. KRISS, S. S. ABYZOV, M. N. LEBEDEVA, I. E. MISHUSTINA, AND I. N. MITSKEVICH

Institute of Microbiology, Academy of Sciences, Moscow

Received for publication February 8, 1960

The geography of plants and animals has been studied extensively, whereas that of microorganisms has until recently received little attention. It was only within the last decade or so that data on geographic regularities of microorganism occurrence in soil have become available (Mishustin, 1947, 1958; Mishustin and Mirzoyeva, 1953).

Microbiological investigations in extensive parts of the world ocean (1954 to 1959) have given data on the abundance and species pattern of microbes (heterotrophs) in geographic areas of both hemispheres, the investigations being carried out within the range of the central Arctic to Antarctic seas (Kriss, 1959). Figure 1 shows the distribution of microbiological stations at which sampling was carried out at different ocean depths in subtropic, tropic, and equatorial, areas of the Pacific Ocean; in equatorial, tropic, subtropic, subantarctic, and antarctic areas of the Indian Ocean; in the Greenland Sea; in the Norwegian Sea; in the Atlantic Ocean (along 30°W) off the Polar Circle down to the south tropic and in the central Arctic region.

METHODS

Sampling was carried out with the aid of a bathometer on standard levels: 0, 10, 25, 50, 75, 100, 150, 200, 250, 300, 400, 500, 600, 800, 1,000, 1,500, 2,000, 2,500, 3,000 m, and at each thousand m below 3,000. In some instances, sampling was carried out on intermediate levels. In the Central Arctic, in the Greenland Sea, in the Norwegian Sea and, with few exceptions, in the Atlantic and Indian Oceans the whole water column was studied at the levels outlined above, from the surface layer down to the floor. In the Pacific Ocean about half the stations were suspended ones; that is, the water was sampled down to 2,000 to 2,500 m in depth.

Selection of samples to be analyzed was made with all microbiological precautions in the laboratories set up either on ships or drifting scientific stations. When the sampling was completed, the water was immediately filtered through membrane ultrafilters in the amounts of 35 to 50 ml. Afterwards, the filters were put with their back side on the surface of nutrient agar (60 per cent tryptic hydrolyzate of fish flour, 40 per cent agar) in petri dishes for growth of microbes sedimented on the filter; the nutrient medium was prepared using ocean water. After 3 to 7 days of incubation at temperatures of 18 to 35 C, the colonies were counted. Representatives of unlike colonies were separated and placed into agar of the same composition. The total number of samples studied was approximately 4,000.

RESULTS

The results suggest a quantitative distribution of microorganisms propagating on albuminous media in diverse geographic zones of the world oceans. Heterotrophic microorganisms of this kind use easily assimilated organic matter and, therefore, can serve as distribution indicators of only slightly transformed organic matter which is still in the initial stage of decay.

In the Pacific Ocean, abundance of heterotrophic microorganisms differs in subtropic and tropic areas (table 1). The lowest content of heterotrophs was observed in the subtropic areas, and the highest, in the equatorial area. In the subtropic areas, 50 ml water samples counting 0 to 9 colonies equalled 44 per cent (40° to 23°W) and 38 per cent (23° to 40°S), but those giving rise to over 100 colonies fluctuated in the range of 14 per cent in the southern subtropic area up to 23 per cent in the northern subtropic area. In the tropic areas, however, water samples with a high content of heterotrophs were predominant; in the equatorial area their percentage was three times greater than the percentage of samples with a low content of heterotrophs (52 per cent of samples with over 100 colonies as contrasted with 17 per cent of samples with 0 to 9 colonies).

TABLE 1

*Quantitative correlation of heterotrophs in diverse geographic zones in Pacific Ocean**

Latitudes	No. of Water Samples Analyzed	No. of Colonies on Filters		
		0–9	10–99	Over 100
40°N–23°N	152	44.4	32.5	23.1
23°N–10°N	165	18.4	41.2	40.4
10°N–10°S	278	16.6	31.3	52.1
10°S–23°S	160	24.7	48.7	26.6
23°S–40°S	272	37.6	48.5	13.9

* The figures show the percentage of water samples, counting a respective number of colonies as correlated with the total number of samples analyzed.

One must take into consideration that summary data on the Pacific Ocean cannot help but obscure the differences caused by dry land closeness, currents, and other factors. We have described sharp differences in the vertical distribution of heterotroph abundance in the water column layers, the differences being conditioned by currents carrying antarctic and arctic waters to the tropic areas and equatorial-tropic waters to high latitudes (Kriss, Lebedeva, Abyzov and Mitskevich, 1958).

The comparison of two sections in the Pacific Ocean (central area) shows quite pronounced differences between them. To the north of the equator, stations on 172°E were located in the region affected by the powerful Kuroshio current, the latter being a branch of the north-equatorial current rich in organic matter. This accounts for the strikingly small number of samples with 0 to 9 colonies and, on the other hand, the great number of samples with over 100 colonies in the northern subtropic and tropic zones on the 172°E section as compared with the stations in the southern part of this section and also with the stations in the same zones on the 174°W section (table 2).

In the eastern section of the Indian Ocean the microbiological stations were located in all geographic zones within the range of Antarctica to the north tropic, whereas in the western part of the Indian Ocean the stations were located to the south of the subtropic convergence area (figure 1).

As in the Pacific Ocean, there was observed a decrease in microbe (heterotroph) abundance with the transition off the equator to Antarctic areas. Table 3 shows that 40 ml samples showing 0 to 9 colonies increased from 7.5 per cent in the equatorial zone up to 73 to 74 per cent in the Antarctic waters; at the same time, the number of samples counting over 100 colonies decreased from 89 per cent around the equatorial area to 13 per cent in the vicinity of Antarctica. It should be pointed out that quite evident differences in the occurrence of heterotroph microorganisms in the water column, place eastern areas apart from western ones in the Indian Ocean to the south of 35°S. As compared with western areas limited by 35°S and 70°S, those in the east showed a lower percentage of samples counting few heterotrophs and a higher percentage of samples counting abundant heterotrophs (table 4).

In the Atlantic Ocean where microbiological

TABLE 2

*Quantitative distribution of heterotrophs on diverse latitudes in Pacific Ocean along 172°E and 174°W**

Latitudes	Section along 172°E				Section along 174°W			
	No. of water samples analyzed	No. of colonies on filters			No. of water samples analyzed	No. of colonies on filters		
		0–9	10–99	Over 100		0–9	10–99	Over 100
40°N–23°N	54	9.3	44.4	46.3	98	79.5	20.5	0
23°N–10°N	111	3.6	37.8	58.6	54	33.3	44.5	22.2
10°N–10°S	196	4.1	32.1	63.8	82	29.2	30.5	40.3
10°S–23°S	96	22.9	49.0	28.1	64	26.6	48.4	25.0
23°S–40°S	123	34.2	48.8	17.0	149	40.9	48.3	10.8

* The figures show the percentage of water samples, counting a respective number of colonies as correlated with the total number of samples analyzed.

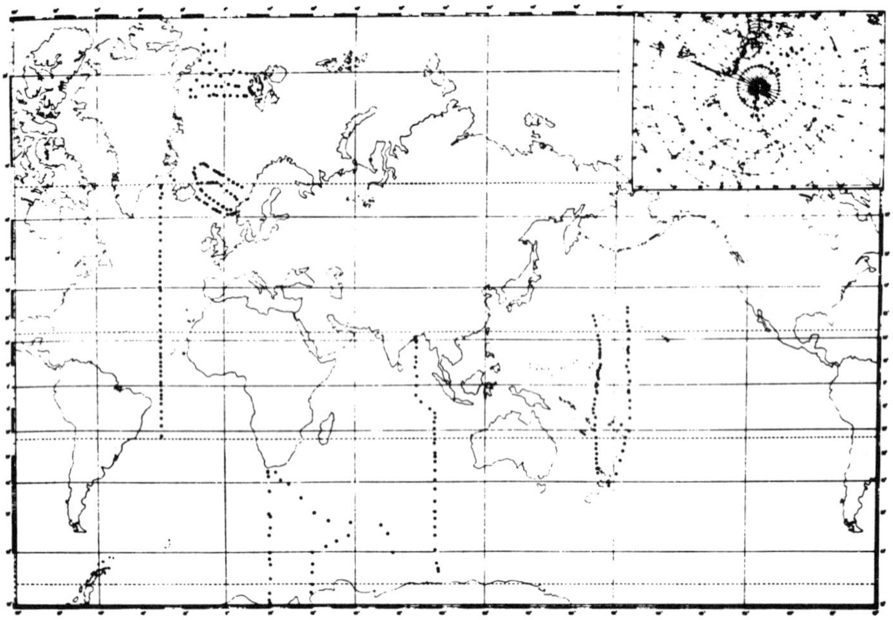

Figure 1. Chart of microbiological stations in the world ocean

TABLE 3

*Quantitative correlation of heterotrophs in diverse geographic zones in Indian Ocean**

Latitudes	No. of Water Samples Analyzed	No. of Colonies on Filters		
		0–9	10–99	Over 100
10°N–21°N	74	1.4	5.4	93.2
10°N–10°S	106	7.5	3.8	88.7
10°S–23°S	89	11.2	14.6	74.2
23°S–35°S	58	17.2	36.2	46.6
35°S–50°S	254	58.2	20.1	21.7
50°S–60°S	263	73.9	12.6	13.5
60°S–70°S	273	72.8	14.1	13.1

* The figures show the percentage of water samples, counting a respective number of colonies as correlated with the total number of samples analyzed.

stations were placed along 30°W from Greenland down to the south tropic, there was a similar decrease in heterotroph abundance with the gradual transition off the equatorial-tropic areas (table 5). In the 40 to 66°N latitudes, the number of samples counting only few heterotrophs reached 90 to 99 per cent. There were no samples with over 100 heterotrophs.

Heterotrophs which use easily assimilated organic matter are scarce in the Greenland Sea and Central Arctic, investigations on heterotroph occurrence in the Arctic Ocean water column being carried out in the area between the North Pole and 83°N. Most samples (79 to 85 per cent) showed quite insignificant heterotroph counts. (table 5).

The Norwegian Sea stands out in heterotroph content as compared with the subarctic area (along 30°W) of the Atlantic Ocean and subantarctic areas. In the Norwegian Sea 65 per cent of the samples showed relatively great heterotroph numbers, thus coming close to the subtropic area of the North Atlantic Ocean. There is a sharp difference between northern and southern areas of the Norwegian Sea: 27 per cent of southern area samples counted over 100 colonies, but only 2 per cent of northern area samples were rich in heterotrophs. Undoubtedly, it is the Gulf Stream that contributes much to the peculiarities of the Norwegian Sea hydrology.

Thus, the results suggest the existence of clear-cut geographic zones in the world ocean distribution of microbes which use easily assimilated organic matter. The significance of these microbic forms in oceanology lies in the

TABLE 4

*Quantitative distribution of heterotrophs on diverse latitudes in eastern and western areas of Indian Ocean**

Latitudes	Longitudes: 80°-90°-100°E				Longitudes: 20°-40°-60°E			
	No. of water samples analyzed	No. of colonies on filters			No. of water samples analyzed	No. of colonies on filters		
		0-9	10-99	Over 100		0-9	10-99	Over 100
35°S-50°S	97	46.5	22.7	30.8	157	70.0	17.3	12.7
50°S-60°S	106	63.1	20.8	16.1	157	84.7	4.5	10.8
60°S-70°S	45	60.0	20.0	20.0	228	85.6	8.3	6.1

* The figures show the percentage of water samples, counting a respective number of colonies as correlated with the total number of samples analyzed.

TABLE 5

*Quantitative correlation of heterotrophs in diverse geographic zones in Atlantic Ocean, Norwegian Sea, Greenland Sea, and Central Arctic**

Latitudes	No. of Water Samples Analyzed	No. of Colonies on Filters		
		0-9	10-99	Over 100
Atlantic Ocean				
22°S-11°S	81	2.5	20.0	77.5
8°S- 8°N	167	3.0	20.0	77.0
10°N-21°N	121	1.6	33.9	64.5
24°N-38°N	104	44.3	38.4	17.3
40°N-48°N	88	90.0	10.0	0
50°N-58°N	78	98.7	1.3	0
60°N-66°N	71	95.7	4.3	0
Norwegian Sea				
60°N-70°N	575	35.0	54.0	11.0
60°N-65°N	203	14.9	57.6	27.5
64°N-70°N	372	46.6	51.6	1.8
Greenland Sea				
78°N-83°N	448	78.8	16.5	4.7
Central Artic				
83°N-90°N	115	85.0	15.0	0

* The figures show the percentage of water samples, counting a respective number of colonies as correlated with the total number of samples analyzed.

fact that their quantitative distribution in seas and oceans affords the measure of distribution of organic matter not transformed into humus and fairly available to the action of hydrolytic ferments.

It seems beyond any doubt that in the high latitude regions of the world ocean (arctic and antarctic areas; subarctic and subantarctic areas) the density of the microbe population is most insignificant. Although these areas are most rich in plant life conditioning ocean water productivity, the metabolic products and the latter's defunct bodies do not produce such food material supply for heterotrophs as occurs in the world ocean tropic area.

Strikingly high was the occurrence of microbe life (heterotroph) in the tropic and especially equatorial areas of the Pacific, Indian, and Atlantic Oceans where the investigations were carried out. With respect to heterotroph abundance these areas stand out sharply as compared with other world ocean geographic zones, thus coming close to numerous small fresh and saline water basins on land in the temperate zone of the northern hemisphere. It seems quite paradoxical that high occurrence of microbes (heterotrophs) is observed in the world ocean areas in which, as compared with high latitudes, plant and animal population is most scarce.

A certain exception falls on the so-called intertrade-drift area in the Pacific Ocean equatorial zone where plankton and necton propagation is more favorably conditioned (Bogoroff, 1959). Still, the area is much more barren of life (except bacterial life) than the subarctic region in the Pacific Ocean. Thus, while heterotroph content in the world ocean water column increases with the transition off the polar areas to the equator, other life forms are characterized by the reverse phenomenon; their abundance increases with the transition off the equatorial-tropic area.

This paradox might be explained by the fact that in the equatorial-tropic areas of the world ocean the chief source conditioning the microbe abundance is slightly transformed, not fully decomposed organic matter of allochthonous nature. There is no doubt that the abundance of this organic matter in the equatorial-tropic areas is much higher than that of analogous autochthonous and allochthonous organic forms found in high latitude waters.

It is believed that the enrichment of equatorial-tropic waters in the Pacific and Indian Oceans by organic matter easily accessible to the action of hydrolytic ferments of microorganisms takes place mainly in the Coral Sea. In this area and in that of the Australasian Archipelago where Indian and Pacific Oceans meet, a great number of islands rich in plant and animal life fill the circumfluent waters with organic matter in forms which are good nutrient materials to heterotrophs and, therefore, condition their active propagation.

Being driven by equatorial and counterequatorial currents, these waters enrich the whole water column of the equatorial-tropic zones in the Pacific and Indian Oceans, particularly Pacific Ocean (western areas) and Indian Ocean (eastern areas). On the 172°E section there was a higher content of heterotrophs than on the 174°W section in the Pacific Ocean (eastern part) (table 2).

In the Indian Ocean, where deep currents drive equatorial-tropic waters to high latitudes as far as the Antarctic coast, subtropic, subantarctic, and antarctic areas have a higher content of heterotrophs in the eastern than in the western half (table 4).

In the Atlantic Ocean the unusually high abundance of slightly transformed organic matter in the water of the equatorial-tropic zone seems to be caused by the discharge of such powerful rivers as the Amazon and Orinoco, Congo and Niger, with their catchment on spacious American and African equatorial areas densely populated by plant and animal life. A certain role, as a source for the allochthonous organic matter, is played by numerous islands in the Caribbean Sea into which branches of the equatorial current enter.

In this respect, Kolbe's findings (1957) of fresh water diatom membranes in the sediments of the Atlantic Ocean equatorial zones are particularly interesting. In bottom samples taken thousands of kilometers off the African coast, he found over 1000 valves of fresh water diatoms per slide. Considering the great distance from land where diatoms were found, the great ocean depths, and the high occurrence of diatom valves, one cannot overestimate the influence of the discharge of equatorial Africa, driven further by trade drifts, on the content of organic and inorganic matter in the water mass of the Atlantic Ocean equatorial-tropic zone. A strong northeaster carrying from the African continent dust clouds of such thickness that they sometimes limit the visibility to 1 to 2 km or even up to 150 m in the Atlantic Ocean area between 15°N and 5°S in close proximity to Africa, is a means of transportation of terrigenous organic substances into the waters of north- and south-equatorial currents which begin at the African coast.

Consideration of the data which characterize the stratification of waters with unusually high heterotroph abundance in diverse areas of the world ocean (Kriss, Lebedeva, Abyzov, and Mitskevich, 1958) makes evident the following:

a. Currents driving equatorial-tropic waters as far as the highest latitudes distribute allochthonous organic matter in the whole water body of the world ocean. Taking into account the quantity in these waters of heterotrophs mineralizing organic matter, one should not underestimate the role of allochthonous organic matter as a source of biogenic matter determining ocean water productivity.

b. Deep currents driving equatorial-tropic waters are not at all slow as it has been believed until recently. Otherwise, it would be difficult to explain the presence in deep layers in temperate zone latitudes and higher latitudes of such abundance of easily decomposing organic matter which (abundance) brings about such high heterotroph content in these layers.

SUMMARY

Microbiological investigations in the Pacific, Indian, Atlantic, Arctic, and Antarctic Oceans, and in the Norwegian and Greenland seas (1954 to 1959) have embraced all geographic zones from the North Pole to the Antarctic seas. The comparative analysis of about 4000 samples taken at diverse depths of the ocean water column has enabled us to form an idea of regularities in microbe population (heterotroph) distribution in the world ocean.

The high latitude regions of the world ocean (arctic and antarctic, subarctic and subantarctic) have the lowest microbial population (heterotroph) density. Remarkably high is the heterotroph content in equatorial-tropic areas of the Pacific, Indian, and Atlantic Oceans.

Plant and animal life increases in abundance in the ocean water body with the transition off the equatorial-tropic area; the microbial distribution, on the other hand, is characterized by a reverse phenomenon: the abundance of heterotrophs increases with the transition off the Polar areas to the equator.

This disagreement can be explained by the fact that in the world ocean equatorial-tropic zone the heterotroph abundance in waters is determined by slightly transformed, not fully decomposed organic matter of allochthonous origin. Besides, it is evident that in the equatorial-tropic areas, the content of this organic matter is greater than that of analogous forms of autochthonous and allochthonous organic matter in high latitude waters.

The enrichment of waters in the equatorial-tropic zone of the Pacific and Indian Oceans by organic matter easily accessible to microorganisms, takes place chiefly in the area of numerous islands rich in life in the Coral Sea and Australasian Archipelago regions; in the Atlantic Ocean this is conditioned by the discharge of the Amazon and Orinoco, Congo and Niger Rivers which have their catchment on spacious areas in equatorial America and Africa.

REFERENCES

Bogoroff, V. 1959 Biologischeskaya structura okeana. [Biological structure of the ocean.] Doklady Akad. Nauk S. S. S. R., **128** (4), 819–822. [In Russian.]

Kolbe, R. 1957 Fresh-water diatoms from Atlantic deep-sea-sediments. Science, **126**, 1053–1056.

Kriss, A. 1959 Morskaya microbiologiya (glubokovodnaya). [Deep-sea microbiology.] Izdaniye Akad. Nauk S. S. S. R., 455 p. [In Russian.]

Kriss, A., M. Lebedeva, S. Abyzov, and I. Mitskevich 1958 Microorganismi kak indicatori gidrologicheskih yavleniy v moryah i okeanah. [Microorganisms as indicators of hydrological phenomena in seas and oceans.] Jurnal Obshchey biologii, **19** (5), 397–413. [In Russian.]

Mishustin, E. 1947 Ekologo-geographicheskaya izmenchivost pochvennih bakteriy. [Ecological-geographic variability of soil bacteria.] Izdaniye Akad. Nauk S. S. S. R., 326 p. [In Russian.]

Mishustin, E. 1958 Geographicheskiy factor i rasprostraneniye pochvennih microorganismov. [The geographic factor and soil microorganism distribution.] Izvest. Akad. Nauk S. S. S. R. Ser. Biol., No. 6, 661–676. [In Russian.]

Mishustin, E., and V. Mirzoyeva 1953 Sootnosheniye osnovnich grupp microorganismov v pochvah raznih tipov. [Correlation of chief microorganism groups in soils of diverse kind.] Pochvovedenie, No. 6, 1–10. [In Russian.]

Reprinted from *Deep-Sea Res. Oceanogr. Abst.*, **18**, 1111–1121 (1971)

Distribution and activity of oceanic bacteria

JOHN MCN. SIEBURTH*

(*Received* 29 *March* 1971; *in revised form* 13 *May* 1971; *accepted* 2 *June* 1971)

Abstract—A survey was conducted on 381 samples collected to depths of 1200 m at 47 offshore stations in the Pacific, Caribbean and Atlantic. All samples were cultured in Agar roll tubes and incubated within 4·5°C of the temperature of the water from which they were obtained. The heterotrophic population thus obtained was tested directly for a few biochemical activities or isolated, tested more extensively and keyed to genus. A marked bacterial flora in the 150 μ thick surface film overlying 'low count' waters was observed only off the west coast of Central America in an area known for upwelling. The ratio of lipolytic:proteolytic:amylolytic bacteria in this flora was 93:92:24 compared to a ratio of 46:53:54 for 'high count' surface and subsurface samples which were obtained sporadically from the Caribbean and Atlantic. Atypical pseudomonads dominated the Pacific samples, while types two and four pseudomonads were twice as high in the Caribbean and Atlantic samples. The bacteria in the more typical 'low count' samples had biochemical activities which differed from the 'high count' samples. The 'low count' flora had similar proteolytic:amylolytic ratios regardless of latitude, depth or oceanic area.

1. INTRODUCTION

THE GENERAL patterns for the distribution of heterotrophic bacteria in the sea were established by studies at the turn of the century (ZOBELL, 1946). KRISS (1963) has summarized his work and that of his colleagues on the distribution of heterotrophic bacteria in the principal oceans and has suggested that bacteria are sensitive indicators of water mass. JANNASCH and JONES (1959) have compared direct microscopic and cultural methods of estimation and have shown that as little as 1/13 to 1/10,000 of the cells present are detected by cultural procedures. The determination of adenosine triphosphate (ATP) in material filtered from suspension (HOLM-HANSEN and BOOTH, 1966) is a sensitive measure of microbial biomass and may indicate heterotrophic biomass in the aphotic zone, but in the photic zone it fails to distinguish between photosynthetic and heterotrophic biomass. Despite their inadequacies, only cultural procedures yield organisms for further study. In addition, these populations may be an index of less fastidious micro-organisms which bloom during periods of organic production, that is, the zymogenous microflora.

Although there is much interest in the physics and chemistry of the sea surface (EWING, 1950; BLANCHARD 1963; JARVIS, GARRETT, SCHIEMAN and TIMMINS, 1967), apparently only two studies have been made in the last decade on the microflora at the air–water interface (SIEBURTH, 1965; TSIBAN, 1967). Despite differences in water temperature up to 30°C from the surface to the depths in tropical waters, KRISS (1963) appears to incubate only at cabin temperatures, while ZOBELL and CONN (1940) found that 18°C gave maximal counts regardless of whether the samples were from cool deep waters or warm shallow waters. The great influence of water temperature on the selection of thermal types and the minimal number of incubation temperatures required to cultivate them has been determined by SIEBURTH (1967). In this study, data are presented in which samples of the surface film and subsurface samples to 1200 m

*Graduate School of Oceanography, University of Rhode Island, Kingston, Rhode Island 02881.

were cultured at temperatures within 4·5°C of the water temperature. Sampling was purposely restricted to depths at which the effect of hydrostatic pressure could be ignored, that is, less than 2000 m or 200 atm (OPPENHEIMER and ZOBELL, 1952). Representative isolates were characterized to genus and for potential biochemical activity against several substrates in an attempt to describe further the distribution, types and activity of cultivable heterotrophs which can occur in oceanic water masses. KRISS, LEBEDEVA and TSIBAN (1966) made a point of the fact that data obtained with the piggy-back sampler (SIEBURTH, FREY and CONOVER, 1963) have not been published. This paper is an attempt to rectify this situation.

MATERIALS AND METHODS

The hydrographic stations occupied and the times of sampling are shown in Fig. 1. The surface samples (150 μ thick) were obtained with the fly-screen sampler of GARRETT (1965) as modified by SIEBURTH (1965). Subsurface samples were collected by an

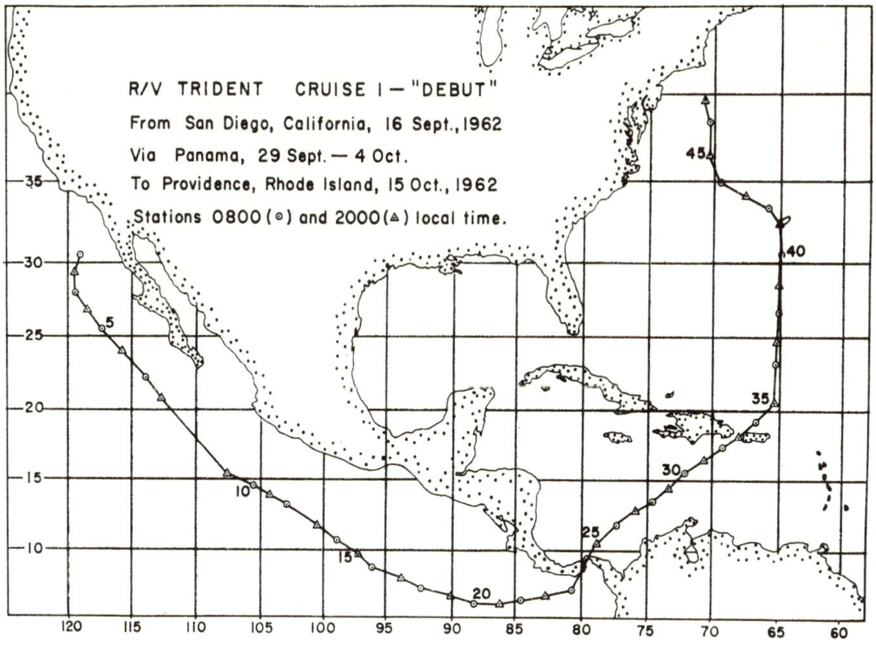

Fig. 1. Station locations and times of sampling during Tr–01.

evacuated bulb sampler attached piggy-back to the Nansen bottles (SIEBURTH, FREY and CONOVER, 1963). One milliliter portions of the sample or serial decimal dilutions of it were pipetted into Astell roll tubes (Colab, Chicago Heights, Ill.). Four milliliters of the tempered medium was added, the tube capped and spun on the Astell spinner while it was chilled and the Agar solidified with a spray of water from the ship's refrigerated drinking fountain. The basal medium (SIEBURTH, 1967) referred to as 0ZR was a modification of the 2216E medium of OPPENHEIMER and ZOBELL (1952) and consisted of Trypticase (Baltimore Biological Laboratories), 1g; Yeast Extract (Difco), 1g; ferric phosphate (Mallinckrodt), 5 mg; Agar (Difco), 15 g; and aged sea water 1 litre. This basal agar was supplemented with either 0·4% gelatin, 0·2% starch of 1% Tween-80. The racks of inoculated and solidified Astell roll tubes were incubated according to the temperature of the water from which the sample was obtained (Table 1). Colony counts were obtained by taking the mean of the three media.

Table 1. *Incubation temperatures and times for the 'environmental temperature counts'. Three heated wall incubators with internal fans were set at* 9°, 18° *and* 27°*C. These and an open container* (0°*C incubator) were placed in the walk-in scientific freezer which was maintained as close as possible to* 0°*C. This scheme provided incubation within* 4·5°*C of the temperature of the water sample.*

Water temp. (°C)	0	to	4·5	to	13·5	to	22·5	to	32·5
Incubation temp. (°C)		0		9		18		27	
Incubation time (days)		12		9		6		3	

Representative colonies were isolated from the 'high count' tubes and keyed to genus by the method of SHEWAN, HOBBS and HODGKISS (1960). Gelatinase was detected by adding 4 g of gelatin (Difco) /l. of OZR to make a modified Frazier gelatin plate which was developed with a mercuric chloride–hydrochloric acid coagulant solution (SOCIETY OF AMERICAN BACTERIOLOGISTS, 1957 pp. 55 and 158). Amylase was detected by adding 2 g of soluble starch (Difco) /l. of OZR agar and developing the plates with either Gram's iodine or dilute Lugol's solution to give a starch iodine reaction which indicated zones of hydrolysis. Oxidation and fermentation of glucose was determined with the O/F medium of HUGH and LEIFSON (1953). Lipase (esterase) activity was detected on OZR agar plates containing 1% Tween-80 (sorbitan mono-oleate). Insoluble oleic acid released by lipase activity formed a distinct halo around active colonies (SIERRA, 1957). Gelatinase and amylase activity of each colony in the 'low count' tubes of the gelatin and starch containing media was determined directly by adding the reagent to the primary cultures.

RESULTS

The distribution of bacteria in the upper 600 m during the Pacific leg of the cruise, was determined from 155 samples obtained at 24 stations (Fig 2). The isotherms are plotted to give an idea of the water strata as well as the incubation temperatures used. The data indicates that 64 out of 155 samples, or some 41%, contained more than five colonies per ml. These higher count samples were mainly distributed in the upper 100 m. The frequency distribution of the bacterial populations in this upper 100 m was previously reported (SIEBURTH, 1965). Counts in the surface film ranged from 1 to > 10,000 per ml, with a mean of 2398 colonies per ml. The distribution and means both indicated a second maximum between 10 and 51 m with a mean of 23 to 26 organisms per ml while a lesser population of 4–8 organisms per ml occurred above and below at 1 and 105 m, respectively.

The bacterial flora occurring in the surface film (Table 2) was characterized by making observations on 200 isolates. Most of the isolates were capable of oxidizing glucose (86%) and hydrolyzing lipids (93%) and protein (91·5%) while only 23·5% were able to attack starch. Atypical pseudomonads (Gram-negative polar flagellated

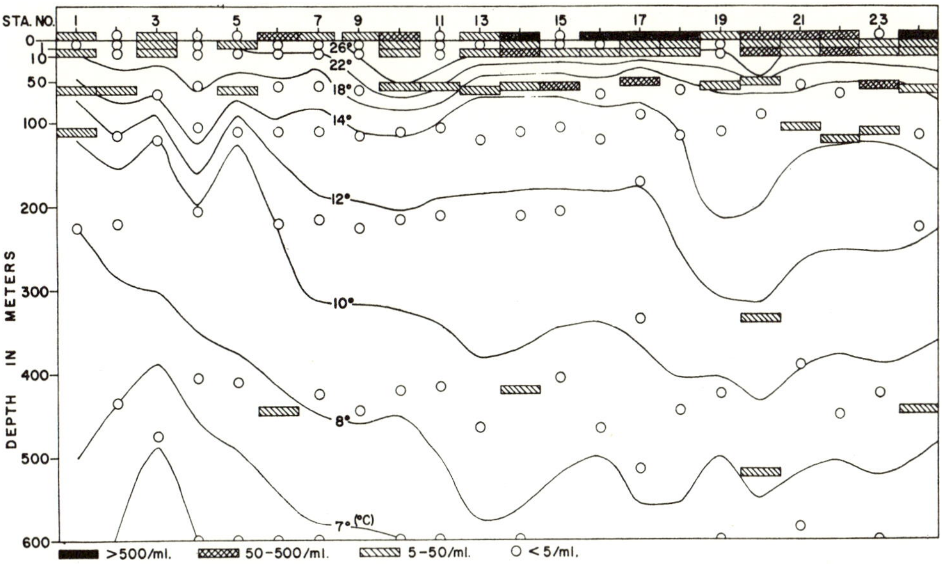

Fig. 2. The populations of heterotrophic bacterial colonies in the upper 600 m during the Pacific leg of Tr–01. Isotherms indicate water structure and incubation temperatures used.

rods which utilized glucose oxidatively but were cytochrome negative) accounted for 66% of the microflora. Types two (cytochrome oxidase positive and oxidative on glucose) and four (cytochrome positive but inactive on glucose) pseudomonads comprised only 23% of the isolates.

Colonies occurring in the 'low count' samples on primary cultivation had a markedly different chemical activity (Table 3). There was no trend with depth with

Table 2. The taxonomic composition and biochemical activity of the dominant bacterial flora of surface films in the Pacific Ocean.

	% Bacterial type					*% Biochemical activity*				
*Station no.**	*Pseudomonas*			*Alcali-*	*Entrobact.*	*Glucose*				
	2	4	*Atypical*	*Achromo.*		*Ferm.*	*Oxid.*	*Lipase*	*Gelatinase*	*Amylase*
10	—	—	100	—	—	—	100	100	100	—
14†	—	—	95	5	—	—	100	100	100	55
16†	30	25	25	20	—	—	60	85	85	25
17†	5	—	90	5	—	—	100	100	95	10
18†	45	25	10	20	—	—	70	90	95	15
19	85	—	10	5	—	—	100	75	100	50
20	15	—	55	—	30	30	60	100	75	20
21	—	—	95	5	—	—	95	90	90	0
22	—	—	100	—	—	—	100	95	90	10
24	—	—	85	5	10	10	80	95	95	50
	18	5	66·5							
Mean	89·5			6·5	4	4	85·6	93	91·5	23·5

*20 isolates per station.
†Greater than 10^8 org./ml.

Table 3. Comparison of biochemical activity with depth for 'low count' samples from the Pacific Ocean.

Mean depth (m)	Per cent activity Gelatinase		Amylase	
0·15mm	55	(333)*	35	(432)
1	57	(164)	29	(181)
10	58	(276)	27	(221)
51	62	(256)	32	(314)
105	47	(122)	35	(77)
Mean	56	(1151)	32	(1225)

*No. of colonies tested.

either substrate. Gelatinase activity ranged from 47 to 62% with a mean of 56% for all 1151 colonies observed while amylase activity ranged from 27 to 35% with a mean of 32% for all 1225 colonies tested. This indicated that the 'low count' flora was less active on proteinaceous substrates and more active against starch than the 'high count' flora.

KRISS, MISHUSTINA and ZEMTSOVA (1962) have reported that there is a marked decrease in the percentage of the bacterial isolates with positive biochemical activities as one approaches the equator. The gelatinase and amylase activity of 'low count' populations in the Pacific, grouped as to latitude, are given in Table 4. For some 2500 colonies tested there was no apparent increase in activity with latitude for either substrate.

Table 4. Comparison of biochemical activity with latitude for 'low count' samples from the Pacific Ocean.

°N lat.	Per cent activity Gelatinase		Amylase	
20–30	44	(293)*	25	(269)
10–20	54	(210)	37	(193)
5–10	55	(742)	28	(844)
Mean	52	(1245)	29	(1306)

*No. of colonies tested.

The distribution of bacteria down to 1100 m during the Caribbean and Atlantic legs of the cruise is shown in Fig. 3. Of the 226 samples obtained at 23 stations, 63 or 27% were 'high count' samples containing more than five colonies per ml. In addition to the higher count samples in the upper 100 m, four stations had populations in excess of 500 colonies per ml which extended from 400 to 1100 m in depth. These higher populations appear to be associated with the edges of basins or different water masses.

The bacterial flora occurring in the 'high count' samples from the Caribbean and Atlantic were characterized by making observations on 400 isolates. The taxonomic affinities and biochemical activities of this flora are given in Table 5. The ability to

*Table 5. The taxonomic composition and biochemical activity of the dominant bacterial flora in the Caribbean and Atlantic. Ten isolates per sample, except where noted (*20/sample).*

Station No.	*Depth* (m)	% *Bacterial type* *Pseudomonas* 2	4	*Atyp.*	*Alcali- Achromo.*	*Enterobact.*	% *Biochemical activity* *Glucose Ferm.*	*Oxid.*	*Lipase*	*Gelatinase*	*Amylase*
	0	10	60	30	—	—	—	20	—	—	—
	1	20	—	80	—	—	—	70	20	80	90
25	10	30	—	70	—	—	—	90	40	40	70
	53	—	—	100	—	—	—	50	10	0	60
	706	80	—	10	—	—	—	100	—	—	—
	0	30	40	30	—	—	—	50	—	10	—
	1	—	—	100	—	—	—	—	90	100	60
	10	80	—	20	—	—	—	100	80	80	70
26	220	—	—	100	—	—	—	100	—	—	70
	439	—	—	50	—	50	50	50	—	40	90
	660	40	60	—	—	—	—	40	100	100	80
	879	90	—	10	—		—	100	90	90	100
	1099	70	—	30	—	—	—	100	100	100	100
27	103	—	—	100	—	—	—	100	100	90	60
32	*0	50	—	30	20	—	—	80	80	80	40
33	*0	95	—	5	—	—	—	100	100	100	100
34	*0	—	—	100	—	—	—	100	—	—	—
	*1	—	—	100	—	—	—	95	—	20	—
	0	40	—	60	—	—	—	100	90	100	30
	1	—	—	10	—	90	90	10	—	—	100
	10	90	10	—	—	—	—	90	100	100	40
	54	100	—	—	—	—	—	100	100	100	100
35	108	—	80	20	—	—	—	—	100	100	60
	216	40	20	40	—	—	—	60	70	90	80
	435	80	20	—	—	—	—	80	100	100	60
	651	80	20	—	—	—	—	80	100	100	60
	869	40	60	—	—	—	—	100	40	50	10
	1087	80	—	20	—	—	—	100	—	—	—
	0	40	—	60	—	—	—	100	100	100	100
	1	—	—	100	—	—	—	100	—	—	—
	10	—	—	100	—	—	—	30	—	40	50
38	55	—	—	100	—	—	—	100	—	40	—
	110	—	—	100	—	—	—	90	—	10	70
	219	—	—	100	—	—	—	100	—	10	100
	439	50	—	50	—	—	—	50	50	50	80
	1097	100	—	—	—	—	—	100	—	—	—
Mean	400 (Isolates)	37	10·4 95·4	48	0·6	4	0·6	76	46	53	54

hydrolyze starch was double while lipase and gelatinase activities were nearly one half that of the Pacific Ocean isolates. This was also reflected by an equal division of the dominant pseudomonad flora between types two and four and the atypical forms lacking cytochrome oxidase activity.

There appears to be a major difference in biochemical activity between 'high and low' count bacterial floras. A comparison is made in Table 6. Gelatinase activity was much greater in the 'high count' than the 'low count' samples from both areas. Amylase activity, however, was only greater in the 'high count' samples from the Caribbean and Atlantic.

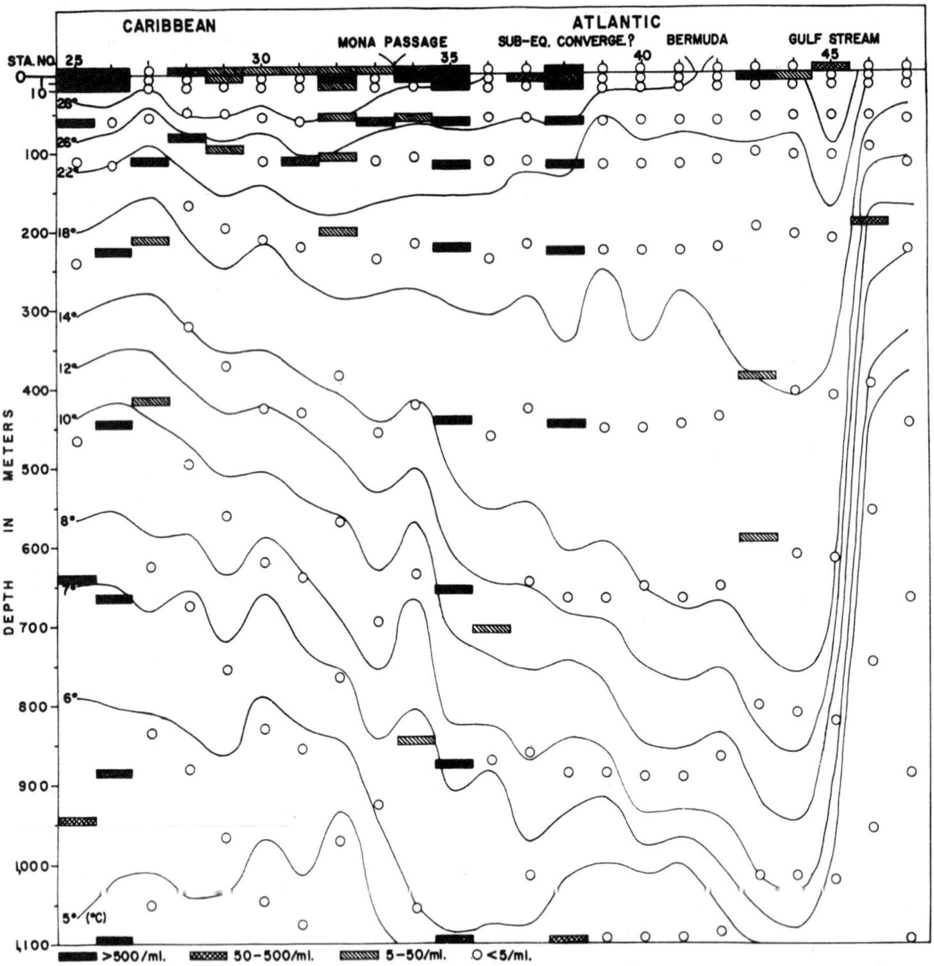

Fig. 3. The populations of heterotrophic bacterial colonies in the upper 600 m during the Caribbean and Atlantic legs of Tr–01. Isotherms, geographical and hydrological features are indicated.

Table 6. A comparison of the biochemical activity of the 'high' and 'low' count populations of the pacific and Caribbean–Atlantic samples.

		Per cent activity			
		Gelatinase		*Amylase*	
Pacific	Low	52	(1245)	29	(1306)
	High	92	(200)	24	(200)
Caribbean–Atlantic	Low	35	(581)	22	(590)
	High	53	(400)	54	(400)

DISCUSSION

The values and distributions obtained in this study are in line with those obtained by earlier workers (ZoBELL, 1946). Superimposed upon the general pattern of a majority of the samples yielding less than five colonies per ml, and most with more being from the upper 100 m, there were sporadically 'high count' samples from the surface film to a depth of 1100 m. In the absence of productivity data one can only surmise that this sporadic occurrence of bacteria is due to spotty productivity. Since the bacterial populations in some 'high count' samples approach population levels (10^3–10^5/ml) encountered in inshore areas (SIEBURTH, 1968), such localized productivity must be intense. An alternative explanation is that high bacterial counts may be due to the microorganisms occurring on a highly populated particle such as a dead copepod captured by chance in that pipetting. The agreement of replicate cultures indicate that such is not the case.

In the much more extensive works of KRISS (1963) and his colleagues (KRISS, LEBEDEVA and MITZKEVITCH, 1960; KRISS, ABYZOV and MITZKEVITCH, 1960; and KRISS, MITZKEVITCH, MISHUSTINA and ABYZOV, 1961) there is a pattern of results which are very different from those reported here. Instead of the usual 'less than' values with a few sporadic 'high' values, Kriss and his colleagues obtain values in between more consistently. SOROKIN (1962) claims that this is due to the adsorption of organic matter on the walls of the Nansen bottle and the development of a microflora with use. BOGOYAVLENSKII (1964) has used the data of LEBEDEVA (1960) to show that it is more than a coincidence that the quantity of heterotrophs obtained correlates directly and consistently with the sampler which was always used at the same depth. When mixups occurred on two occasions, the results reflected this. Data from KRISS (1963) were similarly cited. There is no assurance that the techniques used in defense of the Nansen bottle (KRISS, LEBEDEVA and TSIBAN, 1966) are the same as those used in the earlier studies.

In addition to the possible artefacts induced by sampling, Kriss can also be criticized for his culture methods. The use of agar media containing some 30 g of nutrients per litre, an order of magnitude greater than required for optimal growth, can decrease the colony counts by an order of magnitude (SIEBURTH, 1967). An even more serious criticism is that Kriss incubated his culture plates at cabin temperature regardless of the temperature of the deep water samples being enumerated. In the tropics and subtropics, a deep sample from 3°C would be incubated at temperatures up to 40°C. SIEBURTH (1967) has shown that incubation temperatures above 20°C would inhibit all but a minor component of the bacterial flora of samples obtained at temperatures below 10–12°C.

The methods and procedures used in this study were also far from ideal, but a deliberate attempt was made to avoid those which would induce artefacts. The screen samplers were washed and autoclaved between each use and they were handled in a manner to minimize contamination. The data gives us no cause to think otherwise. For the subsurface samples, sterile and collapsed rubber bulbs attached to the Nansen bottles had their sterile plugs removed as the Nansen bottles were tripped (SIEBURTH, FREY and CONOVER, 1963). In our hands, this sampler worked quite well mechanically. The rubber bulbs were washed and sterilized immediately after use. JANNASCH and MADDUX (1967) have shown that the hydrographic wire and the sampler itself can be a source of contamination. Their use of a sterile inlet tube which remains sheathed until the sampler is tripped could be easily added to the piggy-back sampler. The only indication of contamination is the presence of members of the Enterobacteriaceae at Stas. 20, 24, 26 and 35. These could have been aeromonads or coliforms from marine animals but the most likely source was probably the ship's toilets despite the precautions taken.

Astell roll tube cultures were used because 1 ml of sample could be cultured, the chill water spray reduced contact time with the melted media to less than 30 sec, and the tubes were easy to handle and store aboard ship. On a subsequent cruise it was found that the microflora concentrated by the Cholodny method (JANNASCH and JONES, 1959) or extracted from the membrane with glass beads (MILLER, 1963) could be used to surface inoculate prepoured plates (BUCK and CLEVERDON, 1960) and avoid some of the problems of membrane filter culture (JANNASCH and JONES, 1959; NIEMELÄ, 1965). Surface moisture on prepoured plates as well as contaminants, however, can be disastrous on a cruise. A comparison of surface inoculated plates with Astell roll tube cultures using a heat-sensitive bacterial flora is indicated.

The four temperature system of incubation used in this study appears superior to the more customary single temperature of incubation. The incubation temperatures and times selected were based on optima indicated in an inshore study (SIEBURTH, 1967). The applicability of these results to open-ocean studies was confirmed by Kjell Eimhjellen (personal communication, 1970).

The biochemical activity of the 'low count' colonies on primary culture with one exception was far less than that of the isolates from the 'high count' samples. This observation tends to confirm the notion of a persistent background of bacteria able to subsist on a minimal diet of refractory organic substances (autochthonous microflora) while during periods of organic production a more vigorous bacterial flora which utilizes more readily available substrates (zymogenous microflora) takes over (WINOGRADSKY, 1925). Such field observations, which may reflect what happens in nature, are equally as convincing as laboratory evidence with pure cultures to the contrary. MORRIS (1960) found that under conditions of limiting glucose, both *Escherichia coli* (zymogenous) and *Arthrobacter globiformis* (autochthonous) produced similar growth.

Organic matter which accumulates only in the surface film would be expected to be very different from that also occurring in subsurface samples. Such differences in the nature of the organic matter should also be reflected in the bacterial flora developing on them. In the Pacific samples, a marked flora occurred only in the surface film and this was characterized by marked lipolytic and proteolytic activity. In contrast, the 'high count' floras developing in the Caribbean and Atlantic which had much greater amylase activity were highly variable in regard to lipolytic and proteolytic activity. The activity and taxonomic affinities of the microflora varied with depth and with station indicating that it was unique to the waters sampled and not an artefact carried from one depth and from one station to the other.

KRISS (1963) and his colleagues (KRISS, LEBEDEVA and MITZKEVITCH, 1960; KRISS, ABYZOV and MITZKEVITCH, 1960; and KRISS, MITZKEVITCH, MISHUSTINA and ABYZOV, 1961) treat bacteria as a conservative property of sea water much as salinity or temperature. They presume that water masses are homogenous in regard to the nature of organic matter and therefore the resulting bacterial populations indicate water strata and circulations. They have apparently made no attempt to substantiate this by characterizing the biochemical activity of the bacterial flora in the different water masses. In fact, the opposite has been done. On the basis of the biochemical activity of 3200 isolates obtained from all major oceans, KRISS, MISHUSTINA and ZEMTSOVA (1962) have made some generalizations based on latitude only. They concluded that as they went from high latitudes towards the equator there was a definite decrease in biochemical activity. I have been unable to detect such differences and my tropical isolates were equally as active as the obligate psychrophiles obtained during the New England winters (SIEBURTH, 1967). The data of KRISS, MISHUSTINA and ZEMTSOVA (1962) appear to reflect the decreasing biochemical activity of a flora surviving on contaminated Nansen bottles and the rigors of cabin temperatures in the tropics rather than those of a bacterial flora active in the cold temperatures of the sea.

The data on distribution and biochemical activity reported in this study are at variance with the 'big picture' as presented by Kriss and his colleagues. Critical details of their methods such as the composition of the medium and temperature and times of incubation are lacking (KRISS, 1960, 1963). Only conversations with Kriss' colleagues during a visit to Moscow has clarified his methods and procedures. The only conclusion that one can come to is that they are conducive to artefacts which would explain the results and their conclusions.

The criticism of the Nansen bottle sampling procedure made by SOROKIN (1962) and BOGOYAVLENSKII (1964) appear valid. In addition, excessively rich media and excessive incubation temperature also appear to be a source of artefacts. It is a pity that the methods and procedures used by Kriss and his colleagues were not critically tested by themselves before using them and publishing the data thus obtained so extensively.

Acknowledgements—Appreciation is expressed to M. R. BRACCI for his faithful assistance afloat and ashore. This study was supported in part by grants from the National Science Foundation and the National Institute of Health and a contract from the Office of Naval Research.

REFERENCES

BLANCHARD, D. C. (1963) The electrification of the atmosphere by particles from bubbles in the sea. *Progr. Oceanogr.*, **1**, 71–202.

BOGOYAVLENSKII, A. N. (1964) On the distribution of heterotrophic microorganisms in the Indian Ocean and in Antarctic waters. *Deep-Sea Res.*, **11**, 105–108.

BUCK, J. D. and R. C. CLEVERDON (1960) The spread place as a method for the enumeration of marine bacteria. *Limnol. Oceanogr.*, **5**, 78–80.

EWING, G. (1950) Slicks, surface films and internal waves. *J. mar. Res.*, **9**, 161–187.

GARRETT, W. D. (1965) Collection of slick-forming materials from the sea surface. *Limnol. Oceanogr.*, **10**, 602–605.

HOLM-HANSEN, O. and C. R. BOOTH (1966) The measurement of adenosine triphosphate in the ocean and its ecological significance. *Limnol. Oceanogr.*, **11**, 510–519.

HUGH, R. and E. LEIFSON (1953) The taxonomic significance of fermentative versus oxidative metabolism of carbohydrates by various Gram-negative bacteria. *J. Bact.*, **66**, 24–26.

JANNASCH, H. W. and G. E. JONES (1959) Bacterial populations in sea water as determined by different methods of enumeration. *Limnol. Oceanogr.*, **4**, 128–139.

JANNASCH, H. W. and W. S. MADDUX (1967) A note on bacteriological sampling. *J. mar. Res.*, **25**, 185–189.

JARVIS, N. L., W. D. GARRETT, M. A. SCHEIMAN and C. O. TIMMINS (1967) Surface chemical characterization of surface active material in sea water. *Limnol. Oceanogr.*, **12**, 88–96.

KRISS, A. E. (1960) Micro-organisms as indicators of hydrological phenomena in seas and oceans—I. Methods. *Deep-Sea Res.*, **6**, 88–94.

KRISS, A. E. (1963) *Marine microbiology* (*Deep-Sea*). Trans. by J. M. Shewan and Z. Kabata. Oliver and Boyd, 536 pp.

KRISS, A. E., S. S. ABYZOV and I. N. MITZKEVITCH (1960) Micro-organisms as indicators of hydrological phenomena in seas and oceans—III. Distribution of water masses in the central part of the Pacific Ocean (according to microbiological data). *Deep-Sea Res.*, **6,** 335–345.

KRISS, A. E., M. N. LEBEDEVA and I. N. MITZKEVITCH (1960) Micro-organisms as indicators of hydrological phenomena in seas and oceans—II. Investigation of the deep circulation of the Indian Ocean using microbiological methods. *Deep-Sea Res.*, **6,** 173–183.

KRISS, A. E., M. N. LEBEDEVA and A. V. TSIBAN (1966) Comparative estimate of a Nansen and microbiological water bottle for sterile collection of water samples from depths of seas and oceans. *Deep-Sea Res.*, **13,** 205–212.

KRISS, A. E., I. E. MISHUSTINA and E. V. ZEMTSOVA (1962) Biochemical activity of micro-organisms isolated from various regions of the world. *J. gen. Microbiol.*, **29,** 221–232.

KRISS, A. E., I. N. MITZKEVITCH, I. E. MISHUSTINA and S. S. ABYZOV (1961) Microorganisms as hydrological indicators in seas and oceans—IV. The hydrological structure of the Atlantic Ocean including the Norwegian and Greenland Seas, based on microbiological data. *Deep-Sea Res.*, **7,** 225–236.

LEBEDEVA, M. N. (1960) Microbiological investigations, the second marine expedition in the research vessel *Ob*, 1956–57. (In Russian). *Inf. Byull. Sov. Antarkt. Eksped.* **7,** 153–163.

MILLER, E. J. (1963) A method for the removal from membrane filters of microorganisms filtered from water and air. *J. appl. Bact.*, **26,** 211–215.

MORRIS, J. G. (1960) Studies on the metabolism of *Arthrobacter globiformis*. *J. gen Microbiol.*, **22,** 564–582.

NIEMELÄ, S. (1965) The quantitive estimation of bacterial colonies on membranes filters. *Suomal. -ugr. Seur. Aikak.* (*A*)., **4,** (90), 63 pp.

OPPENHEIMER, C. H. and C. E. ZOBELL (1952) The growth and viability of sixty three species of marine bacteria as influenced by hydrostatic pressure. *J. mar. Res.*, **11,** 10–18.

SHEWAN, J. M., G. HOBBS and W. HODGKISS (1960) A determinative scheme for the identification of certain genera of gram-negative bacteria, with special reference to the Pseudomonadaceae. *J. appl. Bact.*, **23,** 379–390.

SIEBURTH, J. MCN. (1965) Bacteriological samplers for air–water and water–sediment interfaces. *Ocean Science and Ocean Engineering*, *Trans. Joint Conf. MTS and ASLO*, Washington D.C., 1064–1068.

SIEBURTH, J. MCN. (1967) Seasonal selection of estuarine bacteria by water temperature. *J. exp. mar. Biol. Ecol.*, **1,** 98–121.

SIEBURTH, J. MCN. (1968) Observations on bacteria planktonic in Narragansett Bay, Rhode Island, a resumé. *Bull. Misaki mar. biol. Inst., Kyoto Univ.*, **12,** 49–64.

SIEBURTH, J. MCN., J. A. FREY and J. T. CONOVER (1963) Microbiological sampling with a piggy-back device during Nansen bottle casts. *Deep-Sea Res.*, **10,** 757–758.

SIERRA, G. (1957) Studies on bacterial esterases—I. Differentiation of a lipase and two aliesterases during the growth of *Pseudomonas aeruginosa* and some observations on growth and esterase inhibition. *Leeuwenhoed ned. Tijdschr.*, **23,** 241–265.

SOCIETY OF AMERICAN BACTERIOLOGISTS, (1957) *Manual of microbiological methods.* McGraw-Hill. 315 pp.

SOROKIN, Y. I. (1962) Problems in the sampling method used for study of marine microflora. (In Russian). *Okeanologiya* (*Moskow*) **2,** 888–897. (Trans. by Joint Pub. Res. Serv., Wash., D.C.)

TSIBAN, A. V. (1967) On an apparatus for the collection of microbiological samples in the near-surface micro-horizon of the sea. (In Russian). *Gidrobiol. Zh.*, **3,** 84–86.

WINOGRADSKY, S. (1925) Etudes sur la microbiologie du sol—I. Sur la methode. *Annls Inst. Pasteur, Paris*, **39,** 299–354.

ZOBELL, C. E. (1946) *Marine microbiology*, Chronica Botanica, 240 pp.

ZOBELL, C. E., and J. E. CONN (1940) Studies on the thermal sensitivity of marine bacteria. *J. Bact.*, **40,** 223–238.

45

Reprinted from pp. 65, 67–68, 70–76 of *Bull. Misaki Mar. Biol. Inst. Kyoto Univ.,* **12,** 65–76 (Feb. 1968)

SOME ECOLOGICAL ASPECTS OF MARINE BACTERIA IN THE KUROSHIO CURRENT

Nobuo TAGA

Ocean Research Institute, University of Tokyo, Nakano, Tokyo

Introduction

The "Kuroshio" is not only the biggest current in the adjacent sea of Japan, but also one of the biggest ones in the world, along with the Gulf Stream in the Atlantic Ocean. A good deal of investigation has been achieved up to the present in regards to physical, chemical and biological aspects of the Kuroshio current, while any approach has not been performed in order to know the basic mode of bacterial life in the Kuroshio water mass.

The present paper reports a few results of the investigation, which were performed to know the ecological mode of heterotrophic bacterial community in the ecosystem of the Kuroshio water mass.

Methods

Cruises for sampling. The sea water samples were collected on both of the cruise in November, 1965 and in April, 1966, by the Research Vessel "Tansei-maru" of the Ocean Research Institute, University of Tokyo. On her cruise of November in 1965, the vessel was particularly drifted on the Kuroshio axis for about 24 hours, in ordor to collect repeatedly the several water samples at every four and half an hour during a day. The drifted course and sampling stations on both cruises are shown in Fig. 1.

Collection of water samples. Three kinds of water sampler were used in this investigation, *i. e.*, the Nansen's Bottle for the oceanographic routine works; the JZ Sampler (ZoBELL, 1946) for collecting the microbiological samples from the surface water layer; the ORIT Sampler for collecting also the microbiological samples from the vertical various depths. The ORIT-type microbiological sampler, as shown in Fig. 2, was a new type of aseptic water sampler, which could be fixed the sterilized rubber bulbs on the Nansen's Bottle and had been developed jointly by the author's laboratory and the Tsurumi-seiki Kosakusho Co., Ltd., Yokohama.

Oceanographic observations. The GEK measurements were carried out several times to find out the position of the axis of Kuroshio current and to make the vessel drifting on the current axis. As to the environmental conditions of the current water mass, the vertical distributions of temperature and chlorinity were mainly measured. In order to know the other environmental characters of the water mass, the vertical changes of the dissolved oxygen (DO) and pH values were only referred from the data, which were measured by the jointed party of

[*Editor's Note:* Pages 66 (Figures 1 and 2) and 69 (Tables 1 and 2) have been omitted.]

Dr. HORIBE of our Institute.

Enumeration of bacterial number. By filtration, bacteria in sea water samples were collected aseptically on the sterilized Millipore® filters (Type: HA-47 mm), as soon as possible on board. The inoculated filters were placed on the agar plates of Medium PPES-II in Petri dishes. These dishes were sealed by the Parafilm "M"® (American Can Co.), in order to minimize the drying of agar plate and the contamination by other aeromicrobes. These inoculated plates were incubated at 20°C for two weeks before bacterial colonies on filters were counted. The Medium PPES-II used in this investigation was a new type of cultural medium, which had been inspected previously the reliability compared with the ZoBELL's Medium 2216E and had been used for other ecological routine works in our laboratory since a few years (TAGA, 1967). The new medium had the following composition: Polypeptone (Daigo Eiyokagaku Co.), 2 g; Proteose-peptone No. 3 (Difco), 1 g; Bacto-soytone (Difco), 1 g; Bacto-yeast extract (Difco), 1 g; Ferric phosphate soluble (Merck), 0.1 g; Marine mud estract, 100 m*l*; Agar (Wako), 15 g; in 900 m*l* aged sea water; pH adjusted to 7.6 to 7.8.

Biochemical tests of bacteria isolated. The bacterial colonies, grown on each of the inoculated Millipore® filters, were purely isolated and examined seperately as to the nine kinds of biochemical properties. The biochemical tests performed in the present work and their abbreviations were as follows: the hydrolysis properties of gelatin (Ge), casein (Ca), fat or tributyrin (Fa), starch (St), chitin (Ch) and alginate (Al); the decomposition properties of carbohydrates, such as the aerobic acid productions from dextrose (De) and mannite (Ma); the utilization ability of inorganic nitrogen source, such as ammonium salt (IN). The test media used for detecting these biochemical properties were those different stocks of ZoBELL's Medium 2216E, in which the objective organic substrates were added seperately in concentrations of 0.5%, with the exception of 0.75% in the alginate test medium. The biochemical tests were performed by inoculating the isolated bacteria seperately on these different kinds of testing agar plates. After incubating the inoculated plates at 20°C for two to seven days, the positive results of the tests were determined by the formation of hydrolysis halo or the color change of added pH indicator, according to the routine methods described previously by SKERMAN (1959), KIMURA (1961), CAMPBELL and WILLIAMS (1951) and SEKI and TAGA (1963). The test medium used for detecting the utilization ability of inorganic nitrogen had the following composition: dextrose, 5 g; $(NH_4)_2SO_4$, 1 g; K_2HPO_4, 1g; $CaCl_2$ (anhydrous), 0.001 g; Ferric phosphate soluble (Merck), 0.1 g; NaCl, 25 g; in 1000 m*l* water; pH adjusted to 7.4. The positive result of the test was determined by the positive growth of the objective bacterium, which was inoculated repeatedly in the fresh liquid medium and incubated at 20°C for three to seven days.

Results

1. *Diurnal variation of bacterial number in sea water at the Kuroshio Axis*

From the ecological point of view, it might be one of the important problems to make clear what is the magnitude of variation of bacterial number in the same water mass of open sea during a day. In this investigation, such a point was observed, by collecting the water samples on the research vessel which had been drifted in the Kuroshio water mass for 24 hours. The result obtained from this observation is summarized as shown in Table 1 and Fig. 3. As will be seen in Table 2, each series of data in Table 1 was also statistically tested by the method of "analysis of variance by range" (TUKEY, 1951), to confirm if there really existed the variation of bacterial number during a day in each of water mass at the depths of 0, 50, 100, 150, 200 and 400 meters.

According to the result of the statistical test, it is found that there exists no significant difference among the diurnal series of bacterial number at each of the same depths. Consequently, it could be also considered from this result that there was, as a general rule, no significant short-term variation of bacterial population in the Kuroshio water mass during a day.

2. *Distribution pattern and standing crop of heterotrophic bacteria in the Kuroshio water mass*

Vertical patterns of bacterial distributions, comparing with the distributions of a few environmental factors, in sea water mass at the Kuroshio axis are shown in Fig. 4. As will be seen in this figure, vertical distribution of bacteria seems to have the general tendency that the dense bacterial biomass distributes usually only in the topmost productive zone at the depth from 0 to 50 meters and it decreases rapidly in the deeper euphotic layer at the depth from 100 to 200 meters, even if it sometimes increases in the deeper zone at the depth of about 300 to 400 meters. On the other hand, the environmental factors except chlorinity, such as water temperature, pH value and amount of dissolved oxygen (DO) in sea water, show also the decreasing tendencies corresponding to the depths. The vertical density of bacterial biomass in the Kuroshio water mass, however, does not seem to be directly governed only by the distributional tendencies of these environmental factors.

In order to presume the approximate standing crop of heterotrophic bacteria in water mass, the values of population mean of bacterial number with 95% confidence interval were statistically estimated by using the data obtained on several cruises around the Kuroshio region. Table 3 gives the result of the estimation. According to the result of the statistical approximations, it could be considered that the density of heterotrophic bacterial number in the water masses of the Kuroshio and its adjacent region was relatively low, *i. e.*, the estimated value of population mean was about 100 to 250 cells per 10 m*l* of sea water in the upper water mass of the Kuroshio.

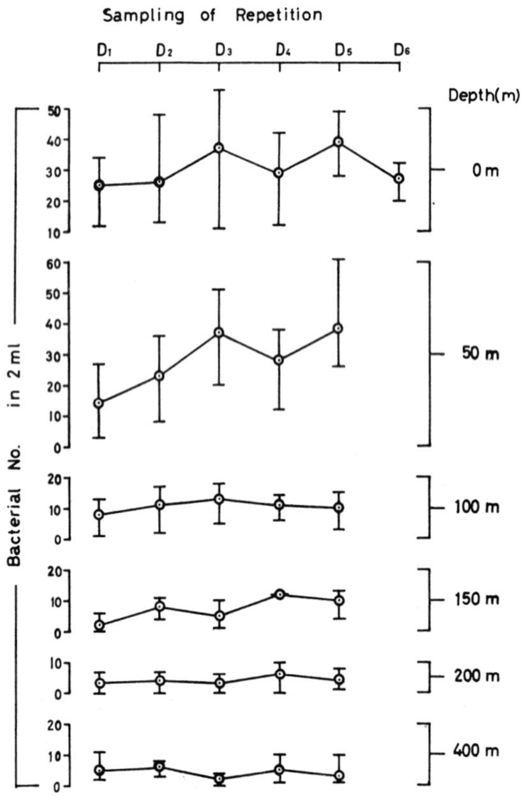

Fig. 3. Diurnal variation of bacterial number in vertical mass of sea water at the Kuroshio Axis.

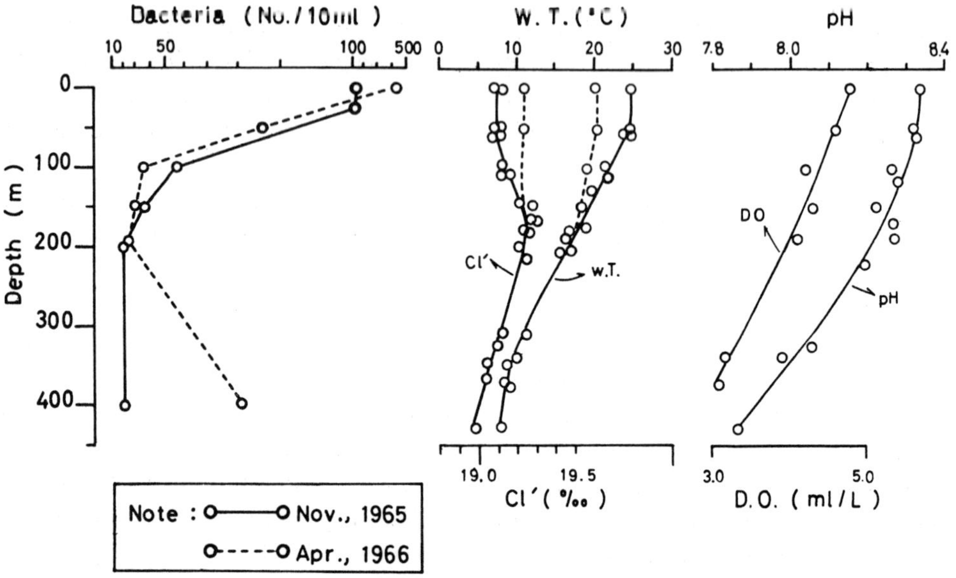

Fig. 4. Vertical pattern of distributions of bacteria and environmental factors in sea water at the Kuroshio Axis.

Table 3. Estimated Value of Population Mean of Heterotrophic Bacterial Number in Sea Water with 95% Confidence Interval

Region (Observed Date)	Water Mass (Meter of Depth)	Population Mean of Bacteria* (No./10 ml)
Kuroshio Axis (Nov., 1965)	(0— 50)	125 (102—153)
	(100—200)	31 (24— 39)
	(400)	19 (13— 27)
Sagami Bay** (Feb., May, Sept., Nov., 1964)	Kuroshio Mass (0—200)	166 (102—247)
	Middle Mass (300—800)	51 (44— 59)
Sagami Bay (Nov., 1964)	Upper Mass (0—200)	161 (66—393)
	Middle Mass (300—800)	113 (72—178)

* Mean values in parenthesis represent the 95% confidence interval.
** Average of overall data in the four months.

3. *Potential biochemical activity of heterotrophic bacterial community in the Kuroshio water mass*

In understanding the mineralization processes of various organic matters in a sea water mass, it would provide an important basic information to analyse the quality of biochemical activities which have potentialized in the heterotrophic bacterial community existing there. To make clear such a problem, it would be required to answer the following points; what is the prevalent type of biochemical properties which have potentialized in the heterotrophic bacterial community, and how much is the overall intensity of their biochemical activities.

In the present investigation, efforts were made only for presuming the quality of the biochemical activities which might be potentially possessed by each ones in the heterotrophic bacterial community in the same water mass. For the purpose of practicing such qualitative analyses, each of the bacterial strains isolated from the same water mass was examined as to the nine kinds of biochemical properties as stated before, thereafter the total number of positive or negative properties to every biochemical test was summed up by the lump of all strains examined, with every water mass. In addition, these positive or negative values, with every biochemical property and with every water mass, were treated statistically with the method by the binormial probability paper, then the average per cent occurrence rate of each biochemical character in heterotrophic bacterial community was estimated with 95% confidence interval. By these estimation processes, for instance, if we obtain the average value of the 75% occurrence rate concerning the gelatin hydrolysis character, it could be qualitatively presumed that the 75% of bacteria in the heterotrophic community had such a biochemical character in a certain sea water mass.

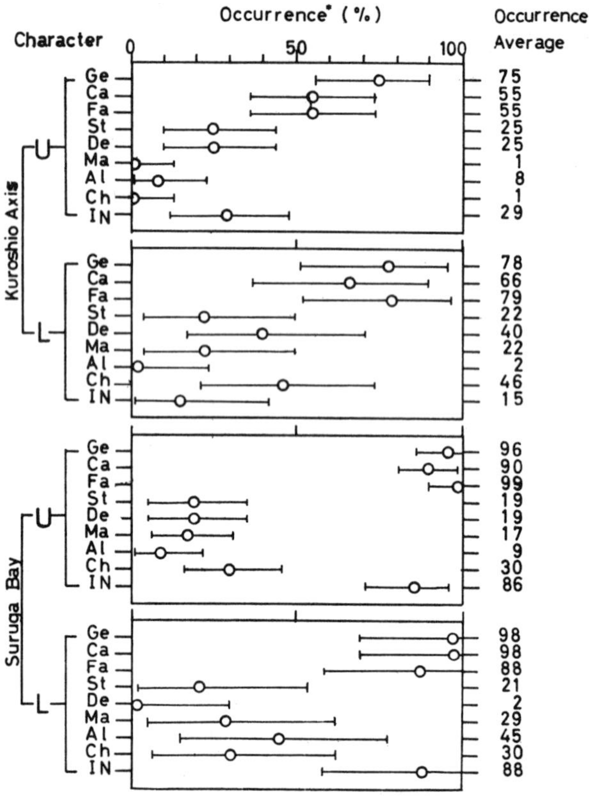

* The estimated mean for the occurrence rate and its 95% confidence interval

Fig. 5. Schematic comparison on the occurrence rate of biochemical characters of heterotrophic bacterial communities in water masses of the Kuroshio Axis and Suruga Bay.

The result obtained concerning the occurrence rate of biochemical characters in waters of the Kuroshio axis and the Suruga Bay can be schematically summarized as shown in Fig. 5. It is obviously found in this figure that the overall distributional patterns of the occurrence rates of biochemical characters are qualitatively different with each other between the bacterial community in upper water mass (U) at the depth from 0 to 150 meters of the Kuroshio axis and that at the depth from 0 to 200 meters of the Suruga Bay, while their patterns, except the inorganic nitrogen utilization (IN) character, are fairly similar with each other between their communities in lower water masses (L) at the depth of 400 meters in both regions. In other words, it might be suggested that the Kuroshio water mass was also microbiologically different from the adjacent neritic water mass in point of the biochemical quality of bacterial community. This

Table 4. Occurrence Rate of Biochemical Character of Heterotrophic Bacterial Community in Sea Water of Various Regions

Region (Observed Date)	Water Mass	Occurrence Rate of Biochemical Characters (%)											
Kuroshio Axis (Apr., 1966)	Upper	Ge,	Ca	≧	Fa	>	St,	De,	Ma,	Al	>	Ch	IN
	(0-150m)	75	55		55		25	25	1	8		1	29
	Lower	Ge,	Ca	≧	Fa	>	St,	De,	Ma,	Al	≦	Ch	IN
	(400m)	78	66		79		22	40	22	2		46	15
Suruga Bay (Apr., 1966)	Upper	Ge,	Ca	≦	Fa	>	St,	De,	Ma,	Al	<	Ch	IN
	(0-200m)	96	90		99		19	19	17	9		30	86
	Lower	Ge,	Ca	>	Fa	>	St,	De,	Ma,	Al	>	Ch	IN
	(400m)	98	98		88		21	2	29	45		30	88
Sagami Bay (May, 1964)	Kuroshio	Ge,	Ca	>	Fa	≧	St,	De,	Ma,	Al	>	Ch	IN
	(0-200m)	64	68		28		32	0	0			0	
	Upper	Ge,	Ca	>	Fa	>	St,	De,	Ma,	Al	>	Ch	IN
	(0-200m)	89	100		78		53	0	0			0	
Sagami Bay (Overall in 1964)	Kuroshio	Ge,	Ca	>	Fa	>	St,	De,	Ma,	Al	≧	Ch	IN
	(0-200m)	73	71				49	34	34			34	
	Upper	Ge,	Ca	>	Fa	>	St,	De,	Ma,	Al	≧	Ch	IN
	(0-200m)	82	81				74	58	54			64	
Order of Overall Occurrence Rate		Protein Hydrolysis > Fat Hydrolysis > Carbohydrate Hydrolysis or Decomposition											

point will be made clearly also by the data summarieed in Table 4.

According to this table, it is found in addition that there exists fairly regular order of their magnitudes among the average occurrence rates of eight biochemical characters except the IN character in any water mass, even if each value of the same character varies to some degree among different series of data. Namely, the tendency of the magnitude of these occurrence rates seems to be on the whole as follows:

Protein hydrolysis>Fat hydrolysis>Carbohydrate hyrolysis or decomposition.

It can be presumed, therefore, that the hydrolytic or decomposing capacity of protein, followed by that of fat, rather than that of carbohydrates, extends and potentializes predominantly in the bacterial community existing in the Kuroshio and its adjacent waters.

Discussion

The purpose of this study was to obtain the basic informations regarding the ecological mode of heterotrophic bacterial community in the water mass of the Kuroshio. For the purpose of carring out such a sort of ecological survey in the sea, successfully were adopted some methods or techniques, such as the water sampling by drifting the vessel for a day, the new type of the ORIT aseptic water sampler, and the bacterial enumeration method by the Millipore filter.

In this investigation, no significant diurnal variation of the bacterial population in the Kuroshio water mass was observed at least during a day. This result might only represent the net bacterial population or standing crop, which was excluded the dead cells and the cells ingested by other organisms from the gross bacterial production in a day, but might not necessarily imply the real growth or production rate of bacteria *in situ* during a day. For the purpose of measuring the latter, another approach should be advanced further by some of the available method.

The vertical distribution pattern of bacteria observed in the Kuroshio water mass was remarkably different from the typical one that had been demonstrated previously by ZoBell (1946). In the water at station off the coast of Southern California, he represented the typical vertical pattern that the distribution of bacterial population generally increased, below the topmost 5 to 10 meters, with depth to 25 or 50 meters and then decreased again to the deeper depths. However, in the present work was not observed such a distributional peak at any intermediate depths from the surface to 100 meters. On the other hand, Saijo (1967) has suggested that the vertical distribution of seston in the Kuroshio water mass showed the same tendency as that of bacteria presented here. Therefore, assuming that there existed some interaction between the both distributions of bacteria and seston, it would be confirmed that the vertical distribution pattern of bacteria, as shown in the present paper, was characteristic in the Kuroshio water mass.

Although a few data are available on the predominant or overall biochemical properties of a group of marine bacteria isolated from each of different regions, some discrepancies among the data are found regarding the occurrence rate or percentage of some biochemical characters based upon the total examined bacteria in each of the regions (ZoBell, 1944; Kriss, *et al.*, 1962; Pfister and Burkholder, 1965).

The result of the present study, that the character of gelatin hydrolysis in the bacterial community was predominated rather than the others, is in good agreement with the results of ZoBell (1944) and Pfister and Burkholder (1965), excepting with that of Kriss, *et al.* (1962). It might be considered from these results that the predominance of the protein hydolysis character in the bacterial community was reflecting the aspect of the most extended process on the overall decomposition or mineralization of organic matters in a certain water mass *in situ*. This matter, on the other hand, particularly when considered in conjunction with the assay of the microbial decomposition of organic matters in sea water, would strongly suggest that the nitrogenous organic matters rather than carbohydrates should be also used as a model substrate in the assay.

Summary

1. The magnitude of diurnal variation of heterotrophic bacterial number in the Kuroshio water mass has been investigated by collecting repeatedly the water samples at every four and half an hour on the research vessel, which was drifted on the Kuroshio axis for about 24 hours. According to the statistical test of the observed data, no significant short-term variation of bacterial population can be found at least during a day in several depths of the Kuroshio water mass.

2. The vertical pattern of bacterial distribution in the Kuroshio water mass has the declining tendency, on the whole, in which the dense bacterial biomass distributes only in topmost zone at the depth from 0 to 50 meters and its density decreases rapidly in the deeper euphotic layer at the depth from 100 to 200 meters.

3. According to the result of the statistical approximation with 95% confidence interval, it is considered that the estimated value of population mean of heterotrophic bacteria is relatively low and about 100 to 250 cells per 10 m*l* of sea water in the upper water mass of the Kuroshio.

4. Biochemical properties of the isolated bacteria have been examined for their ability to hydrolyze or decompose the organic substrates, such as gelatin, casein, fat (tributyrin), starch, chitin, alginate, dextrose and mannite. According to these examinations, it is presumed on the whole that the hydrolytic or decomposing capacity of protein, followed by that of fat, rather than that of carbohydrates, extends and potentializes predominantly in the bacterial community existing in the Kuroshio and its adjacent waters. In addition, it may be suggested that the Kuroshio water mass is microbiologically different from the adjacent neritic water mass in point of the overall biochemical quality potentialized in the bacterial community.

Acknowledgements

The author appreciates the aids of Capt. I. Tadama, the other officers and crew of the R/V Tanseimaru of the Ocean Research Institute, University of Tokyo. His thanks are due to Mr. K. Ohwada, graduate student of University of Tokyo, for his generous assistance in carrying out the microbiological examinations on board, and to Miss Kiyoko Shiozawa and Miss Kazu Hoshino for their technical assistances. He wishes also to express his thanks to Dr. Y. Horibe of the Ocean Research Institute for having kindly supplied the data of chemical observations in the same cruises, and to Dr. T. Ishii of the same Institute for his many valuable advices on the statistical treatment of the data.

References

Campbell, L. L. and Williams, O. B. (1951): A study of chitin-decomposing micro-organisms of marine origin. *J. gen. Microbiol.*, **5,** 894-905.

Kimura, T. (1961): A method for rapid detection of alginic acid-digesting bacteria. *Bull. Faculty of Fisheries, Hokkaido Univ.*, **12**(1), 41-48.

Kriss, A. E., Mishustina, I. E. and Zemtsova, E. V. (1962): Biochemical activity of micro-organisms isolated from various regions of the world ocean. *J. gen. Microbiol.*, **29,** 221-232.

Pfister, R. M. and Burkholder, P. R. (1965): Numerical taxonomy of some bacteria isolated from Antarctic and tropical seawaters. *J. Bacteriol.*, **90,** 863-872.

Saijo, Y. (1967): Private communication.

Seki, H. and Taga, N. (1963): Microbiological studies on the decomposition of chitin in marine environment—1. Occurrence of chitinoclastic bacteria in the neritic region. *J. Oceanogr. Soc. Japan*, **19,** 101-108.

Skerman, V. B. D. (1959): A guide to the identification of the genera of bacteria. Williams and Wilkins Co., Baltimore, 217 p..

Taga, N. (1967): The media for cultivation of marine bacteria (unpublished data).

ZoBell, C. E. and Upham, H. (1944): A list of marine bacteria including descriptions of sixty new species. *Bull. Scripps Inst. Oceanogr.*, **5,** 239-292.

ZoBell, C. E. (1946): Marine Microbiology. Chronica Botanica Co., Waltham, 240 p..

AUTHOR CITATION INDEX

SUBJECT INDEX

About the Editor

CAROL D. LITCHFIELD is Associate Professor in the Department of Microbiology and the Marine Science Center of Rutgers University. In addition to a lecture course in marine microbiology, she has developed a field and laboratory course on marine microbial ecology. Following receipt of her M.S. degree in bacteriology from the University of Cincinnati, she worked as a research scientist in the Department of Oceanography at Texas A & M University and later received her Ph.D. in biochemistry from the same institution. She became interested in the microbiology of marine sediments during a postdoctoral yeat at the University College of North Wales, Marine Science Laboratories. She has been active in the New Jersey branch of the American Society for Microbiology and has served as vice-chairman and chairman of the Aquatic and Terrestrial Division of the national American Society for Microbiology and is a member of the Environmental Microbiology Committee of the A.S.M.